Vortical Flows

Jie-Zhi Wu · Hui-Yang Ma
Ming-De Zhou

Vortical Flows

Springer

Jie-Zhi Wu
State Key Laboratory for Turbulence and Complex Systems, College of Engineering
Peking University
Beijing
China

Ming-De Zhou
Department of Aerospace and Mechanical Engineering
University of Arizona
Tucson, AZ
USA

Hui-Yang Ma
Department of Physics
University of Chinese Academy of Sciences
Beijing
China

ISBN 978-3-662-47060-2 ISBN 978-3-662-47061-9 (eBook)
DOI 10.1007/978-3-662-47061-9

Library of Congress Control Number: 2015938755

Springer Heidelberg New York Dordrecht London

Printed on acid-free paper

Springer-Verlag GmbH Berlin Heidelberg is part of Springer Science+Business Media
(www.springer.com)

Preface

Vortical flows are flows with *vortices* as their skeleton structures. Vortices are seen everywhere in our universe and on the earth: from spiral galaxies, atmospheric and oceanic circulations to hurricanes and typhoons, tornadoes to bath stub vortices; from volcanoes' erupted smoke rings and mushroom clouds of nuclear explosions to vortex rings ejected from the mouth of dolphin and smoker, or formed in a heart downstream of the mitral valve that separates the left atrium and left ventricle; from tip vortices of aircraft, rotor blade, and turbo fan to complicated ring-like structures in the wake of birds, insects and fishes; from well organized laminar vortices to coherent turbulent structures.

This book provides a systematic introduction to the physical theory of vortical flows at graduate level. It grew from our monograph *Vorticity and Vortex Dynamics* (Springer 2006), but has been thoroughly rewritten. Some advanced topics in the monograph have been removed, and more basic topics have been added. Recent advances since 2006 in the field of fundamental interest are included. Nevertheless, two basic characteristics of the monograph are inherited and further enhanced, which make both the monograph and the present book differ from other existing books on the subject:

(1) We consider the theory of vortical flows as a branch of fluid dynamics focusing on *shearing process* in fluid motion, measured by *vorticity*. A vortex is defined as a fluid body with high vorticity concentration. The evolution of vorticity field is governed by *vorticity dynamics*. Coexisting with this process is the compression–expansion process (*compressing process* for short) measured by dilatation, pressure, or other thermodynamic variables, of which the main structure is shock waves where *entropy process* is naturally involved. The three fundamental processes in fluid motion are coupled with each other both inside the flow field and at solid boundary. We believe that only on the basis of this broad background can the physics of vortical flows be fully understood.

(2) We study vortical flows according to their natural evolution stages, from being generated to dissipated. As preparation, the first three chapters of the book provide background knowledge for entering vortical flows. Due to the coupling of shearing process with other processes, this knowledge appears wider and more

profound than common books on vortical flows. Chapter 1 reviews standard fundamental kinematics and dynamics of generic viscous and compressible flow, including some elementary results of process identification and decomposition. The whole of Chap. 2 is devoted to the basic theory of fundamental processes in fluid motion, their splitting and coupling. Chapter 3 discusses general theory and physics of vorticity dynamics. Although later chapters will be mainly confined to incompressible flow, Chaps. 1–3 cover much broader materials with a hope to facilitate future exploration of more complicated compressible vortical flows.

The rest of the book deals with vortices and vortical flows. Of various vortices the primary form is *layer-like vortices* or *shear layers*, and secondary but stronger form is *axial vortices* mainly formed by the rolling up of shear layers. Thus, Chap. 4 is on attached shear layer (namely *boundary layer*) and free shear layers. As Reynolds number approaches infinity, these layers become asymptotically attached and free vortex sheets, which are the subject of Chap. 5. This chapter ends with vortex-sheet rolling up and initial formation of axial vortices, so it is naturally followed by Chaps. 6 and 7 on typical solutions of columnar vortices and vortex rings, respectively. Chapter 8 studies flow separation first, which is a key localized dynamic process turning a simple attached flow to complex, namely to become global separated flow with concentrated vortices that is studied next. Chapter 9 is an introduction to total force and moment acting to a body moving through the fluid, in terms of various vortical structures.

Chapters 10 and 11 discuss the instability and breakdown of axial vortices, and vortical structures in transitional and turbulent shear flows, respectively. Both chapters require some elementary knowledge of flow instability and turbulence, which are placed (somewhat artificially) in the beginnings of Chaps. 10 and 11, respectively. Finally, A general theory of vector and tensor field is presented in the Appendix for readers' convenience.

Problems are given at the end of each chapter and Appendix, some for helping to understand the basic theories, and some involving specific applications; but the emphasis of both is always on physical thinking. Problems with asterisk may need more effort.

The reader of this book is assumed to have learned undergraduate fluid mechanics or aerodynamics in majors of mechanics, aerospace and mechanical engineering, and be familiar with physics, advanced calculus and differential equations. Better background of these fields will make it easier to understand the present book. Most part of the book materials has been used as Lecture Notes and were taught by J.Z. at Peking University over the past 15 years as a one-semester graduate course of advanced fluid dynamics. The course has been proved acceptable by most students with warm and inspiring feedback.

August 2014

Jie-Zhi Wu
Hui-Yang Ma
Ming-De Zhou

Acknowledgments

We are indebted to many colleagues and students involved in the courses based on this book, whose encouragement has been the major source of our thrust. In particular, we owe much to Profs. Shiyi Chen, Zhensu She, Cun-Biao Li, Xi-Yun Lu, Xie-Yuan Yin, Chui-Jie Wu, Wei-Dong Su and Yi-Peng Shi, with whom our close cooperations and numerous brainstorming in teaching and research have strongly motivated the preparation of the book.

Our special sincere thanks go to Prof. Ronard L. Panton, who reviewed carefully most chapters and provided very insightful comments; to Prof. Yasuhide Fukumoto, who helped improve the contents of Chap. 7; to Prof. Tianshu Liu, who helped update and enrich the contents of Chap. 8; and to Prof. Bartek Protas, who made useful comments on Sect. 2.4. Our special thanks also go to Dr. Zhen Li, who read repeatedly the early drafts of Chaps. 1–7 and offered thoughtful comments on the structures and contents.

The writing of the book has been proceeded through intensive interactions with many of our former and current graduate students. Prof. Hui Zhao, Drs. Yan-Tao Yang, Ri-Kui Zhang, Li-Jun Xuan and Feng Mao, as well as Mrs. Jin-Yang Zhu, Luo-Qin Liu, Shu-Fan Zou, An-Kang Gao, and Lin-Lin Kang have provided extensive support in reviewing the drafts, designing and answering problems, and editing texts. Ms. Feng-Rong Zhu has provided excellent support to figure drawing. We owe a lot to all of them.

The authors are also very grateful for the continuous support from the National Natural Science Foundation of China (Nos. 10332040, 10572005, 10532010, 90405007, 10921202), Ministry of Science and Technology (Project No. 2009CB724100), and internal funding of State Key Laboratory for Turbulence and Complex Systems, College of Engineering, Peking University.

Contents

Chapter 1
Fundamentals of Fluid Dynamics

1.1 Basic Fluid Kinematics

1.1.1 Description and Visualization of Fluid Motion

Fluid dynamics studies the motion of continuous media with fluidity. A fluid then has a dual feature and can be described in two ways. On the one hand, the fluid consists of continuously distributed *material elements* or *particles*, each of which retains its identity all the time so that one can trace the fluid motion and evolution by tracking each element. We may define the identity or "label" of a fluid particle by a known particle's location $\boldsymbol{x}_0$ at $t = t_0$, say $\boldsymbol{x}_0 = \boldsymbol{X}$. As it moves smoothly as t increases, its new position $\boldsymbol{x}$ at $t > t_0$ is a differentiable function of $\boldsymbol{X}$ and t:

$$\boldsymbol{x} = \boldsymbol{f}(\boldsymbol{X}, t). \tag{1.1.1a}$$

Note that as the label of a fixed particle, $\boldsymbol{X}$ does not change with time; namely there must be $\partial \boldsymbol{X}/\partial t = \boldsymbol{0}$. Inversely, we may write

$$\boldsymbol{X} = \boldsymbol{F}(\boldsymbol{x}, t). \tag{1.1.1b}$$

Initially separated particles cannot merge to a single particle at later time, even though they could be tightly squeezed together. Meanwhile, a single particle cannot be split into two or more particles at later time. This implies that the mapping between $\boldsymbol{x}$ and $\boldsymbol{X}$ is one-to-one. Their transformation Jacobians are never zero nor infinity. Figure 1.1 sketches the mapping of a material fluid body from the $\boldsymbol{X}$-space (a reference space) to the $\boldsymbol{x}$-space (the physical space) and its reverse.

Now, following the movement of a particle $\boldsymbol{X}$, its position $\boldsymbol{x}$ becomes a function of t, so that the particle has a velocity

$$\boldsymbol{u} = \frac{\partial \boldsymbol{x}}{\partial t}(\boldsymbol{X}, t) \quad \text{or} \quad u_i = \frac{\partial x_i}{\partial t}, \quad i = 1, 2, 3, \tag{1.1.2}$$

J.-Z. Wu et al., *Vortical Flows*, DOI 10.1007/978-3-662-47061-9_1

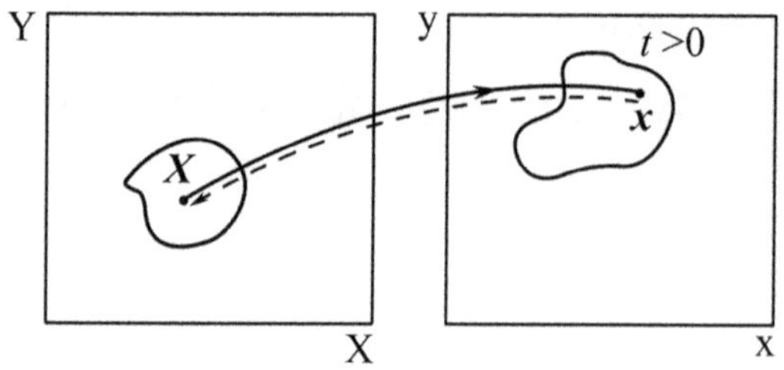

Fig. 1.1 One-to-one mapping between the $\boldsymbol{X}$-space and $\boldsymbol{x}$-space for a material fluid body

with X_i being parameters. This way of description, known as *particle or Lagrangian description*, is a direct extension of Newton's particle kinematics.

On the other hand, fluid motion can be treated by a *field theory* where, like in an electromagnetic field, the spatial position $\boldsymbol{x}$ and time t are independent variables. The fields of velocity $\boldsymbol{u}$, pressure p, and other derived physical quantities are all functions of $(\boldsymbol{x}, t)$ and will be assumed sufficiently smooth except on certain surfaces of discontinuity. If the fluid is unbounded, except otherwise stated it is assumed to be at rest at infinity or, by a Galilean transformation, have uniform motion. If the flow has a boundary, the boundary is assumed to be piecewise smooth. The collection of these fields constitutes a *flow field*. This way of description is known as *field or Eulerian description*. The two descriptions enrich each other.

To see this duality of material and field in fluid motion, consider an elementary manifestation of the fluidity: the velocity difference at two neighboring points $\boldsymbol{x}_0$ and $\boldsymbol{x} = \boldsymbol{x}_0 + \delta\boldsymbol{x}$ with $|\delta\boldsymbol{x}| = \delta r \to 0$. In the field description, a use of Taylor expansion gives

$$\delta\boldsymbol{u} = \boldsymbol{u}(\boldsymbol{x}) - \boldsymbol{u}(\boldsymbol{x}_0) = \delta\boldsymbol{x} \cdot \nabla\boldsymbol{u}(\boldsymbol{x}_0) + O(\delta r^2). \tag{1.1.3}$$

In contrast, in the material description, after δt the particles at the two ends of $\delta\boldsymbol{x}$, $\boldsymbol{x}_0$ and $\boldsymbol{x}$, will move to

$$\begin{aligned}\boldsymbol{x}_0 &\to \boldsymbol{x}_0 + \boldsymbol{u}(\boldsymbol{x}_0)\delta t,\\ \boldsymbol{x} &\to \boldsymbol{x} + \boldsymbol{u}(\boldsymbol{x})\delta t = \boldsymbol{x} + \boldsymbol{u}(\boldsymbol{x}_0)\delta t + \delta\boldsymbol{x} \cdot \nabla\boldsymbol{u}(\boldsymbol{x}_0)\delta t + O(\delta r^2).\end{aligned}$$

Hence, there is

$$\delta\boldsymbol{x}(\boldsymbol{x}, t + \delta t) = \delta\boldsymbol{x}(\boldsymbol{x}, t) + \delta\boldsymbol{u}(\boldsymbol{x}, t)\delta t, \quad \delta\boldsymbol{u}(\boldsymbol{x}, t) = \delta\boldsymbol{x} \cdot \nabla\boldsymbol{u}(\boldsymbol{x}_0, t),$$

so that

$$\frac{d\delta\boldsymbol{x}}{dt} = \delta\boldsymbol{u} = \delta\boldsymbol{x} \cdot \nabla\boldsymbol{u} \equiv \frac{D\delta\boldsymbol{x}}{Dt}, \tag{1.1.4}$$

which is the same as (1.1.3) but enriches the latter's implication by comparing the velocities at neighboring points and identifying $\delta\boldsymbol{u}$ as the rate of change of a *material line element* $\delta\boldsymbol{x}$. The operator D/Dt is used to emphasize that the derivative is taken by following the particle, and is called *material derivative*.

In general, the field description and material description are not fully equivalent. The former does not care which specific particle is moving through a field point $\boldsymbol{x}$ at time t, but the latter does and has to ensure the identification of all infinitely many particles at all time. In other words, the two descriptions will become fully equivalent only if to the field description we add a vectorial constraint $\partial\boldsymbol{X}/\partial t = \boldsymbol{0}$ or

$$\frac{D\boldsymbol{X}}{Dt} = \boldsymbol{0}. \tag{1.1.5}$$

For most of fluid dynamics problems it suffices to stay on the simpler field description without constraint (1.1.5), as we do in the major portion of the book.[1] But, as will be seen in Sect. 1.2, in developing the formulation of fluid dynamics in terms of the field description, tracking material fluid elements is still necessary because the link between the flow field evolution and its causes, i.e., the forces acting on the fluid, is provided by the Newton mechanics which is formulated for material fluid body and particles.

It is appropriate here to introduce four different types of lines in a flow field, defined based on the above two descriptions. First, a curve tangent to the velocity $\boldsymbol{u}(\boldsymbol{x}, t)$ everywhere at a time t is a *streamline* at this time. Let it be represented as $\boldsymbol{x}(s)$ in terms of parameter s. Then its equation follows from eliminating dt in the component form of (1.1.2):

$$\frac{d\boldsymbol{x}}{ds} = \boldsymbol{u}(\boldsymbol{x}, t) \quad \text{or} \quad \frac{dx_1}{u_1(\boldsymbol{x}, t)} = \frac{dx_2}{u_2(\boldsymbol{x}, t)} = \frac{dx_3}{u_3(\boldsymbol{x}, t)}, \tag{1.1.6}$$

of which the solution passing a given $\boldsymbol{x}(s_0)$ is the required streamline. The concept of streamlines does not distinguish different particles and belongs to the field description. In an experiment, if we spread the tracer particles in the flow and take a photo with *very short time exposure*, then we see a set of short line segments, of which a smooth connection can represent a family of instantaneous streamlines.

Next, a *particle-path line* or *pathline* is the curve created by the motion of a particle $\boldsymbol{X}$ as time goes on, which can be obtained by solving the ordinary differential equation (1.1.2) in time:

$$\frac{\partial \boldsymbol{x}}{\partial t} = \boldsymbol{u}(\boldsymbol{x}, t), \quad \boldsymbol{x}(\boldsymbol{X}, t_0) = \boldsymbol{X}. \tag{1.1.7}$$

The concept of pathlines belongs to the material description. In flow visualization, if we introduce a tracer particle into the fluid and photograph its motion by a *long time exposure*, we obtain a pathline.

[1] Later in Chap. 3 we shall see that for a special class of flows the two descriptions become equivalent, and the constraint (1.1.5) can be dropped.

To visualize a flow, instead of taking a long-time exposure to trace a single particle, it is more informative to introduce some dyed fluid continuously at a fixed point $\boldsymbol{x}_0$ and take a *snapshot* at a later time t_0. The photo shows a curve consisting of the spatial positions at t_0 of *all* fluid particles which have passed $\boldsymbol{x}_0$ at any time $\tau \le t_0$ and continue to move ahead, where $\tau \in (-\infty, t_0]$ is a parameter for identifying different particles. Such a curve is called a *streakline*, the third type of lines. Mathematically, the positions of these particles at t_0 can be obtained by applying (1.1.7) not to a single particle but all particles $\boldsymbol{X}(\tau)$:

$$\frac{\partial \boldsymbol{x}}{\partial t} = \boldsymbol{u}(\boldsymbol{x}, t), \quad \boldsymbol{x}(\boldsymbol{X}(\tau), \tau) = \boldsymbol{x}_0, \quad -\infty < \tau \le t_0. \tag{1.1.8}$$

Changing t_0 continuously will yield an animation of the streak-line evolution.

While a streakline involves a time-sequence of particles $\boldsymbol{X}(\tau)$ passing a single $\boldsymbol{x}_0$, the visualization can be extended to releasing dyed particles from different points of a line $\boldsymbol{X}(s)$, say, and then take a snapshot. Furthermore, one may insert a thin metal wire $\boldsymbol{X}(s)$ in a moving fluid (e.g. water) and introduce a pulsating current of frequency f through it. The wire will electrolyze the water and release hydrogen bubbles on-and-off at discrete time $\tau_n = nT = n/f$ with $n = 0, 1, 2, \ldots$, which can be illuminated. Thus, each current-on action produces a bright-dark strip or column of short pathlines of the hydrogen bubbles. The strips are initially parallel to the wire but then advected by local flow velocity, and hence exhibit approximately the local velocity profiles including vortical structures. These pulsating strips and short bubble traces therein are called *time lines*, the fourth type of lines. Photos or animations of the velocity profiles may clearly exhibit the flow pattern and its evolution. They are defined by $\boldsymbol{x}_n(\boldsymbol{X}(s, \tau_n), t)$ that are governed by

$$\frac{\partial \boldsymbol{x}_n}{\partial t} = \boldsymbol{u}(\boldsymbol{x}_n, t), \quad \boldsymbol{x}_n(\boldsymbol{X}(s, \tau_n), \tau_n) = \boldsymbol{X}(s). \tag{1.1.9}$$

As a simple example for the behavior of these lines, consider a two-dimensional (2D) and incompressible unsteady velocity field $\boldsymbol{u} = (u, v)$ in the (x, y)-plane:

$$u = -x, \quad v = y + t.$$

Then by (1.1.6)–(1.1.8), the parametric expressions for the streamline, pathline, and streakline are found to be:

$$\begin{aligned}
&\text{Streamline}: \quad x = x_0 e^{-(s-s_0)}, \quad y = (y_0 + t)e^{(s-s_0)} - t,\\
&\text{pathline}: \quad x = x_0 e^{-(t-t_0)}, \quad y = (y_0 + t_0 + 1)e^{t-t_0} - t - 1,\\
&\text{streakline}: \quad x = x_0 e^{-(t-\tau)}, \quad y = (y_0 + \tau + 1)e^{t-\tau} - t - 1, \quad \tau \in (-\infty, t_0).
\end{aligned}$$

It is easily verified that at a given t_0, the streamline passing $\boldsymbol{x}_0$, the pathline of a particle locating at $\boldsymbol{x}_0$, and the streakline passing $\boldsymbol{x}_0$ have a common tangent vector at $\boldsymbol{x}_0$. Thus, over a very short time interval streaklines, pathlines, and instantaneous

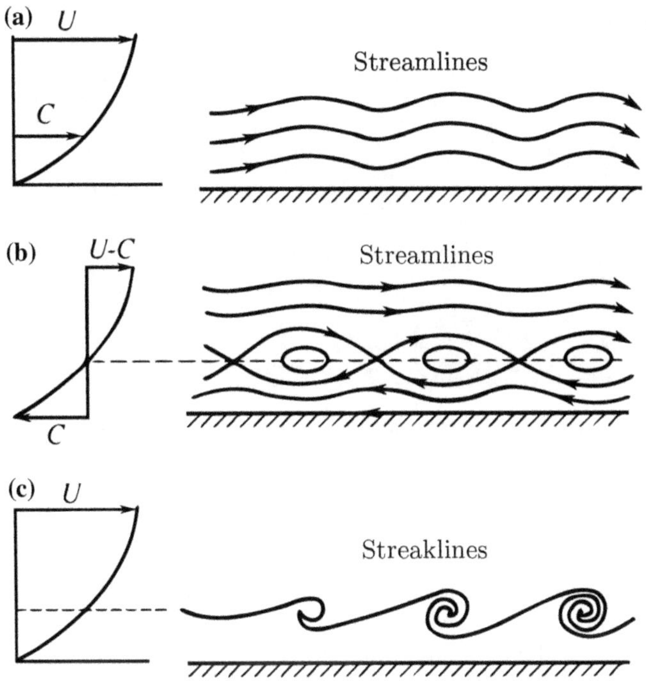

Fig. 1.2 Schematic streamlines (viewed in different frames) and streaklines in a boundary layer with travelling instability waves. C is the wave speed. Reproduced from Taneda (1985)

streamlines are identical. When the flow is steady, i.e., in (1.1.2) $\boldsymbol{u}$ is independent of t, the three curves coincide at any time. For more general unsteady flows, however, the three curves are entirely different. A pathline or a streakline can intersect itself, but a streamline cannot. The behavior of streamlines and pathlines vary drastically as the observer changes from a fixed frame of reference to a moving one, but the streaklines will remain the same.

In the past, most experimental visualizations of vortical flows exhibited streaklines, and most numerical visualizations plot exhibited streamlines. It has now been realized that observing more kinds of lines can lead to clearer understanding of a complex flow field, but their interpretation needs great care since vortical flows are inherently more or less unsteady. Figure 1.2 sketches the unsteady streamlines and streaklines due to the instability travelling waves in a flat-plate boundary layer, viewed from different frames of reference. Note that the streamlines in the frame moving with the wave exhibit some vortex-like structure (so-called "cat-eyes"), but whether or not these cat-eyes can be classified as vortices should be judged by the degree of concentration of the vorticity rather than merely by the frame-dependent streamlines.

Figure 1.3 shows both streamlines and streaklines due to the unsteady vortex shedding from a circular cylinder, where their difference is obvious. However, while

Fig. 1.3 Streamlines and streaklines in unsteady vortex shedding from a *circular cylinder*. From Taneda (1985)

streaklines can tell where the vorticity resides in a flow, it tells very little about the surrounding fluid and the entrainment process. In this regard instantaneous streamlines in an unsteady flow are still useful. A correct understanding of some highly time-dependent vortical flows, of which the evolution is essentially a material process, should use jointly all three kinds of lines to avoid misinterpretation that could happen if only streaklines are visualized (cf. Kurosaka and Sundaram 1986). Meanwhile, Lagrangian formulation may be of crucial importance. These issues will be further addressed in Sect. 8.5 when we study *unsteady flow separation*. For example, streamlines, pathlines, and streak lines of the same unsteady flow are simultaneously displayed in Fig. 8.28, along with a discussion of their respective roles.

Figure 1.4 is a time-line photo of the flow over a circular cylinder. The metal wire is a vertical straight line at the far left of the photo.

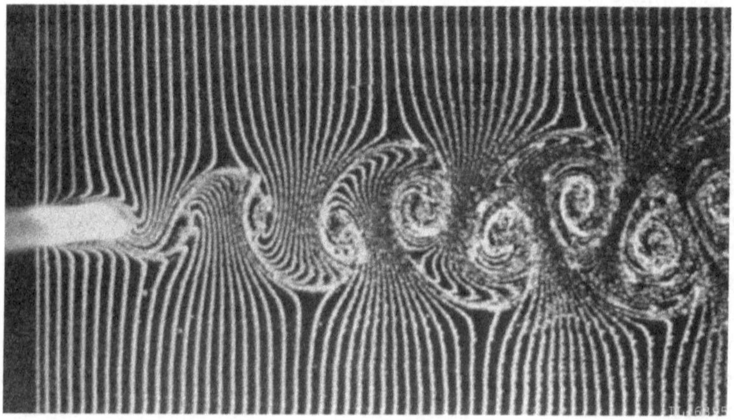

Fig. 1.4 Time lines behind a *circular cylinder* at Reynolds number 152. From Taneda (1985)

1.1.2 Dilatation and Vorticity

While the velocity field $\boldsymbol{u}(\boldsymbol{x}, t)$ discussed in the preceding subsection is the most primary vector field in any flows, our major concern is various *flow structures* like those seen in Figs. 1.3 and 1.4, which are highly localized and occupy only a very small portion of the flow domain, but play a role as the "organizers" of the entire flow. Structures come from the *variation* of the velocity in space and time rather than the velocity itself. To understand these structures, we have to consider the spatial derivatives of the velocity field and their temporal evolution.

In a flow domain, the structures are locally characterized by various products of the gradient operator ∇ with $\boldsymbol{u}$, because the vector ∇ measures both the direction along which the variation is steepest and the magnitude of variation in that direction per unit length. The mathematical foundation of our study is the general vector-field theory, of which a systematic introduction is given in Appendix,[2] of which the results will be simply cited in the main text.

The most primary derived fields that describe the local spatial variation of a velocity field $\boldsymbol{u}$ are its divergence, a scalar field called *dilatation*, and its curl, an axial vector called *vorticity*:

$$\vartheta \equiv \nabla \cdot \boldsymbol{u} = \frac{\partial u}{\partial x} + \frac{\partial v}{\partial y} + \frac{\partial w}{\partial z}, \tag{1.1.12a}$$

$$\boldsymbol{\omega} \equiv \nabla \times \boldsymbol{u} = \left(\frac{\partial w}{\partial y} - \frac{\partial v}{\partial z}, \frac{\partial u}{\partial z} - \frac{\partial w}{\partial x}, \frac{\partial v}{\partial x} - \frac{\partial u}{\partial y}\right). \tag{1.1.12b}$$

Intuitively, ϑ measures the isotropic expansion or compression of the fluid, while $\boldsymbol{\omega}$ measures the rotation of fluid particles. Their physical meanings can be more clearly understood by considering their volume integrals. By the generalized Gauss theorem (A.2.8), we obtain

$$\vartheta = \lim_{V \to 0} \frac{1}{V} \int_{\partial V} \boldsymbol{n} \cdot \boldsymbol{u} dS, \tag{1.1.13a}$$

$$\boldsymbol{\omega} = \lim_{V \to 0} \frac{1}{V} \int_{\partial V} \boldsymbol{n} \times \boldsymbol{u} dS, \tag{1.1.13b}$$

indicating that their net contributions are solely from the *normal* and *tangential* components of $\boldsymbol{u}$ on the boundary, respectively, as sketched in Fig. 1.5 with V being a small sphere. Clearly, the dilatation represents a net isotropic *outflow* where only the velocity normal to sphere's surface plays a role. This is an outcome of the compressing process. In contrast, the vorticity represents a non-isotropic "*curl up*" property where only the velocity tangent to sphere's surface is involved. This is an outcome of the shearing process. Note that the velocity curling-up associated with vorticity can be

[2] This theory is also applicable to all other vector fields to be encountered in our study. Before moving on, the reader is strongly recommended to get a full familiarity of the materials in Appendix as a necessary preparation.

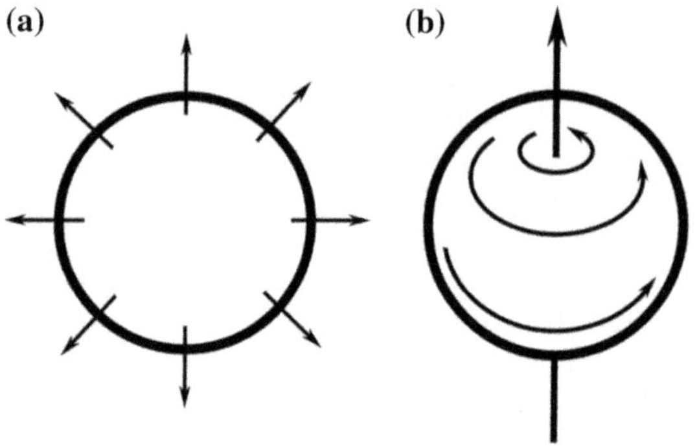

Fig. 1.5 The velocities associated with dilatation (**a**) and vorticity (**b**) in a small sphere

further revealed by the *circulation* of $\boldsymbol{u}$ along a closed loop C, which by the Stokes theorem gives

$$\oint_C \boldsymbol{u} \cdot d\boldsymbol{x} = \int_S \boldsymbol{\omega} \cdot \boldsymbol{n} dS, \tag{1.1.14}$$

where S is any directional surface spanned by C. A further taste of the behavior of dilatation and vorticity can be felt by considering a simple plane-wave fluctuating velocity field $\boldsymbol{u} = \hat{\boldsymbol{u}}(t)e^{i(\boldsymbol{k}\cdot\boldsymbol{x}-nt)}$, where $\boldsymbol{k} = k\boldsymbol{e}_k$ is the wave vector with $k = 2\pi/\lambda$ being the wave number of $\boldsymbol{u}$ and $\boldsymbol{e}_k$ the unit vector in the wave propagation direction, $n = 2\pi f$ is the circular frequency, and $\hat{\boldsymbol{u}}$ is a uniform amplitude. For this wave the differentiation operation by ∇ is reduced to the multiplication by $i\boldsymbol{k}$, so that

$$\nabla \cdot \boldsymbol{u} = i\boldsymbol{k} \cdot \boldsymbol{u}, \quad \nabla \times \boldsymbol{u} = i\boldsymbol{k} \times \boldsymbol{u}. \tag{1.1.15}$$

Thus, the ϑ-wave is a *longitudinal* wave propagating along the velocity direction, while the $\boldsymbol{\omega}$-wave is a *transverse* wave propagating perpendicular to the velocity direction. This is in consistency with (1.1.13), with $\boldsymbol{n}$ there being analogous to $i\boldsymbol{k}$ here. Note that in an arbitrary bounded domain the geometric relations between wave oscillating and propagating directions may not be as simple as that given by (1.1.15). In general, we may split the velocity field into two parts, $\boldsymbol{u} = \boldsymbol{U}(\boldsymbol{x}) + \boldsymbol{u}'(\boldsymbol{x}, t)$, where $\boldsymbol{U}(\boldsymbol{x})$ is an arbitrary steady basic flow which can have both divergence and curl, and $\boldsymbol{u}'(\boldsymbol{x}, t)$ describes an unsteady velocity wave. Then, if $\nabla \times \boldsymbol{u}' = \boldsymbol{\omega}' = \boldsymbol{0}$, we say the wave is longitudinal; while if $\nabla \cdot \boldsymbol{u}' = \vartheta' = 0$, we say it is transverse. Later we shall see that the two waves are qualitatively different; their propagation speeds are determined by different dynamic mechanisms and have different values. Physics fields that have these wave behaviors are usually called *longitudinal field* and *transverse field*, respectively (cf. Morse and Feshbach 1953); we now see that in fluid motion the longitudinal and transverse fields are specified to compressing and shearing fields.[3]

[3]In this book the two pair of names will be used alternatively.

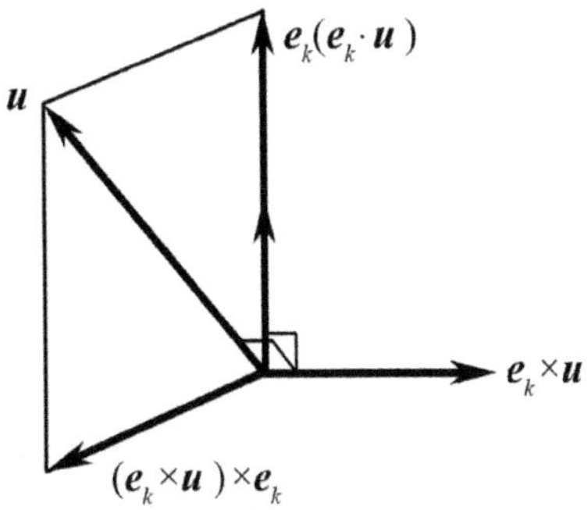

Fig. 1.6 Geometric orthogonal decomposition of $\boldsymbol{u}$ with respect to wave vector $\boldsymbol{k}$

If a $\boldsymbol{u}$-fluctuation emits simultaneously both longitudinal and transverse waves, the fluctuation $\boldsymbol{u}$-field as a whole must be a composite wave. For the above plane wave the composition is simple, since (ignoring factor i) the two waves in (1.1.15) are just orthogonal. Indeed, since $\boldsymbol{k} = \boldsymbol{e}_k k$, $\boldsymbol{e}_k(\boldsymbol{e}_k \cdot \boldsymbol{u})$ is a component vector of $\boldsymbol{u}$ along $\boldsymbol{k}$; while $\boldsymbol{e}_k \times \boldsymbol{u}$ differs from the perpendicular components of $\boldsymbol{u}$ (which is $\boldsymbol{u} - \boldsymbol{e}_k(\boldsymbol{e}_k \cdot \boldsymbol{u})$) by a 90° on the plane normal to $\boldsymbol{k}$. The former can be turned back to the latter simply by its one more cross product with $\boldsymbol{e}_k$, see Fig. 1.6. Therefore, we construct the composite $\boldsymbol{u}$ as

$$|\boldsymbol{k}|^2\boldsymbol{u} = \boldsymbol{k}(\boldsymbol{k} \cdot \boldsymbol{u}) - \boldsymbol{k} \times (\boldsymbol{k} \times \boldsymbol{u}). \tag{1.1.16}$$

Then, recalling the rule $\nabla \to i\boldsymbol{k}$, from (1.1.16) we arrive at a differential identity

$$\nabla^2\boldsymbol{u} = \nabla(\nabla \cdot \boldsymbol{u}) - \nabla \times (\nabla \times \boldsymbol{u}) = \nabla\vartheta - \nabla \times \boldsymbol{\omega}, \tag{1.1.17}$$

which clearly reveals the functional relation between $\boldsymbol{u}$, ϑ, and $\boldsymbol{\omega}$. The above elementary observations suggest that the dilatation and vorticity fields are measures of two physically distinct processes, which are mutually independent of and complementary to each other in many aspects.

1.1.3 Velocity Gradient and Its Decompositions

We now move to the next product of the gradient operator ∇ and $\boldsymbol{u}$: the dyad $\nabla\boldsymbol{u}$ known as the *velocity gradient tensor*. Once again, our interest is the intrinsic decompositions of $\nabla\boldsymbol{u}$.[4]

Like any matrix, the second-order tensor $\nabla\boldsymbol{u}$ can be split into a symmetric part and an antisymmetric part:

---

[4] Whenever written in component form, throughout this book we use the convention that the (i, j)th component of $\nabla\boldsymbol{u}$ is $\nabla_i u_j = \partial_i u_j = u_{j,i}$.

$$\nabla \boldsymbol{u} = \mathbf{D} + \boldsymbol{\Omega}, \tag{1.1.18}$$

where, with the superscript T denoting transpose,

$$\mathbf{D} = \frac{1}{2}[\nabla \boldsymbol{u} + (\nabla \boldsymbol{u})^T] \tag{1.1.19a}$$

$$\boldsymbol{\Omega} = \frac{1}{2}[\nabla \boldsymbol{u} - (\nabla \boldsymbol{u})^T] \tag{1.1.19b}$$

are the symmetric *strain-rate tensor* and antisymmetric *vorticity tensor* (or *spin tensor*), respectively. These tensors are associated with dilatation and vorticity via the following relations:

$$\vartheta = \nabla \cdot \boldsymbol{u} = D_{ii}, \tag{1.1.20a}$$

$$\omega_i = \epsilon_{ijk}\Omega_{jk}, \quad \Omega_{jk} = \frac{1}{2}\epsilon_{ijk}\omega_i, \tag{1.1.20b}$$

where ϵ_{ijk} are the components of the permutation tensor (Appendix A.1.4). Thus, $\boldsymbol{\omega}$ is equivalent to $\boldsymbol{\Omega}$. Note that for any vector $\boldsymbol{b}$ there is identity

$$2\boldsymbol{b} \cdot \boldsymbol{\Omega} = \boldsymbol{\omega} \times \boldsymbol{b} = -\boldsymbol{b} \times \boldsymbol{\omega}, \tag{1.1.21a}$$

so there also is

$$2\nabla \cdot \boldsymbol{\Omega} = -\nabla \times \boldsymbol{\omega}. \tag{1.1.21b}$$

From (1.1.20) we see a remarkable difference of the two fundamental processes measured by ϑ and $\boldsymbol{\omega}$ or $\boldsymbol{\Omega}$: *the compressing process can exist in any of one-, two-, or three-dimensional flow, but the shearing process does not appear in one-dimensional flow at all, appears merely in part in two-dimensional flow like a scalar (its direction is always perpendicular to the flow plane), and exhibits its full behavior in and only in three-dimensional flow*. Moreover, as a divergence-free vector, $\boldsymbol{\omega}$ has just two independent components in three dimensions. This geometric difference of the two processes is transformed to a more intuitive difference in their integrals (1.1.13), where the value of ϑ and the two independent components of $\boldsymbol{\omega}$ produce the normal component and two tangential components of the velocity at boundary, respectively. Similar contrast appears in (1.1.15) as well, where the normal vector $\boldsymbol{n}$ is replaced by the wave vector $\boldsymbol{k}$.

As a very basic application of (1.1.18), let us revisit the velocity difference of two neighboring points $\boldsymbol{x}$ and $\boldsymbol{x}_0$ given by (1.1.4), which can now be decomposed to

$$\delta \boldsymbol{u} = \frac{D\delta \boldsymbol{x}}{Dt} = \delta \boldsymbol{x} \cdot \mathbf{D}_0 + \frac{1}{2}\boldsymbol{\omega}_0 \times \delta \boldsymbol{x}, \tag{1.1.22}$$

where $\delta \boldsymbol{x} = \boldsymbol{x} - \boldsymbol{x}_0$ is a material line element and suffix 0 denotes taking values at fixed $\boldsymbol{x}_0$. Thus, like rigid-body rotation, the role of $\boldsymbol{\omega}_0$ is to rotate the line element $\delta \boldsymbol{x}$ around $\boldsymbol{x}_0$ with angular velocity $\boldsymbol{\omega}_0/2$. Namely, *the vorticity can be understood as twice of the angular velocity of a fluid element.*

In contrast, the first term of (1.1.22) is unique to deformable body and deserves a detailed examination. Let $\delta s = |\delta \boldsymbol{x}|$, the inner product of (1.1.22) and $\delta \boldsymbol{x}$ yields[5]

$$\frac{1}{2}\frac{D}{Dt}(\delta s^2) = \delta \boldsymbol{x} \cdot \mathbf{D}_0 \cdot \delta \boldsymbol{x} \equiv 2\phi, \tag{1.1.23}$$

and hence the relative velocity solely "induced" by pure straining,

$$\delta \boldsymbol{u}_\phi \equiv \delta \boldsymbol{x} \cdot \mathbf{D}_0 = \nabla \phi, \tag{1.1.24}$$

is *irrotational* or has a *potential*. It is along the *normal* to surfaces $\phi =$ constant, which are *quadric center surfaces* (ellipsoid, hyperboloid, and paraboloid, etc. centered at $\boldsymbol{x}_0$), known as the *tensor surfaces* of $\mathbf{D}$ (Appendix A.1.3).

How the specific elements of D_{ij} affect $\delta \boldsymbol{u}_\phi$ can be further clarified.[6] First, consider the diagonal components of D_{ij} with $i = j$. Since

$$\phi_{,ii} = \delta x_{j,i} D_{0ji} = D_{0ii} = \vartheta_0,$$

the trace of $\mathbf{D}$ simply represents an isotropic expansion/compression, which vanishes if the flow is incompressible. Then, similar to the derivation of (1.1.23), the rate of change of δs reads

$$\frac{1}{\delta s}\frac{D}{Dt}(\delta s) = D_{0ij}\frac{dx_i}{ds}\frac{dx_j}{ds}. \tag{1.1.25}$$

In particular, if $\delta \boldsymbol{x}$ is along the x_1-axis, say, such that $dx_i/ds = \delta_{i1}$ and $dx_j/ds = \delta_{j1}$, (1.1.25) implies

$$\frac{1}{\delta s}\frac{D}{Dt}(\delta s) = D_{0ij}\delta_{i1}\delta_{j1} = D_{011}. \tag{1.1.26}$$

Thus, the diagonal elements of D_{ij} are responsible for the relative stretching rate of a line element parallel to x_i-axis. They are called *normal components of the strain-rate*.

---

[5] Since only material line element $\delta \boldsymbol{x}$ is involved, operator D/Dt can be replaced by d/dt, see (1.1.4).

[6] See, e.g. Aris (1962), Zhuang et al. (2009), and Panton (2013).

Next, for understanding the off-diagonal elements of D_{ij}, consider two material line elements $\delta\boldsymbol{x}$ and $\delta\boldsymbol{x}'$ initiated from $\boldsymbol{x}_0$ with angle θ, such that $\delta\boldsymbol{x}\cdot\delta\boldsymbol{x}' = \delta s\delta s'\cos\theta$. Then one finds

$$2D_{0ij}\frac{dx_i}{ds}\frac{dx_j'}{ds'} = \cos\theta\left[\frac{1}{\delta s}\frac{d}{dt}(\delta s) + \frac{1}{\delta s'}\frac{d}{dt}(\delta s')\right] - \sin\theta\frac{d\theta}{dt}. \tag{1.1.27}$$

In particular, when $\delta\boldsymbol{x}$ and $\delta\boldsymbol{x}'$ are along the x_1- and x_2-axes, respectively, we simply have

$$D_{012} = -\frac{1}{2}\frac{d\theta}{dt}. \tag{1.1.28}$$

Namely, the off-diagonal elements of D_{ij} measure half of the decrease rate of change of the angle of two material line elements originally along the ith and jth axes, respectively. These off-diagonal elements of D_{ij} are called *shearing components of the strain-rate*.

The strain-rate tensor $\mathbf{D}$ has three real principal values. $\mathbf{D}$ has a *principal-axis coordinate system* in which with $D_{ij} = 0$ for $i \neq j$ (Appendix A.1.3). Then the above results show that each instantaneous *principal axis* keeps straight. It experiences a stretching (shrinking) along its direction and rotating with angular velocity $\boldsymbol{\omega}/2$, but with no tilting.

In summary, the velocity at $\boldsymbol{x} = \boldsymbol{x}_0 + \delta\boldsymbol{x}$ is

$$\boldsymbol{u}(\boldsymbol{x}) = \boldsymbol{u}_0 + \nabla\phi + \frac{1}{2}\boldsymbol{\omega}_0 \times \delta\boldsymbol{x}. \tag{1.1.29}$$

This result is known as the *Cauchy-Stokes theorem* or **fundamental theorem of deformation kinematics** (Truesdell 1954): *The instantaneous state of a fluid motion at every point is the superposition of a uniform translation, an irrotational stretching or shrinking along three orthogonal principal axes, and a rigid rotation around an axis*. Figure 1.7 illustrates this theorem schematically.

Example: Simple shear flow. Consider a unidirectional shear flow on the (x, y)-plane, $\boldsymbol{u} = (ky, 0, 0)$ with constant shear rate k. Its velocity-gradient tensor and strain-rate tensor are

$$\{u_{j,i}\} = \begin{pmatrix} 0 & 0 & 0 \\ k & 0 & 0 \\ 0 & 0 & 0 \end{pmatrix}, \quad \{D_{ij}\} = \begin{pmatrix} 0 & k/2 & 0 \\ k/2 & 0 & 0 \\ 0 & 0 & 0 \end{pmatrix},$$

respectively, while by (1.1.12) there is $\vartheta = 0$ and $\boldsymbol{\omega} = (0, 0, -k)$. This is a simplified prototype of boundary layer and free shear layer to be addressed in Chap. 4, in which both the strain rate and vorticity are very strong. Here, D_{ij} has principal values $\pm k/2$ and can be reduced to diagonal form by rotating the axes counter-clockwise through an angle $\pi/4$. This strain rate represents a uniform elongation in one principle direction and a uniform foreshortening in the second one at right angle to the first,

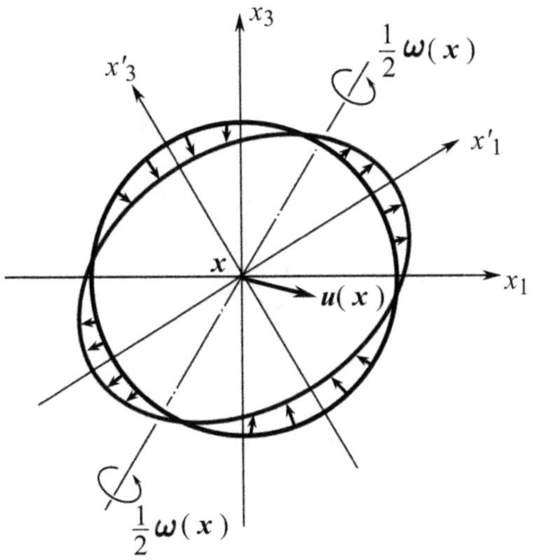

Fig. 1.7 The deformation of small fluid sphere. Only the pattern on the (x_1, x_3) plane is shown

so that the solenoidality condition $\vartheta = 0$ is ensured. Thus, a small sphere of radius ϵ will deform to an ellipsoid during a time interval dt, with the lengths of its semi-axes being

$$\epsilon(1 + kdt/2), \ \epsilon(1 - kdt/2), \ \epsilon.$$

However, to maintain the simple shear motion, the simultaneous rotation around the z axis with angular velocity $\omega/2 = -k/2$ just produces a clockwise turning of the principal axes back to their original directions. For example, after a dt time the rotation makes the actual angle between the first principal axis and the x-axis become $\pi/4 - kdt/2$. The situation is shown in Fig. 1.8, which shows that *in order to maintain the shear flow pattern it is essential for the strain rate to be accompanied by the vorticity*.

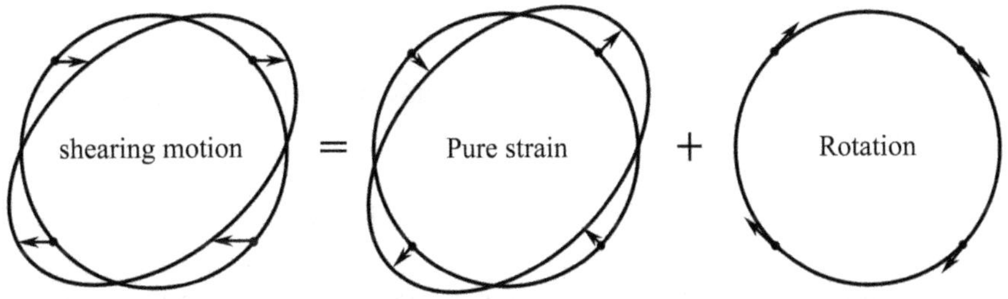

Fig. 1.8 Decomposition of a simple shear motion into a pure strain and a rotation. From Lighthill (1986)

Having seen the double decomposition (1.1.18) of $\nabla \boldsymbol{u}$, we recall that its divergence is the familiar $\nabla^2 \boldsymbol{u}$ that has decomposition (1.1.17), or

$$\nabla \cdot \nabla \boldsymbol{u} = \nabla \cdot (\vartheta \mathbf{I} + 2\boldsymbol{\Omega}),$$

indicating that while $\nabla \boldsymbol{u}$ has nine independent components in a three-dimensional space, its divergence has only three. This fact suggests that from $\nabla \boldsymbol{u}$ one may separate a divergence-free tensor to let the remaining part be solely expressed by ϑ and $\boldsymbol{\omega}$. Indeed, since $\mathbf{D} = \mathbf{D}^T$ and $\boldsymbol{\Omega} = -\boldsymbol{\Omega}^T$, we may write

$$(\nabla \boldsymbol{u})^T = \mathbf{D} - \boldsymbol{\Omega} + \vartheta \mathbf{I} - \vartheta \mathbf{I},$$

so that the strain-rate tensor and, by (1.1.18) the velocity gradient tensor, have *intrinsic triple decompositions*

$$\mathbf{D} = \vartheta \mathbf{I} + \boldsymbol{\Omega} - \mathbf{B}, \tag{1.1.30a}$$

$$\nabla \boldsymbol{u} = \vartheta \mathbf{I} + 2\boldsymbol{\Omega} - \mathbf{B}, \tag{1.1.30b}$$

where

$$\mathbf{B} \equiv \vartheta \mathbf{I} - (\nabla \boldsymbol{u})^T \quad \text{with} \quad \nabla \cdot \mathbf{B} = \mathbf{0} \tag{1.1.31}$$

is the divergence-free tensor we are seeking for, which is known as the *surface-deformation tensor* due to its physical implication to be explained later in Sect. 1.1.4. The divergence of (1.1.30b) recovers (1.1.17) at once, with no contribution from $\mathbf{B}$. This triple decomposition is very useful as one studies the velocity field near a material surface, and in dynamic problems for which the traceless property of $\mathbf{B}$ can bring considerable simplification.

Interestingly, similar to (1.1.30), a triple decomposition can be found for the double inner-product $\mathbf{D} : \mathbf{D}$, which as will be seen in Sect. 1.2.2 plays a key role in viscous flow dissipation into heat. In fact, $\nabla \boldsymbol{u} : \nabla \boldsymbol{u} = u_{j,i} u_{i,j}$ may be alternatively decomposed to

$$\begin{aligned} u_{j,i} u_{i,j} &= (u_{j,i} u_i)_{,j} - u_i \vartheta_{,i} = (u_{j,i} u_i - \delta_{ij} u_i \vartheta)_{,j} + \vartheta^2; \\ &= (D_{ij} + \Omega_{ij})(D_{ji} + \Omega_{ji}) = D_{ij} D_{ji} - \frac{1}{2}\omega^2. \end{aligned}$$

Then, a comparison of these two expressions yields the desired decomposition identity at once:

$$\mathbf{D} : \mathbf{D} = \vartheta^2 + \frac{1}{2}\omega^2 - \nabla \cdot (\mathbf{B} \cdot \boldsymbol{u}). \tag{1.1.32}$$

But since this is a nonlinear product, coupling among different constituents of the velocity must appear as can be seen in $\mathbf{B} \cdot \boldsymbol{u}$.

1.1.4 Local and Global Material Derivatives

In fluid kinematics we study not only the spatial relations of the flow quantities as we did in the preceding subsections, but also their temporal variation in a universal way, namely without concerning specific physical cause and effect of these temporal variation.[7] Of the temporal variations the most important kind is the time rate of change of flow quantities, which is examined now. We shall make a combined use of the material and field descriptions to derive the governing equations of fluid motion. Readers are assumed to have been familiar with the procedure and results, and the focus here is a neater formulation based on tensor analysis (Appendix A.1) and deeper physical understanding thereof.

Consider any field quantity $\mathcal{F}(\boldsymbol{x}, t)$ carried by a material element located at $\boldsymbol{x}$ at time t. Since the element motion makes $\boldsymbol{x} = \boldsymbol{x}(t)$ as in (1.1.1a), we have $\mathcal{F}(\boldsymbol{x}, t) = \mathcal{F}(\boldsymbol{x}(t), t)$; thus, the material rate of change of $\mathcal{F}$ is

$$\frac{D}{Dt}\mathcal{F}(\boldsymbol{x}(t), t) = \frac{\partial \mathcal{F}}{\partial t} + \frac{\partial \mathcal{F}}{\partial x_i}\frac{dx_i}{dt},$$

which by (1.1.2) implies

$$\frac{D\mathcal{F}}{Dt} = \frac{\partial \mathcal{F}}{\partial t} + \boldsymbol{u} \cdot \nabla \mathcal{F}. \tag{1.1.34}$$

Thus, when acting on any field quantities, the material-derivative operator D/Dt is split to a local time-variation $\partial/\partial t$ and a variation by advection $\boldsymbol{u} \cdot \nabla$. In particular, the acceleration $\boldsymbol{a}$ of a material element reads

$$\boldsymbol{a} = \frac{D\boldsymbol{u}}{Dt} = \frac{\partial \boldsymbol{u}}{\partial t} + \boldsymbol{u} \cdot \nabla \boldsymbol{u}. \tag{1.1.35}$$

Here again appears the velocity gradient tensor $\nabla \boldsymbol{u}$. In addition to producing the strain rate $\mathbf{D}$, the vorticity $\boldsymbol{\omega}$, and the dilatation ϑ, by various operations one can generate some other important quantities from this tensor. In fact, not only the inner products of $\nabla \boldsymbol{u}$ with $\delta \boldsymbol{x}$ and $\boldsymbol{u}$ *from left* yield the rate of change of $\delta \boldsymbol{x}$ and that of $\boldsymbol{u}$ by advection, respectively, as we have seen, but also its inner product with $\boldsymbol{u}$ *from right* is meaningful: $\nabla \boldsymbol{u} \cdot \boldsymbol{u} = \boldsymbol{u} \cdot (\nabla \boldsymbol{u})^T = \nabla(|\boldsymbol{u}|^2/2)$ is the gradient of kinetic energy.

Then, since $\nabla \boldsymbol{u} = \nabla \boldsymbol{u} - (\nabla \boldsymbol{u})^T + (\nabla \boldsymbol{u})^T$, by (1.1.18) and (1.1.21a) we can split $\boldsymbol{u} \cdot \nabla \boldsymbol{u}$ into two terms, and hence refine (1.1.35) to

$$\boldsymbol{a} = \frac{\partial \boldsymbol{u}}{\partial t} + \boldsymbol{\omega} \times \boldsymbol{u} + \nabla\left(\frac{1}{2}q^2\right), \quad q = |\boldsymbol{u}|. \tag{1.1.36}$$

[7] Some authors use the term "kinematics" more restrictively, only to the spatial relations of the relevant quantities at a single time instance.

Namely, the advective acceleration has two causes: the gradient of kinetic energy and a vorticity cause in the direction perpendicular to both $\boldsymbol{u}$ and $\boldsymbol{\omega}$. Vector $\boldsymbol{\omega} \times \boldsymbol{u}$ is known as the *Lamb vector*, which implies a *transverse force* to the fluid motion, and as will be seen in Chap. 9 it is responsible for the lift of an aircraft.

So far we have traced the local variation of a field quantity by following a material fluid element. We may now similarly trace the global variation of a field quantity in an arbitrary domain $D(t)$, of which the boundary may move and change shape over time with velocity $\boldsymbol{v_b}$. To this end we recall the Newton-Leibniz formula in elementary calculus,

$$\frac{d}{dt}\int_{a(t)}^{b(t)} f(x,t)dx = \int_a^b \frac{\partial f}{\partial t}dx + \frac{db}{dt}f(b,t) - \frac{da}{dt}f(a,t). \tag{1.1.37}$$

This formula can be extended to multi-dimensional space, where we replace the speed of moving bounds, da/dt and db/dt, by the normal velocity $\boldsymbol{n}\cdot\boldsymbol{v_b}$ of the moving boundary surface $\partial D(t)$. This yields

$$\frac{d}{dt}\int_{D(t)} \mathcal{F}(\boldsymbol{x},t)dV = \int_{D(t)} \frac{\partial \mathcal{F}}{\partial t}dV + \int_{\partial D(t)} \boldsymbol{n}\cdot\boldsymbol{v}_b\mathcal{F}dS. \tag{1.1.38}$$

In particular, we want to trace the global variation of $\mathcal{F}$ in a *material volume* $\mathcal{V}$, which consists of the same fluid body and whose boundary velocity is the flow velocity $\boldsymbol{u}$. Let $D(t) = \mathcal{V}$, and notice that the time derivative on the right-hand side of (1.1.38) has been shifted inside the volume integral, we may replace $\mathcal{V}$ by a fixed *control volume* V that is instantaneously coincide with $\mathcal{V}$. Thus, (1.1.38) yields the *material derivative* of the integral:

$$\frac{d}{dt}\int_{\mathcal{V}} \mathcal{F}dv = \int_V \frac{\partial \mathcal{F}}{\partial t}dV + \int_{\partial V} \boldsymbol{n}\cdot\boldsymbol{u}\mathcal{F}dS \tag{1.1.39a}$$

$$= \int_V \left[\frac{\partial \mathcal{F}}{\partial t} + \nabla\cdot(\boldsymbol{u}\mathcal{F})\right]dV \tag{1.1.39b}$$

$$= \int_V \left(\frac{D\mathcal{F}}{Dt} + \vartheta\mathcal{F}\right)dV. \tag{1.1.39c}$$

These formulas give alternative expressions of the time rate of the material-volume integral of the $\mathcal{F}$-field. Equation (1.1.39a) is known as the **Reynolds transport theorem**. It indicates that the time variation of $\mathcal{F}$ in $\mathcal{V}$ has two parts: one is due to the local time derivative of $\mathcal{F}$ in V, and the other due to the moving of $\mathcal{V}$ which brings some $\mathcal{F}$ across the boundary ∂V with the rate $u_n\mathcal{F}$ per unit area, see Fig. 1.9. Equation (1.1.39b) is from (1.1.39a) by the Gauss theorem (Appendix A.2.1), which can in turn be cast to (1.1.39c) by using the fact $\nabla\cdot(\boldsymbol{u}\mathcal{F}) = \boldsymbol{u}\cdot\nabla\mathcal{F} + \vartheta\mathcal{F}$ and (1.1.34).

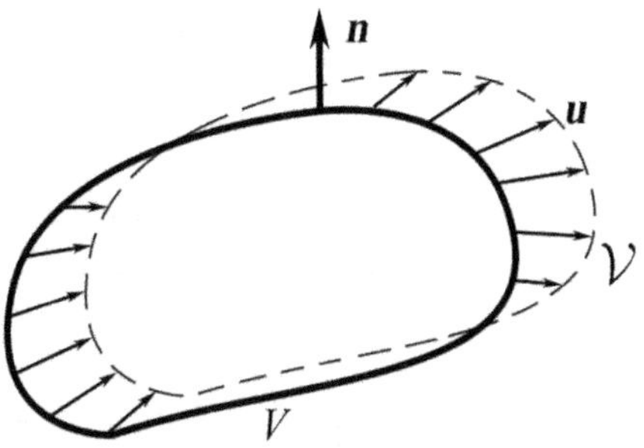

Fig. 1.9 Fluid motion in a control volume V that leads to the Reynolds transport theorem

The local and global variations of any quantity $\mathcal{F}$ in a flow field, (1.1.34) and (1.1.39), are the basis of deriving basic equations of fluid dynamics in the next section.

In addition to the above formulas, we shall also deal with the rate of change of integrals over material lines and surfaces. Since the fluid particles forming a material line, surface, or volume do not change as time, the operator d/dt in front of any of such integrals can be shifted into the integration symbol to become D/Dt therein. But line element $d\boldsymbol{x}$ and surface element[8] $d\boldsymbol{S} = \boldsymbol{n}dS$ in these integrals all vary as time. Thus we first need to list their material rates of change (the rate of change of dv is also included for completeness):

$$\frac{D}{Dt}(d\boldsymbol{x}) = d\boldsymbol{x} \cdot \nabla \boldsymbol{u}, \tag{1.1.40}$$

$$\frac{D}{Dt}(d\boldsymbol{S}) = d\boldsymbol{S} \cdot \mathbf{B}, \tag{1.1.41}$$

$$\frac{D}{Dt}(dv) = \vartheta dv, \tag{1.1.42}$$

where $\mathbf{B}$ is the divergence-free surface-deformation tensor introduced by (1.1.31). Equation (1.1.40) follows directly from (1.1.4) by setting $d\boldsymbol{r} = d\boldsymbol{x}$, while (1.1.42) follows from (1.1.39c) by setting $\mathcal{F} = 1$. To derive (1.1.41), we construct a volume element $dv = d\boldsymbol{x} \cdot d\boldsymbol{S}$ and then use (1.1.40) and (1.1.42) (Problem 1.7). Therefore, in addition to (1.1.39c), we can write down the general rules of the rate of change of the integrals of any quantity $\mathcal{F}$ over material line C and surface S:

$$\frac{d}{dt}\int_C d\boldsymbol{x} \circ \mathcal{F} = \int_C \left\{ d\boldsymbol{x} \circ \frac{D\mathcal{F}}{Dt} + (d\boldsymbol{x} \cdot \nabla \boldsymbol{u}) \circ \mathcal{F} \right\}, \tag{1.1.43}$$

$$\frac{d}{dt}\int_S d\boldsymbol{S} \circ \mathcal{F} = \int_S \left\{ d\boldsymbol{S} \circ \frac{D\mathcal{F}}{Dt} + (d\boldsymbol{S} \cdot \mathbf{B}) \circ \mathcal{F} \right\}, \tag{1.1.44}$$

[8] A surface element is a vector consisting of its normal direction $\boldsymbol{n}$ and area dS.

where $\circ$ denotes any meaningful tensor product for generality, including scalar product, inner product, cross product, tensor product, etc.

As an important example of (1.1.43), we apply it to consider the rate of change of the *circulation* along a material loop $\mathcal{C}$, defined by (1.1.14) and denoted by $\Gamma_{\mathcal{C}}$. Set $\mathcal{F} = \boldsymbol{u}$ in (1.1.43). Since $d\boldsymbol{x} \cdot \nabla\boldsymbol{u} = d\boldsymbol{u}$ and $d\boldsymbol{u} \cdot \boldsymbol{u} = d(q^2/2)$ with $q = |\boldsymbol{u}|$, its integral over a material loop vanishes because q^2 is single-valued function of $\boldsymbol{x}$. Thus, we immediately obtain the famous *Kelvin circulation formula*

$$\frac{d\Gamma_{\mathcal{C}}}{dt} = \frac{d}{dt}\int_S \boldsymbol{\omega} \cdot \boldsymbol{n}dS = \oint_{\mathcal{C}} \boldsymbol{a} \cdot d\boldsymbol{x} = \int_S (\nabla \times \boldsymbol{a}) \cdot \boldsymbol{n}dS \tag{1.1.45}$$

for a closed *material* curve $\mathcal{C}$, where $\boldsymbol{a}$ is the fluid acceleration.

A few words are needed for the less familiar surface-deformation process. For a surface element of unit area, by (1.1.41) there is

$$\frac{1}{dS}\frac{D}{Dt}(d\boldsymbol{S}) = \frac{D\boldsymbol{n}}{Dt} + \frac{\boldsymbol{n}}{dS}\frac{D}{Dt}dS \tag{1.1.46a}$$

$$= \boldsymbol{n} \cdot \mathbf{B} = -(\boldsymbol{n} \times \nabla) \times \boldsymbol{u}. \tag{1.1.46b}$$

Thus, the change of $d\boldsymbol{S}$ consists of a rotation of its direction $\boldsymbol{n}$ and a change of its area; and, because $\boldsymbol{n} \times \nabla$ in (1.1.46b) involves only tangential gradient, at any point *on a material surface the rate of change of* $d\boldsymbol{S}$ *depends solely on the velocity on that surface*.

As a kinematic application of (1.1.46a) and the triple decomposition (1.1.30), consider the expressions of $\nabla\boldsymbol{u}$ and $\mathbf{D}$ of a viscous fluid element sticking to a stationary solid wall with $\boldsymbol{u} = \mathbf{0}$ due to viscous adherence (see Sect. 1.3.1). Let $\boldsymbol{n}$ be the wall normal pointing out of the fluid, and denote the tangential gradient by ∇_π. Then, due to the adherence, there is $\nabla_\pi\boldsymbol{u} = \mathbf{0}$ at the stationary wall. Meanwhile, since the surface element $d\boldsymbol{S}$ on the wall does not change as time in either area or direction, by (1.1.46a) there also is $\boldsymbol{n} \cdot \mathbf{B} = \mathbf{0}$. Therefore, we simply have $\nabla\boldsymbol{u} = \boldsymbol{n}(\boldsymbol{n} \cdot \nabla\boldsymbol{u})$, and hence by (1.1.18) and (1.1.30) obtain a pair of elegant formulas

$$\nabla\boldsymbol{u} = \boldsymbol{n}\boldsymbol{n}\vartheta + \boldsymbol{n}(\boldsymbol{\omega} \times \boldsymbol{n}), \tag{1.1.47a}$$

$$2\mathbf{D} = 2\boldsymbol{n}\boldsymbol{n}\vartheta + \boldsymbol{n}(\boldsymbol{\omega} \times \boldsymbol{u}) + (\boldsymbol{\omega} \times \boldsymbol{n})\boldsymbol{n}, \tag{1.1.47b}$$

in which only three independent components of $\nabla\boldsymbol{u}$ and $\mathbf{D}$, i.e., ϑ and $\boldsymbol{\omega}$, are relevant. Equation (1.1.47b) is due to Caswell (1967). These formulas are important in understanding near-wall fluid kinematics. They can be extended to rigid rotating boundary, and more generally, arbitrarily moving and deforming flexible boundary. In these cases the surface deformation tensor $\mathbf{B}$ plays an important role.

1.2 Dynamic Equations of Fluid Motion

For a general fluid, the local conservative equations for mass, momentum and energy are derived from their corresponding integral conservation laws, which can be conveniently obtained by assigning $\mathcal{F}$ with density ρ, momentum $\rho\boldsymbol{u}$, and total energy $\rho E = \rho(q^2/2 + e)$ ($q = |\boldsymbol{u}|$ and e is the specific internal energy) in turn in the general integral formulas (1.1.39). While the procedure of deriving these governing equations is well stated in standard textbooks, we shall pay special attention to their underlying physics.

1.2.1 Dynamic Equations for General Fluids

We first consider a general fluid.

1. Mass conservation. Setting $\mathcal{F} = \rho$ in (1.1.39) and using the mass conservation law, we obtain

$$\int_{\mathcal{V}} \left(\frac{D\rho}{Dt} + \rho\vartheta \right) dv = 0,$$

from which the differential form of the *continuity equation* follows due to the arbitrariness of $\mathcal{V}$:

$$\frac{D\rho}{Dt} + \rho\vartheta = \frac{\partial\rho}{\partial t} + \nabla\cdot(\rho\boldsymbol{u}) = 0, \tag{1.2.1}$$

where the second expression takes the so-called *conservative form*, implying that it is directly from (1.1.39a) for a fixed control volume.

Now, write $\mathcal{F} = \rho\widehat{\mathcal{F}}$ such that $\widehat{\mathcal{F}}$ is a quantity of unit mass, as a corollary of the mass conservation, from (1.1.39c) and (1.2.1) we obtain a very convenient formula

$$\frac{d}{dt}\int_{\mathcal{V}} \rho\widehat{\mathcal{F}} dv = \int_{V} \rho\frac{D\widehat{\mathcal{F}}}{Dt} dV, \tag{1.2.2}$$

of which the combination with (1.1.39b) yields an alternative form of the material derivative of $\mathcal{F}$:

$$\rho\frac{D\widehat{\mathcal{F}}}{Dt} = \frac{\partial}{\partial t}(\rho\widehat{\mathcal{F}}) + \nabla\cdot(\rho\boldsymbol{u}\widehat{\mathcal{F}}). \tag{1.2.3}$$

These results will be very useful below.

2. Momentum balance and stress tensor. As an extension of Newton's second law to fluid, the rate of change of fluid momentum of a material volume $\mathcal{V}$ must be balanced by the total body force exerted over $\mathcal{V}$ and *surface stress* exerted on its boundary $\partial\mathcal{V}$. Assigning $\widehat{\mathcal{F}}$ as velocity $\boldsymbol{u}$, by (1.1.39) and (1.2.2) the integral momentum balance reads

$$\int_V \rho \frac{D\boldsymbol{u}}{Dt} dv = \int_V \rho \boldsymbol{f} dV + \int_{\partial V} \boldsymbol{t} dS, \tag{1.2.4}$$

where $\boldsymbol{f}$ is external body force of unit mass.

Notice that the surface stress $\boldsymbol{t}$ is a vector function of both $\boldsymbol{x}$ and the orientation of the surface element $d\boldsymbol{S} = \boldsymbol{n} dS$ as well as time t. For strictly inviscid fluid the dependence of $\boldsymbol{t}$ on $\boldsymbol{n}$ is simply $\boldsymbol{t} = -\boldsymbol{n}p$, only a normal force, where p is the *static pressure*. This comes from a kinetic assumption that at every time the energy distribution among different molecules and different degrees of freedom of the same molecules is always in equilibrium as maintained by molecular collisions.

In real fluid, however, between collisions molecules drift to new positions, and take a short but nonzero time (a few collisions) before taking up the mean energy and momentum characteristic of their new positions. These finite-time processes cause the slight departure from the equilibrium distribution of energy and lead to the phenomena of *viscosity*, *heat conduction*, and *diffusion*.[9] Here, the viscosity makes the stress gain tangential components as well as normal one. For an infinitesimal surface element $d\boldsymbol{S}$, $\boldsymbol{t}$ can only depend on $\boldsymbol{n}$ linearly, but the dependence must be extended to a general linear combination of their components, namely

$$t_i(\boldsymbol{n}, \boldsymbol{x}, t) = T_{ij}(\boldsymbol{x}, t) n_j(\boldsymbol{x}, t)$$

with T_{ij} being the coefficients. Then since both $\boldsymbol{t}$ and $\boldsymbol{n}$ are vectors, by the *tensor identification theorem* (see Appendix A.1.3) T_{ij} must be the components of a tensor $\mathbf{T}$, namely $\boldsymbol{t} = \mathbf{T} \cdot \boldsymbol{n}$. This simple yet rigorous argument proves the existence of the *stress tensor*.

Then, let $\boldsymbol{r} = \boldsymbol{x} - \boldsymbol{x}_0$ be the position vector relative to any fixed point $\boldsymbol{x}_0$, the *integral angular momentum balance* is *postulated* as

$$\frac{d}{dt} \int_{\mathcal{V}} \rho \boldsymbol{r} \times \boldsymbol{u} dv = \int_V \rho \boldsymbol{r} \times \boldsymbol{f} dV + \int_{\partial V} \boldsymbol{r} \times \boldsymbol{t} dS, \tag{1.2.5}$$

where again $\boldsymbol{t} = \mathbf{T} \cdot \boldsymbol{n}$. Then by this equation one may easily prove that $T_{ij} = T_{ji}$, e.g. $\mathbf{T}$ must be a symmetric tensor, and hence

$$\boldsymbol{t}(\boldsymbol{x}, \boldsymbol{n}) = \boldsymbol{n} \cdot \mathbf{T}(\boldsymbol{x}) = \mathbf{T}(\boldsymbol{x}) \cdot \boldsymbol{n}. \tag{1.2.6}$$

---

[9] See, e.g. Lighthill (1956).

Once this is acknowledged in the momentum and energy equations, the angular-momentum balance is no longer an independent equation and needs not to be considered.[10]

We now substitute $\boldsymbol{t} = \boldsymbol{n} \cdot \mathbf{T}$ into (1.2.4) and use the generalized Gauss theorem. Owing to the arbitrariness of V, this yields the differential form of the momentum balance, known as *Cauchy motion equation* and applicable to any continuous media:

$$\rho \frac{D\boldsymbol{u}}{Dt} = \rho \boldsymbol{f} + \nabla \cdot \mathbf{T}. \tag{1.2.7}$$

Here, the physical meaning of $\nabla \cdot \mathbf{T}$ can be clearly understood by considering a small volume V_ϵ:

$$\nabla \cdot \mathbf{T} = \lim_{V_\epsilon \to 0} \left(\frac{1}{V_\epsilon} \int_{\partial V_\epsilon} \boldsymbol{t} dS \right), \tag{1.2.8}$$

which is the *resultant surface stress* over the boundary of infinitesimal unit volume.

3. Total-energy balance. Finally, assigning $\widehat{\mathcal{F}}$ as total energy $E = e + q^2/2$ gives the integral total-energy balance, showing that the rate of change of total-energy of volume $\mathcal{V}$ is caused by the work done by external body force and surface stress, as well as heat addition and heat conduction:

$$\int_V \rho \frac{DE}{Dt} dv = \int_V \rho(\boldsymbol{f} \cdot \boldsymbol{u} + \dot{Q}) dV + \int_{\partial V} (\boldsymbol{t} \cdot \boldsymbol{u} - \boldsymbol{n} \cdot \boldsymbol{q}) dS, \tag{1.2.9}$$

where $\rho \boldsymbol{f} \cdot \boldsymbol{u}$ and $\boldsymbol{t} \cdot \boldsymbol{u}$ are the work rate done by the body force and surface stress, respectively; $\dot{Q}$ is the time rate of heat addition per unit mass; and $\boldsymbol{q}$ is the heat flux across the boundary, with the minus sign indicating that the flux into V causes an internal-energy increase therein. Similar to the momentum equation, the local differential form of the total-energy balance reads

$$\rho \frac{DE}{Dt} = \rho \boldsymbol{f} \cdot \boldsymbol{u} + \nabla \cdot (\mathbf{T} \cdot \boldsymbol{u}) - \nabla \cdot \boldsymbol{q} + \rho \dot{Q}. \tag{1.2.10}$$

Here, like (1.2.8),

$$\nabla \cdot (\mathbf{T} \cdot \boldsymbol{u}) = \lim_{V_\epsilon \to 0} \left(\frac{1}{V_\epsilon} \int_{\partial V_\epsilon} \boldsymbol{t} \cdot \boldsymbol{u} dS \right), \tag{1.2.11}$$

which is the *resultant work rate done by surface stress* over the boundary of infinitesimal unit volume.

[10]The postulation (1.2.5) holds for most common fluids but is not universally true. In those exceptional cases the angular-momentum balance is an independent law and stress tensor is no longer symmetric.

1.2.2 Constitutive Relations and Thermodynamics

1. Constitutive relations. To make (1.2.1), (1.2.7), and (1.2.10) solvable, we have to fix the specific expression for the stress tensor. In strictly inviscid fluid or *ideal fluid*, there is $\mathbf{T} = -p\mathbf{I}$ and (1.2.7) reduces to the *Euler equation*:

$$\rho\frac{D\boldsymbol{u}}{Dt} = \rho\boldsymbol{f} - \nabla p. \tag{1.2.12}$$

Then for real viscous fluid we write

$$\mathbf{T} = -p\mathbf{I} + \mathbf{V}, \quad \boldsymbol{t} = -p\boldsymbol{n} + \boldsymbol{n}\cdot\mathbf{V}, \tag{1.2.13}$$

where the symmetric *viscous stress tensor* $\mathbf{V}$ comes from the aforementioned slightly non-equilibrium molecular interactions, which causes inner friction when there is a relative motion between various parts of the fluid. If the relative velocity is small, we may assume the inner friction depends only on the first-order velocity derivatives, and V_{ij} is a linear function of $u_{j,i}$. Then, for constant $\boldsymbol{u}$ or rigid rotation there should be no friction; thus V_{ij} should be a linear combination of $u_{j,i} + u_{i,j} = 2D_{ij}$ since it vanishes for constant $\boldsymbol{u}$ or rigid rotation. This is also consistent with the symmetry of V_{ij}. The coefficients of this linear relation are called *viscous coefficients* or simply *viscosities*. For homogeneous and isotropic media like water and air, which are our concern, it can be proved that the viscous coefficients are scalars. This kind of fluids is known as *Newtonian Fluids*. Thus, the desired linear relation takes the following alternative forms, known as the *mechanical constitutive relation*:

$$\begin{aligned}
\mathbf{V} &= \lambda\vartheta\mathbf{I} + 2\mu\mathbf{D} && (1.2.14a)\\
&= \zeta\vartheta\mathbf{I} + 2\mu\mathbf{D}', \quad \zeta \equiv \lambda + \frac{2}{3}\mu, && (1.2.14b)\\
&= \mu_\theta\vartheta\mathbf{I} + 2\mu\boldsymbol{\Omega} - 2\mu\mathbf{B}, \quad \mu_\theta \equiv \lambda + 2\mu = \zeta + \frac{4}{3}\mu, && (1.2.14c)
\end{aligned}$$

where

$$\mathbf{D}' = \mathbf{D} - \frac{1}{3}\vartheta\mathbf{I} \quad \text{with} \quad \mathrm{tr}\mathbf{D}' = 0 \tag{1.2.15}$$

is the *deviator* of $\mathbf{D}$ and represents a pure deformation from a sphere, say, to ellipsoid of the same volume by squeezing or "straining". In the above expressions of viscous stress tensor, (1.2.14a) is the original form postulated as an analogy with the Lamé constants in elasticity, with μ and λ being named the *shear viscosity* and second viscosity, respectively. But this form is not the clearest expression and one uses more often the other two forms derived from (1.2.14a). Equation (1.2.14b) comes from (1.2.15) with ζ being called the *bulk viscosity* by analogy with the bulk modulus in elasticity. Tensor $2\mu\mathbf{D}'$ is actually the whole viscous stress tensor for *monatomic gases*, of which the molecules have only translational energy. Namely,

$$\mathbf{V} = 2\mu\mathbf{D}' \quad \text{for monatomic gas.} \tag{1.2.16}$$

Comparing (1.2.16) with (1.2.14b) gives

$$3\lambda + 2\mu = 0 \quad \text{for monatomic gas,} \tag{1.2.17}$$

known as the *Stokes relation*. But for *diatomic gases* the molecules have two rotational degrees of freedom of which the energy change slightly lags behind that of translational energy, making the mean of the isotropic part of $\mathbf{T}$ slightly differ from the pressure. In this case the remaining isotropic part $\vartheta\mathbf{I}/3$ of $\mathbf{D}$ should be merged into the dilatation term of (1.2.14a), leading to (1.2.14b). But the bulk viscosity ζ has not yet exhausted the full viscous resistance to the isotropic expansion/compression. The latter should be $\mu_\theta\vartheta$ as revealed by (1.2.14c), which is from the triple decomposition (1.1.30a). In fact, for one-dimensional flow with vanishing $\boldsymbol{\Omega}$ and $\mathbf{B}$, μ_θ is the sole viscosity. At higher dimensions two or three, the respective roles of μ (assumed constant here) and μ_θ can be seen most clearly by substituting (1.2.14c) into (1.2.7):

$$\nabla \cdot \mathbf{V} = \nabla(\mu_\theta\vartheta) - \nabla \times (\mu\boldsymbol{\omega}), \tag{1.2.18}$$

indicating that μ_θ and μ are involved dynamically in viscous compressing and shearing processes, respectively.

The stress $\boldsymbol{t} = \boldsymbol{n} \cdot \mathbf{T}$ can now be decomposed according to (1.2.13) and (1.2.14). On an arbitrary surface, either inside the fluid or at a deforming-moving boundary, the general stress formula is

$$\boldsymbol{t} = (-p + \mu_\theta\vartheta)\boldsymbol{n} + \mu\boldsymbol{\omega} \times \boldsymbol{n} + \boldsymbol{t}_s, \tag{1.2.19a}$$

where

$$\boldsymbol{t}_s = -2\mu\boldsymbol{n} \cdot \mathbf{B} = 2\mu(\boldsymbol{n} \times \nabla) \times \boldsymbol{u} \tag{1.2.19b}$$

is the stress due to surface-deformation rate, which is a viscous resistance to the variation of the direction and area of the surface element $\boldsymbol{n}dS$. Unlike the normal and shear stresses, when μ is constant the resultant force due to $\boldsymbol{t}_s$ acting on an open surface S has a simple expression

$$\int_S \boldsymbol{t}_s dS = 2\mu \oint_{\partial S} d\boldsymbol{x} \times \boldsymbol{u}, \tag{1.2.20}$$

which vanishes on a closed surface without boundary line.

Similar to the mechanical constitutive relation between $\mathbf{T}$ and $\mathbf{D}$, a *thermodynamic constitutive relation* is assumed between the heat conduction vector $\boldsymbol{q}$ and temperature T, known as the *Fourier law*:

$$\boldsymbol{q} = -k\nabla T, \tag{1.2.21}$$

where k is heat conductivity. For the purpose of this book it suffices to treat μ, k, and ζ constant.[11] μ and k can be related by the dimensionless *Prandtl number*

$$Pr = \frac{\mu c_p}{k}, \tag{1.2.22}$$

that measures the relative importance of dissipation and heat conduction and is a constant for each gas. For water and gases $Pr = O(1)$, and hence unless $|\nabla T|$ is extremely high, if one of μ and k is to be neglected so should be the other. This is the case for ideal fluids. In a viscous flow, the viscous forces in certain regions may be self-balanced and hence those regions behave as if the viscosity is zero. However, in other regions of the same flow viscous forces may become important.

2. Energy dissipation. Having obtained constitutive relations, we now make an important observation by considering the rate of change of kinetic energy $q^2/2$, derived as a corollary of the momentum equation (1.2.7). Its inner product with $\boldsymbol{u}$ leads to

$$\begin{aligned}\rho\frac{D}{Dt}\left(\frac{1}{2}q^2\right) &= \rho\boldsymbol{f}\cdot\boldsymbol{u} + \boldsymbol{u}\cdot(\nabla\cdot\mathbf{T})\\ &= \rho\boldsymbol{f}\cdot\boldsymbol{u} + \nabla\cdot(\mathbf{T}\cdot\boldsymbol{u}) + p\vartheta - \mathbf{V}:\mathbf{D},\end{aligned} \tag{1.2.23}$$

where $p\vartheta$ is the work rate done by pressure as it compresses the fluid. Note that one cannot stop at the first line but has to shift $\boldsymbol{u}$ in its last term into the operator ∇, which as seen in (1.2.11) represents the resultant work rate done by surface stress over the small-volume boundary. It is this step that produces an extra term, which by (1.2.13) reads

$$\Phi \equiv \mathbf{V}:\mathbf{D} = \lambda\vartheta^2 + 2\mu\mathbf{D}:\mathbf{D} \tag{1.2.24a}$$
$$= \zeta\vartheta^2 + 2\mu\mathbf{D}':\mathbf{D}' \tag{1.2.24b}$$
$$= \mu_\theta\vartheta^2 + \mu\omega^2 - 2\mu\nabla\cdot(\mathbf{B}\cdot\boldsymbol{u}) \tag{1.2.24c}$$

is the *dissipation rate*, where the third expression comes from (1.1.32). Therefore, due to the appearance of viscous stress tensor $\mathbf{V}$, the mechanical work done by body force and surface stress does not completely become kinetic energy. Rather, a part

[11] In general, μ and k are at most functions of ρ and T as the physical properties of a fluid; but ζ has been found to depend on not only ρ, T but also frequency of the gas motion, and can be enhanced enormously at very high frequencies (cf. Landau and Lifshitz 1959).

of it must be transformed irreversibly into heat. This is where one finds the necessity to invoke thermodynamics and consider the balance of total energy E, both kinetic and internal.

3. Local-equilibrium thermodynamics. Due to the involvement of thermodynamics, some basic concepts and equations are needed. In addition to state variables p, ρ and *absolute temperature* T, three more state variables are often used: *specific internal energy* e, *specific enthalpy* h, and *specific entropy* s. For common gases, e and h can be defined by

$$e = c_v T, \quad h = e + \frac{p}{\rho} = c_p T, \tag{1.2.25}$$

respectively, with c_v and c_p being specific heats at constant volume and pressure, respectively. The fundamental differential equation of thermodynamics for $s(e, \rho)$ or $s(h, p)$ is

$$T ds = de + p d\left(\frac{1}{\rho}\right) = dh - \frac{1}{\rho} dp, \tag{1.2.26}$$

where the differential operator d can be replaced by ∇, D/Dt, or $\partial/\partial t$, depending on the context.

Now, the above physical identification of Φ can be confirmed by considering the governing equation for internal energy e by subtracting the kinetic-energy equation (1.2.23) from the total-energy equation (1.2.10):

$$\rho \frac{De}{Dt} = -p\vartheta + \Phi - \nabla \cdot \boldsymbol{q} + \rho \dot{Q}, \tag{1.2.27}$$

which can in turn be cast to the *entropy equation* by a combination of (1.2.26) and the continuity equation,

$$\rho \frac{De}{Dt} = \rho T \frac{Ds}{Dt} - p\vartheta,$$

yielding

$$\rho T \frac{Ds}{Dt} = \Phi - \nabla \cdot \boldsymbol{q} + \rho \dot{Q}. \tag{1.2.28}$$

Thus, as expected, the dissipation represents precisely the part of work rate done by viscous surface stress that is irreversibly transferred to internal energy and becomes a source of entropy. Another source of entropy is heat conduction. Note that $p\vartheta$ causes a *reversible* change of the internal energy, and is absent in the entropy equation.

Of the work rate done and dissipation rate produced by various stress constituents, that by $\boldsymbol{t}_s$ is very special. When μ is constant, for any volume V there evidently is

$$\int_{\partial V} \boldsymbol{t}_s \cdot \boldsymbol{u} dS = -2\mu \int_{\partial V} \boldsymbol{n} \cdot \mathbf{B} \cdot \boldsymbol{u} dS = -2\mu \int_V \nabla \cdot (\mathbf{B} \cdot \boldsymbol{u}) dV,$$

which is precisely the integral of the last term of (1.2.24c) and hence cancelled in (1.2.23). Therefore, *for constant viscosity the work done by the viscous resistance to surface deformation never changes the kinetic energy but is directly and totally dissipated into heat* (Wu et al. 1999).

A remark on the feature and role of thermodynamics in fluid dynamics is in order here (e.g. Lagerstrom 1964). Classical thermodynamics deals with very general systems, but is restricted to the study of equilibrium states. Its concern is relation between initial and final equilibrium states even if the transition process is irreversible. In both states the thermodynamic variables have uniform distribution. The transient process between the two states is beyond the scope of equilibrium thermodynamics, but has to be taken into account in fluid dynamics. To reach a rational combination of thermodynamics and fluid dynamics, therefore, the state variables of equilibrium thermodynamics must be generalized so that they have meanings in nonequilibrium states, in the sense that a *local equilibrium* is still reached in every fluid element during the transient process. Meanwhile, concepts such as viscous stress tensor and heat conduction vector must be introduced, which have no analogy in equilibrium thermodynamics.

Among various thermodynamic state variables, the physical implications of pressure p are quite rich. For ideal fluid, it enters the stress tensor $\mathbf{T}$ as its isotropic part, namely by setting $\boldsymbol{t} = -p\boldsymbol{n}$ in (1.2.8) there is

$$-\nabla p = \lim_{V_\epsilon \to 0} \left(\frac{1}{V_\epsilon} \int_{\partial V_\epsilon} -p\boldsymbol{n} dS \right).$$

This *mechanic pressure* is then identified to be the same as the *thermodynamic pressure*. But for a viscous fluid there is an extra isotropic normal stress $\mu_\theta \vartheta \boldsymbol{n}$, so the pressure p becomes only the inviscid part of this stress. Moreover, since e and ρ are well defined in nonequilibrium states, the thermodynamic relation (1.2.26) is considered an equality for both equilibrium and irreversible thermodynamics, but for the latter the concept of p is extended again, as explained by Lighthill (1956): "*the thermodynamic pressure has been redefined so that at each stage its relation to other thermodynamic variables is the same as in equilibrium. The difference is because with this definition of pressure the work done by the fluid is really less than* $p d\rho^{-1}$; *hence* $T ds$ *still exceeds the total energy supplied.*"

The generalization of equilibrium thermodynamics in fluid dynamics is also seen in the total-energy equation (1.2.10), where has been implicitly assumed that e is defined in nonequilibrium state. Therefore, this equation is an extension rather than a consequence of the first law of equilibrium thermodynamics. Accordingly, the second law of thermodynamics has to be extended as well. The entropy equation (1.2.28) implies that the extended second law will be ensured for any flow if

$$\Phi \geq 0, \quad \boldsymbol{q} \cdot \nabla T \leq 0,$$

which by (1.2.21) and (1.2.24) in turn requires

$$\zeta \geq 0, \quad \mu \geq 0, \quad k \geq 0. \tag{1.2.30}$$

Once ζ, μ, and k satisfy these inequalities in viscous and heat-conducting flows (they do from their measured values), the second law is fully satisfied.

1.2.3 Navier-Stokes Equations and Perfect Gas

From now on we assume the shear viscosity μ, longitudinal viscosity μ_θ, and heat conductivity k are constant, and the rate of heat addition $\dot{Q}$ is zero. Then by (1.2.14c), (1.2.21), and (1.2.18), plus the state equation of gas, the complete set of fundamental equations can be chosen as

$$\frac{D\rho}{Dt} + \rho\vartheta = 0, \tag{1.2.31a}$$

$$\rho\frac{D\boldsymbol{u}}{Dt} = \rho\boldsymbol{f} + \nabla(-p + \mu_\theta\vartheta) - \nabla \times (\mu\boldsymbol{\omega}), \tag{1.2.31b}$$

$$\rho T\frac{Ds}{Dt} = \Phi + k\nabla^2 T, \tag{1.2.31c}$$

$$f(p, \rho, T) = 0, \tag{1.2.31d}$$

where Φ is the dissipation function given by (1.2.24), and (1.2.31d) is the equation of state depending on the specific fluid. The momentum equation (1.2.31b) is the famous *Navier-Stokes equation* (N-S equation for short), and the whole set of equations in (1.2.31) are often called the Navier-Stokes equations. Under proper initial and boundary conditions to be discussed in Sect. 1.3.1, the Navier-Stokes equations solve three velocity components and two independent thermodynamic variables (say, p, ρ or T, s) by five scalar equations, and the rest of state variables are deduced from (1.2.31d).

In fluid dynamics it is often convenient to express the energy equation in terms of enthalpy $h = e + p/\rho$ or *total* enthalpy $H = h + q^2/2$. To derive their equations, we first notice that by (1.2.31a) and the definition of h we have

$$p\vartheta = p\rho\frac{D}{Dt}\left(\frac{1}{\rho}\right), \quad \rho\frac{Dh}{Dt} = \rho\left[\frac{De}{Dt} + p\frac{D}{Dt}\left(\frac{1}{\rho}\right)\right] + \frac{Dp}{Dt}.$$

Substituting these into in (1.2.27) yields the *enthalpy equation*

$$\rho\frac{Dh}{Dt} = \frac{Dp}{Dt} + \nabla \cdot (k\nabla T) + \Phi. \tag{1.2.33}$$

Then, by using (1.2.7) and (1.2.13) there is

$$\begin{aligned}\frac{Dp}{Dt} &= \frac{\partial p}{\partial t} + \boldsymbol{u} \cdot \nabla p = \frac{\partial p}{\partial t} - \rho \boldsymbol{u} \cdot \frac{D\boldsymbol{u}}{Dt} + \boldsymbol{u} \cdot (\nabla \cdot \mathbf{V}) \\ &= \frac{\partial p}{\partial t} - \rho \frac{D}{Dt}\left(\frac{1}{2}q^2\right) + \nabla \cdot (\boldsymbol{u} \cdot \mathbf{V}) - \Phi.\end{aligned}$$

Thus, (1.2.33) is cast to the *total-enthalpy equation*

$$\rho \frac{DH}{Dt} = \frac{\partial p}{\partial t} + \nabla \cdot (\boldsymbol{u} \cdot \mathbf{V}) + \nabla \cdot (k \nabla T), \tag{1.2.34}$$

which is especially convenient for steady flow.

In this book we use *perfect gas* as the model of compressible fluid, of which some basic properties are outlined here. Its equation of state is

$$p = \rho RT \quad \text{or} \quad pv = RT, \quad R = c_p - c_v = c_v(\gamma - 1), \tag{1.2.35}$$

where $v = 1/\rho$ is the *specific volume*, $R > 0$ is a constant, and $\gamma = c_p/c_v$ is the ratio of specific heats. A perfect gas is said to be *polytropic* or *calorically perfect* if c_p, c_v, and γ are assumed constant, to which we shall be exclusively confined. Its entropy can be inferred from two of p, ρ and T by (1.2.26) and (1.2.35),

$$\frac{ds}{c_v} = \frac{dT}{T} + \frac{pdv}{c_v T} = \gamma \frac{dT}{T} - \frac{vdp}{c_v T},$$

so that

$$\begin{aligned}\frac{ds}{c_v} &= \gamma d(\ln T) - (\gamma - 1)d(\ln p) = d\ln(T^{\gamma}/p^{\gamma-1}) \\ &= d(\ln T) - (\gamma - 1)d(\ln \rho) = d\ln(T/\rho^{\gamma-1}) \\ &= d(\ln p) - \gamma d(\ln \rho) = d\ln(p/\rho^{\gamma}).\end{aligned} \tag{1.2.36}$$

A flow with two independent thermodynamic variables is called *baroclinic*, while that with only one independent thermodynamic variable is called *barotropic*. For example, an isentropic flow of *polytropic gas* is barotropic, for which (1.2.36) yields isentropic relations of thermodynamic variables between two states 1 and 2 by

$$\frac{p_2}{p_1} = \left(\frac{\rho_2}{\rho_1}\right)^{\gamma} = \left(\frac{T_2}{T_1}\right)^{\gamma/(\gamma-1)}. \tag{1.2.37}$$

In Sect. 1.1.2 we have learned that compressibility is associated with longitudinal waves, of which the weak form is sound wave. The propagation speed a of sound wave is an important quantity, defined by

$$a^2 = \left(\frac{\partial p}{\partial \rho}\right)_s. \tag{1.2.38}$$

For perfect gas this gives, by (1.2.36),

$$a^2 = \gamma RT = (\gamma - 1)h = \frac{\gamma p}{\rho}. \tag{1.2.39}$$

Thus a^2 is proportional to T and its magnitude may serve as a measure of the internal energy in the flow field.

1.2.4 Dominant Non-dimensional Parameters

We now consider the dimensionless form of the N-S equations. While it is natural to nondimensionalize kinematic and dynamic variables by characteristic velocity U, length L, and time period τ or frequency $f = 1/\tau$, it is more natural to nondimensionalize thermodynamic variables by their constant reference values, say at infinity (Lagerstrom 1964). Thus we have two groups of dimensionless variables, denoted by asterisk and tilde, respectively:

$$\begin{aligned}
&x_i^* = \frac{x_i}{L}, \quad t^* = ft, \quad \boldsymbol{u}^* = \frac{\boldsymbol{u}}{U}, \quad p^* = \frac{p - p_\infty}{\rho_\infty U^2}, \quad \Phi^* = \frac{L^2 \Phi}{U^2 \mu_\infty}, \\
&\tilde{p} = \frac{p}{p_\infty}, \quad \tilde{T} = \frac{T}{T_\infty}, \quad \tilde{\rho} = \frac{\rho}{\rho_\infty}, \quad \tilde{h} = \frac{h}{h_\infty}, \quad \tilde{s} = \frac{s}{c_p}, \\
&\tilde{\mu} = \frac{\mu}{\mu_\infty} = O(1), \quad \tilde{\mu}_\theta = \frac{\mu_\theta}{\mu_\infty} = O(1),
\end{aligned} \tag{1.2.40}$$

where μ and μ_θ have been assumed constant. If the Stokes relation $3\lambda + 2\mu = 0$ or $\zeta = 0$ is used, we simply have $\tilde{\mu}_\theta = 4/3$; and, for polytropic gas we simply have $\tilde{h} = \tilde{T}$. p^* and $\tilde{p}$ appear simultaneously because pressure has both dynamic and thermodynamic implications.

Then, consider a polytropic gas and assume the external body force $\boldsymbol{f}$ and heat addition $\dot{Q}$ are absent. Multiplying the continuity equation (1.2.31a), momentum equation (1.2.31b), energy equation (1.2.31c), and state equation (1.2.35) by $L/\rho_\infty U$, $L/\rho_\infty U^2$, $L/U h_\infty \rho_\infty$, and $1/p_\infty$, respectively, we obtain dimensionless N-S equations

$$St \frac{\partial \tilde{\rho}}{\partial t^*} + \nabla^* \cdot (\tilde{\rho} \boldsymbol{u}^*) = 0, \tag{1.2.41a}$$

$$\tilde{\rho}\left(St \frac{\partial \boldsymbol{u}^*}{\partial t^*} + \boldsymbol{u}^* \cdot \nabla^* \boldsymbol{u}^*\right) = -\nabla^* p^* + \frac{1}{Re}[\nabla^*(\tilde{\mu}_\theta \vartheta^*) - \nabla^* \times \boldsymbol{\omega}^*], \tag{1.2.41b}$$

$$\tilde{\rho}\tilde{T}\left(St \frac{\partial \tilde{s}}{\partial t^*} + \boldsymbol{u}^* \cdot \nabla^* \tilde{s}\right) = \frac{1}{Re}\left[\frac{1}{Pr}\nabla^{*2}\tilde{h} + (\gamma - 1)M_\infty^2 \Phi^*\right], \tag{1.2.41c}$$

$$\tilde{p} = \tilde{\rho}\tilde{h}, \tag{1.2.41d}$$

where ϑ and $\boldsymbol{\omega}$ are scaled by U/L. In these equations there appear three major dynamic parameters, the *Strouhal number*, the *Reynolds number* and the *Mach number*:

$$St = \frac{fL}{U}, \quad Re = \frac{\rho_\infty UL}{\mu_\infty}, \quad M_\infty = \frac{U}{a_\infty}, \tag{1.2.42a}$$

and two fluid-property parameters

$$Pr = \left(\frac{\mu c_p}{k}\right)_\infty, \quad \gamma = \frac{c_p}{c_v}. \tag{1.2.42b}$$

Let L, f, U have been properly chosen such that all dimensionless variables in (1.2.41) are of the same $O(1)$. Then only the three dimensionless parameters can be of varying orders of magnitude. While St is a general measure of flow unsteadiness with a characteristic frequency,[12] Re and M are the key parameters for the two fundamental dynamic processes we are discussing.

The Reynolds number measures the relative importance of the inertial force compared with viscous force, and dominates the characteristic pattern of all shearing processes. In particular, it governs the vorticity evolution and diffusion patterns. For $Re \ll 1$, the vorticity diffuses to far field in a quite isotropic manner (the corresponding flow is called the *Stokes flow*). As Re increases, more vorticity is advected downstream by increasingly large velocity field, so that the fluid carrying strong vorticity is confined to a more and more narrow parabola-like (in two dimensions) wake region as sketched in Fig. 1.10. When $Re \gg 1$, the parabolic region becomes a very thin *boundary layer* attaching to the body surface and a thin wake downstream the body. Thus, the overall vortical flow pattern changes dramatically from $Re \ll 1$ to $Re \gg 1$.

The present book will not address flow theory at small Reynolds numbers, for which the reader may consult e.g. Batchelor (1967), Rosenhead (1963), Panton (2013), and other textbooks. By contrast, almost the entire vortex dynamics theory is devoted to the flows at large Re, which have very close relevance to human life on the Earth where the major moving fluids are air and water, both having very small viscosities. In particular, it is in this large-Re regime where there appear various highly localized *vortical structures* due to shearing. As suggested by Fig. 1.10, the primary vortical structures are *shear layers*, including firstly the just mentioned boundary layers which are the inevitable and direct product of the shearing process and no-slip condition at solid wall at large Re; and secondly the *free shear layers* which form after boundary layers leave the body and enter the interior of the flow. Then sufficiently thin free shear layers have a natural tendency to roll into even further localized *axial vortices* as the secondary vortical structures. It is these layer-like and axial vortices that play a key role as the sinews and muscles or organizers of

[12] For unsteady flow without characteristic frequency, one may use L and U to define the dimensionless time $t^* = Ut/L$, which corresponds to setting $St = 1$.

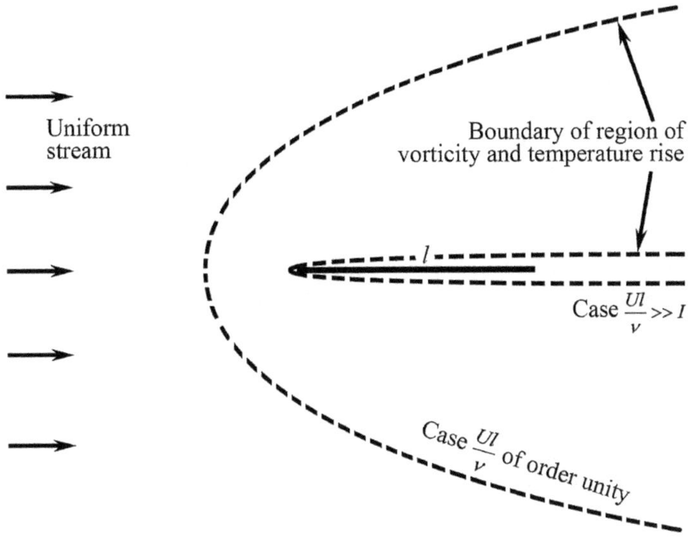

Fig. 1.10 Diffusion and convection of both vorticity and heat from a plate in stream. After Batchelor (1967)

complex flows. The dynamics of shear layers and axial vortices will be presented in Chaps. 4–7, following a general theory of vorticity dynamics.

A far-reaching property of large Reynolds-number flows is the instability of various shear layers and vortices, which causes transition from laminar flow to *turbulence*—the most complicated existence of shearing process—full of chaotic vortical structures (Chaps. 10 and 11). It will be seen that the occurrence and manner of vortical-flow instability and transition depend strongly on specific body geometries and specific values of Re; thus, flow patterns at $Re \gg 1$ do not simply become uninfluenced by viscosity and so independent of Re, but rather are markedly different at Re around 10^m for $m = 2, 3, 4, 5, 6, 7$ as stressed by Lighthill (1963). Therefore, *Reynolds number "is the liveliest of all the non-dimensional parameters"*.

Here we just mention that, in aerodynamics, one is mainly concerned with the mean behavior of strongly fluctuating turbulent flows, i.e., taking $\boldsymbol{u} = \boldsymbol{U} + \boldsymbol{u}'$ with zero mean fluctuation, $\overline{\boldsymbol{u}'} = \boldsymbol{0}$. Then, the mean of nonlinear advection in the Navier-Stokes equation produces a *turbulent force* $f_j = -\partial_i(\overline{\rho u'_i u'_j})$. This force could be modelled as $\nu_t \nabla^2 \boldsymbol{U}$, where ν_t (or something alike in more advanced models) behaves as a "turbulent viscosity" that is about of $O(10^3)$ of the molecular viscosity ν. In aerodynamics the turbulent Reynolds number $Re_{\text{tub}} = UL/\nu_t$ is still very large.

In contrast to the Reynolds number, (1.2.42a) shows that the primary meaning of M is the measure of how fast the fluid motion is compared to its sound wave, i.e., the propagation speed of small longitudinal disturbance. Moreover, for polytropic gas, by (1.2.39) there is

$$\frac{U^2}{\gamma RT} = M_\infty^2, \tag{1.2.43}$$

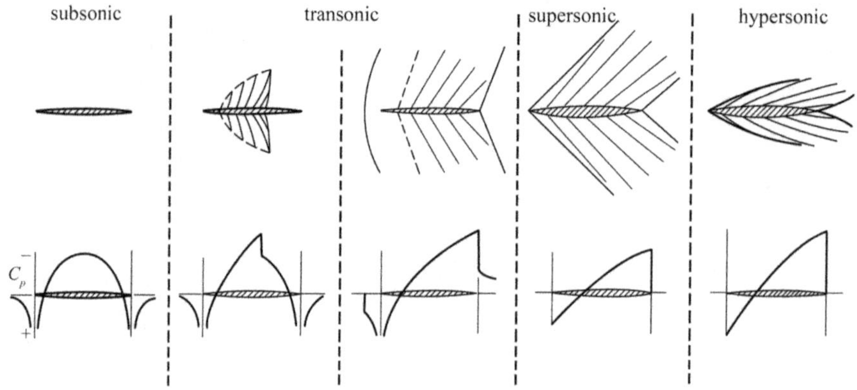

Fig. 1.11 Wave patterns around a thin airfoil at different flow regimes. In the *upper row*, *lines* denote Mach waves and shocks, and *dashed lines* in transonic flow regime mark the boundary of supersonic regions. From Liepmann and Roshko (1957)

indicating that the Mach number also measures how much the internal energy is excited by the fluid motion and interacts the latter. When $M_\infty^2 \ll 1$, the kinetic energy is much smaller than the internal energy, implying that the fluid motion can only excite very little internal energy, and hence the dissipation Φ^* can be ignored. Meanwhile, by (1.2.39) there is also

$$\tilde{p} = \gamma M^2 p^* + 1, \tag{1.2.44}$$

indicating that the thermodynamic role of the pressure, $\tilde{p} - 1$, decreases as M^2 reduces.

Similar to the critical influence the Reynolds number has on the shear-flow patterns sketched in Fig. 1.10, so does the Mach number on the compressible flow patterns, see Fig. 1.11.

1.3 Wall-Bounded Flows

1.3.1 Boundary Conditions

Specific solutions of the Navier-Stokes equations (1.2.31) depend on proper initial and boundary conditions. The thermodynamic boundary conditions at a solid wall are usually specified as either the temperature distribution or the heat flux $\boldsymbol{n} \cdot \boldsymbol{q} = -k\partial T/\partial n$ on the wall. Below we briefly outline mechanical boundary conditions.

In external flow problems, the fluid extends to infinity and we need a far-field condition. Typically, for a solid body that may have arbitrary motion and deformation in the fluid, we set $\boldsymbol{u} = \boldsymbol{0}$ at infinity. It is sometimes convenient to fix the frame of reference to the moving body so that the far-field velocity has a prescribed distribution. Thus, if the body performs a translational motion through the fluid with velocity $\boldsymbol{U} = -\boldsymbol{e}_x U$, in the frame fixed to the body the far-field boundary condition will be $\boldsymbol{u} = \boldsymbol{e}_x U$.

Our main concern is the boundary condition on a solid-body surface, denoted by ∂B. A solid surface can be defined by an equation of surface $\partial B : F(\boldsymbol{x}, t) = 0$. Since ∂B remains to be a material surface during the fluid motion, as a *kinematic boundary condition* there must be

$$\frac{DF}{Dt} = \frac{\partial F}{\partial t} + \boldsymbol{u} \cdot \nabla F = 0 \text{ on } \partial B. \tag{1.3.1}$$

For viscous flows some further conditions on ∂B have to be satisfied. Let $\boldsymbol{u}_B(\boldsymbol{x}, t)$ be the specified velocity of ∂B and $[\![\mathcal{F}]\!]$ denote the jump of any quantity $\mathcal{F}$ right across ∂B. Then the boundary condition should ensure the continuity of velocity, i.e., $[\![\boldsymbol{u}]\!] = (\boldsymbol{u} - \boldsymbol{u}_B) = \boldsymbol{0}$ on ∂B, which can be decomposed to the *no-through condition* and *no-slip condition*:

$$\boldsymbol{n} \cdot [\![\boldsymbol{u}]\!] = 0, \tag{1.3.2a}$$

$$\boldsymbol{n} \times [\![\boldsymbol{u}]\!] = \boldsymbol{0}. \tag{1.3.2b}$$

An important corollary of (1.3.2) is the continuity of material acceleration $\boldsymbol{a} \equiv D\boldsymbol{u}/Dt$:

$$\boldsymbol{n} \cdot [\![\boldsymbol{a}]\!] = 0, \tag{1.3.3a}$$

$$\boldsymbol{n} \times [\![\boldsymbol{a}]\!] = \boldsymbol{0}. \tag{1.3.3b}$$

Note that if (1.3.2) holds at $t = 0$ then it is ensured by (1.3.3) for all $t > 0$.

1.3.2 Fluid Reaction to Solid Boundaries

Having set up the boundary conditions for a viscous fluid, it is straightforward to obtain the total force and moment acting on boundary by the fluid. Let us consider a typical external-flow problem: a material body of volume B moves arbitrarily in a viscous fluid. To be sufficiently general, we permit bodies having arbitrarily deformable boundary as in cases of fish swimming and insect flight in external biofluiddynamics, of nonlinear fluid-solid coupling, and of flow control by flexible walls, etc. Thus, we assume the body surface ∂B has specified velocity distribution $\boldsymbol{u} = \boldsymbol{b}(\boldsymbol{x}, t)$. The fluid volume V_f is bounded internally by the material surface ∂B and externally by a control surface Σ. The latter may have arbitrary velocity $\boldsymbol{v}(\boldsymbol{x}, t)$ or extend to infinity where the fluid is at rest or in uniform translation. The flow domain is sketched in Fig. 1.12.

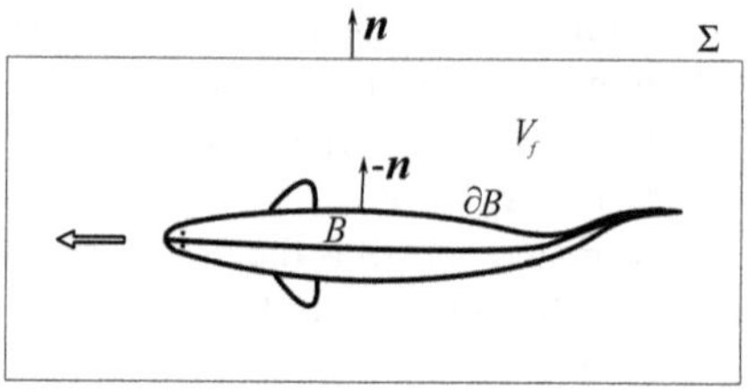

Fig. 1.12 Flow domain to be analyzed and notations

Since (1.2.20) implies that the integral of $\boldsymbol{t}_s$ over a closed boundary vanishes, in calculating the total force we may drop $\boldsymbol{t}_s$. Then by Newton's third law the total force exerted to the body by the fluid is

$$\boldsymbol{F} = -\int_{\partial B} (-\Pi \boldsymbol{n} + \boldsymbol{\tau}) dS, \tag{1.3.4a}$$

where $\Pi = p - \mu_\theta \vartheta$ and $\boldsymbol{\tau} = \mu \boldsymbol{\omega} \times \boldsymbol{n}$. This primary form of the force formula distinguishes the normal and tangent components of the stress. However, it keeps silence on what flow structures generated by the body motion are major causes of the force, despite huge amount of flow visualization results obtained either experimentally or numerically have revealed the crucial importance of these structures for the total force. To save room for this additional information, we rewrite (1.3.4a) by either integrals of local momentum balance or the rate of change of total momentum in a generic control volume (Problem 1.10):

$$\boldsymbol{F} = -\int_{V_f} \rho \boldsymbol{a} dV + \int_{\Sigma} (-\Pi \boldsymbol{n} + \boldsymbol{\tau}) dS \tag{1.3.4b}$$

$$= -\frac{\partial}{\partial t} \int_{V_f} \rho \boldsymbol{u} dV + \int_{\Sigma} [-\Pi \boldsymbol{n} + \boldsymbol{\tau} - \rho \boldsymbol{u}(u_n - v_n)] dS, \tag{1.3.4c}$$

where $\boldsymbol{a} = D\boldsymbol{u}/Dt$ is the fluid acceleration.

In particular, consider an externally unbounded flow past a body with uniform upstream velocity $\boldsymbol{U}$. In this case it is quite often that in body-fixed frame of reference the flow is *steady* in a fixed control surface Σ. Then the total momentum is time-independent, and (1.3.4c) is reduced to

$$\boldsymbol{F} = \int_{\Sigma} (-\Pi \boldsymbol{n} + \boldsymbol{\tau} - \rho \boldsymbol{u}\boldsymbol{u} \cdot \boldsymbol{n}) dS. \tag{1.3.5}$$

If the Reynolds number is sufficiently large such that the viscous stress on Σ can be neglected, (1.3.5) is simplified to

$$\boldsymbol{F} = -\int_{\Sigma} (p\boldsymbol{n} + \rho \boldsymbol{u}\boldsymbol{u} \cdot \boldsymbol{n}) dS. \tag{1.3.6}$$

Similarly, the total moment also has three alternative formulas, in which the integrals of $\boldsymbol{x} \times \boldsymbol{t}_s$ over ∂B and Σ are nonzero but can be expressed by the total vorticity in B and $V = V_f + B$, respectively (Problem 1.10):

$$\boldsymbol{M} = -\int_{\partial B} \boldsymbol{r} \times (-\Pi \boldsymbol{n} + \boldsymbol{\tau}) dS - 2\mu \int_B \boldsymbol{\omega} dV \tag{1.3.7a}$$

$$= -\int_{V_f} \boldsymbol{r} \times \rho \boldsymbol{a} dV + \int_\Sigma \boldsymbol{r} \times (-\Pi \boldsymbol{n} + \boldsymbol{\tau}) dS - 2\mu \int_V \boldsymbol{\omega} dV \tag{1.3.7b}$$

$$= -\frac{\partial}{\partial t} \int_{V_f} \boldsymbol{r} \times \rho \boldsymbol{u} dV + \int_\Sigma \boldsymbol{r} \times [-\Pi \boldsymbol{n} + \boldsymbol{\tau} - \rho \boldsymbol{u}(u_n - v_n)] dS$$
$$- 2\mu \int_V \boldsymbol{\omega} dV. \tag{1.3.7c}$$

Later in Sect. 3.3.1 we shall see that the total vorticity in B or V can be further expressed by the total circulation in two dimensions and by the moment of normal vorticity component at ∂B and Σ in three dimensions.

Another formula for the *rate of work done by a rigid body* for maintaining its motion can be derived from the kinetic-energy balance (1.2.23). The general motion of a rigid body is given by

$$\boldsymbol{u}_B = \boldsymbol{U}(t) + \boldsymbol{W}(t) \times \boldsymbol{r}, \tag{1.3.8}$$

where $\boldsymbol{W}$ is the angular velocity of the body and $\boldsymbol{r} = \boldsymbol{x} - \boldsymbol{x}_0$ with origin $\boldsymbol{x}_0$ being at the instantaneous rotating center. Then, assume the fluid is externally unbounded and at rest at infinity, and denote the force and moment acting to fluid by $\boldsymbol{F}_f$ and $\boldsymbol{M}_f$, since the rate of work done to the fluid by the body-surface stress $\boldsymbol{t}$ is

$$\int_{\partial B} (\boldsymbol{W} \times \boldsymbol{r}) \cdot \boldsymbol{t} dS = \boldsymbol{W} \cdot \int_{\partial B} \boldsymbol{r} \times \boldsymbol{t} dS = \boldsymbol{W} \cdot \boldsymbol{M}_f,$$

there is

$$\int_{\partial B} \boldsymbol{t} \cdot \boldsymbol{u}_B dS = \boldsymbol{U} \cdot \boldsymbol{F}_f + \boldsymbol{W} \cdot \boldsymbol{M}_f = -(UD + WM),$$

where (D, M) are the drag and resistance torque experienced by the body, respectively. Hence, by (1.2.23) we simply have

$$UD + WM = \frac{d}{dt} \int_V \frac{1}{2} \rho q^2 dv + \int_V \Phi dV - \int_V (\rho \boldsymbol{f} \cdot \boldsymbol{u} + p\vartheta) dV. \tag{1.3.9}$$

Thus, the rate of work done by the body translation and rotation is consumed to change the fluid kinetic energy, to overcome the rate of work done by body force and pressure force to the fluid, and to cause dissipation. In particular, for a steady incompressible flow with uniform incoming velocity $\boldsymbol{U}$ over a stationary body without external force, the total kinetic energy is also time-independent; so by (1.2.24c) and the adherence condition we simply have

$$D = \frac{1}{U}\int_V \Phi dV = \frac{\mu}{U}\int_V \omega^2 dV. \tag{1.3.10}$$

1.4 Problems for Chapter 1

1.1. Consider a one-dimensional travelling wave with x-dependent Eulerian amplitude and phase,

$$u(x,t) = \frac{dx}{dt} = a(x)\cos\gamma(x,t), \quad \gamma = nt - kx.$$

In addition to the conventional velocity $u = dx/dt$ in the Eulerian description, the wave will cause fluid particles to have migration described by their Lagrangian motion with velocity u_L. Show that when $a(x) \ll 1$, the Lagrangian velocity $u_L(x_0, t)$ of a fluid particle which was at $x_0 = 0$ when $t = 0$ will differ from the field velocity $u(x_0, t)$ at $t > 0$ by

$$(u_L - u)(x_0, t) = \frac{1}{2n}\left[a\frac{da}{dx}\sin 2\gamma + ka^2(1 - \cos 2\gamma)\right]. \tag{1.4.1}$$

In addition, show that in this purely oscillatory wave field the fluid particle has a nonzero mean velocity $\overline{u}_L = ka^2/2n$, known as the *Stokes drift*.

1.2. Consider a time-independent velocity field $\boldsymbol{u} = (u, v, w)$ with

$$u = -ay, \quad v = ax, \quad w = b(x^2 + y^2). \tag{1.4.2}$$

(a) Compute and draw the flow pattern.

(b) Solve the streamline differential equation (1.1.6) to show that there are two integration constants

$$\psi_1 = \sqrt{x^2 + y^2}, \quad \psi_2 = z - \frac{b}{a}(x^2 + y^2)\tan^{-1}\left(\frac{y}{x}\right),$$

such that $\psi_1 =$ const. and $\psi_2 =$ const. form two families of *stream surfaces*, and their intersection at point $\boldsymbol{x}_0$ is a streamline passing $\boldsymbol{x}_0$.

(c) Find the vorticity field $\boldsymbol{\omega}(\boldsymbol{x})$ and show that $\boldsymbol{u} \cdot \boldsymbol{\omega} = 0$. This feature is known to hold if and only if

$$\boldsymbol{u} = \lambda \nabla \mu \tag{1.4.3}$$

for some scalar fields λ and μ (this kind of flows is *complex lamella flow*, see Sect. 3.1.2). Then, for the flow (1.4.2), show that in (1.4.3) we have

$$\lambda = x^2 + y^2, \quad \mu = \tan^{-1}\left(\frac{y}{x}\right) + \frac{b}{a}z,$$

where λ is an integrating factor and $\mu = \text{const.}$ represents a family of surfaces perpendicular to velocity, like the velocity potential in irrotational flow.

Hint. A surface which is perpendicular to the velocity field everywhere satisfies the differential equation

$$\boldsymbol{u} \cdot d\boldsymbol{x} = udx + vdy + wdz = 0 \tag{1.4.4}$$

with $d\boldsymbol{x}$ being any displacement vector on the surface. Using the result of (b) to cast this equation to a perfect differential for (1.4.2).

Remark The system of surfaces $\psi_1 = \text{const.}$, $\psi_2 = \text{const.}$, and $\phi = \text{const.}$ form a set of natural generalized coordinate surfaces (not mutually orthogonal) for the flow (1.4.2). The values of (ψ_1, ψ_2) at a point determine which streamline goes through the point, and that of μ determines the potential of the point along the streamline (Morse and Feshbach 1953, p. 15).

1.3. Consider a rigidly rotating fluid with $\boldsymbol{u} = \boldsymbol{u}_0 + \boldsymbol{\Omega} \times \boldsymbol{r}$, where $\boldsymbol{u}_0$ and $\boldsymbol{\Omega}$ are constant vectors. Assume $\boldsymbol{u}_0$ is aligned to the direction of $\boldsymbol{\Omega}$. Show that in cylindrical coordinates (r, θ, z), the streamlines are given by

$$x = r\cos\theta, \quad y = r\sin\theta, \quad z = \frac{u_0}{\Omega}\theta + C,$$

and

$$\boldsymbol{\omega} \times \boldsymbol{u} = \nabla\Phi, \quad \Phi = [(\Omega \cdot \boldsymbol{r})^2 - \Omega^2 r^2].$$

1.4. Derive (1.1.24), (1.1.25), and (1.1.27).

1.5. Prove that for incompressible simple shear flow in the (x, y)-plane, $\boldsymbol{u} = (ky, 0)$ above a stationary flat plate at $y = 0$, the maximum stretching rate of a material line element is along the direction of 45° from the x-axis.

1.6. Prove that for the simple shear flow the surface deformation tensor is $\mathbf{B} = -k\boldsymbol{e}_x\boldsymbol{e}_y$. Consider a surface element $d\boldsymbol{S} = \boldsymbol{n}dS$ in this flow, with $\boldsymbol{n} \cdot \boldsymbol{e}_x = \cos\theta$. Prove that the material rate of change of this surface element, per unit area, is characterized by

$$\frac{D\boldsymbol{n}}{Dt} = -k\cos^2\theta(\boldsymbol{e}_z \times \boldsymbol{n}), \quad \frac{1}{dS}\frac{D}{Dt}dS = -k\sin\theta\cos\theta. \tag{1.4.5}$$

Draw a sketch to explain this result for typical values of θ.

1.7. Complete the derivation of (1.1.41) and (1.1.46b), and show that on a rotating rigid-body surface there is

$$\frac{1}{dS}\frac{D}{Dt}dS = \boldsymbol{W} \times \boldsymbol{n}, \tag{1.4.6}$$

where $\boldsymbol{W}$ is the body's angular velocity. Then, extend (1.1.47) to a rotating rigid wall.

1.8. Consider the transformation of vorticity viewed from an inertial frame of reference Σ to a non-inertial frame of reference Σ', of which the velocity seen in Σ is

$$\boldsymbol{u}_0 = \boldsymbol{U}(t) + \boldsymbol{\Omega}(t) \times \boldsymbol{r}, \quad \boldsymbol{r} = \boldsymbol{x} - \boldsymbol{x}_0,$$

where $\boldsymbol{U}$ and $\boldsymbol{\Omega}$ are uniform translational and angular velocities, respectively. For both three- and two-dimensional flows, prove *separately* that the vorticity $\boldsymbol{\omega}$ viewed in Σ and $\boldsymbol{\omega}'$ in Σ' are related by

$$\boldsymbol{\omega} = \boldsymbol{\omega}' + 2\boldsymbol{\Omega}. \tag{1.4.7}$$

1.9. Let $\boldsymbol{r} = \boldsymbol{x} - \boldsymbol{x}_0$ be the position vector relative to a fixed point $\boldsymbol{x}_0$. Prove that if (1.2.5) is postulated then the stress tensor $\mathbf{T}$ must be symmetric.

1.10. Derive the total force and moment formulas in (1.3.4) and (1.3.7). Give an example to explain why (1.3.4b) or (1.3.4c) is sometimes more useful than (1.3.4a) although it involves more complicated volume integral.

1.11. Extend (1.1.47) to the case where the solid wall has arbitrary rotation at angular velocity $\boldsymbol{W}$, and prove that the wall shear stress is (the general expression of the skin friction)

$$\boldsymbol{\tau}_w = \mu \boldsymbol{\omega}_{\mathrm{re}} \times \hat{\boldsymbol{n}},$$

where μ is the shear viscosity, $\hat{\boldsymbol{n}} = -\boldsymbol{n}$ is the unit normal of the wall pointing into the fluid, and $\boldsymbol{\omega}_{\mathrm{re}} \equiv \boldsymbol{\omega} - 2\boldsymbol{W}$ is the relative vorticity.

Chapter 2
Fundamental Processes in Fluid Motion

It has been stated in *Preface* of this book that the physics of vortical flows can be fully understood only if the subject is put on a broad background about the splitting and coupling of different fundamental dynamic processes in fluid motion. The concept of process splitting has been repeatedly touched upon in Chap. 1, both kinematically and dynamically, when we discuss dilatation and vorticity, decomposition of velocity gradient, strain-rate tensor, dynamic constitutive relation and stress tensor, the Navier-Stokes equations, and dominant dimensionless parameters. Accordingly, the stress force is split into normal and tangent components, and velocity boundary condition is split to normal no-through and tangent no-slip conditions. In this chapter we present a systematic theory on this subject, starting from a physical explanation of fundamental processes in fluid motion, in non-mathematical language.

2.1 Preliminary Observations

Conceive a piston suddenly put in slight motion at $t = 0$, in a one-dimensional pipe flow which was originally at rest. Then the fluid at one side of the piston is compressed with higher pressure and density, while that at the other side is rarefied with lower pressure and density. If the piston suddenly stops after a short time, opposite processes will occur. Like an elastic body, the fluid cannot be indefinitely compressed and has a *normal elasticity* (for which a more common term is *compressibility*) to resist the compressing, which causes a propagating wave in which the fluid is alternatively compressed and expanded. This is *longitudinal* acoustic wave or sound wave. Owing to the involvement of thermodynamics, many weak compressing waves may squeeze together to form a stronger wave and even an almost discontinuity of fluid properties across the wavefront, namely a shock wave. In this compressing process all fluid elements are acted by a pressure force (modified by a viscous normal stress) like many micro-pistons. Compressing process is one of the basic forms of existence of the fluid motion.

J.-Z. Wu et al., *Vortical Flows*, DOI 10.1007/978-3-662-47061-9_2

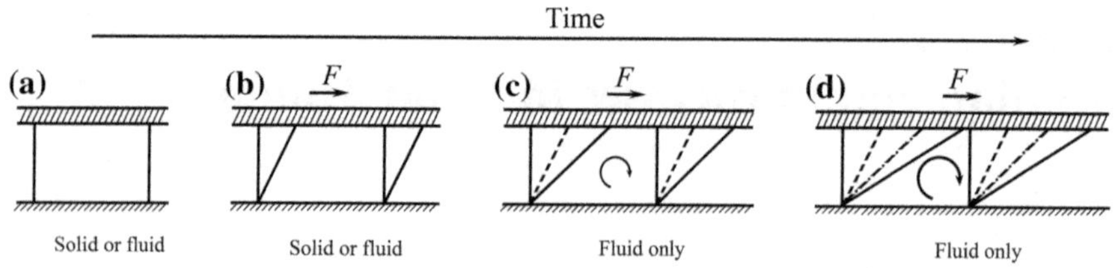

Fig. 2.1 Shearing process between parallel plates and driven by upper plate. **a** and **b** Finite deformation of elastic body. **a–d** Indefinitely deforming of fluid and formation of shear layer, in which vorticity is enhancing as time goes on. Based on Pritchard (2011)

Conceive now a viscous fluid and an elastic solid test substance between two parallel plates, and let the lower plate be stationary and upper plate start to move suddenly in its plane. If the test substance is an elastic solid, the imposed shear force deforms slightly the solid and causes a shear stress proportional to the displacement that stops the motion (Fig. 2.1a, b). In sharp contrast, if the test substance is a viscous fluid, although the shear force exerted by upper plate also drives a thin adjacent fluid layer to move which in turn drives the next thin fluid layer and so on, the fluid has no shear elasticity at all but just responses the force by indefinite deformation till the force is balanced by a shear stress proportional to the normal gradient of velocity (Fig. 2.1a through d), in which fluid elements have to rotate and thereby gain *vorticity*. This flow pattern exemplifies another fundamental process or the basic form of existence in fluid motion, known as *transverse process* or *shearing process*. Any fluid more or less has a shear viscosity, and thus the shearing is also a universal process in fluid motion.

The lack of shear elasticity in fluid is closely related to its *fluidity*. When an aircraft flies in the air or a fish swims through the water, it not only separates the surrounding fluid, which moves away from its path and forms a variable pressure distribution on the surface of aircraft (or fish); but also, it "rubs" the fluid to form a *shear layer* attached to its surface. A shear layer is the primary structure in vortical flows, also known as *layer-like vortex* (this is a more general use of the term "vortex"). If the wall has a downstream end, the attached shear layer sheds off to become a free shear layer. A sufficiently thin free shear layer may automatically roll into a concentrated *axial vortex*, which is the secondary (and stronger) structures of vortical flows. Therefore, vorticity and vortices are the major product of the shearing process and exist only in fluids. In this regards the very insightful assertion of Shi-Jia Lu, the only female disciple of Ludwig Prandtl, that the present authors have cited in many occasions, deserves to be cited again:

The essence of fluid is vortices. A fluid cannot stand rubbing; once you rub it there appear vortices.

Fluid dynamics would still be simple if the compressing and shearing processes could always evolve independent of each other. This is however not true: generically they are coupled as sketched in Fig. 2.2. For example, owing to the viscosity, in the pipe-piston system the fluid elements near the pipe wall must be rubbed by tangent

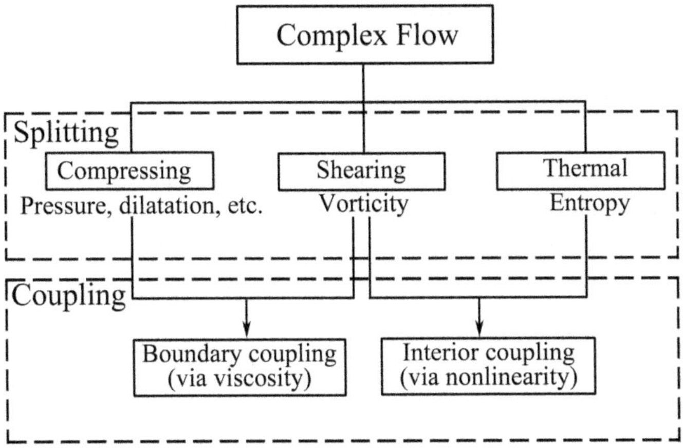

Fig. 2.2 Schematic diagram of decomposition and coupling of fundamental dynamic processes in a generic complex flow

pressure gradient, the longitudinal sound wave or shock wave, and the flow cannot be strictly one-dimensional and irrotational. On the other hand, the motion and interaction of vortices must alter the surrounding pressure distribution, including producing sound. Note that even in incompressible flow model decoupled from thermodynamics, the compressing process still exists but is solely represented by a *mechanical pressure*, of which the key role in vortical flow is to generate new vorticity at a wall via its tangent gradient. The generated vorticity is soon diffused into the interior of fluid. This event is a major product of boundary coupling between shearing and compressing.

The theory on the splitting and coupling of these three fundamental processes may serve as a sharp scalpel to investigate the extremely rich and complex flow patterns seen in nature and technology. It also guides one to best understand the inherent connections and distinctions of various disciplines of fluid dynamics.

In Sect. 2.2 we introduce the most powerful mathematic tool of vector decomposition, the *Helmholtz decomposition*, and apply it to the momentum balance to derive the dynamic equations for vorticity and dilatation. Section 2.3 addresses the coupling of the processes, both inside the flow field and at boundary. In Sect. 2.4 we consider far-field dynamic asymptotics of vorticity and dilatation in unbounded domain and thereby derive the velocity far-field behavior. Sections 2.5 and 2.6 introduce two widely used decoupled and minimally coupled model flows, the inviscid compressible flow and viscous incompressible flow, respectively. The latter will be the major model in our study of vortical flows in the rest of the book.

2.2 Intrinsic Decomposition of Fundamental Processes

2.2.1 Helmholtz Decomposition

The *Helmholtz decomposition*[1] states that for any continuous vector field $\boldsymbol{f}$ in a bounded or unbounded domain there must exist scalar field ϕ and solenoidal vector field $\boldsymbol{A}$, known as the scalar and vector *Helmholtz potentials* of $\boldsymbol{f}$, respectively, such that $\boldsymbol{f}$ can be globally decomposed to

$$\boldsymbol{f} = \nabla\phi + \nabla \times \boldsymbol{A}, \quad \nabla \cdot \boldsymbol{A} = 0. \tag{2.2.1}$$

For a given vector field $\boldsymbol{f}$, ϕ and $\boldsymbol{A}$ can be obtained by solving Poisson equations derived from (2.2.1):

$$\nabla^2\phi = \nabla \cdot \boldsymbol{f}, \quad \nabla^2\boldsymbol{A} = -\nabla \times \boldsymbol{f}. \tag{2.2.2}$$

For a full discussion of this decomposition see Appendix A.3.2. For example, the decomposition (1.1.17) of $\nabla^2\boldsymbol{u}$,

$$\nabla^2\boldsymbol{u} = \nabla(\nabla \cdot \boldsymbol{u}) - \nabla \times (\nabla \times \boldsymbol{u}) = \nabla\vartheta - \nabla \times \boldsymbol{\omega}, \tag{2.2.3}$$

is a kinematic Helmholtz decomposition, indicating that the vector field $\nabla^2\boldsymbol{u}$ consists of (or can be intrinsically split into) the gradient of scalar ϑ, which is a curl-free vector, and the curl of vector $\boldsymbol{\omega}$, which is a solenoidal or divergence-free vector.

Then, let $\boldsymbol{a} = D\boldsymbol{u}/Dt$, the Navier-Stokes equation in the form of (1.2.31b) for constant μ also appears as a *dynamic* Helmholtz decomposition of the total body force (inertial force $-\rho\boldsymbol{a}$ plus external body force $\rho\boldsymbol{f}$)[2]:

$$\rho(\boldsymbol{f} - \boldsymbol{a}) = \nabla\Pi + \nabla \times (\mu\boldsymbol{\omega}), \tag{2.2.4}$$

where $\Pi = p - \mu_\theta\vartheta$ and $\mu\boldsymbol{\omega}$ are the scalar and vector Helmholtz potentials of the total body force, respectively. Some of the underlying physics of this decomposition has been discussed in Chap. 1. Its exact existence lies in the fact that the Helmholtz decomposition is a linear operation and so is the constitutive relations of Newtonian fluid. According to our terminology defined in Sect. 1.1.2, the fields of ϑ and Π are longitudinal and that of $\boldsymbol{\omega}$ is transverse.

The Helmholtz decomposition can also be applied to velocity field itself, say

$$\boldsymbol{u} = \nabla\phi + \nabla \times \boldsymbol{\psi}, \quad \nabla \cdot \boldsymbol{\psi} = 0. \tag{2.2.5}$$

---

[1] According to Truesdell (1954), this decomposition should bear the name of G. G. Stokes who first introduced it in 1851. Here we just follow the convention.

[2] The notation $\boldsymbol{f}$ here should not be confused with the general vector field introduced in (2.2.1). This form of the N-S equation was first observed by Poisson in 1829 and re-emphasized by Truesdell (1954).

A difference between (2.2.3) or (2.2.4) and (2.2.5) is worth noticing. The first two are *natural* because their potentials themselves are physical fields. By contrast, (2.2.5) is a somewhat artificial treatment where the potentials are auxiliary functions and have to be calculated by solving (2.2.2). Nevertheless, to this end a comparison of (2.2.1) and (2.2.3) can slightly simplify our task. We just introduce an auxiliary vector field $\boldsymbol{F}$ such that $\boldsymbol{F}$ satisfies a single vector Poisson equation with source $\boldsymbol{u}$:

$$\boldsymbol{u} = \nabla^2 \boldsymbol{F} = \nabla(\nabla \cdot \boldsymbol{F}) - \nabla \times (\nabla \times \boldsymbol{F}) = \nabla\phi + \nabla \times \boldsymbol{\psi}.$$

Thus, $\boldsymbol{F}$ can be solved for given $\boldsymbol{u}$ under proper boundary condition, and then $(\phi, \boldsymbol{\psi})$ can be obtained from $\nabla \cdot \boldsymbol{F}$ and $\nabla \times \boldsymbol{F}$. The procedure is given in Appendix A.3.2. Then one may further proceed to finding $\boldsymbol{u}$ for given ϑ and $\boldsymbol{\omega}$. This is the inversion of (1.1.12), as the equations therein are viewed as differential equations for $\boldsymbol{u}$ with given $(\vartheta, \boldsymbol{\omega})$ field. The result is the famous Biot-Savart formula to be discussed in Chap. 3.

In unbounded domain, if $\boldsymbol{f}$ approaches zero fast enough as $|\boldsymbol{x}| \to \infty$, the solutions of (2.2.2) exist and are unique, and hence so are the Helmholtz potentials; but in a bounded domain this is not the case without prescribing boundary conditions. In this regard, we have the **Helmholtz-Hodge theorem** proved in Sect. A.3.2:

A finite and continuous vector field $\boldsymbol{f}(\boldsymbol{x})$ in a bounded domain V can always be uniquely split into a longitudinal part $\nabla\phi$ and a transverse part $\mathbf{f}_\perp$, where $\nabla \cdot \mathbf{f}_\perp = 0$ and

$$\boldsymbol{n} \cdot \boldsymbol{f}_\perp = 0 \quad \text{or} \quad \frac{\partial \phi}{\partial n} = \boldsymbol{n} \cdot \boldsymbol{f} \quad \text{at } \partial V. \tag{2.2.6}$$

In this case $\mathbf{f}_\perp$ and $\nabla\phi$ are functionally orthogonal in V in the sense that

$$\int_V \nabla\phi \cdot \boldsymbol{f}_\perp dV = 0. \tag{2.2.7}$$

Similar to the geometric orthogonality, being functional orthogonality implies physical decoupling of these two vector fields. Note that in (2.2.1) one can further separate a harmonic function ζ from scalar ϕ with $\nabla^2\zeta = 0$, such that $\nabla\zeta$ is also functionally orthogonal to both $\nabla\phi$ and $\boldsymbol{f}_\perp$. Thus, strictly, $\boldsymbol{f}$ has a *triple orthogonal decomposition*. The harmonic part is necessary for satisfying boundary conditions and thereby influences both.

Whether the Helmholtz-Hodge theorem can be applied to a specific vector field in V depends critically on the physical behavior of this field. For example, for an incompressible flow over a stationary body, the no-through condition at the body surface implies that the velocity $\boldsymbol{u}$ is a transverse vector and satisfies (2.2.6). But in dynamic decomposition (2.2.4) the solenoidal part $\nabla \times \boldsymbol{\omega}$ is generally not parallel to the boundary, since $\boldsymbol{n} \cdot (\nabla \times \boldsymbol{\omega}) = (\boldsymbol{n} \times \nabla) \cdot \boldsymbol{\omega}$ can be nonzero in viscous flow as long as $\boldsymbol{\omega}$ varies along the boundary. Rather, the full adherence condition stated in Sect. 1.3.1 implies that there will appear a pair of *boundary coupling relations* between two Helmholtz potentials. The boundary dynamic coupling of Π and $\mu\boldsymbol{\omega}$ will be extensively discussed latter in Sects. 2.3.2 and 3.4.4, which implies the violation of the orthogonality condition (2.2.7). Consequently, Π and $\mu\boldsymbol{\omega}$ are globally coupled.

2.2.2 Dynamic Equations for Vorticity and Dilatation

In this subsection we seek a thorough natural decomposition of the momentum equation, which enables deriving the dynamic equations for vorticity and dilatation. While the dynamic Helmholtz decomposition (2.2.4) is already a very informative formulation of the momentum balance, this equation alone is insufficient to fully understand the properties of fundamental processes and their coupling mechanisms. It is still necessary to go through all equations in (1.2.31). Besides, (2.2.4) is for fluid elements per unit volume, where the inertial force $-\rho\boldsymbol{a} = -\rho D\boldsymbol{u}/Dt$ is taken as a whole; but in the field description the acceleration itself should be further decomposed, and vorticity and dilatation are defined for unit mass. Therefore, we should consider the momentum equation per unit mass:

$$\boldsymbol{a} = -\frac{1}{\rho}\nabla p + \boldsymbol{\eta} = -\nabla h + T\nabla s + \boldsymbol{\eta}, \tag{2.2.8a}$$

where we have set external body force $\boldsymbol{f} = \boldsymbol{0}$ for simplicity. The replacement of pressure by enthalpy via (1.2.26) makes the entropy gradient appear explicitly, which is expected to be the root of the coupling of dynamic processes and irreversible thermodynamics. The vector $\boldsymbol{\eta}$ represents all viscous effects, which by (1.2.18) reads

$$\boldsymbol{\eta} \equiv \frac{1}{\rho}\nabla\cdot\mathbf{V} = \nu_\theta\nabla\vartheta - \nu\nabla\times\boldsymbol{\omega}, \tag{2.2.8b}$$

where the shearing and longitudinal kinematic viscosities

$$\nu = \frac{\mu}{\rho}, \quad \nu_\theta = \frac{\mu_\theta}{\rho} \tag{2.2.9}$$

must be variable for compressible flow. In this case, unlike (2.2.4), on both sides of (2.2.8a) no term except ∇h can be naturally decomposed. We have to treat them one by one.

First, we re-express the acceleration by (1.1.36),

$$\boldsymbol{a} = \frac{\partial\boldsymbol{u}}{\partial t} + \boldsymbol{\omega}\times\boldsymbol{u} + \nabla\left(\frac{1}{2}q^2\right) \tag{2.2.10}$$

to display a natural longitudinal term. Thus (2.2.8a) becomes

$$\frac{\partial\boldsymbol{u}}{\partial t} + \boldsymbol{\omega}\times\boldsymbol{u} = -\nabla H + T\nabla s + \boldsymbol{\eta}, \quad H = h + \frac{1}{2}q^2, \tag{2.2.11}$$

known as the *Crocco-Vazsonyi equation*, where H is the total enthalpy.

Next, let ν_0 and $\nu_{\theta 0}$ be constant reference values of ν and ν_θ, respectively, we write

$$\nu = \frac{\mu}{\rho} = \nu_0(1+\tilde{\nu}'), \quad \nu_\theta = \frac{\mu_\theta}{\rho} = \nu_{\theta 0}(1+\tilde{\nu}'_\theta), \tag{2.2.12a}$$

such that

$$\boldsymbol{\eta} = \nabla(\nu_{\theta 0}\vartheta) - \nabla\times(\nu_0\boldsymbol{\omega}) + \boldsymbol{\eta}', \tag{2.2.12b}$$

$$\boldsymbol{\eta}' = \nu_{\theta 0}\tilde{\nu}'_\theta\nabla\vartheta - \nu_0\tilde{\nu}'\nabla\times\boldsymbol{\omega}. \tag{2.2.12c}$$

Since $\boldsymbol{\eta}'$ is generically of smaller order unless $\tilde{\nu}'$ and $\tilde{\nu}'_\theta$ have exceptionally strong variation, we ignore it in the following discussion.

With these preparations, (2.2.11) is cast to

$$\frac{\partial \boldsymbol{u}}{\partial t} + \nabla(H - \nu_{\theta 0}\vartheta) + \nabla\times(\nu_0\boldsymbol{\omega}) = -\boldsymbol{L}, \tag{2.2.13a}$$

where

$$\boldsymbol{L} \equiv \boldsymbol{\omega}\times\boldsymbol{u} - T\nabla s \tag{2.2.13b}$$

is genuinely nonlinear that has both divergence and curl. We call $\boldsymbol{L}$ the *generalized Lamb vector*.

Equation (2.2.13a) exhibits as much the "natural" Helmholtz decomposition as we can reach. We now substitute the velocity decomposition (2.2.5) into (2.2.13a) as the only "artificial" element of our decomposition:

$$\nabla\left(\frac{\partial\phi}{\partial t} + H - \nu_{\theta 0}\vartheta\right) + \nabla\times\left(\frac{\partial\boldsymbol{\psi}}{\partial t} + \nu_0\boldsymbol{\omega}\right) = -\boldsymbol{L}. \tag{2.2.14}$$

Naturally, to study each of shearing and compressing processes as well as their coupling, we can take the curl and divergence of (2.2.14).[3] On the one hand, its curl yields

$$\frac{\partial\boldsymbol{\omega}}{\partial t} - \nu_0\nabla^2\boldsymbol{\omega} = -\nabla\times\boldsymbol{L}, \quad \boldsymbol{\omega} = -\nabla^2\boldsymbol{\psi}, \tag{2.2.15}$$

which is the basic equation of vorticity dynamics for the shearing process and will be discussed extensively throughout this book. On the other hand, the divergence of (2.2.14) leads to an equation for the compressing process:

$$\frac{\partial\vartheta}{\partial t} + \nabla^2(H - \nu_{\theta 0}\vartheta) = -\nabla\cdot\boldsymbol{L}, \quad \vartheta = \nabla^2\phi, \tag{2.2.16}$$

---

[3]Equivalently but without raising the order of equation, one may project (2.2.14) onto a solenoidal vector space to remove its longitudinal part, see, e.g. Chorin and Marsden (1992).

which is however merely an intermediate product due to the involvement of another longitudinal variable H. In fact, for compressible flow with nonzero ϑ or Mach number, the continuity Equation (1.2.31a) and energy equation, say (1.2.31c), have to be combined with (2.2.16) to jointly describe both compressing and thermodynamic processes. This procedure has been conducted by Mao et al. (2011). After a lengthy algebra, they obtained a general viscous dilatation equation for polytropic gases:

$$\frac{\partial^2 \vartheta}{\partial t^2} - \nabla^2 \left[\left(a^2 + \nu_\theta \frac{D}{Dt} \right) \vartheta \right] = Q_\theta, \quad \text{with} \tag{2.2.17a}$$

$$\begin{aligned} Q_\theta &= -\nabla^2 \left[\left(\frac{D}{Dt} + \frac{\partial}{\partial t} \right) \left(\frac{1}{2} q^2 \right) \right] - \left(\frac{\partial}{\partial t} + \nu \nabla^2 \right) \nabla \cdot (\boldsymbol{\omega} \times \boldsymbol{u}) - \nu \nabla^2 \omega^2 \\ &+ \frac{1}{(\gamma - 1) c_v} \nabla \cdot \left[\frac{1}{\gamma} \frac{\partial}{\partial t} (a^2 \nabla s) + \frac{1}{\gamma} \nabla (a^2 \boldsymbol{u} \cdot \nabla s) - \nabla \left(a^2 \frac{Ds}{Dt} \right) \right]. \end{aligned} \tag{2.2.17b}$$

The basic structure of this complicated equation is seen from the left-hand side of (2.2.17a). It is a *parabolic* equation for viscous longitudinal wave, of one order higher than (2.2.15) and (2.2.16) due to the removal of H from the latter. The sound speed a is to be calculated from the energy equation (recall $a^2 = (\gamma - 1)h$). When $\nu_\theta = 0$, (2.2.17) degenerates to an inhomogeneous *hyperbolic* wave equation as it should. Q_θ collects nonlinear terms from the Lamb vector, variation of kinetic energy, vorticity, and entropy, whose physical roles will be classified in the next subsection.

Equations (2.2.15) and (2.2.17) may be viewed as a pair of general coupled dynamic equations for the shearing and compressing equations. They are both of parabolic type, but with distinct physical-mathematical properties reflecting the very different characteristics of flow structures in shearing and compressing processes. When the two processes coexist, one should carefully observe the respective distributions and motions of these structures and thereby trace their physical origins.

To exemplify the complex coexistence of the two processes, Fig. 2.3 shows one of a sequence of photos as a vertical plane shock strikes a finite 25° wedge. As the shock wave passes the wedge base, the flow separates to form thin vortex layers that roll up into discrete vortices (these mechanisms will be discussed in later chapters). Further interaction produces increasingly complex pattern of shock waves, vortex layers, and vortices [for the whole sequence see Van Dyke (1982)]. Figure 2.4 shows a fighter plane flying at $M_\infty \simeq 1$. At the wing's upper surface the flow is supersonic and full of expansion waves made visible by moisture condensation, which coexists with wingtip vortices.

Figure 2.5 is a slice of the $(\boldsymbol{\omega}, \vartheta)$ distribution for a compressible isotropic turbulence at a large Reynolds number and turbulent Mach number $M_t = u'/\langle a \rangle = 1.03$, where u' is the root-mean-square (rms) turbulence velocity and $\langle a \rangle$ is the mean speed of sound. While the vorticity field exhibits intermittent filament-like structures, the dilatation field is highly peaked at some sheet-like shock waves with extremely

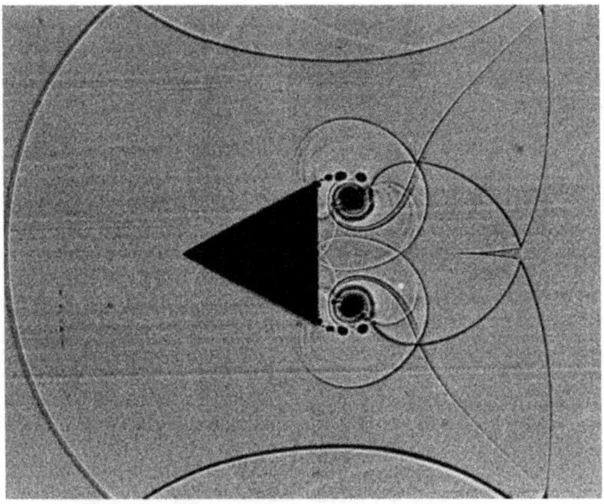

Fig. 2.3 Shock-vortex interaction during the diffraction of a shock wave by a finite wedge. From Van Dyke (1982), Fig. 241

Fig. 2.4 Coexistence of wingtip vortices and supersonic waves around a fighter flying at nearly sound speed. Courtesy of http://www.sky-flash.com/EJ.v.Koningsveld

strong $\vartheta < 0$ called shocklets (in their near neighborhood the vortical structures may become sheet-like as well). Statistically, Wang et al. (2012, 2013) have observed that the vortical structures obey the same law as those in incompressible flow, while the dilatation structures obey a law similar to one-dimensional shocklets (recall that compressing process exists for one- to three-dimensional flows).

2.3 Coupling and Splitting of Fundamental Processes

The Helmholtz decomposition is a linear operator, but the Navier-Stokes equation contains various nonlinear terms that cause process coupling inside the fluid. Besides, a linear and viscous coupling occurs at boundary due to no-slip condition. Both coupling mechanisms are discussed in this section.

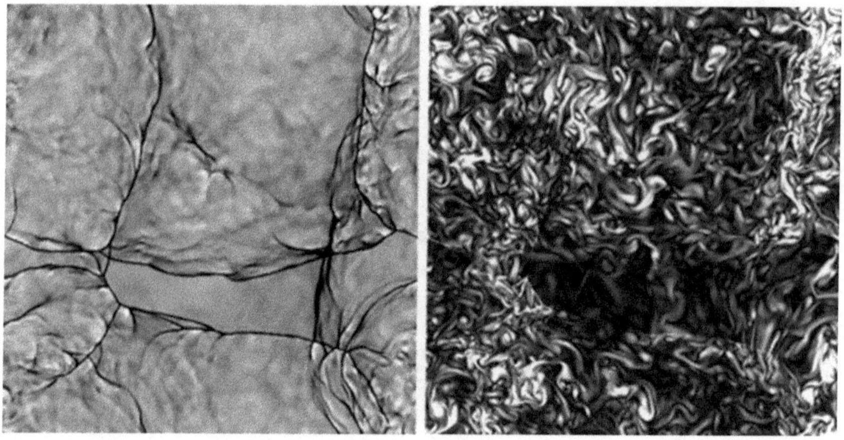

Fig. 2.5 The distribution of dilatation (*left*) and vorticity (*right*) on a slice of a three-dimensional and high Mach-number isotropic turbulence. Direct numerical simulation with grid 512^3 at turbulent Mach number $M_t = 1.03$. Courtesy of Shiyi Chen

2.3.1 Process Nonlinearity and Coupling Inside the Flow

In the interior of the fluid motion the coupling mechanisms of different processes can be identified from the nonlinear terms of (2.2.15) and (2.2.17). To see the respective role of each nonlinear term in each process, we first follow Mao et al. (2011) to consider a pair of coupled nonlinear model equations for functions f and g in a domain $(x, t) \in \mathcal{D}$:

$$f_t + f_x^2 - \epsilon g_x f = 1, \tag{2.3.1a}$$

$$g_t - g g_x - f f_x = 0, \tag{2.3.1b}$$

where subscripts denote partial derivatives. The nonlinear terms in these equations fall into three types:

(1) *Cross generation*: A nonlinear mechanism from motion g (or f) that serves as the inhomogeneous source of another motion f (or g), so that it can *produce* motion f (or g) from nothing. Of this type is $-ff_x$ in (2.3.1b), since for example if $g = 0$ at $t = 0$ then after a small time interval δ there will be

$$g = \int_0^\delta f f_x dt + O(\delta^2).$$

(2) *Cross modulation*: A nonlinear mechanism in the evolution of motion f (or g) by which motion g (or f) modulates the evolution of existing f (or g) but not creates it from nothing. Of this type is $-\epsilon g_x f$ in (2.3.1a).

(3) *Self-nonlinearity*: A nonlinear mechanism that takes place within each motion. Of this type are f_x^2 in (2.3.1a) and $-gg_x$ in (2.3.1b). For example, if ϵ is negligibly small in a subdomain $\mathcal{D}_1$, there will be

$$f_t + f_x^2 = 1 \quad \text{in } \mathcal{D}_1.$$

This self-nonlinearity will affect the coupling with process g indirectly through types (1) and (2).

Let us now revisit (2.2.15) and (2.2.17). First, of all nonlinear terms therein, the generalized Lamb vector $\boldsymbol{L} = \boldsymbol{\omega} \times \boldsymbol{u} - T\nabla s$ stands at the crossroad of the two dynamic processes. It appears in both equations.[4] This vector is the most commonly encountered mechanism of the compressing-shearing coupling in the interior of a flow field. More specifically, write $\boldsymbol{u} = \boldsymbol{v} + \nabla\phi$ such that $\boldsymbol{v}$ is the transverse part (including the harmonic part induced by the vorticity field, so that ϕ always represents the compressible potential flow induced by ϑ), the Lamb vector is split into $\boldsymbol{\omega} \times \boldsymbol{v}$ and $\boldsymbol{\omega} \times \nabla\phi$. Since $\boldsymbol{\omega} = \nabla \times \boldsymbol{v}$, the former is purely transverse; but the latter implies cross modulation.

Next, for clarity, denote terms for self nonlinearity, cross generation, and cross modulation by SN, CG, and CM, respectively. Then in the shearing equation (2.2.15), $\nabla \times \boldsymbol{L}$ is the *only* nonlinear term and can be expanded to

$$\nabla \times \boldsymbol{L} = \underbrace{\nabla \times (\boldsymbol{\omega} \times \boldsymbol{v})}_{SN} + \underbrace{\nabla \times (\boldsymbol{\omega} \times \nabla\phi)}_{CM} - \underbrace{\nabla T \times \nabla s}_{CG}. \tag{2.3.2}$$

Here, the first term is a self nonlinearity of the shearing process, which will be discussed in detail later; the third term is an external source from entropy process that appears in baroclinic flow; and the second term exhibits how a compressing process modulates an existing shearing process which, by denoting $\boldsymbol{u}_\phi = \nabla\phi$, can be written as

$$\nabla \times (\boldsymbol{\omega} \times \boldsymbol{u}_\phi) = \boldsymbol{u}_\phi \cdot \nabla\boldsymbol{\omega} - \boldsymbol{\omega} \cdot \nabla\boldsymbol{u}_\phi + \vartheta\boldsymbol{\omega}.$$

In contrast, in compressing process governed by (2.2.17), we see that physical sources of the ϑ-wave are contained in the three explicitly time-dependent terms on the right-hand side: the divergence of Lamb vector,[5] the kinetic energy, and the entropy gradient, of which the first two are major nonlinear terms and we split them to

$$\nabla \cdot (\boldsymbol{\omega} \times \boldsymbol{u}) = \underbrace{\nabla \cdot (\boldsymbol{\omega} \times \boldsymbol{v})}_{CG} + \underbrace{(\nabla \times \boldsymbol{\omega}) \cdot \nabla\phi}_{CM}, \tag{2.3.3a}$$

$$\frac{1}{2}\nabla^2|\boldsymbol{u}|^2 = \underbrace{\frac{1}{2}\nabla^2|\boldsymbol{v}|^2}_{CG} + \underbrace{\frac{1}{2}\nabla^2|\nabla\phi|^2}_{SN} + \underbrace{\nabla^2(\boldsymbol{v} \cdot \nabla\phi)}_{CM}. \tag{2.3.3b}$$

---

[4] In Chap. 3 we shall see some special types of flows where the Lamb-vector effect disappears completely or partly, although both $\boldsymbol{\omega}$ and $\boldsymbol{u}$ are nonzero.

[5] The longitudinal waves produced by this mechanism is called *vortex sound*.

Except a self-nonlinearity $\nabla^2(|\nabla\phi|^2/2)$, there are two external sources, $\nabla\cdot(\boldsymbol{\omega}\times\boldsymbol{v}-T\nabla s)$ and $\nabla(|\boldsymbol{v}|^2/2)$, and two cross-modulation mechanisms, $\nabla^2(\boldsymbol{v}\cdot\nabla\phi)$ and $(\nabla\times\boldsymbol{\omega})\cdot\nabla\phi$, where the former represents the convection of ϕ by the shearing velocity $\boldsymbol{v}$, as is evident from (2.2.16) in which

$$\partial_t\vartheta+\nabla^2(\boldsymbol{v}\cdot\nabla\phi)=\nabla^2[(\partial_t+\boldsymbol{v}\cdot\nabla)\phi].$$

Note that *the shearing process can be a source of compressing process, but not vise versa.*[6]

2.3.2 Process Linear Coupling on Boundaries

The compressing and shearing processes are also coupled on flow boundary ∂V but by a different mechanism. To see this we first notice that (2.2.15) and (2.2.16) are one order higher than the original N-S equation (2.2.4), indicating that their boundary conditions are not only the velocity adherence (1.3.2) but also one-order higher, the acceleration adherence (1.3.3), i.e., $[\![\boldsymbol{a}]\!]=\boldsymbol{0}$, for otherwise there may appear spurious solutions due to the raising of the order. Dynamically, this requires applying the N-S equation itself to ∂V.

Then, because (1.3.3) takes $\boldsymbol{a}$ as a whole on the material boundary, it is convenient to start from the Navier-Stokes equation (2.2.4) rather than (2.2.13a). As observed in Sect. 2.2.1, the Helmholtz-Hodge theorem is not applicable; there appears a boundary coupling mechanism between Π and $\mu\boldsymbol{\omega}$. Below we illustrate the physical implication of this coupling by considering a high-Re steady and two-dimensional flow over a stationary airfoil C on the (x, y) plane, of which the streamline pattern is sketched in Fig. 2.6.

For this flow (2.2.4) is reduced to

$$\boldsymbol{u}\cdot\nabla\boldsymbol{u}=-\frac{1}{\rho}\nabla p-\nu\nabla\times\boldsymbol{\omega},\tag{2.3.4}$$

where ρ and ν are constant, and $\boldsymbol{\omega}=\omega\boldsymbol{e}_3$. Let $\boldsymbol{n}$ be the unit normal of any streamline (including the airfoil C), and $\boldsymbol{u}=\boldsymbol{t}q$ be the velocity with $\boldsymbol{t}$ being the unit tangent vector. Assume $\boldsymbol{t}$ moves counterclockwise viewed on the fluid side but clockwise viewed on the body side. Then $(\boldsymbol{n},\boldsymbol{t},\boldsymbol{e}_3)$ form a right-hand orthonormal triad. Denote the arclength of the streamline by s so that

$$\frac{\partial\boldsymbol{t}}{\partial s}=\kappa\boldsymbol{n},\tag{2.3.5}$$

[6]The above splitting of the nonlinear source terms in (2.2.17) indicates that they still contain some elements of compressing process itself but not expressible by dilatation, unless introducing integral operator.

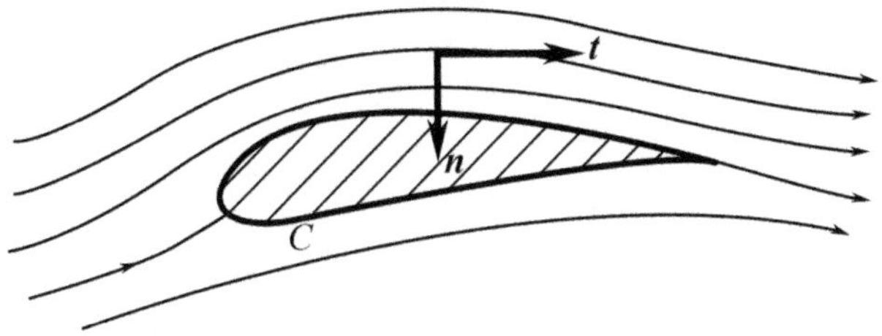

Fig. 2.6 Two-dimensional steady streamlines over an airfoil and intrinsic coordinates

where κ is the curvature of the streamline. Then there is

$$\boldsymbol{u} \cdot \nabla \boldsymbol{u} = q\partial_s q\boldsymbol{t} + \kappa q^2 \boldsymbol{n}, \quad -\boldsymbol{t} \cdot (\nabla \times \boldsymbol{\omega}) = \partial_n \omega, \quad -\boldsymbol{n} \cdot (\nabla \times \boldsymbol{\omega}) = -\partial_s \omega.$$

Substituting these into (2.3.4) yields

$$\frac{1}{\rho}\frac{\partial p}{\partial s} = -q\frac{\partial q}{\partial s} + \nu\frac{\partial \omega}{\partial n}, \tag{2.3.6a}$$

$$\frac{1}{\rho}\frac{\partial p}{\partial n} = -\kappa q^2 - \nu\frac{\partial \omega}{\partial s}. \tag{2.3.6b}$$

Obviously, in the effectively inviscid region where $\nu|\nabla \times \boldsymbol{\omega}| \ll 1$, the streamwise pressure gradient is balanced by the streamwise variation of the kinetic energy, while the normal pressure gradient is balanced by the turning of the streamlines if $q \neq 0$ (for a given normal pressure gradient, the smaller q must be associated with sharper turning).

As we move from the effectively inviscid region toward the airfoil C (a special streamline) and finally reach C, however, the momentum-balance mechanism changes dramatically. The no-slip condition requires $q = 0$ there, so by (2.3.6) the gradient of pressure has to be solely balanced by those of vorticity:

$$\frac{1}{\rho}\frac{\partial p}{\partial s} = \nu\frac{\partial \omega}{\partial n}, \tag{2.3.7a}$$

$$\frac{1}{\rho}\frac{\partial p}{\partial n} = -\nu\frac{\partial \omega}{\partial s}. \tag{2.3.7b}$$

This pair of equations represents *a strong viscous and linear coupling of the two fundamental processes at boundaries, and implies a global* $(\boldsymbol{\omega}, p)$ *coupling*.

The most important observation of this boundary coupling is that the vorticity diffusion flux $\nu\partial\omega/\partial n$ at the solid wall as seen in (2.3.7a) is radically different from that inside the fluid as seen in (2.3.6a). In the latter case the vorticity is simply diffused from one side of the streamline to the other side; but since no vorticity exists inside the stationary airfoil, in the former case $\nu\partial\omega/\partial n$ implies that the vorticity is diffused only at the fluid side, either toward or away from the wall. Lighthill (1963) points

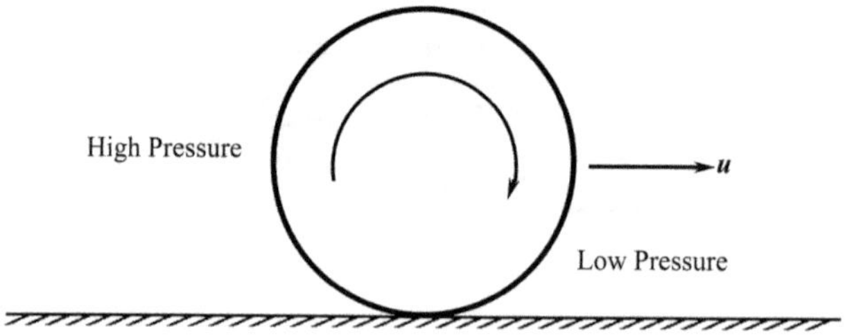

Fig. 2.7 Schematic illustration of vorticity generation by pressure gradient and no-slip condition

out that, this *boundary vorticity flux* measures how much vorticity is created at the wall due to the no-slip condition and sent into the fluid by diffusion in per unit area and unit time. Therefore, (2.3.7a) represents a mechanism that *the tangent pressure gradient at the wall produces vorticity due to adherence.*

To picture this mechanism, Fig. 2.7 shows a small fluid ball right on the wall. Assume at $t = 0$ there suddenly appears a tangent pressure gradient $\partial_s p < 0$, which forces the ball to move right. The ball cannot slide at the wall but only rolls, and hence gains an angular velocity or vorticity ω. Because for $n > 0$ (inside the wall) and $t < 0$ there is $\omega = 0$, the vorticity in the rolling ball must be *newly created* by the joint action of $\partial_s p$ and no-slip condition. The created $\omega < 0$ first occurs in the fluid layer adjacent to the wall and is diffused to the region $n < 0$, with decreasing magnitude at larger distance above the wall. The vorticity will be positive if $\partial_s p > 0$. This mechanism is precisely described by the right-hand side of (2.3.7a), which is a normal vorticity diffusion flux. A general theory of this boundary coupling and vorticity generation will be addressed later in Sect. 3.4.4.

Example 1: Water hammer. Consider a very long water-pipe flow with a valve as sketched in Fig. 2.8. When the valve is being closed, its upstream flow will be stagnated with an increase of pressure p, while the flow downstream will firstly gain a low pressure that propagates at sound speed (the water is slightly compressible) to and is reflected back at the far-downstream open end of the pipe, to become a high-pressure peak that then runs upstream and hits the valve like a hammer. This process happens periodically as the p-wave runs back and forth repeatedly, which may cause severe damage to the valve and nearby hydraulic facilities. Thus, the water-hammer prediction and control are a crucial issue in various hydraulic pipeline systems.

Fig. 2.8 Sketch of water hammer flow

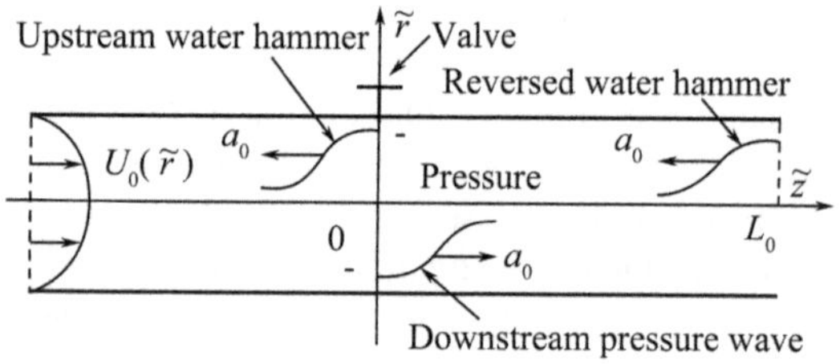

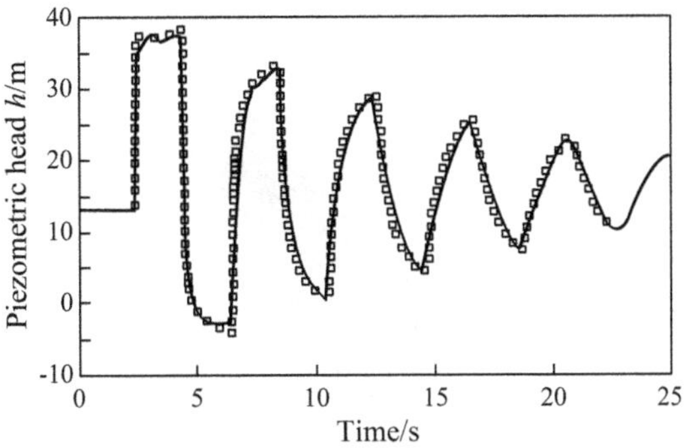

Fig. 2.9 A comparison of theoretical (*squares*) and experimental (*solid line*) wave propagation of water hammer. From Xuan et al. (2012)

In practical hydraulic systems the ratio of pipe radius to the length is typically of $O(10^{-4})$ or smaller, so one had long been satisfied with one-dimensional linearized transient flow model. The wall friction must have accumulated effect on the p-wave during a multi-period time interval, which in that model can only be added empirically. But all efforts within such a model failed to capture the strength, period, and peculiar wave pattern of the water hammer. What had been missing in such models is the *closed-loop coupling* between the compressing and shearing processes, by which the strong pressure wave generates strong vorticity wave at pipe wall via (2.3.7), known as Stokes wave (for details see Sects. 4.1.3 and 4.3), and the latter in turn strongly modulates the former. To capture this key physics one has to consider at least a viscous axisymmetric transient flow model with coupled equations for axial and radial velocities. The problem can then be reduced to a linear integral-differential equation for p, of which an almost-analytic solution was found by Xuan et al. (2012). Remarkably, a simple enlargement of the constant molecular viscosity already enables predicting turbulent water hammer in excellent agreement with experiments, see Fig. 2.9. The theory has also led to a very effective water-hammer control principle physically.

Example 2: Flow instability in combustion chamber. If in the above example the ratio of pipe radius to length is not small, the transient flow can no longer be linearized. Internal self-nonlinearities of three processes and their nonlinear coupling as well as linear boundary coupling will all occur. This transient flow may happen in a rocket combustor, where combustion instability may trigger strong pressure waves propagating back and forth in the combustion chamber, which "rub" vorticity waves at chamber wall that in turn modulate the pressure waves. The closed-loop interactions of the processes may reach a resonance that could damage the combustor. A successful predictive theory has been developed by Flandro et al. (2006), which includes all major interactions. Figure 2.10 compares the measured and predicted time history of pressure wave in a model combustion chamber. The highly nonlinear feature of the p-wave, including the sudden increment of mean pressure (DC shift), is captured by the theory.

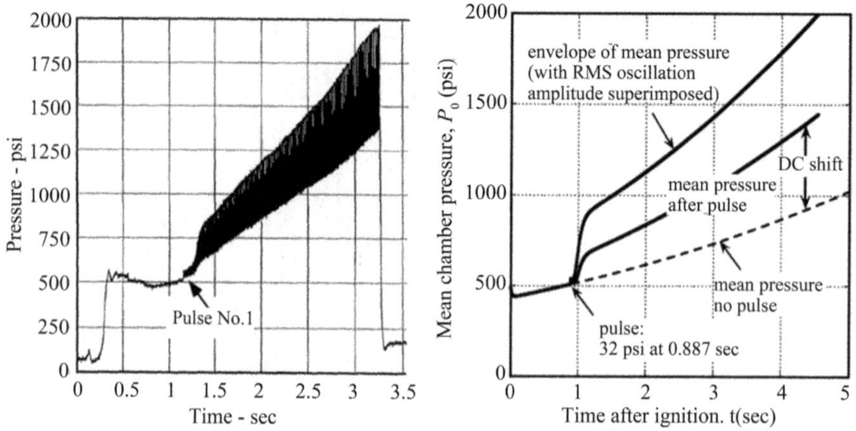

Fig. 2.10 Measured (*left*) and predicted (*right*) pressure history in a model combustion chamber. From Flandro et al. (2006)

2.3.3 Linearized Process Splitting in Unbounded Space

Having discussed how the fundamental processes in fluid motion are coupled, in what situation they can be fully split is also evident: If our concern is confined to unbounded flow, then the boundary coupling of fundamental processes of fluid motion disappears; and if the processes are just quite small disturbances of magnitudes of $O(\epsilon) \ll 1$ to a uniform flow field so that their governing equations can be linearized, then their nonlinear coupling can also be neglected. This linearized case, although simple but of fundamental significance, has been addressed by Wu (1956) and revisited by Mao (2010).

Consider a linearized viscous and heat-conducting fluid motion of arbitrary Reynolds number and Mach number, on a uniform flow background denoted by suffix ∞. Define the disturbed quantities by

$$\begin{aligned}
&\boldsymbol{u} = U_0(\epsilon \boldsymbol{u}' + \cdots), \quad \rho = \rho_\infty(1 + \epsilon\rho' + \cdots), \quad h = c_p T_\infty(1 + \epsilon h' + \cdots),\\
&s = c_p(\epsilon s' + \cdots), \quad \nu = \nu_\infty(1 + \epsilon\nu' + \cdots),\\
&\nu_\theta = \nu_{\theta\infty}(1 + \epsilon\nu' + \cdots), \quad k = k_\infty(1 + \epsilon k' + \cdots),
\end{aligned}$$

along with $(\vartheta, |\boldsymbol{\omega}|) = O(\epsilon)$. The prime disturbance variables are dimensionless. Then to the order of $O(\epsilon)$, (2.2.15) and (2.2.17) are reduced to a pair of linear equations

$$(\partial_t - \nu_\infty \nabla^2)\boldsymbol{\omega} = \boldsymbol{0}, \tag{2.3.8a}$$

$$(\partial_t^2 - a_\infty^2 \nabla^2)\vartheta = \nu_{\theta\infty}\nabla^2\partial_t\vartheta - \epsilon a_\infty^2 \nabla^2 \partial_t s'. \tag{2.3.8b}$$

Evidently, the transverse and longitudinal processes, associated with velocities $\boldsymbol{v}$ and $\nabla\phi$, respectively, are decoupled. It can be easily shown that the splitting $\boldsymbol{u} = \boldsymbol{v} + \nabla\phi$ is unique if the initial values of $\boldsymbol{v}$ and $\nabla\phi$ are given.

Then, to transform the entropy term in (2.3.8b), we use the linearized version of entropy equations (1.2.31c) and (2.2.16), now written as:

$$\partial_t s' = \frac{\gamma - 1}{\gamma a_\infty^2}\kappa\nabla^2 h',$$
$$(\partial_t - \nu_\theta\nabla^2)\vartheta = \epsilon(-\nabla^2 h' + h_\infty\nabla^2 s'),$$

where $\kappa \equiv \gamma\nu_\infty/Pr$ is known as the *thermometric conductivity*. Combining these equations yields

$$\frac{(\gamma - 1)\kappa}{\gamma a_\infty^2}\partial_t\vartheta = -\epsilon\left(\partial_t - \frac{\kappa}{\gamma}\nabla^2\right)s'.$$

We further assume that the viscosity and heat conductivity are small:

$$\epsilon \ll \nu, \nu_\theta, k = O(\delta) \ll 1. \tag{2.3.9}$$

In this case, observe that s' is of also $O(\delta)$, smaller than other variables, and hence $\kappa/\gamma\nabla^2 s' = O(\delta^2)$ should be neglected. Therefore, substituting the above ϑ-s relation into (2.3.8b) simplifies the latter to

$$(\partial_t^2 - a_\infty^2\nabla^2)f = b\nabla^2\partial_t f, \quad b \equiv \kappa - \frac{\kappa}{\gamma} + \nu_{\theta\infty}, \tag{2.3.10}$$

where f can stand for ϕ, ϑ, or p. The entropy also satisfies (2.3.10) if terms of both $O(\epsilon\delta)$ (its leading order) and $O(\epsilon\delta^2)$ are retained. Thus, the longitudinal process has been further split to independent compressing process and entropy process. Note that while (2.3.8a) is a standard *parabolic* (diffusion) equation (2.3.10) is a third-order parabolic equation that will degenerate to the standard hyperbolic wave equation if $b = 0$ as it should. Since the dissipation has been neglected as a nonlinear effect, the role of viscosity-conductivity in (2.3.10) is to make the longitudinal waves *dispersive* that are eventually flattened.

The above results can be stated as (Wu 1956; Mao et al. 2010).

Linear Splitting Theorem. *In an unbounded domain where the fluid is otherwise at rest, a linearized viscous and heat-conducting disturbance flow* ($\mathbf{u}$, p, s) *may be expressed as the sum of a transverse (shearing) process and a longitudinal process governed by parabolic equations* (2.3.8a) *and* (2.3.8b), *respectively, and* $\mathbf{u} = \mathbf{v} + \nabla\phi$. *This splitting is unique if the initial values of* $\mathbf{v}$ *and* $\nabla\phi$ *are given. Moreover, for small viscosity and heat conductivity given by* (2.3.9), *the longitudinal process can be further decoupled to "sound mode" and "entropy mode", both governed by* (2.3.10).

We remark that the solutions of the above linear equations can also be superposed to a known basic flow field to study how the latter affects the propagation of linear viscous vortical, acoustic, and entropy waves. In this case the basic flow can be symbolically treated as the sources of mass, external body force, and heat addition, which makes (2.3.8a) and (2.3.8b) or (2.3.10) inhomogeneous as given by Mao et al. (2010).

2.4 Far-Field Asymptotics in Unbounded Flow

A fundamental issue in all studies of externally unbounded flows is the asymptotic behavior of vorticity and dilatation field as $r \equiv |\boldsymbol{x}| \to \infty$. This is a necessary prerequisite for not only prescribing far-field boundary conditions for external-flow problems, but also ensuring the convergence of relevant integrals of $\boldsymbol{\omega}$ and ϑ, including their "induced" velocity field, over the entire space or arbitrarily large external boundary. Having developed the general theory of process splitting and coupling, we can now address this issue in more detail than the velocity condition at infinity mentioned briefly in Sect. 1.3.1.

Throughout this book, whenever the fluid is externally unbounded, we start from an infinitely extended space (free space) V_∞ with fluid at rest and having uniform properties at infinity:

$$\boldsymbol{u} = \boldsymbol{0}, \quad (\boldsymbol{\omega}, \vartheta) = (\boldsymbol{0}, 0), \quad (p, \rho, s) = (p_\infty, \rho_\infty, s_\infty) \quad \text{at} \quad r \equiv |\boldsymbol{x}| = \infty. \tag{2.4.1}$$

We assume that the flow is created by a moving body (bodies) of finite size. The dynamics of the flow is Galilean invariant, and we may superimpose a constant velocity $\boldsymbol{U}$ to $\boldsymbol{u}$, with the understanding that (2.4.1) holds for the disturbance part of the composite velocity field $\boldsymbol{u} + \boldsymbol{U}$.

2.4.1 Vorticity and Dilatation Far Fields

No matter how strong, nonlinear, and coupled can various dynamic processes be in a volume $V_{NL} = O(L^n)$ around the body, with n being spatial dimension, (2.4.1) requires all disturbances to die out at infinity. Hence, there must be a far-field region V_L between V_{NL} and infinity where all processes have decayed to a weak level of $O(\epsilon) \ll 1$ and satisfy *linearized equations* derived in Sect. 2.3.3. This permits us to determine the far-field behaviors of $\boldsymbol{\omega}$ and ϑ separately.

Before proceed, we make a few general observations. Firstly, conventionally the asymptotics of parabolic equations is referred to as the state $t \to \infty$. This limiting approach cannot be used here, since it would lead to a far field contradicting (2.4.1). Instead, we should require finite $t < \infty$ but allow $r \to \infty$. For example, when a wing starts to move at $t = 0$, as will be explained in the next chapter, its generated vortex system forms a closed loop, of which the streamwise extension, say $L(t)$, is continuously elongated as t. Thus for studying the far-field behavior of the vortex loop the flow domain has to be correspondingly expanded. As long as $t < \infty$, there always exists a sufficiently large sphere of radius $r_0 \gg L(t)$ such that for $r \gg r_0$ the loop still looks like a small vortex ring near the origin.

Secondly, as is familiar in heat equation, a parabolic wave generated at $t = 0$ can propagate instantly to infinity but decays *exponentially* as distance. The former property implies that physically parabolic equations do not describe deterministic phenomena but are a statistic model for disturbances propagated by molecules, since there is no upper bound on the possible velocities of the molecules. The latter property implies that a parabolic field can never be mathematically compact with a finite support outside which the field is exactly zero. However, there always exists a finite domain outside which the parabolic field is exponentially small and hence negligible. Throughout the book we shall use the word "*compact*" in this ordinary sense, despite mathematically the field is still not strictly compact.

Thirdly, for a remote observer at $r = |\boldsymbol{x}| \gg L$, the body or even the whole region V_{NL} may be viewed as a point-like disturbance to the fluid at the origin $r = 0$. Moreover, to satisfy the upstream and downstream conditions, the disturbance cannot be of step-function type but has to be of doublet type, namely a *pulse signal*, which is of course a compact disturbance. The body motion can only reach a finite distance for any $t < \infty$, and as $r \to \infty$ this motion appears increasingly slow as if we look at a remote airplane. Thus, the flow in n-dimensional space can well be assumed symmetric with respect to x for $n = 1$, axisymmetric for $n = 2$, and spherically symmetric for $n = 3$. Consequently, it suffices to estimate the asymptotic behaviors of (ω, ϑ)-field by using the symmetric fundamental solutions of (2.3.8a) and (2.3.10), with r being the only spatial variable. For $n = 1$, 2, and 3, r represents rectangular coordinator $|x|$, polar radius, and radius, respectively.

The above observations can be demonstrated by the fundamental solutions of (2.3.8a) and (2.3.10). We drop the suffix ∞ of the reference values for neatness. First, as a standard heat equation, (2.3.8a) has fundamental solution G_n of unified form for $n = 1, 2, 3$:

$$G_n(\boldsymbol{x}, t; \boldsymbol{x}', t') = \frac{H(\tau)}{(4\pi\nu\tau)^{n/2}} \exp\left(-\frac{r^2}{4\nu\tau}\right), \tag{2.4.2}$$

where $\tau = t - t'$ and $H(\tau)$ is the Heaviside step function. Therefore, for any $t < \infty$ the vorticity field must decay exponentially like e^{-r^2} as $r \to \infty$.

Similar to (2.4.2), the longitudinal wave decays faster algebraically as n increases. Thus it suffices to consider the solution of (2.3.10) with $n = 1$ for velocity $u(r, t)$, say:

$$u_{tt} - a^2 u_{xx} = b u_{xxt}.$$

The pulse signal has initial condition

$$u(0, t) = 0 \quad \text{for } t < 0 \text{ and } t > \delta,$$
$$u(0, t) = \frac{u_0}{\delta} \quad \text{for } t \in [0, \delta],\ \delta \to 0.$$

The solution cannot be expressed in closed form, but its asymptotic behavior is clear. An outgoing pulse signal evolves as (the detailed algebra is skipped)

$$|\boldsymbol{u}(r,t)| = O\left(\frac{e^{-\beta r}}{r^{n/2}}\right), \quad r \to \infty, \quad t < \infty, \tag{2.4.3}$$

where β is a positive constant, which decays also exponentially but slower than vorticity. The estimate of ϑ is the same as (2.4.3). Therefore, we may state (Liu et al. 2014)

Vorticity-Dilatation Compactness. *In an unbounded compressible fluid at rest at infinity in n-dimensional space with* $n = 2, 3$, *originally compact vorticity and dilatation fields must remain compact in a sufficiently large domain for any finite* $t < \infty$.

The compactness of the vorticity and dilatation field ensures the finiteness of their mth tensorial moment integrals over the space for finite integers m:

$$\left|\left|\int_{V_\infty} \boldsymbol{xx}\ldots\boldsymbol{x}\boldsymbol{\omega} dV\right|\right| < \infty, \quad \left|\left|\int_{V_\infty} \boldsymbol{xx}\ldots\boldsymbol{x}\vartheta dV\right|\right| < \infty. \tag{2.4.4}$$

Owing to this *physical* compactness, the flow field in a neighborhood of infinity can only be irrotational and incompressible, which we shall always assume to have *single-valued and smooth* velocity potential.

2.4.2 Velocity Far Field

The preceding dynamic analysis of the $(\boldsymbol{\omega}, \vartheta)$ far field enables us to determine the far-field velocity induced by vorticity and dilatation through a kinematic analysis.

In addition, the assumed boundary condition (2.4.1) at infinity ensures that for any $t < \infty$ there is:

$$\int_{V_\infty} \vartheta dV = \int_{\partial V_\infty} \boldsymbol{n}\cdot\boldsymbol{u} dS = 0, \tag{2.4.5a}$$

$$\int_{V_\infty} \boldsymbol{\omega} dV = \int_{\partial V_\infty} \boldsymbol{n}\times\boldsymbol{u} dS = \boldsymbol{0}. \tag{2.4.5b}$$

Specifically, one may conceive ∂V_∞ as the boundary of the complementary fluid volume V_{out} exterior to V_∞ and including infinity, in which as we assumed the flow has smooth and single-valued velocity potential with $\vartheta = 0$. Then (2.4.5a) follows. On the other hand, (2.4.5b) comes from a pair of general results of vorticity kinematics to be addressed in Sect. 3.3.1 below: the *Föppl total-vorticity theorem* if the spatial dimension of V_∞ is $n = 3$, and the *total-circulation theorem* if $n = 2$, both being the direct corollaries of the vorticity compactness.

Then, as seen in Sect. 2.2.1, the velocity decomposition (2.2.5) leads to Poisson equations for the scalar potential ϕ and vector potential $\boldsymbol{\psi}$ of velocity:

$$\nabla^2\phi = \vartheta, \tag{2.4.6a}$$

$$\nabla^2\boldsymbol{\psi} = -\boldsymbol{\omega}, \quad \nabla\cdot\boldsymbol{\psi} = 0, \tag{2.4.6b}$$

so that for a given $(\vartheta, \boldsymbol{\omega})$ distribution in an unbounded flow, one can solve the $(\phi, \boldsymbol{\psi})$ field as (Appendix A.3)

$$\phi = \int_{V_\infty} G\vartheta' dV', \quad \boldsymbol{\psi} = -\int_{V_\infty} G\boldsymbol{\omega}' dV', \tag{2.4.7}$$

where the two-point function $G(\boldsymbol{x}, \boldsymbol{x}')$ is the fundamental solution of the Poisson equation, representing the field at $\boldsymbol{x}$ generated by a pointwise disturbance of unit strength at $\boldsymbol{x}'$ (Fig. 2.11):

$$G(\boldsymbol{r}) = \begin{cases} \dfrac{1}{2\pi}\log r & \text{if } n = 2, \\ -\dfrac{1}{4\pi r} & \text{if } n = 3, \end{cases} \tag{2.4.8}$$

with $r = |\boldsymbol{r}|$ and $\boldsymbol{r} = \boldsymbol{x} - \boldsymbol{x}'$.

Now, while $\boldsymbol{x}'$ runs over a compact domain of $\vartheta \neq 0$ and $\boldsymbol{\omega} \neq \boldsymbol{0}$, the field point $\boldsymbol{x}$ can approach infinity with $x = |\boldsymbol{x}| \to \infty$. When $x > x' = |\boldsymbol{x}'|$, the Taylor expansion of G around $\boldsymbol{x}' = \boldsymbol{0}$ converges:

$$G(\boldsymbol{x} - \boldsymbol{x}') = G_0 - x_i'(G_{,i})_0 + \frac{1}{2}x_i'x_j'(G_{,ij})_0 - \cdots, \tag{2.4.9}$$

where the suffix 0 denotes evaluation at $\boldsymbol{x} = \boldsymbol{0}$, making the functions dependent on $\boldsymbol{x}$ only. Owing to (2.4.5), as we substitute $G(r)$ into (2.4.7) the first term has no contribution and the leading-order approximation comes from the second term depending linearly on $\boldsymbol{x}'$. Thus, let

$$\boldsymbol{I}_\theta \equiv \int \boldsymbol{x}'\vartheta' dV' \tag{2.4.10}$$

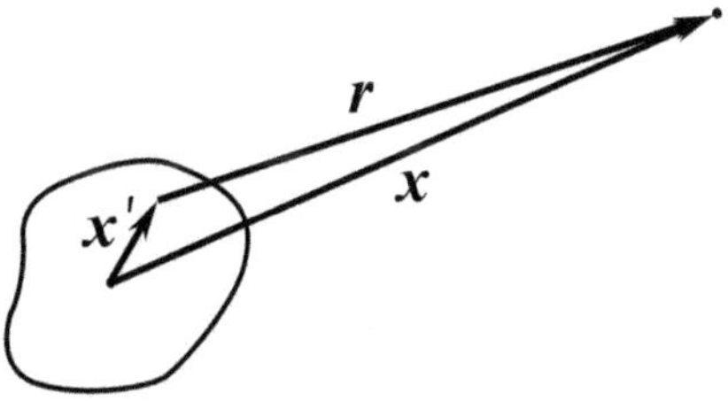

Fig. 2.11 Geometry of position vectors $\boldsymbol{x}$ and $\boldsymbol{x}'$ for observation point and field point, respectively. $\boldsymbol{r} = \boldsymbol{x} - \boldsymbol{x}'$

which by (2.4.4) must be finite, the ϑ-induced velocity has the leading-order expansion

$$\boldsymbol{u}_\theta = \nabla\phi \simeq -\nabla[\boldsymbol{I}_\theta \cdot (\nabla G)_0] = -\boldsymbol{I}_\theta \cdot (\nabla\nabla G)_0, \tag{2.4.11}$$

indicating $\boldsymbol{u}_\theta$ is irrotational as it should. It is also divergence-free, since

$$\nabla \cdot \boldsymbol{u}_\theta \simeq -I_{\theta j} G_{,jii} = 0.$$

Similarly, for the $\boldsymbol{\omega}$-induced velocity there is

$$\boldsymbol{u}_v = \nabla \times \boldsymbol{\psi} \simeq \nabla \times \int \boldsymbol{x}' \cdot (\nabla G)_0 \boldsymbol{\omega}' dV' = \nabla \times \left(\left(\int \boldsymbol{\omega}' \boldsymbol{x}' \cdot (\nabla G)_0 dV' \right) \right).$$

Since for $n = 3$

$$\int (\omega_i x_j + x_i \omega_j) dV = \int \nabla \cdot (x_i x_j \boldsymbol{\omega}) dV = 0, \tag{2.4.12}$$

only the anti-symmetric part of $\omega_i x_j$, i.e., $(\omega_i x_j - x_i \omega_j)/2$, has contribution to $\boldsymbol{u}_v$, namely

$$\frac{1}{2}(\boldsymbol{\omega}'\boldsymbol{x}' - \boldsymbol{x}'\boldsymbol{\omega}') \cdot (\nabla G)_0 = -\frac{1}{2}(\nabla G)_0 \times (\boldsymbol{x}' \times \boldsymbol{\omega}').$$

But (2.4.12) does not hold for $n = 2$; rather, since $\boldsymbol{\omega} = \omega \boldsymbol{e}_3$ with $\boldsymbol{\omega} \cdot \nabla G = 0$ we simply have

$$\boldsymbol{\omega}'\boldsymbol{x}' \cdot (\nabla G)_0 = -(\nabla G)_0 \times (\boldsymbol{x}' \times \boldsymbol{\omega}').$$

Hence, let

$$\boldsymbol{I}_v \equiv \frac{1}{n-1} \int \boldsymbol{x}' \times \boldsymbol{\omega}' dV' \tag{2.4.13}$$

which by (2.4.4) must also be finite, the $\boldsymbol{\omega}$-induced velocity has the leading-order expansion

$$\boldsymbol{u}_v \simeq -\nabla \times [(\nabla G)_0 \times \boldsymbol{I}_v] = -\boldsymbol{I}_v \cdot (\nabla\nabla G)_0 = -\nabla[\boldsymbol{I}_v \cdot (\nabla G)_0], \tag{2.4.14}$$

which is also irrotational as well as divergence-free.

By (2.4.8), for $n = 2$ and 3 the explicit forms of $(\nabla G)_0$ and $(\nabla\nabla G)_0$ are

$$(G_{,i})_0 = \frac{1}{2(n-1)\pi} \frac{e_i}{x^{(n-1)}}, \tag{2.4.15a}$$

$$(G_{,ij})_0 = \frac{1}{2(n-1)\pi} \frac{\delta_{ij} - n e_i e_j}{x^n}, \tag{2.4.15b}$$

where $\boldsymbol{e} = \boldsymbol{x}/x$ is the unit vector along $\boldsymbol{x}$. Therefore, we may summarize the above results as follows:

Velocity Far-Field Behavior. *In an unbounded compressible fluid at rest at infinity in n-dimensional space with* $n = 2, 3$, *if there is no net mass source nor total vorticity, then the far-field velocity induced by the vorticity and dilatation is dominated by*

$$\boldsymbol{u} = -(\boldsymbol{I}_\theta + \boldsymbol{I}_\vartheta) \cdot (\nabla\nabla G)_0 = O(x^{-n}), \tag{2.4.16}$$

which is an incompressible potential flow with $\phi = O(x^{-(n-1)})$.

2.4.3 Far-Field Asymptotics for Steady Flow

In this book, whenever we say "V extends to the entire space V_∞", we will be always talking about the "true" infinity, namely not only the fluid is at rest at infinity so that (2.4.1) is satisfied, but also V_∞ contains the whole vorticity and dilatation field so that (2.4.5a) and (2.4.5b) are satisfied. In this case the far-field flow becomes all-over irrotational so that the velocity has asymptotic behavior $|\boldsymbol{u}| = O(|\boldsymbol{x}|^{-n})$ as $\boldsymbol{x} \to \infty$ for $t < \infty$, given by (2.4.16).

Conceive now a finite body B centered at $\boldsymbol{x}_0$ starts motion in an otherwise still viscous fluid in V_∞ at $t = 0$, and then turns to move at constant velocity $\boldsymbol{U}_B = -U\boldsymbol{e}_x$. For definiteness assume the flow is incompressible. The moving body creates a disturbance flow field with nonzero vorticity. As long as the Reynolds number based on body size is not very small, the body carries a part of vorticity along with it and leaves the rest behind it, forming a *vortical wake*. The downstream end of the wake is the "starting vortical structure" generated at the earliest time $t = 0$, which stays around $\boldsymbol{x}_0$. As the body keeps moving to $x(t) = x_0 - Ut$, the wake extends from $x(t)$ to x_0, with a continuously increasing length $L(t) \sim Ut$. The flow in V_∞ is evidently time-dependent.

In the frame of reference fixed to the body, let the velocity field be $\boldsymbol{u} = \boldsymbol{U} + \boldsymbol{v}$ with $\boldsymbol{U} = -\boldsymbol{U}_B = U\boldsymbol{e}_x$, one sees

$$\boldsymbol{u} \to U\boldsymbol{e}_x \text{ and } \boldsymbol{v} \to \boldsymbol{0} \text{ as } |\boldsymbol{x}| \to \infty, \quad \boldsymbol{u} = \boldsymbol{0} \text{ for } \boldsymbol{x} \in \partial B.$$

At sufficiently large t after start, the starting vortical structure and its unsteady motion are very far away from the solid body, so their influence on the near-body flow is well negligible. Namely, the velocity field in a subspace surrounding the body, say V_{st}, may become (not always) time-independent or *steady* with $\boldsymbol{u} = \boldsymbol{u}(\boldsymbol{x})$. Meanwhile, one may set $V_\infty = V_{st} + V_{nst}$, where V_{nst} is a downstream subspace containing the starting vortical structure that retreats at speed $\boldsymbol{U}$, so the flow there is still inherently unsteady. This situation is exemplified in Fig. 2.12 for two- and three-dimensional wing flows, where the starting vortical structure is known as *starting vortex* (for details see Sects. 8.5.2 and 9.1.2).

In Fig. 2.12, the common boundary of V_{st} and V_{nst} must cut through the wake (as a *wake plane*) to exclude the unsteady growth of the wake in V_{nst}. Of course the wake plane, say W, should be located sufficiently far from the active unsteady-

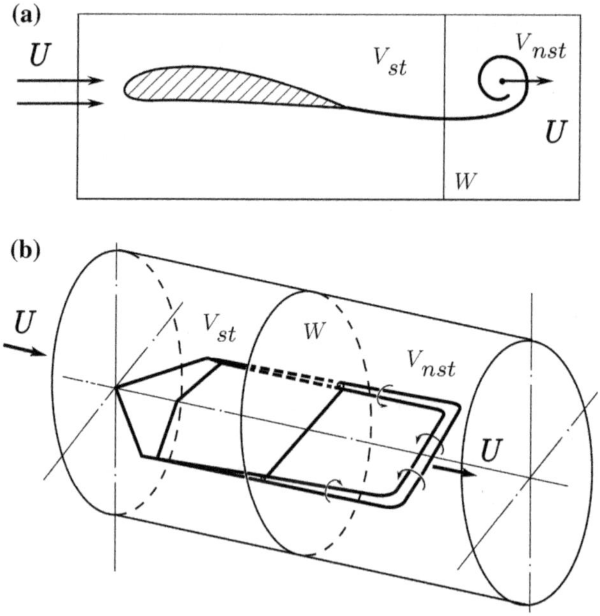

Fig. 2.12 Wing wake including starting vortex in, **a** two dimensions and **b** three dimensions. The flow in $V_{\rm st}$ is steady and that in $V_{\rm nst}$ is unsteady. The common boundary of $V_{\rm st}$ and $V_{\rm nst}$ is a wake plane W

flow region in $V_{\rm nst}$ so that at W all flow unsteadiness cannot be felt, where one can set time-independent downstream boundary conditions for the flow in $V_{\rm st}$. In $V_{\rm st}$, then, one sees an *open* wake, like an "infinitely" long paraboloid shown in Fig. 1.10. Evidently, this type of viscous steady flow must have different far-field asymptotics as $|\boldsymbol{x}| \to$ "∞". Since having two distinct infinite spaces V_∞ and $V_{\rm st}$ with different boundary conditions at infinity could tempt one to forget their inherent physical relation, we prefer to put them into a unified picture by assuming the steady flow to occur at arbitrarily large but finite t, so that $V_{\rm st}$ and $V_{\rm nst}$ are two true subset of V_∞ shown in Fig. 2.12. Therefore, the integral condition (2.4.5b) no longer holds over $V_{\rm st}$, but we now have

$$\int_{V_{\rm st}} \omega dV = -\int_{V_{\rm nst}} \omega dV, \quad n = 2, 3. \tag{2.4.17}$$

Namely, owing to (2.4.5b), the vortical structures in $V_{\rm st}$ and $V_{\rm nst}$ must coexist, one implying another.

Accordingly, the far-field estimate of velocity (2.4.16) in $V_{\rm st}$ has to be revised due to two distinct reasons. The first revision comes from the special property of two-dimensional flow, which also holds for unsteady flow. For any three-dimensional flow over a finite body, the flow domain $V_f = V_\infty - B$ is singly-connected and (2.4.5b) always holds. By contrast, in two dimensions V_f is *doubly-connected*. Even if $\boldsymbol{u} = \nabla\phi$ throughout V_{st}, the potential ϕ can be multi-valued and gains an arbitrary nonzero circulation Γ along any loop surrounding the body once. Such a potential is called *cyclic* or *circulatory*. For this case (2.4.5b) and (2.4.14) are invalid in any domain enclosing the body. Instead, in the far-field Taylor expansion (2.4.9) of Green's function $G(\boldsymbol{x}-\boldsymbol{x}_0)$, the first term will be G_0 rather than $x_i'(G_{,i})_0$, with the coefficient being that Γ. Consequently, if at far field the two-dimensional flow is *irrotational*, the expansion of potential ϕ has an extra leading term:

$$\phi = \frac{\Gamma}{2\pi}\theta + c_i\partial_i(\log r) + c_{ij}\partial_i\partial_j(\log r) + \cdots ,$$

where θ is the polar angle and we have assumed no source of mass as before. Thus, for $\boldsymbol{x} \in V_{\text{st}} \subset V_\infty$ and as $|\boldsymbol{x}| \to$ "∞", to the leading order there is

$$\boldsymbol{v} = \frac{\Gamma}{2\pi}\nabla\theta = \boldsymbol{e}_\theta\frac{\Gamma}{2\pi r} = O(r^{-1}). \tag{2.4.19}$$

The second and more delicate revision comes from the "infinitely" extended paraboloidal wake in viscous vortical flow, for which neither (2.4.19) nor (2.4.16) is applicable. Rather, instead of (2.3.8a), the linearized far-field vorticity equation for steady flow reads

$$\left(\nabla^2 - 2k\frac{\partial}{\partial x}\right)\boldsymbol{\omega} = \boldsymbol{0}, \quad k \equiv \frac{U}{2\nu}, \tag{2.4.20}$$

known as the *Oseen equation*, which is uniformly effective at any $Re = Ua/\nu$, where a is the body length scale. For example, for steady flow over a sphere, in spherical coordinates (r, ϕ, θ) with $x = r\cos\theta$, the vorticity $\boldsymbol{\omega} = \omega\boldsymbol{e}_\phi$ is along the azimuthal direction, and to the leading order of $a/r \ll 1$ (2.4.20) has approximate solution (e.g., Lamb 1932; Milne-Thomson 1968)

$$\omega = A(1+kr)\frac{\sin\theta}{r^2}e^{-kr(1-\cos\theta)} \quad \text{for } r \gg a, \tag{2.4.21}$$

where constant A is to be determined by the solution near the sphere at $R = O(a)$. The exponential factor in (2.4.21) describes a paraboloidal wake

$$r(1-\cos\theta) = L > 0, \quad r = |\boldsymbol{x}|,$$

outside which the vorticity decays exponentially but inside which it decays much slowly. Consequently, the far-field behavior of velocity is also different from (2.4.16). A rigorous estimate (Galdi 2011) has shown that outside the wake only the transverse components of $\boldsymbol{u}(\boldsymbol{x})$ decay according to (2.4.16) for $n = 3$ or $n = 2$ with $\Gamma = 0$, or (2.4.19) for $n = 2$ with $\Gamma \neq 0$; while for $\boldsymbol{v}\cdot\boldsymbol{e}_x = v_x(\boldsymbol{x})$ there is

$$|v_x(\boldsymbol{x})| = O(r^{-1/2}) \text{ as } r \to \text{“}\infty\text{”}, \quad \boldsymbol{x} \in \text{wake}, \tag{2.4.22a}$$
$$|v_x(\boldsymbol{x})| = O(r^{-1/2-\epsilon}) \text{ as } r \to \text{“}\infty\text{”}, \quad \boldsymbol{x} \notin \text{wake}, \tag{2.4.22b}$$

for some $\epsilon > 0$.

2.5 A Decoupled Model Flow: Inviscid Gas Dynamics

Owing to the big difficulty in theoretical studies of general viscous and compressible flows with coupled shearing and compressing processes, various simplified flow models have been developed and extensively explored to minimize the process coupling. The basis of these models is the fact that the behavior of shearing and compressing processes are governed by the Reynolds number Re and Mach number M, respectively, and hence their decoupling is possible if one of these numbers becomes infinity or zero. Two most widely used model flows are thereby identified: inviscid gas dynamics with $Re \to \infty$, which is briefly reviewed in this section; and viscous incompressible flow with $M \to 0$, which is the subject of the next section.

2.5.1 *Basic Equations*

We have seen in Sect. 1.2.4 that, for flow over a streamlined body at large Reynolds number Re, the shearing process is confined in a very thin *boundary layer* and its wake, outside which the global flow can be treated as effectively inviscid and irrotational. Therefore, one can consider the *asymptotic flow model at* $Re \to \infty$, neglect the thickness of the boundary layer, assume that the global flow satisfies only the no-through condition (1.3.2a) at the body surface. Then the linear viscous boundary coupling between the compressing and shearing processes disappears, and one can solve the inviscid global flow field first by the following equations:

$$\frac{D\rho}{Dt} + \rho\vartheta = 0, \tag{2.5.1a}$$
$$\rho\frac{D\boldsymbol{u}}{Dt} = -\nabla p, \tag{2.5.1b}$$
$$\rho T\frac{Ds}{Dt} = 0, \tag{2.5.1c}$$
$$p = \rho R T. \tag{2.5.1d}$$

The Euler equation (2.5.1b) can be replaced by the *Crocco equation* (2.2.11) for per unit mass:

$$\frac{\partial \boldsymbol{u}}{\partial t} + \boldsymbol{\omega} \times \boldsymbol{u} = -\nabla H + T\nabla s. \tag{2.5.2}$$

On the other hand, it is often convenient to replace (2.5.1c) by the total-enthalpy equation (1.2.34), which now reads

$$\rho \frac{DH}{Dt} = \frac{\partial p}{\partial t}. \tag{2.5.3}$$

These equations form the basis of *inviscid gas dynamics theory*, e.g., Liepmann and Roshko (1957). The no-slip condition (1.3.2b) will be imposed only when the viscous shear flow in boundary layer is to be solved for calculating the skin friction. This ingenious strategy invented by Prandtl in 1904 has been proved to be extremely successful and applied widely in practice.

2.5.2 Unsteady Potential Flows

If the flow remains attached and curved shock-waves therein, if any, are weak so that the entropy gradient behind the shocks is small, the flow away from boundary can be further treated irrotational with $\boldsymbol{u} = \nabla\phi$,[7] so that one obtains a fully decoupled model flow in which only one- two-, or three-dimensional longitudinal (compressing and entropy) process exists, with $Ds/Dt = 0$ following fluid particles' motion except across shock waves. This *isentropic* condition can be further strengthened to *homoentropic* condition $s = s_\infty$ if in far upstream the flow is uniform with constant s_∞. Then for arbitrary inviscid fluid (2.5.2) is reduced to

$$\nabla\left(\frac{\partial\phi}{\partial t} + H\right) = 0,$$

which can be integrated once to yield the unsteady Bernoulli equation

$$\frac{\partial\phi}{\partial t} + h + \frac{1}{2}|\nabla\phi|^2 = \frac{\partial\phi}{\partial t} + \int\frac{dp}{\rho} + \frac{1}{2}|\nabla\phi|^2 = 0, \tag{2.5.4}$$

where the integration constant $C(t)$ has been absorbed into ϕ without loss of generality. Then, by (2.5.1a), (2.5.1c), and thermodynamic relation (1.2.26), (2.5.4) is cast to

$$\nabla^2\phi = -\frac{1}{\rho}\frac{D\rho}{Dt} = -\frac{1}{a^2}\frac{Dh}{Dt} = \frac{1}{a^2}\frac{D}{Dt}\left(\frac{\partial\phi}{\partial t} + \frac{1}{2}|\nabla\phi|^2\right),$$

in which $D/Dt = \partial_t + \nabla\phi\cdot\nabla$. Hence, we obtain the *unsteady velocity potential equation* for homoentropic flow of any fluid:

$$\frac{\partial^2\phi}{\partial t^2} + \frac{\partial}{\partial t}|\nabla\phi|^2 + \nabla\phi\cdot\nabla\nabla\phi\cdot\nabla\phi - a^2\nabla^2\phi = 0; \tag{2.5.6a}$$

---

[7]If a shock is strong, the curl of (2.5.2) indicates that $\nabla T \times \nabla s$ will produce vorticity, see Sect. 3.4.2.

or, in more compact form,

$$\frac{D^2\phi}{Dt^2} - \frac{D}{Dt}\left(\frac{1}{2}|\nabla\phi|^2\right) - a^2\nabla^2\phi = 0. \tag{2.5.6b}$$

These equations are the basis of *nonlinear acoustics theory*.

2.5.3 Steady Isentropic Flow

Consider now a uniform incoming flow $\boldsymbol{U}$ over a stationary body, where the inviscid flow can often be treated steady. Then it is easily seen that (2.2.17) can be integrated twice and reduced to the first-order nonlinear *velocity equation* for inviscid isentropic flow, well known in gas dynamics (Problem 2.1):

$$\boldsymbol{u} \cdot \mathbf{D} \cdot \boldsymbol{u} - a^2\nabla \cdot \boldsymbol{u} = 0, \quad \text{or} \tag{2.5.7a}$$

$$\boldsymbol{u} \cdot \nabla\left(\frac{1}{2}q^2\right) - a^2\nabla \cdot \boldsymbol{u} = 0, \tag{2.5.7b}$$

where $\mathbf{D}$ is the symmetric strain-rate tensor. This equation is the first of two basic equations for inviscid, isentropic, and steady flow. To see its component form, it suffices to consider two-dimensional flow on (x, y)-plane with $\boldsymbol{u} = (u, v)$. Denote partial derivatives by subscripts, (2.5.7) yields

$$\left(1 - \frac{u^2}{a^2}\right)u_x + \left(1 - \frac{v^2}{a^2}\right)v_y - \frac{uv}{a^2}\left(v_x + u_y\right) = 0. \tag{2.5.8}$$

Obviously, the velocity equation will be most useful if $\boldsymbol{u}$ can be expressed in terms of a single scalar variable, e.g. by ϕ for potential flow or by ψ for two-dimensional flow. Then (2.5.7) or (2.5.8) becomes an equation for that scalar alone. For example, in terms of ϕ, (2.5.8) becomes the second-order *full velocity-potential equation* (cf. Problem 2.6).

Equation (2.5.7) alone is not yet closed as it involves variable sound speed a. This variable comes from the total-enthalpy equation (2.5.3) that simply reads $\boldsymbol{u} \cdot \nabla H = 0$, yielding the *energy integral*

$$H = \frac{1}{2}q^2 + h = \frac{1}{2}q^2 + c_pT = C(\psi), \tag{2.5.9a}$$

where ψ is a symbolic notation of a streamline. For polytropic gas this equation is specified to

$$\frac{1}{2}q^2 + \frac{\gamma}{\gamma - 1}\frac{p}{\rho} = \frac{1}{2}q^2 + \frac{a^2}{\gamma - 1} = C(\psi), \tag{2.5.9b}$$

or

$$a^2 = a_\infty^2 + \frac{\gamma - 1}{2}(U^2 - q^2), \tag{2.5.10}$$

which is the second of two basic equations for inviscid, isentropic, and steady flow. Equations (2.5.7) and (2.5.10) constitute the very basis of *high-speed steady aerodynamics* from subsonic to supersonic flows (for supersonic flow the jump relations across shock waves need to be added). This subject is beyond the concern of the present book (but see Problems 2.6 and 2.7).

2.6 Minimally-Coupled Model: Incompressible Flow

Alternative to the above asymptotic model with $Re \to \infty$, the asymptotic model with $M \to 0$ leads to *viscous incompressible flow* with $\vartheta = \nabla \cdot \boldsymbol{u} = 0$, in which the compressing-shearing coupling is minimized. In practice one does not require this velocity solenoidality condition to hold exactly, but will consider the flow as incompressible if the following dimensionless numbers are all small (Lighthill 1963):

$$\frac{nL}{a_\infty}, \frac{U}{a_\infty}, \frac{T_w - T_\infty}{T_\infty}, \frac{gL}{a_\infty^2} \ll 1, \tag{2.6.1}$$

where n is characteristic frequency, T_w the wall temperature, g the gravitational acceleration. These conditions can be inferred from inspecting (1.2.41c) and (1.2.44). Then thermodynamics is not involved either. Note that while (1.2.41a) and (1.2.44) suggest that the compressibility effect is proportional to M_∞^2, very high-frequency density fluctuation may result in $\vartheta \neq 0$ and this effect is proportional to M_∞. Indeed, comparing the Strouhal number St_a based on a_∞ as in (2.6.1) and that based on U yields $St_a/St = M_\infty$. Obviously, the incompressible flow is the best simplified model for the study of vortical flows. Most contents of this book will be developed within this model. Before entering the vorticity and vortex dynamics, therefore, it is appropriate to review the general theoretical foundation of incompressible flow.

2.6.1 Momentum Formulation versus Vorticity Formulation

For incompressible flow, the compressing process is degenerated to and only represented by the mechanical pressure field p. What remain nontrivial in the general Navier-Stokes equations are only the continuity equation (which becomes a kinematic constraint) and momentum equation, sufficient for solving $\boldsymbol{u}$ and p:

$$\nabla \cdot \boldsymbol{u} = 0, \tag{2.6.2a}$$

$$\begin{aligned} \rho \frac{D\boldsymbol{u}}{Dt} &= -\nabla p + \mu \nabla^2 \boldsymbol{u} \\ &= -\nabla p - \mu \nabla \times \boldsymbol{\omega}. \end{aligned} \tag{2.6.2b}$$

Alternative to (2.6.2b), we may write

$$\frac{\partial \boldsymbol{u}}{\partial t} + \boldsymbol{\omega} \times \boldsymbol{u} = -\nabla H - \nu \nabla \times \boldsymbol{\omega} \tag{2.6.3}$$

for per unit mass, where $H = p/\rho + q^2/2$.

For flow over a solid boundary, it is well known that (2.6.2) and velocity adherence condition (1.3.2),

$$[\![\boldsymbol{u}]\!] = \boldsymbol{0} \quad \text{at boundary}, \tag{2.6.4}$$

along with external flow condition, form a well-posed problem, which we call *momentum formulation*. But due to the $(\boldsymbol{u}, p)$ coupling inside the fluid, pressure has also to be solved, from the divergence of (2.6.2b) or (2.6.3):

$$\nabla^2 p = -\rho \nabla \cdot (\boldsymbol{u} \cdot \nabla \boldsymbol{u}) = -\rho \nabla \boldsymbol{u} : \nabla \boldsymbol{u}, \tag{2.6.5a}$$

$$\nabla^2 H = -\nabla \cdot (\boldsymbol{\omega} \times \boldsymbol{u}). \tag{2.6.5b}$$

These Poisson equations indicate that p and H are variables depending on the entire instantaneous flow field with no progressing wave.

Actually, since all flow structures of incompressible flow come solely from the shearing process, pressure behavior in the interior of a flow is not our concern. It can be removed by taking the curl of (2.6.2b) or (2.6.3) (assuming no non-conservative external body force):

$$\frac{\partial \boldsymbol{\omega}}{\partial t} + \nabla \times (\boldsymbol{\omega} \times \boldsymbol{u}) = \frac{D\boldsymbol{\omega}}{Dt} - \boldsymbol{\omega} \cdot \nabla \boldsymbol{u} = \nu \nabla^2 \boldsymbol{\omega}, \tag{2.6.6}$$

which is known as the *Helmholtz vorticity equation*. In the interior of the flow the shearing has only a self-nonlinearity. In contrast to (2.6.5), this equation involves *time evolution* and *viscous diffusion* of vorticity. In fact, one could well calculate the $(\boldsymbol{u}, \boldsymbol{\omega})$-field first by solving (2.6.6) along with $\boldsymbol{\omega} = \nabla \times \boldsymbol{u}$, of which the inversion is the Biot-Savart formula to be addressed in the next chapter, and then calculate pressure p.

Under the velocity adherence condition (2.6.4), however, unlike (2.6.2b), (2.6.6) *does not* form a well-posed problem because it is one order higher than (2.6.2b) and possible spurious solution could appear. The same is true for (2.6.5). Additional boundary conditions are necessary, of which the natural (and optimal) choice is *dynamic boundary conditions of Neumann type*, derived from (2.6.2b) itself applied on boundary with the use of acceleration adherence (1.3.3).

For example, consider two-dimensional flow on the (x, y) plane over a wall with $\boldsymbol{u} = (u, v)$, $\boldsymbol{\omega} = \boldsymbol{e}_z\omega$, and acceleration $\boldsymbol{a}_B = (a_{sB}, a_{nB})$. In this case (2.6.6) degenerates to

$$\frac{D\omega}{Dt} = \nu\nabla^2\omega, \tag{2.6.7}$$

and the additional conditions for vorticity and pressure are a simple extensions of (2.3.7a) and (2.3.7b), respectively:

$$\nu\frac{\partial\omega}{\partial n} = -\frac{1}{\rho}\frac{\partial p}{\partial s} + a_{sB}, \tag{2.6.8a}$$

$$\frac{1}{\rho}\frac{\partial p}{\partial n} = -\nu\frac{\partial\omega}{\partial s} - a_{nB}. \tag{2.6.8b}$$

As remarked in Sect. 1.3.1, as long as (2.6.4) holds at $t = 0$, it is ensured by (2.6.8a). We thus call (2.6.7) and (2.6.8a), as well as their extension to three dimensions, the *vorticity formulation*. The interior $(\boldsymbol{u}, p)$ coupling becomes a viscous (ω, p) coupling at boundary via viscosity and no-slip condition.

Vortical structures appear in flows with $Re \gg 1$ only. In this case $\omega = O(Re^{1/2})$ as will be seen in boundary layer and free shear layer (Chap. 4). Since $\partial/\partial s = O(1)$, if $Re \gg 1$ and $a_{nB} = 0$ the right-hand side of (2.6.8b) is small and so is $\partial_n p$. But the right-hand side of (2.6.8a) remains of $O(1)$, and hence so must be the left-hand side even if $Re \to \infty$. Therefore, (2.6.8a) is the key relation where, as said before, $\nu\partial\omega/\partial n \equiv \sigma$ is the *boundary vorticity flux* (BVF) that measures the vorticity creation rate at the wall and diffuses the vorticity into the fluid.

Mathematically and physically, the above analysis indicates that *in viscous wall-bounded flow a completely decoupling of shearing process from compressing process is impossible.* Their coupling always exists but is *minimized* in incompressible flow.

Numerically, however, the calculation of p can be bypassed by fractional-step vorticity-based methods (cf. Wu et al. 2006), where (2.6.8a) is the key condition to be satisfied, which can be implicitly imposed. This observation has motivated many studies in fluid dynamics and applied mathematics communities to design vorticity-based numerical methods, either grid-free or with grids. Besides, in two dimensions the velocity decomposition $\boldsymbol{u} = \nabla\phi + \nabla\times\boldsymbol{\psi}$ has vector potential $\boldsymbol{\psi} = (0, 0, \psi)$ and $\boldsymbol{\psi}$ is reduced to a single scalar ψ known as the *stream function*. We thus have

$$\boldsymbol{u} = \nabla\phi + \nabla\psi\times\boldsymbol{e}_z, \quad \nabla^2\phi = 0, \quad \nabla^2\psi = -\omega. \tag{2.6.9}$$

Thus, (2.6.7) may be replaced by a fourth-order equation for ψ, fully decoupled from p.

Although practice has shown that momentum-based numerical schemes are more convenient in use than vorticity-based schemes and have been the mainstream of CFD, the latter sheds deeper and unique insight to vortical-flow physics. Especially, if the flow field has been solved by whatever formulation and the distributions of $(\boldsymbol{u}, \boldsymbol{\omega}, p)$ are all available, the vorticity formulation offers a very powerful means for understanding the physics. For example, the role of pressure in the flow evolution in vorticity formulation has much clearer physical meaning than its role in momentum formulation. In particular, since (2.6.6) is homogeneous in ω, *the BVF is the only source in incompressible flow; no BVF, no vortical flows.* In contrast, the origin of vorticity in incompressible flows is hidden in momentum formulation.

2.6.2 Incompressible Potential Flow

If we take double limits, not only $M \to 0$ but also $Re \to \infty$, then we arrive at the "inviscid" and incompressible flow model, of which the real meaning is that the flow is still viscous with $\mu \neq 0$ but $\mu \to 0$. In this model the vortical regions shrink to infinitely thin "*vortex sheets*" (Chap. 5), away from which there is $\boldsymbol{u} = \nabla\phi$. Since the vortex sheets attached to solid body surface do not alter the body geometry, we may impose the no-through condition (1.3.2a) at the body surface and thereby solve the Laplace equation $\nabla^2\phi = 0$. The viscous term in (2.6.2b) automatically disappears. This is the simplest flow model in the entire fluid dynamics, yet still an inevitable part of vortical flow theory. For example the flow "induced" by a two-dimensional *point-vortex* at origin $r = 0$ is irrotational for all $r > 0$. It is also necessary for satisfying some boundary conditions. Therefore, in the rest of this section we highlight the main content of incompressible potential-flow theory.

Consider an externally unbounded fluid domain V_f which is at rest at infinity and in which a moving body B causes a single-valued and *acyclic* (or non-circulatory) velocity potential ϕ, which is solved from the kinematic problem

$$\nabla^2\phi = 0 \;\text{ in }\; V_f, \tag{2.6.10a}$$

$$\frac{\partial\phi}{\partial n} = \boldsymbol{n}\cdot\boldsymbol{u}_B \;\text{ on }\; \partial B, \quad \phi \to 0 \;\text{ as }\; x = |\boldsymbol{x}| \to \infty, \tag{2.6.10b}$$

where $\boldsymbol{u}_B$ is the velocity of ∂B. In particular, in two-dimensional flow on the (x, y) plane, the scalar stream function ψ satisfies irrotational condition $\nabla^2\psi = 0$. Thus, in complex plane $Z = x + iy$ the complex velocity potential $W(Z) = \phi + i\psi$ is analytic, making the powerful analytic-function theory and conformal mapping technique applicable.

In any case, once $\phi(\boldsymbol{x})$ is solved, the pressure can be computed from the Bernoulli integral for incompressible potential flow, which is the only dynamic equation at hand:

$$\frac{\partial\phi}{\partial t} + \frac{1}{2}|\nabla\phi|^2 + \frac{p}{\rho} = 0, \tag{2.6.11}$$

where the time-dependent integration constant is again absorbed into ϕ. An important consequence of (2.6.11) is that it leads to a very simple relation between the total pressure force acting on the body surface ∂B and the rate of change of the integral of $\phi \boldsymbol{n}$ over ∂B:

$$\rho \frac{d}{dt} \int_{\partial B} \phi \boldsymbol{n} dS = - \int_{\partial B} p \boldsymbol{n} dS. \tag{2.6.12}$$

Alternative to using (2.6.11), this relation may directly follow from integrating the Euler equation $\rho(D/Dt)\nabla\phi = -\nabla p$ over the body volume B.

If a solid body B moves through a fluid at rest at infinity and causes a potential flow, since $\nabla^2\phi = 0$ and hence $\nabla\phi \cdot \nabla\phi = \nabla \cdot (\phi\nabla\phi)$, by (2.6.10b) the total kinetic energy of the flow is

$$K = \frac{\rho}{2} \int_{V_f} \nabla\phi \cdot \nabla\phi dV = \frac{\rho}{2} \int_{\partial B} \phi \frac{\partial \phi}{\partial n} dS = \frac{\rho}{2} \int_{\partial B} \phi \boldsymbol{n} \cdot \boldsymbol{u}_B dS. \tag{2.6.13}$$

Evidently, the flow has no memory of its history but completely depends on the current motion of boundary. Since $K = 0$ implies $\nabla\phi = \boldsymbol{0}$ everywhere, if ∂V_f is suddenly brought to rest then the entire flow stops instantaneously. Therefore, *if there is a fluid flow without moving boundary, it must be vortical or compressible, or both.*

If one adds any disturbance $\boldsymbol{u}'$ to the velocity field with kinetic energy $K' > 0$, such that $\boldsymbol{u}_1 = \nabla\phi + \boldsymbol{u}'$, and if $\boldsymbol{u}' \cdot \boldsymbol{n} = 0$ on ∂V_f, then

$$\begin{aligned} K_1 &= \frac{\rho}{2} \int_{V_f} (\nabla\phi + \boldsymbol{u}') \cdot (\nabla\phi + \boldsymbol{u}') dV \\ &= K + K' + \rho \int_{\partial V_f} \boldsymbol{n} \cdot \boldsymbol{u}' \phi dS = K + K' > K. \end{aligned} \tag{2.6.14}$$

This is the famous **Kelvin's minimum kinetic energy theorem**: *Among all incompressible flows satisfying the same normal velocity boundary condition, the potential flow has minimum kinetic energy.*

Consider now a body moving with velocity $\boldsymbol{U}(t)$ through a three dimensional fluid at rest at infinity. Alternative to (2.6.13) that expresses the total kinetic energy K by the potential at body surface, we may express K by the far-field velocity obtained in Sect. 2.4.2. In the fluid domain V_f surrounding a body of volume V_B we have identity

$$\int_{V_f} q^2 dV = \int_{V_f} U^2 dV + \int_{V_f} (\boldsymbol{u} + \boldsymbol{U}) \cdot (\boldsymbol{u} - \boldsymbol{U}) dV. \tag{2.6.15}$$

Here, let $V = V_f + V_B$ be the volume of entire space, the first term is simply $U^2(V - V_B)$, while by using $\nabla \cdot \boldsymbol{u} = 0$ the second term is cast to

$$\int_{V_f} \nabla(\phi + \boldsymbol{U} \cdot \boldsymbol{x}) \cdot (\boldsymbol{u} - \boldsymbol{U}) dV = \int_{\partial V_f} (\phi + \boldsymbol{U} \cdot \boldsymbol{x})(\boldsymbol{u} - \boldsymbol{U}) \cdot \boldsymbol{n} dS.$$

Now ∂V_f consists of body surface ∂B and external boundary Σ, say, and on the former $(\boldsymbol{u} - \boldsymbol{U}) \cdot \boldsymbol{n} = 0$ as required by (2.6.10b). Thus the surface integral is only over Σ, which can be taken as a sphere S_R of radius $R \gg 1$ with $\boldsymbol{n} = \boldsymbol{e}$. Then, by (2.4.14), at large r the flow appears to be induced by a *dipole*:

$$\phi = -\boldsymbol{A} \cdot \nabla G = -\frac{\boldsymbol{A} \cdot \boldsymbol{e}}{r^2}, \tag{2.6.16a}$$

$$\boldsymbol{u} = \nabla\phi = -(\boldsymbol{A} \cdot \nabla)\nabla G = -\frac{\boldsymbol{A} - 3(\boldsymbol{A} \cdot \boldsymbol{e})\boldsymbol{e}}{r^3}, \tag{2.6.16b}$$

where $\boldsymbol{A}$ is proportional to $\boldsymbol{I}_v$ defined by (2.4.13), which is to be determined by the specific body shape and motion via solving (2.6.10). The constant factor $1/4\pi$ in (2.4.15) has been absorbed into $\boldsymbol{A}$. Now (2.6.15) can be cast to

$$\begin{aligned}\int_V q^2 dV &= U^2\left(\frac{4}{3}\pi R^3 - V_B\right) \\ &+ \int_{S_R}\left[3(\boldsymbol{A} \cdot \boldsymbol{e})(\boldsymbol{U} \cdot \boldsymbol{e}) - R^3(\boldsymbol{U} \cdot \boldsymbol{e})^2 - \frac{2(\boldsymbol{A} \cdot \boldsymbol{e})^2}{R^3}\right] d\Omega,\end{aligned}$$

where $d\Omega$ is the solid angle element. All three terms of the surface integral are of the form

$$a_i b_j \int_{S_R} e_i e_j d\Omega,$$

which amounts to an angle average over all directions. Since in spherical coordinates $\boldsymbol{e} = (e_x, e_y, e_z) = (\sin\theta\cos\phi, \sin\theta\sin\phi, \cos\theta)$ and $d\Omega = \sin\theta d\theta d\phi$, the integration yields

$$\int_{S_R} e_i e_j d\Omega = \frac{4\pi}{3}\delta_{ij}. \tag{2.6.17}$$

Thus, after neglecting the third term that vanishes as $R \to \infty$ and dropping the cancelling terms $(4/3)\pi R^3 U^2$, we finally obtain

$$K = \frac{1}{2}\rho \int_{V_f} q^2 dV = \frac{1}{2}\rho(4\pi \boldsymbol{A} \cdot U - V_B U^2), \tag{2.6.18}$$

where the second term is the kinetic energy of the virtual fluid displaced by the moving body. We see again that once the body stops ($\boldsymbol{U} = \boldsymbol{0}$) so does the fluid immediately.

2.6.3 Accelerated Body Motion and Virtual Mass

We now take a closer look at the flow caused by body's uniform motion. Since both (2.6.10a) and (2.6.10b) are linear in ϕ, this potential must depend linearly on the components of $\boldsymbol{U}$. Thus we may write

$$\phi = U_i\hat{\phi}_i, \quad \nabla^2\hat{\phi}_i = 0 \text{ in } V_f, \quad \frac{\partial\hat{\phi}_i}{\partial n} = n_i, \quad i = 1, 2, 3, \tag{2.6.19}$$

such that each $\hat{\phi}_i$ is the velocity potential caused by the body motion with a unit velocity in the ith direction. By their boundary condition, $\hat{\phi}_i$ are only functions of the body-surface geometry and can be obtained once for all for a given shape. Correspondingly, the dipole $\boldsymbol{A}$ and total kinetic energy K of the flow must be a linear and a quadratic forms of U_i, respectively:

$$A_i = c_{ij}U_j, \quad K = \frac{1}{2}m_{ij}U_iU_j. \tag{2.6.20}$$

By (2.6.18) and (2.6.13) it follows that

$$m_{ij} = \rho(4\pi c_{ij} - V_B\delta_{ij}), \tag{2.6.21a}$$

$$c_{ij} = \frac{1}{4\pi}\left(\int_{\partial B}\hat{\phi}_i\frac{\partial\hat{\phi}_j}{\partial n}dS + V_B\delta_{ij}\right). \tag{2.6.21b}$$

Here, it can be ensured c_{ij} is a symmetric tensor, and have so is m_{ij}. Moreover, since K is related to the total fluid momentum $\boldsymbol{P}$ by $dK = \boldsymbol{U}\cdot d\boldsymbol{P}$, there also is

$$P_i = m_{ik}U_k, \quad \text{namely} \quad \boldsymbol{P} = 4\pi\rho\boldsymbol{A} - \rho V_B\boldsymbol{U}. \tag{2.6.22}$$

Therefore, the total force exerted to the body by the potential flow takes exactly the form of Newton's second law:

$$F_i = -\frac{dP_i}{dt} = -m_{ik}\frac{dU_k}{dt}. \tag{2.6.23}$$

which will be known for a given motion once m_{ik} are calculated. The matrix m_{ik} is called *added mass*, *apparent mass*, or *virtual mass* and should be added to the body's own mass when calculating its motion.

If the body has a rotation as well as uniform translation such that

$$\boldsymbol{u}_B(\boldsymbol{x}, t) = \boldsymbol{U}(t) + \boldsymbol{\Omega}(t)\times\boldsymbol{r}, \quad \boldsymbol{r} = \boldsymbol{x} - \boldsymbol{x}_0,$$

we may simply add a potential $\phi_\Omega = \Omega_k \hat{\phi}_k$, with boundary condition

$$\frac{\partial \hat{\phi}_k}{\partial n} = (\boldsymbol{r} \times \boldsymbol{n})_k, k = 4, 5, 6.$$

2.6.4 *Force on a Body in Steady Flow*

We now consider the force exerted on a body B in a steady flow $\boldsymbol{u} = \nabla\phi$ with uniform incoming velocity $\boldsymbol{U}$. This issue is of great importance since it leads to two far-reaching consequences that had strongly stimulated the development of the whole fluid dynamics. One is the *d'Alembert Paradox* (d'Alembert in 1768) that had excited enormous efforts over a century for filling the very basic gap between mathematical theory and practical observations, a gap that was finally resolved by Prandtl's *boundary layer theory* (Chap. 4). The other is the *Kutta-Joukowski theorem* (Kutta in 1902 and Joukowski in 1906) that laid down the first cornerstone of modern aerodynamics.[8]

In this steady and potential flow, the total force formula (1.3.6) applies. Assume the flow at a sufficiently larger fixed control surface Σ is irrotational, the pressure can be replaced by $-|\boldsymbol{u}|^2/2$ due to the Bernoulli integral (2.6.11):

$$\boldsymbol{F} = \rho \int_\Sigma \left(\frac{1}{2}|\boldsymbol{u}|^2 \boldsymbol{n} - \boldsymbol{u}\boldsymbol{u} \cdot \boldsymbol{n} \right) dS. \tag{2.6.24}$$

Let $\boldsymbol{u} = \boldsymbol{U} + \boldsymbol{v}$ with $\boldsymbol{v}$ being the disturbance velocity that approaches zero at infinity, such that

$$\begin{aligned} \frac{1}{2}|\boldsymbol{u}|^2 n_i - u_i u_j n_j &= \frac{1}{2}|\boldsymbol{v}|^2 n_i + U_j v_j n_i + \frac{1}{2}U^2 n_i \\ &\quad - v_i v_j n_j - U_i v_j n_j - v_i U_j n_j - U_i U_j n_j. \end{aligned}$$

Substitute this into (2.6.24) and notice that quadratic terms of $\boldsymbol{U}$ can be taken out of the integral to leave zero integrals $\int \boldsymbol{n} dS$ over Σ, and the term $U_i \int v_j n_j dS$ also integrates to zero due to $\nabla \cdot \boldsymbol{v} = 0$. To handle the remaining terms, let Σ be large enough so that quadratic terms of $\boldsymbol{v}$ in the integral are negligible. Thus (2.6.24) is reduced to

$$\boldsymbol{F} = \rho \boldsymbol{U} \cdot \int_\Sigma (\boldsymbol{v}\boldsymbol{n} - \boldsymbol{n}\boldsymbol{v}) dS = \rho \boldsymbol{U} \times \int_\Sigma \boldsymbol{n} \times \boldsymbol{v} dS. \tag{2.6.25}$$

Then, as Σ further retreats toward infinity, if condition (2.4.5) holds such that by (2.4.16) there is $|\boldsymbol{v}| = O(r^{-n})$ for spatial dimension $n = 2, 3$, the above integral approaches zero as $O(r^{-1})$. Therefore, we arrive at

[8] On the historical development of these crucial results see an inspirative book by Darrigol (2009).

The d'Alembert Paradox. *A body at constant translational motion through an unbounded incompressible fluid at rest at infinity will experience no force if the fluid has no net vorticity*

We have seen in Sect. 2.4.3, however, that for two-dimensional flow over a body, if there is a circulation Γ over $V = V_f + B$, the far-field velocity estimate becomes (2.4.19), indicating that the integral of $\boldsymbol{n} \times \boldsymbol{v}$ over Σ is of $O(1)$ no matter how large Σ is. Thus, by the generalized Gauss theorem there is

$$\int_{\Sigma} \boldsymbol{n} \times \boldsymbol{v} dS = \int_{V} \boldsymbol{\omega} dV = \boldsymbol{e}_z \Gamma,$$

where $V = V_f + B$, and we immediately obtain

The Kutta-Joukowski Theorem. *A two-dimensional unbounded, steady, and incompressible cyclic potential flow over a body with circulation Γ exerts a transverse force to the body*

$$\boldsymbol{F} = -\rho U \Gamma \boldsymbol{e}_y, \tag{2.6.26}$$

where $\Gamma < 0$ if $F_y > 0$. The determination of Γ will be studied in Chap. 9.

The above derivation of (2.6.26) is a simplified version of Joukowski's original one. All one needs in proving both d'Alembert paradox and Kutta-Joukowski theorem is that the flow is inviscid and steady in V, and irrotational at Σ, satisfying corresponding far-field conditions. However, this approach is not without flaw. Later in Sect. 9.1 we shall revisit this issue and show that (2.6.26) is an exact result for two-dimensional and steady viscous flow at the limit $Re \to \infty$, which in turn is a special case of a three-dimensional *vortex-force theory* that can be further generalized to finite-Re flow.

2.7 Problems for Chapter 2

2.1. For steady viscous and compressible flow in a free space, show that (2.2.17) is reduced to

$$(a^2 + \nu_\theta \boldsymbol{u} \cdot \nabla)\vartheta - \frac{1}{2}\boldsymbol{u} \cdot \nabla q^2 + \nu \boldsymbol{u} \cdot (\nabla \times \boldsymbol{\omega}) - a^2 \boldsymbol{u} \cdot \nabla s^* = 0, \tag{2.7.1}$$

where $s^* = s/c_p$. Then using this equation to derive (2.5.7).

2.2. Discuss the physical role of viscosity and heat-conductivity in the far-field asymptotic analysis of vorticity and dilatation fields. What would happen if they are set identically zero?

2.3. Give a detailed derivation of (2.3.7).

2.4. Estimate the order of magnitude of the normal pressure gradient at the wall based on (2.3.7b), and its Re-dependence. Compare the result with the order of magnitude of boundary vorticity flux.

2.5*.[9] It has been proved in Sect. 2.4 that, in an n-dimensional unbounded free space with the fluid at rest at infinity, $n = 2, 3$, as $x = |\boldsymbol{x}| \to \infty$ the pressure $p - p_\infty$ decays exponentially and the flow is asymptotically irrotational and incompressible, with $\phi = O(x^{-n+1})$. Assume ϕ is time-dependent there, so that on a big sphere S of radius R, from the Bernoulli equation with neglecting q^2 it follows that

$$-\int_S (p - p_\infty)\boldsymbol{n}dS = \rho \int_S \frac{\partial \phi}{\partial t}\boldsymbol{n}dS = O(1)$$

since S increases as $O(R^{n-1})$. But this assertion contradicts the exponential decay behavior of $p - p_\infty$. Please resolve this apparent paradox.

2.6. Consider a two-dimensional steady, inviscid, and homoentropic flow of polytropic gas. Let the uniform free-stream velocity be $\boldsymbol{U} = U\boldsymbol{e}_x$ and denote the disturbance flow by $\boldsymbol{u} = (u, v, w)$. Assume that

$$\frac{|\boldsymbol{u}|}{U} \ll 1, \quad \frac{|\boldsymbol{u}|}{a} \ll 1.$$

Expend (2.5.7) and (2.5.10) up to the second order of $|\boldsymbol{u}|/U$ and $|\boldsymbol{u}|/a$, and show that the perturbation equation for (u, v, w) reads (subscripts denote derivatives)

$$\left[1 - M_\infty^2 - (\gamma+1)M_\infty^2 \frac{u}{U}\right] u_x + \left[1 - (\gamma-1)M_\infty^2 \frac{u}{U}\right] v_y = M_\infty^2 \frac{v}{U}(u_y + v_x), \tag{2.7.2}$$

where $M_\infty = U/a_\infty$. Observe that in the linearized version the perturbation equation is elliptic if $M_\infty < 1$ and hyperbolic if $M_\infty > 1$.

2.7. In transonic regime there can be $|1 - M^2| \ll 1$, where $M = |\boldsymbol{u}|/a$ is the local Mach number. Show that in this case

$$1 - M^2 = 1 - M_\infty^2 - (\gamma+1)M_\infty^2 \frac{u}{U} + \text{higher order terms.}$$

Then determine the appropriate form of transonic perturbation equation.

[9] Throughout the book, problems with asterisk are optional.

Chapter 3
Vorticity Dynamics

From now on we focus on the behavior of shearing process and its coupling with compressing process. This chapter presents the basic theory of vorticity dynamics (kinematics and kinetics). We start from kinematic properties of the vorticity, including its spatial properties and time evolution, locally and globally.[1] Follows the spirit of Truesdell (1954), a pioneering monograph on vorticity dynamics, in most of treatment we do not need to be concerned with any specific physical causes and effects of the motion, so the results are within the framework of kinematics and universal. The kinetics will be treated at the end of this chapter.

3.1 Kinematic Properties of Vorticity Field

This section introduces the most primary kinematic properties of the vorticity field, which are immediate corollaries of the definition of this vector, $\boldsymbol{\omega} = \nabla \times \boldsymbol{u}$. These results belong to *spatial* properties of vorticity.

3.1.1 Vorticity Tube and Circulation

The most primary differential property of the vorticity field, due solely and directly to its definition $\boldsymbol{\omega} = \nabla \times \boldsymbol{u}$, is that $\boldsymbol{\omega}$ is a divergence-free (solenoidal) field. In three-dimensional flow, the integral form of this feature is

$$\int_{\partial V} \boldsymbol{\omega} \cdot \boldsymbol{n} dS = 0. \tag{3.1.1}$$

[1] In this book, phenomenological descriptions of a motion are categorized into *kinematics*, like $v = at$ and $x = x_0 + \frac{1}{2}at^2$ for a particle's motion at constant acceleration a. *Kinetics* reveals physical causes of the motion, like $F = ma$. Kinematics plus kinetics as a whole is *dynamics*.

J.-Z. Wu et al., *Vortical Flows*, DOI 10.1007/978-3-662-47061-9_3

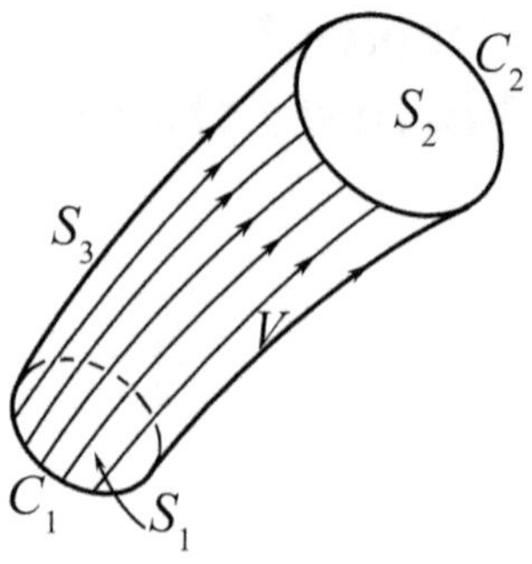

Fig. 3.1 A segment of a vorticity tube

A curve in a flow field tangent to the vorticity $\boldsymbol{\omega}$ at its every point is called a *vorticity line*. Let C be a closed loop which itself is not a vorticity line and can shrink to a point inside the fluid. Then all vorticity lines passing through C defines a tube-like surface called a *vorticity tube*.[2] Let two sectional surfaces S_α with boundary loops C_α ($\alpha = 1, 2$) cut the tube to form a volume V, and let S_3 be its side boundary, see Fig. 3.1. Since by definition $\boldsymbol{\omega} \cdot \boldsymbol{n} = 0$ on S_3, and $\boldsymbol{n}_2$ is directed toward the inside of V, from (3.1.1) it follows that

$$\int_{S_1} \boldsymbol{\omega} \cdot \boldsymbol{n} dS = \int_{S_2} \boldsymbol{\omega} \cdot \boldsymbol{n} dS. \tag{3.1.2}$$

These sectional integrals are called *vorticity-tube strength*. By the Stokes theorem, it is evident that (3.1.2) is equivalent to the circulation relation

$$\oint_{C_1} \boldsymbol{u} \cdot d\boldsymbol{x} = \oint_{C_2} \boldsymbol{u} \cdot d\boldsymbol{x}. \tag{3.1.3}$$

Therefore, we arrive at a fundamental kinematic theorem due to Helmholtz (1858) and Kelvin (1869):

The First Helmholtz Vorticity-Tube Theorem. *The strength of a vorticity tube or its circulation is invariant along the tube.*

Owing to the theorem, *a vorticity tube with nonzero circulation cannot terminate inside a fluid*, for otherwise the vorticity there must be infinite.

The behavior of a vorticity tube adjacent to a flow boundary is of particular importance. For a viscous fluid, the no-slip boundary condition (1.3.2b), $\boldsymbol{n} \times [\![\boldsymbol{u}]\!] = \boldsymbol{0}$, implies that the *normal-vorticity must be continuous across the boundary*:

$$\boldsymbol{n} \cdot [\![\boldsymbol{\omega}]\!] = (\boldsymbol{n} \times \nabla) \cdot [\![\boldsymbol{u}]\!] = 0. \tag{3.1.4}$$

[2] Ever since Helmholtz (1858) and even earlier, in almost all literature on vortex motion one sees the term "vortex tube" rather than "vorticity tube". This terminology has caused some confusion because so far there has no commonly accepted rational definition for a vortex, which itself is a complicated issue; cf. Wu et al. (2006), Sect. 6.6.

Therefore, a vorticity tube with nonzero circulation must form a closed loop inside the fluid or extend through a boundary to another rotational continuous medium.

For example, at a non-rotating solid boundary of normal $\boldsymbol{n}$ there must be $\boldsymbol{\omega} \cdot \boldsymbol{n} = 0$ and the vorticity tube cannot terminate there either. In reality, as a vorticity tube approaches a non-rotating boundary, since the fluid velocity must approach zero, (3.1.2) will force the tube to expand like a horn to keep its strength constant. Note that the termination on a non-rotating wall would be allowable if the fluid is ideal and only impermeable condition $\boldsymbol{n} \cdot [\![\boldsymbol{u}]\!] = 0$ is imposed. But then the most important source of vortices, see Sect. 2.3.2 as well as Sect. 3.4.4, would be lost.

Alternatively, if the solid boundary rotates at angular velocity $\boldsymbol{W}$, (3.1.4) implies that the vorticity lines normal to the boundary will be continued into the solid with $\omega_n = 2W_n$. But since the tangential vorticity involves normal gradient of the velocity, it is in general discontinuous across the boundary, namely $\boldsymbol{n} \times (\boldsymbol{\omega} - 2\boldsymbol{W}) \neq \boldsymbol{0}$. This situation is shown in Fig. 3.2.

Owing to this continuation of normal vorticity as well as that of velocity across boundary, for a flow over a solid body or another fluid (e.g., a water drop in the air) in arbitrary motion, we may conveniently consider the fluid plus the solid or the other fluid as a single *unbounded continuous system*, in which the velocity is everywhere continuous, and across the boundary the *normal vorticity* is also continuous. The first Helmholtz theorem still applies to the whole unbounded system in which all vorticity tubes with nonzero circulations remain closed. When a fluid vorticity tube hits a rotating boundary ∂B, say a gyro or a helicopter rotor blade, it must be split into a normal tube and a tangent tube. The former goes through ∂B with the solid being its continuation, while the latter has to go around ∂B.

Three examples sketched in Fig. 3.3 may illustrate how to apply the first Helmholtz theorem to construct qualitative vortical flow patterns. In Fig. 3.3a, a *tornado-like vortex* must expand as it approaches a non-rotating wall. Thus, if one creates a vertical vortex in a cup of water with a spoon, its vorticity tubes must turn to the direction along the bottom, and then climb up along the side wall, forming a boundary

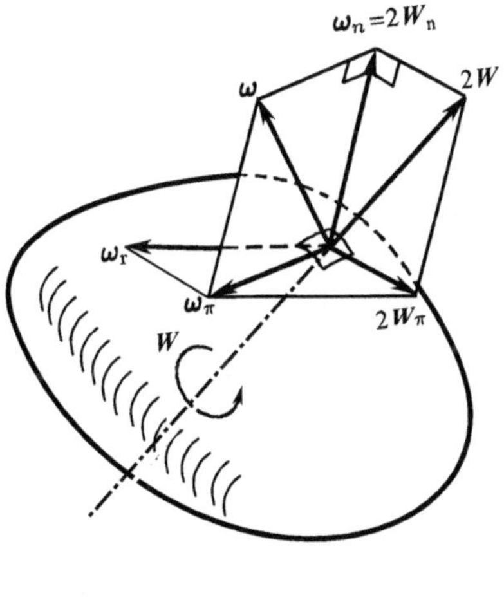

Fig. 3.2 The continuity of normal vorticity and discontinuity of tangent vorticity across a rotating body surface. The relative vorticity $\boldsymbol{\omega}_r = \boldsymbol{\omega} - 2\boldsymbol{W}$ has only tangent components

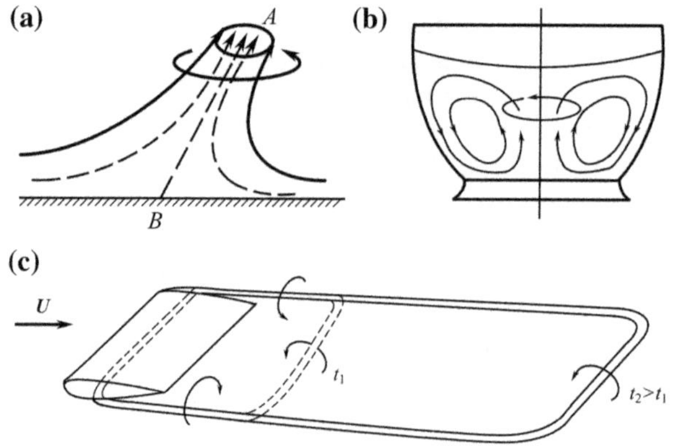

Fig. 3.3 **a** Vorticity lines in a tornado-like vortex. **b** Vorticity lines in a cup of rotating water. **c** Vortex system associated with a flying wing

layer with vorticity of the sign opposite to that of the main vortex, as in Fig. 3.3b. Figure 3.3c is another commonly seen example, where from the wing tips of an aircraft a pair of vortices shed off, known as *trailing vortices* or wing-tip vortices. A vorticity tube in this vortex system must also form a closed loop. Its upstream segment consists of the boundary layers of upper and lower surfaces of the wing, while its downstream segment is called the *starting vortex system* which is about at rest relative to the surrounding air. Note that the vortex loop in this figure is of course merely a highly idealized sketch. The trailing vortices are actually unstable and will break and reconnect to form a series of vortex rings (Sect. 10.2.4). Thus, rather than all the way connecting with the starting vortices, the real downstream end of the wing vortex system is at the location where the first vortex ring is formed.

In a two-dimensional flow a vorticity field only has incomplete appearance. All vorticity tubes there must be simply perpendicular to the flow plane rather than forming closed loops. In a three-dimensional space this simplified pattern is merely an idealized approximation.

3.1.2 Geometric Relation of Velocity and Vorticity

Vorticity and velocity are close related in space which deserves a detailed observation in its different aspects as they characterize various vortical flow patterns. We start from the geometric relation between $\boldsymbol{\omega}$ and $\boldsymbol{u}$.

1. Vorticity components in intrinsic streamline coordinates. We first express $\boldsymbol{\omega} = \nabla \times \boldsymbol{u}$ in terms of intrinsic (natural) *streamline coordinates*. Namely, at each point of a streamline C we introduce an orthonormal triad consisting of tangent vector $\boldsymbol{t}$, principal normal $\boldsymbol{n}$ (toward the center of curvature), and binormal $\boldsymbol{b} = \boldsymbol{t} \times \boldsymbol{n}$, see Fig. 3.4. The variations of $(\boldsymbol{t}, \boldsymbol{n}, \boldsymbol{b})$ along C are given by the *Frenet-Serret formulas*

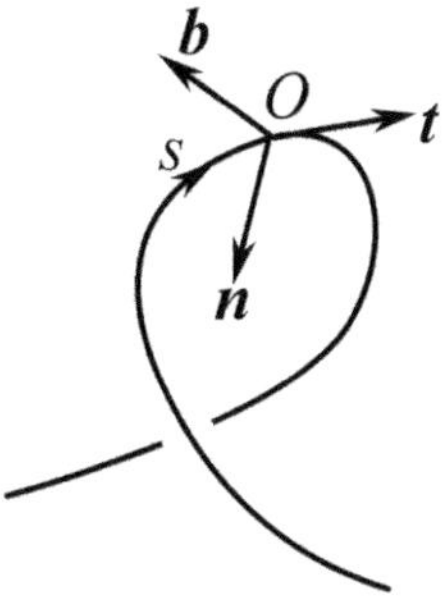

Fig. 3.4 Intrinsic orthogonal streamline coordinates

of classical differential geometry (e.g., Aris 1962):

$$\frac{\partial \boldsymbol{t}}{\partial s} = \kappa \boldsymbol{n}, \quad \frac{\partial \boldsymbol{n}}{\partial s} = -\kappa \boldsymbol{t} + \tau \boldsymbol{b}, \quad \frac{\partial \boldsymbol{b}}{\partial s} = -\tau \boldsymbol{n}, \tag{3.1.5}$$

where κ and τ are respectively the *curvature* and *torsion* of C. The curvature radius is $r = 1/\kappa$ with $dr = -dn$. The torsion τ measures how much a curve deviates from a plane curve; it is the curvature of the projection of C onto the $(\boldsymbol{n}, \boldsymbol{b})$ plane. For two-dimensional flow on the (x, y)-plane, $\boldsymbol{b}$ degenerates to the constant vector $\boldsymbol{e}_z$ normal to the flow plane and $\tau = 0$. With this orthonormal triad there is $\nabla = \boldsymbol{t}\partial_s + \boldsymbol{n}\partial_n + \boldsymbol{b}\partial_b$, so by (3.1.5)

$$\nabla \times \boldsymbol{t} = \kappa \boldsymbol{b} + (\boldsymbol{n} \times \partial_n \boldsymbol{t} + \boldsymbol{b} \times \partial_b \boldsymbol{t}),$$

where

$$\boldsymbol{t} \times (\boldsymbol{n} \times \partial_n \boldsymbol{t} + \boldsymbol{b} \times \partial_b \boldsymbol{t}) = \frac{1}{2}(\boldsymbol{n}\partial_n + \boldsymbol{b}\partial_b)|\boldsymbol{t}|^2 = 0$$

due to $|\boldsymbol{t}| = 1$. Thus

$$\nabla \times \boldsymbol{t} = \xi \boldsymbol{t} + \kappa \boldsymbol{b}, \tag{3.1.6a}$$

where the scalar coefficient

$$\xi \equiv \boldsymbol{t} \cdot (\nabla \times \boldsymbol{t}) = \boldsymbol{b} \cdot \partial_n \boldsymbol{t} - \boldsymbol{n} \cdot \partial_b \boldsymbol{t} \tag{3.1.6b}$$

involves not only a curve C but also a bundle of neighboring curves; it is known as the *torsion of neighboring vector lines* (Truesdell 1954).

With the above geometric preparation, we can now write down the vorticity expression in the streamline frame:

$$\boldsymbol{\omega} = \nabla \times (q\boldsymbol{t}) = \xi q \boldsymbol{t} + (\partial_b q)\boldsymbol{n} + (\kappa q - \partial_n q)\boldsymbol{b}. \tag{3.1.7}$$

Thus, ω_n and ω_b will appear if q varies from curve C to a neighboring C' along the binormal and normal directions, respectively, and ω_s will appear if $\xi \neq 0$. In two-dimensional flow only the last term of (3.1.7) remains nonzero:

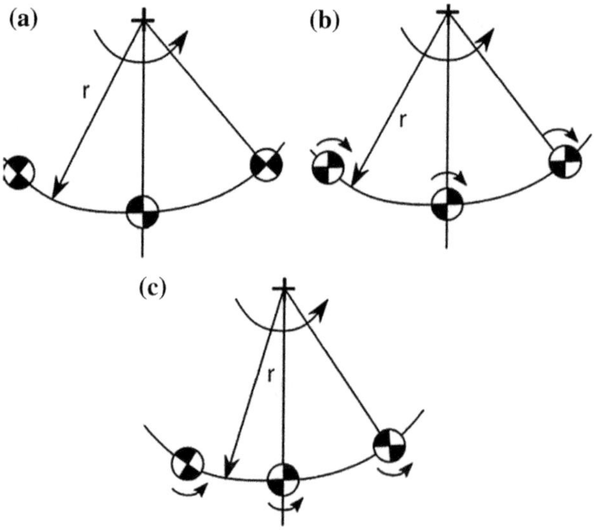

Fig. 3.5 Trajectory rotation and spin of a fluid element. **a** Rigidly rotating flow. **b** Irrotational flow. **c** Generic rotational flow

$$\omega = \kappa q - \partial_n q = \frac{q}{r} + \frac{\partial q}{\partial r} = \frac{1}{r}\frac{\partial}{\partial r}(rq). \tag{3.1.8}$$

In an axisymmetric circular motion with $\boldsymbol{u} = q\boldsymbol{e}_\theta$ in polar coordinates (r, θ), (3.1.8) gives its vorticity. The two terms there just represent a trajectory rotation around the center and a spin, respectively; Fig. 3.5 shows their three typical combinatory effects in the circular motion of a fluid element.

We remark that generically it is impossible to imbed the streamline coordinates into a (nominally general) global curvilinear orthonormal coordinate system in three-dimensional space, as listed in most textbooks, by setting the streamline as a coordinate line of the latter. In fact, it is easily found that in such a global system $\omega_s = \xi q$ would have to vanish identically, conflicting many real flows. This apparent dilemma comes from the fact that such a global orthonormal coordinate system *does not* generally exist but under very restrictive condition, which is violated once $\xi \neq 0$.

2. Lamb vector and helicity density. As a continuation of preceding discussion on the $\boldsymbol{\omega}$-$\boldsymbol{u}$ geometric relation, we denote $q = |\boldsymbol{u}|$ and $\omega = |\boldsymbol{\omega}|$, and by vector algebra obtain a geometrically orthogonal triangle relation

$$q^2\omega^2 = |\boldsymbol{\omega} \times \boldsymbol{u}|^2 + |\boldsymbol{\omega} \cdot \boldsymbol{u}|^2. \tag{3.1.9}$$

Here, the Lamb vector $\boldsymbol{\omega} \times \boldsymbol{u}$ has appeared in (1.1.36) where it was identified as a transverse force, and in Sect. 2.2 where it was further identified as a key nonlinear dynamic coupling mechanism of the shearing and compressing processes. By contrast, the scalar $\boldsymbol{\omega} \cdot \boldsymbol{u}$ is known as *helicity density*. If in a local flow ω and q are assumed fixed, then an increase of the angle between $\boldsymbol{\omega}$ and $\boldsymbol{u}$ implies an increase of Lamb vector and decrease of the helicity density. In particular, in two-dimensional flow $\boldsymbol{\omega} \cdot \boldsymbol{u} \equiv 0$ and the Lamb vector is always maximized, implying a maximum local transverse force for fixed q and ω. On the other hand, although helicity density does not appear in dynamic equations, it can be either positive or negative depending on the rotating direction is right-hand or left-hand sided that characterizes the *polarity* or *handedness* of a transverse vector field (not only flow field). It can be significant near the center of a strong vortex with axial velocity. Such a vortex is called *swirling vortex*. For example, the helicity density in tornadoes and updraft regions in rotating thunderstorms may reach 10 m/s^2 and 0.1 m/s^2, respectively (Lautenschlager et al. 1988).

Then, in the streamline coordinates both helicity density and Lamb vector are expressible by the kinetic energy $q^2/2$ and its gradients, along with the geometric properties of the streamline:

$$\boldsymbol{\omega} \cdot \boldsymbol{u} = \xi q^2, \tag{3.1.10a}$$

$$\boldsymbol{\omega} \times \boldsymbol{u} = \left[\kappa q^2 - \partial_n \left(\frac{1}{2}q^2\right)\right]\boldsymbol{n} - \partial_b \left(\frac{1}{2}q^2\right)\boldsymbol{b}. \tag{3.1.10b}$$

Some special behaviors of the Lamb vector and helicity density lead to certain simplified vortical flows. Below we consider three types of such flows without involving initial and boundary conditions.

3. Complex lamellar flow. In general, a vector field $\boldsymbol{f}$ is said to be a *complex lamellar field* if $\boldsymbol{f} \cdot (\nabla \times \boldsymbol{f}) = 0$. It has been proved that this condition holds if and only if $\boldsymbol{f} = g\nabla h$, say, and there exists a set of surfaces h = constants which are orthogonal to $\boldsymbol{f}$ vector everywhere. Assign $\boldsymbol{f} = \boldsymbol{u}$, we obtain *complex lamellar flow* as the first special flow (this name came from some similarity with potential flow which used to be called lamellar flow), in which $\boldsymbol{\omega} \neq \boldsymbol{0}$ but the helicity density vanishes identically:

$$\boldsymbol{\omega} \cdot \boldsymbol{u} = \boldsymbol{u} \cdot (\nabla \times \boldsymbol{u}) \equiv 0, \tag{3.1.11}$$

i.e., the streamlines are orthogonal to vorticity lines. Such a flow exists if and only if

$$\boldsymbol{u} = \lambda \nabla \mu, \quad \boldsymbol{\omega} = \nabla \lambda \times \nabla \mu. \tag{3.1.12}$$

In this flow there exists a set of equi-potential surfaces μ = constants orthogonal to streamlines everywhere, and potential flow is the special case with a constant λ (see Problem 1.2).

Now, if instead of assigning $\boldsymbol{f} = \boldsymbol{u}$ we set $\boldsymbol{f} = \boldsymbol{\omega} \times \boldsymbol{u}$, we obtain a complex-lamellar Lamb vector field with

$$(\boldsymbol{\omega} \times \boldsymbol{u}) \cdot [\nabla \times (\boldsymbol{\omega} \times \boldsymbol{u})] = 0, \tag{3.1.13}$$

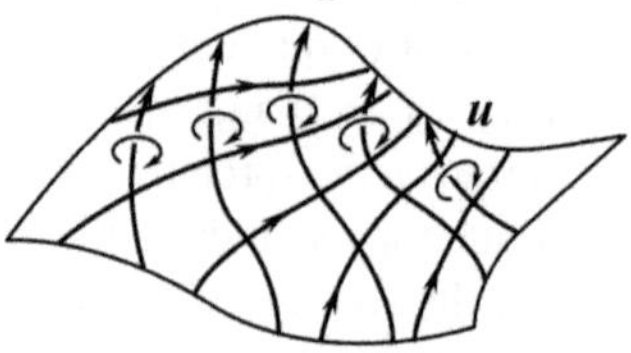

Fig. 3.6 A Lamb surface

then there exists a set of *Lamb surfaces* $h = \text{const.}$, such that both $\boldsymbol{u}$ and $\boldsymbol{\omega}$ are everywhere tangent to a Lamb surface. Namely, a Lamb surface is both stream surface and vorticity surface (Fig. 3.6). The existence of Lamb surfaces is important regarding the *integrability* and *chaos* of the flow. For a system, if under a given initial condition of finite accuracy its motion in subsequent times can be determined with finite accuracy, then the system is said to be *integrable*. Otherwise the system is *non-integrable*, extremely sensible to the initial condition. An uncertainty of the initial condition will cause the system's motion indeterminate or impossible to determine, and the motion will exhibit chaos. It can be shown that the pathline equations for a steady flow on an arbitrary two-dimensional surface is always integrable; thus, *steady flow with Lamb surfaces can in no way be chaotic.*

4. Beltrami flow. Opposite to the complex lamellar flow, the other extreme situation is $\boldsymbol{\omega} \times \boldsymbol{u} = \boldsymbol{0}$ and $\boldsymbol{\omega} \cdot \boldsymbol{u} \neq 0$. This happens if the streamlines are always parallel to vorticity lines, i.e., $\boldsymbol{\omega} = \xi \boldsymbol{u}$. This type of flow is called *Beltrami flow* or *helical flow*. By (3.1.10), the condition of vanishing Lamb vector is equivalent to

$$\partial_n \left(\frac{1}{2}q^2\right) = \kappa q^2, \quad \partial_b \left(\frac{1}{2}q^2\right) = 0. \tag{3.1.14}$$

For Beltrami flow, the vorticity form of the acceleration formula (1.1.36) is reduced to

$$\frac{D\boldsymbol{u}}{Dt} = \frac{\partial \boldsymbol{u}}{\partial t} + \nabla \left(\frac{1}{2}q^2\right),$$

the same as for potential flow. Then, for steady compressible Beltrami flow or any incompressible Beltrami flow, the continuity equation $\nabla \cdot (\rho \boldsymbol{u}) = 0$ implies

$$\nabla \cdot \left(\frac{\rho \boldsymbol{\omega}}{\xi}\right) = \nabla \left(\frac{\rho}{\xi}\right) \cdot \boldsymbol{\omega} = 0.$$

Thus, by (3.1.10a), the surfaces

$$\frac{\xi}{\rho} = \frac{\boldsymbol{\omega} \cdot \boldsymbol{u}}{\rho q^2} = \frac{\omega}{\rho q} = \text{const.} \tag{3.1.15}$$

are both vorticity surfaces and stream surfaces.

Each of the complex-lamellar flow and helical flow shares some of the properties of potential flow (Truesdell 1954). But these special flow models are not of equal importance or simplicity. We have seen that the Lamb vector is the major source of the vortical flow nonlinearity and major mechanism of shearing-compressing coupling in the interior of the flow field. In particular, once an incompressible flow becomes Beltramian the Helmholtz equation (2.6.6) will be linearized, and the nonlinear shearing-compressing coupling only remains in the compressing process through the vorticity-induced kinetic energy (Sect. 2.2.1) which is subjected to condition (3.1.14).

A realistic vortical flow may be approximately treated Beltramian in certain subregions, but can never be globally so. Later we will see that the total kinetic energy of incompressible flow can be expressed in terms of an integrated scalar moment of the Lamb vector; and a thin boundary layer on a wing surface must gain a strong Lamb-vector peak, of which the integration is responsible for the total lift of the wing.

The Beltrami flow is of great significance in some different senses, though. Firstly, the linear equation $\nabla \times \boldsymbol{u} = \xi \boldsymbol{u}$ implies that $\boldsymbol{u}$ is an eigenvector of the curl operator with eigenvalue ξ. Recall that $\xi = \boldsymbol{t} \cdot (\nabla \times \boldsymbol{t})$ along a streamline depends only on the geometry of a bundle of neighboring streamlines (which for Beltrami flow are also vorticity lines). These vector lines could then be generated everywhere in the entire flow domain, in terms of which the velocity and vorticity fields can be described in an intrinsic manner. More specifically, the positive or negative sign of ξ implies that $\boldsymbol{u}$ and $\boldsymbol{\omega} = \xi \boldsymbol{u}$ form a right- or left-hand *helical wave*. Physically, it is known that this *handedness* or *polarity* is a property inherent to all transverse (solenoidal) vector field.

Mathematically, it has been proven that for a domain of arbitrary geometric shape all the eigenvectors of the curl operator form a complete set of orthogonal (in inner-product space) basis vectors, onto which the projection of any transverse vector field constituents a *helical-wave decomposition* of that field into components with different wave numbers and polarities. For a review of relevant theory and examples see, e.g., Wu et al. (2006), Yang et al. (2010), and references therein.

Moreover, as mentioned before chaos cannot happen in steady flows having Lamb surfaces. It is thus natural to see if steady Beltrami flow, for which Lamb surfaces do not exist, can be chaotic. Then, if $\nabla \xi \neq \mathbf{0}$, the velocity will be on a set of surfaces with $\nabla \xi$ as normal, and hence still integrable. Therefore, the only possibility for incompressible steady flow to be chaotic is a Beltrami flow with constant ξ, say $\xi = 1$ and $\boldsymbol{\omega} = \boldsymbol{u}$. This assertion was confirmed by a Beltrami flow (u, v, w) defined in a periodic box $0 \leq (x, y, z) \leq 2\pi$:

$$u = A \sin z + C \cos y, \quad v = B \sin x + A \cos z, \quad w = C \sin y + B \cos x. \tag{3.1.16}$$

The flow indeed exhibits chaotic yet steady streamlines for certain value ranges of parameters A, B, and C. This "*ABC flow*" has close relation with the aforementioned helical-wave decomposition (Wu et al. 2006).

5. Generalized Beltrami flow. While the Beltrami flow represents a too idealized vortical flow model, a much wider type of vortical flows can be found if we relax the condition of $\boldsymbol{\omega} \times \boldsymbol{u} = \mathbf{0}$ to

$$\nabla \times (\boldsymbol{\omega} \times \boldsymbol{u}) = \mathbf{0} \quad \text{or} \quad \boldsymbol{\omega} \times \boldsymbol{u} = \nabla \chi. \tag{3.1.17}$$

A flow satisfying (3.1.17) is called a *generalized Beltrami flow*. The inviscid and incompressible steady flow is an example, which by (2.6.3) is governed by

$$\boldsymbol{\omega} \times \boldsymbol{u} = -\nabla H. \tag{3.1.18}$$

A generalized Beltrami flow has some simple properties. First, (3.1.13) is ensured, so any generalized Beltrami flow, steady or unsteady, must have Lamb surfaces. Second, (3.1.17) makes the Helmholtz equation (2.6.6) linearized, and thus opens a door to analytically obtainable exact vortex flow solutions, some are very useful and will be presented in Chap. 6.

3.1.3 Two-Dimensional and Axisymmetric Vortical Flows

Vortical flows with various spatial symmetries can serve as examples of the above special flows. We have exemplified Beltrami flow by the steady ABC flow (3.1.16), which has cubic-box symmetry. We now consider two wide classes of vortical flows with spatial symmetries: two-dimensional flow and axisymmetric flow. A subclass of the later is the *rotationally symmetric flow* generated by a flow on the meridian plane rotating around the z-axis, with zero circumferential velocity. These two classes of flow models are very commonly used in theoretical analysis of typical vortical flows, i.e., shear layers (Chap. 4) and axial vortices (Chap. 6).

Both two-dimensional and rotationally symmetric flows are complex lamellar, since the vorticity must be perpendicular to the flow plane. When the flow is incompressible, both permit for introducing a scalar stream function ψ, which is the only nonzero component of the vectorial potential $\boldsymbol{\psi}$ of the velocity. In two-dimensional flow $\psi(x, y)$ is the third component of $\boldsymbol{\psi}$ or stream function, which satisfies $\boldsymbol{u} = \nabla\psi \times \boldsymbol{e}_z$ or

$$u = \frac{\partial \psi}{\partial y}, \quad v = -\frac{\partial \psi}{\partial x}. \tag{3.1.19}$$

In the streamline coordinates with unit vectors $(\boldsymbol{t}, \boldsymbol{n})$, since

$$\boldsymbol{u} = q\boldsymbol{t} = \boldsymbol{t}\frac{\partial \psi}{\partial n} - \boldsymbol{n}\frac{\partial \psi}{\partial s},$$

there is $\partial_s \psi = 0$ or $\psi =$ constant along every streamline, and

$$q = q(\psi, t) = \frac{\partial \psi}{\partial n}. \tag{3.1.20}$$

Moreover, the kinematic relation $\nabla^2 \boldsymbol{\psi} = -\boldsymbol{\omega}$ is reduced to a scalar Poisson equation

$$\nabla^2 \psi = -\omega. \tag{3.1.21}$$

In rotationally symmetric flow observed in cylindrical coordinates (r, θ, z) there is $\boldsymbol{\psi} = (0, \psi_\theta, 0)$, so

$$\nabla \times \boldsymbol{\psi} = \frac{\boldsymbol{e}_z}{r} \frac{\partial (r\psi_\theta)}{\partial r} - \frac{\boldsymbol{e}_r}{r} \frac{\partial (r\psi_\theta)}{\partial z}.$$

Thus, let $\psi(r, z) = r\psi_\theta$ be the *Stokes stream function*, for $\boldsymbol{u} = (u, 0, w)$ there is

$$u = -\frac{1}{r}\frac{\partial \psi}{\partial z}, \quad w = \frac{1}{r}\frac{\partial \psi}{\partial r}. \tag{3.1.22}$$

Once again, in streamline coordinates with $\boldsymbol{u} = \boldsymbol{t}q$ there is $\psi =$ constant along every streamline, and instead of (3.1.20) we have

$$q = q(\psi, t) = \frac{1}{r}\frac{\partial \psi}{\partial n}. \tag{3.1.23}$$

Correspondingly, the vorticity is $\boldsymbol{\omega} = (0, \omega_\theta, 0)$, having only one component as well. Thus the Poisson equation $\nabla^2 \boldsymbol{\psi} = -\boldsymbol{\omega}$ is reduced to

$$\frac{\partial}{\partial r}\left(\frac{1}{r}\frac{\partial \psi}{\partial r}\right) + \frac{1}{r}\frac{\partial^2 \psi}{\partial z^2} = -\omega_\theta. \tag{3.1.24}$$

For incompressible flow, these scalar stream functions automatically ensure the continuity equation $\nabla \cdot \boldsymbol{u} = 0$ and can fully describe the velocity and vorticity field.

The use of scalar stream functions can be somewhat extended. Firstly, for *steady compressible flow* the continuity equation is reduced to $\nabla \cdot (\rho \boldsymbol{u}) = 0$, and hence (u, v) and (u, w) in (3.1.19) and (3.1.22) can be replaced by $(\rho u, \rho v)$ and $(\rho u, \rho w)$, respectively. Secondly, if in a three-dimensional flow the third velocity component w is independent of z, then it does not appear in the continuity equation and (3.1.19) still holds. The same is true for a nonzero circumferential velocity v in an axisymmetric flow, where (3.1.22) still holds. But in these extended cases a single ψ cannot fully describe the flow field.

The complex lamellar flow and generalized Beltrami flow have an interesting intersection set, when a single scalar stream function ψ can fully describe the velocity and vorticity, see (3.1.19)–(3.1.24). For two-dimensional flow we have $\boldsymbol{\omega} \times \boldsymbol{u} = \omega \nabla \psi$ and

$$\nabla \times (\boldsymbol{\omega} \times \boldsymbol{u}) = \nabla \omega \times \nabla \psi = \boldsymbol{e}_z \frac{\partial(\omega, \psi)}{\partial(x, y)}, \tag{3.1.25}$$

where

$$\frac{\partial(a, b)}{\partial(x, y)} \equiv \frac{\partial a}{\partial x}\frac{\partial b}{\partial y} - \frac{\partial a}{\partial y}\frac{\partial b}{\partial x} \tag{3.1.26}$$

denotes the Jacobian of the mapping from (a, b) to (x, y). The Jacobian vanishes if and only if a and b depend solely on each other (their dependence on t is permitted). Therefore, a two-dimensional flow is generalized Beltramian if and only if

$$\omega = f(\psi, t). \tag{3.1.27a}$$

This condition reduces (3.1.21) to

$$\nabla^2 \psi = -f(\psi, t). \tag{3.1.27b}$$

Similarly, for rotationally symmetric flow there is $\boldsymbol{\omega} \times \boldsymbol{u} = (\omega_\theta / r)\nabla\psi$ and

$$\nabla \times (\boldsymbol{\omega} \times \boldsymbol{u}) = \boldsymbol{e}_\theta \frac{\partial(\omega_\theta / r, \psi)}{\partial(z, r)}. \tag{3.1.28}$$

Hence a rotationally symmetric flow is generalized Beltramian if and only if ω_θ / r and ψ depend solely on each other, i.e.,

$$\omega_\theta = r f(\psi, t). \tag{3.1.29a}$$

This condition reduces (3.1.24) to

$$\frac{\partial}{\partial r}\left(\frac{1}{r}\frac{\partial \psi}{\partial r}\right) + \frac{1}{r}\frac{\partial^2 \psi}{\partial z^2} = -r f(\psi, t). \tag{3.1.29b}$$

Then, for both two-dimensional and rotationally symmetric flows it can be easily verified that (for the latter ω stands for ω_θ / r)

$$\boldsymbol{\omega} \times \boldsymbol{u} = \omega \nabla \psi = f(\psi, t)\nabla\psi = \nabla \int f(\psi, t) d\psi, \tag{3.1.30}$$

which gives an explicit expression for the scalar potential χ in (3.1.17).

3.1.4 Biot-Savart Formulas

In addition to the geometric relations between $\boldsymbol{\omega}$ and $\boldsymbol{u}$ discussed above, their *spatially functional relation* is more essential. This issue amounts to solving the velocity field

$\boldsymbol{u} = \nabla\phi + \nabla \times \boldsymbol{\psi}$ from an $(\boldsymbol{\omega}, \vartheta)$ distribution defined in a domain V:

$$\nabla \cdot \boldsymbol{u} = \nabla^2\phi = \vartheta, \quad \nabla \times \boldsymbol{u} = -\nabla^2\boldsymbol{\psi} = \boldsymbol{\omega}, \tag{3.1.31a}$$

which is the inversed operation of that defines dilatation and vorticity for given velocity field. The solvability is ensured by compatibility conditions

$$\int_V \vartheta dV = \oint_{\partial V} u_n dS, \quad \nabla \cdot \boldsymbol{\omega} = 0, \tag{3.1.31b}$$

along with proper boundary condition of $\boldsymbol{u}$ on ∂V. Our main concern is the existence and uniqueness of the solution. The uniqueness problem is simple: if *either* $\boldsymbol{n} \cdot \boldsymbol{u}$ *or* $\boldsymbol{n} \times \boldsymbol{u}$ at ∂V is given then the solution of (3.1.31a), if exists, must be unique (Problem 3.2). But uniqueness does not assure existence. The latter can be answered by finding the explicit integral expression of the velocity field in terms of the given $(\vartheta, \boldsymbol{\omega})$ field in a domain V.

1. General formulation. The desired integral expression is just a simple differentiation of the integral expression for the Helmholtz potentials $(\phi, \boldsymbol{\psi})$ of the velocity field, given by the following formulas derived in Appendix A.3.3:

$$\phi = \int_V G\vartheta' dV' - \int_{\partial V} G\boldsymbol{n}' \cdot \boldsymbol{u}' dS', \tag{3.1.32a}$$

$$\boldsymbol{\psi} = -\int_V G\boldsymbol{\omega}' dV' + \int_{\partial V} G\boldsymbol{n}' \times \boldsymbol{u}' dS', \tag{3.1.32b}$$

where primed quantities are evaluated at $\boldsymbol{x}'$. However, as remarked in Appendix A.3.3, (3.1.32) represents only one of infinitely many pairs of the possible potentials; one still needs to choose a single pair that satisfies the velocity boundary condition.

Consider first an unbounded domain V_∞ in n-dimensional space with the fluid at rest at infinity. Then since $|\boldsymbol{u}| = O(r^{-n})$ as $|\boldsymbol{x}| \to \infty$ (Sect. 2.4.2), (3.1.32) is reduced to

$$\phi(\boldsymbol{x}) = \int_{V_\infty} G(\boldsymbol{r})\vartheta' dV', \quad \boldsymbol{\psi}(\boldsymbol{x}) = -\int_{V_\infty} G(\boldsymbol{r})\boldsymbol{\omega}' dV'. \tag{3.1.33}$$

The integrals converge due to the exponential decay of ϑ and $\boldsymbol{\omega}$ (Sect. 2.4.1). Hence, the gradient of ϕ and curl of $\boldsymbol{\psi}$ gives a unique velocity field at once:

$$\begin{aligned}\boldsymbol{u}(\boldsymbol{x}) &= \int_{V_\infty} (\vartheta' \nabla G + \boldsymbol{\omega}' \times \nabla G) dV' \\ &= \frac{1}{2(n-1)\pi} \int_{V_\infty} \frac{\vartheta' \boldsymbol{r} + \boldsymbol{\omega}' \times \boldsymbol{r}}{r^n} dV',\end{aligned} \tag{3.1.34}$$

which is called the *generalized Biot-Savart formula*. Note that the velocities induced by $\vartheta' dV'$ and $\boldsymbol{\omega}' dV'$ are always in and perpendicular to the $\boldsymbol{r}$-direction, respectively. This is yet another example showing the different effects of compressing and shearing processes.

When the domain V is bounded and singly-connected, for a generic boundary shape the solution of Poisson equations in (3.1.31a) under specified boundary condition can only be carried out numerically, which is inconvenient in applications. Instead, one wishes to extend the analytic formula (3.1.34) to bounded domain, but then the formulation is complicated by a few issues (e.g., Serrin 1959; Wu et al. 2006; Tong et al. 2009).

This being the case, within the framework of kinematics, we would rather bypass altogether these troublesome issues and use the continuation artifice introduced in the context of Fig. 3.2: taking the fluid plus solid as a single continuum system in an unbounded domain. Then the simple formula (3.1.34) is sufficient. In later applications we will further confine ourselves to incompressible flow, so (3.1.34) is reduced to

$$\boldsymbol{u}(\boldsymbol{x}) = \frac{1}{2(n-1)\pi} \int_{V_\infty} \frac{\boldsymbol{\omega}' \times \boldsymbol{r}}{r^n} dV'. \tag{3.1.35}$$

2. Velocity induced by vortex sheet. At large Reynolds numbers, the vorticity is usually confined in a thin *shear layer* of small thickness δ. As Re increases, δ decreases but $\boldsymbol{\omega}$ therein increases. In the limiting case of $Re \to \infty$ so that $\delta \to 0$, the shear layer becomes surface-like called a *vortex sheet*, for which the distributed vorticity across the layer is a delta function, but the integral

$$\boldsymbol{\gamma} = \lim_{\delta \to 0} \int_0^\delta \boldsymbol{\omega} dn \tag{3.1.36}$$

remains finite. Vector $\boldsymbol{\gamma}$ defined on the vortex-sheet surface is known as *vortex sheet strength*. Then in (3.1.35) we have $\boldsymbol{\omega} dV = \boldsymbol{\gamma} dS$ and

$$\boldsymbol{u}(\boldsymbol{x}) = \frac{1}{2(n-1)\pi} \int_S \frac{\boldsymbol{\gamma}(\boldsymbol{x}') \times \boldsymbol{r}}{r^n} dS(\boldsymbol{x}'), \tag{3.1.37}$$

which is a basis of *vortex-sheet kinematics* to be addressed in Chap. 5.

3. Velocity induced by vortex filament. If $\boldsymbol{\omega}$ is confined in a very thin vorticity tube C called *vortex filament* (Fig. 3.7), then by the first Helmholtz theorem the tube has a single strength or circulation Γ and must be closed in the unbounded space. Thus, let $d\boldsymbol{x} = \boldsymbol{t} dl$ be a line element at $\boldsymbol{x}' \in C$, in (3.1.35) there is $\boldsymbol{\omega} dV = \Gamma \boldsymbol{t} dl$ and hence

$$\boldsymbol{u}(\boldsymbol{x}) = \frac{\Gamma}{2(n-1)\pi} \oint_C \frac{d\boldsymbol{x}' \times \boldsymbol{r}}{r^n}. \tag{3.1.38}$$

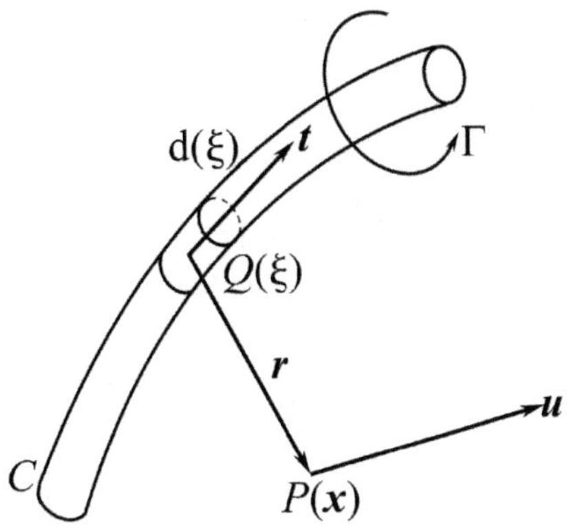

Fig. 3.7 The velocity at point $P(\mathbf{x})$ induced by a vortex filament of circulation Γ

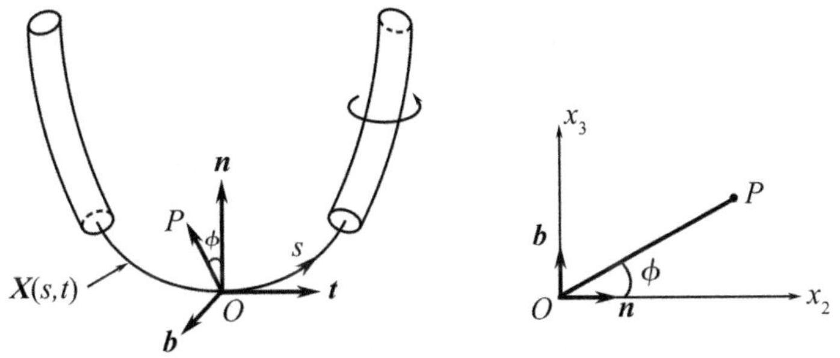

Fig. 3.8 Definition sketch for the induced velocity near a vortex filament

This formula has the same form as the Biot-Savart formula for the static magnetic field induced by a line current, following which it is often said that $\boldsymbol{u}$ in (3.1.38) is "induced" by the vortex filament.[3]

The simplest and most familiar example of (3.1.38) is the two-dimensional circumferential velocity "induced" by a single point vortex of circulation Γ:

$$u_\theta = \frac{\Gamma}{2\pi r}. \tag{3.1.39}$$

Note that a point vortex does not induce any motion of the vortex itself.

In three dimensions, vortex filaments are generically curves and (3.1.39) is not applicable. One of the complexities of curved vortex filaments is the fact that it can have *self-induced motion*. To see this, we look at the extension of (3.1.39) to such a curved filament C, focusing on the motion at a point on C induced by its neighboring segments of C.

[3]This explains the name of generalized Biot-Savart formula. But in fluid mechanics "induction" is a misnomer, since in the kinematic relation between the vorticity and velocity no causality can be identified. With this reservation in mind, the word "induction" will still be used for simplicity.

Let O be a point on C and s be the arc-length starting from O, and let P be a nearby point away from C with position vector $\boldsymbol{x} = \sigma \boldsymbol{e}_\sigma$ in the cross-sectional polar coordinates (σ, ϕ), see Fig. 3.8. We wish to calculate the induced velocity at P by the filament. Introduce a moving intrinsic triad $(\boldsymbol{t}, \boldsymbol{n}, \boldsymbol{b})$ at P as the tangent, normal, and binormal unit vectors as shown in the figure. Since $\partial \boldsymbol{x}/\partial s = \boldsymbol{t}$, we may use the Frenet-Serret formulas (3.1.5) to approximate the position vector $\boldsymbol{x}'$ of a neighboring moving point along the line vortex and its tangent vector $\boldsymbol{t}'$ as

$$\begin{aligned}\boldsymbol{x}' &= \left(\frac{\partial \boldsymbol{x}}{\partial s}\right)_0 s + \frac{1}{2}\left(\frac{\partial^2 \boldsymbol{x}}{\partial s^2}\right)_0 s^2 + O(s^3) \simeq s\boldsymbol{t} + \frac{1}{2}\kappa s^2 \boldsymbol{n},\\ \boldsymbol{t}' &= \boldsymbol{t} + \left(\frac{\partial \boldsymbol{t}}{\partial s}\right)_0 s + O(s^2) \simeq \boldsymbol{t} + \kappa s \boldsymbol{n},\end{aligned}$$

where suffix 0 denotes evaluation at O and κ is the curvature there. Then in (3.1.38) with $d\boldsymbol{x}' = \boldsymbol{t}' ds$ we have

$$\begin{aligned}\boldsymbol{t}' \times (\boldsymbol{x} - \boldsymbol{x}') &= (\boldsymbol{t} + \kappa s \boldsymbol{n}) \times \left(\sigma \boldsymbol{e}_\sigma - s\boldsymbol{t} - \frac{1}{2}\kappa s^2 \boldsymbol{n}\right)\\ &= \sigma \boldsymbol{e}_\phi + \sigma \kappa s \boldsymbol{n} \times \boldsymbol{e}_\sigma + \frac{1}{2}\kappa s^2 \boldsymbol{b},\end{aligned}$$

where $\boldsymbol{e}_\phi = \boldsymbol{t} \times \boldsymbol{e}_\sigma$ is the unit vector along the ϕ-direction. Hence, by (3.1.38), the velocity at P induced by a segment of the line vortex at $-L \le s \le L$ is (cf. Tong et al. 2009)

$$\boldsymbol{u} \simeq \frac{\Gamma\sigma}{4\pi}\boldsymbol{e}_\phi \int_{-L}^{L} \frac{ds}{r^3} + \boldsymbol{n} \times \boldsymbol{e}_\sigma \frac{\Gamma \sigma \kappa}{4\pi} \int_{-L}^{L} \frac{s ds}{r^3} + \frac{\Gamma}{8\pi}\kappa \boldsymbol{b} \int_{-L}^{L} \frac{s^2 ds}{r^3},$$

where $r = |\boldsymbol{x} - \boldsymbol{x}'| \simeq \sqrt{\sigma^2 + s^2} = \sigma\sqrt{1+\eta^2}$ with $\eta = s/\sigma$, and the second integral vanishes due to symmetry. Thus the above expression is cast to

$$\begin{aligned}\boldsymbol{u} &= \frac{\Gamma}{4\pi\sigma}\boldsymbol{e}_\phi \int_{-L/\sigma}^{L/\sigma} \frac{d\eta}{(1+\eta^2)^{3/2}} + \frac{\Gamma}{8\pi}\kappa \boldsymbol{b} \int_{-L/\sigma}^{L/\sigma} \frac{\eta^2 d\eta}{(1+\eta^2)^{3/2}} + \text{bounded terms}\\ &= \frac{\Gamma}{2\pi\sigma}\boldsymbol{e}_\phi + \frac{\Gamma\kappa}{4\pi}\boldsymbol{b} \log\frac{L}{\sigma} + \text{bounded terms}.\end{aligned} \tag{3.1.40}$$

Equation (3.1.40) is called *local-induction approximation*, which is the very basis of *vortex-filament dynamics* which studies the movement and deformation of complex vortex filaments. For details of the theory, including the extension of (3.1.40), see Wu et al. (2006).

Obviously, in (3.1.40) the first term is a singular circulatory motion about the filament, the same effect as (3.1.39) induced by a straight vortex filament on a two-dimensional cross plane. This term causes a rotation of fluid element at point P around the filament but does not cause any motion of the filament itself. In contrast, the second term indicates that *a curved vortex filament has a self-induced motion along its binormal direction with speed of magnitude* $\log(L/\sigma)$ *that becomes logarithmic infinity as* $\sigma \to 0$. Therefore, a curved line vortex of zero core size can never be a realistic vortex model. Any generic curved thin vortex filament, including thin vortex ring, has to have a small but finite core.

3.2 Vorticity Kinetic Vector and Circulation-Preserving Flow

When we derived the dynamic equations for vorticity and dilatation in Sect. 2.2.2, the key step is taking the curl and divergence of the momentum equation. The former leads to the curl of acceleration, $\nabla \times \boldsymbol{a}$, which is to be balanced by the curl of various forces. Thus, the vector $\nabla \times \boldsymbol{a}$ bridges vorticity kinematics and kinetics, which Truesdell (1954) calls the vorticity diffusion vector. Perhaps a more appropriate name is *vorticity kinetic vector*. But if we keep this vector as a general notation as will be done in this and the next sections, various temporal kinematic aspects of vortical flow can still be examined in a universal manner. The kinetic contents of $\nabla \times \boldsymbol{a}$ will be discussed in Sect. 3.4.

3.2.1 General Evolution Formulas

To examine the temporal evolution of vorticity field, we start from the Lagrangian formula for material acceleration:

$$\boldsymbol{a} = \frac{D\boldsymbol{u}}{Dt} = \frac{\partial \boldsymbol{u}}{\partial t} + \boldsymbol{\omega} \times \boldsymbol{u} + \nabla\left(\frac{1}{2}q^2\right), \tag{3.2.1}$$

of which the curl may take a few equivalent forms:

$$\begin{aligned} \nabla \times \boldsymbol{a} &= \frac{\partial \boldsymbol{\omega}}{\partial t} + \nabla \times (\boldsymbol{\omega} \times \boldsymbol{u}) &\quad (3.2.2\text{a})\\ &= \frac{\partial \boldsymbol{\omega}}{\partial t} + \nabla \cdot (\boldsymbol{u}\boldsymbol{\omega} - \boldsymbol{\omega}\boldsymbol{u}) &\quad (3.2.2\text{b})\\ &= \frac{D\boldsymbol{\omega}}{Dt} - \boldsymbol{\omega} \cdot \nabla \boldsymbol{u} + \vartheta \boldsymbol{\omega}. &\quad (3.2.2\text{c}) \end{aligned}$$

Here, the compressibility effect $\vartheta\boldsymbol{\omega}$ can be replaced by the density-variation effect via the continuity equation (1.2.31a). This casts (3.2.2c) to the *Beltrami equation*

$$\frac{D}{Dt}\left(\frac{\boldsymbol{\omega}}{\rho}\right) = \frac{\boldsymbol{\omega}}{\rho} \cdot \nabla \boldsymbol{u} + \frac{1}{\rho}\nabla \times \boldsymbol{a}. \tag{3.2.3}$$

Owing to taking the curl of acceleration, the vorticity kinetics covers less contents than the general fluid kinetics, while the vorticity kinematics cover much wider contents than the general fluid kinematics.

The roles of vorticity kinetic vector $\nabla \times \boldsymbol{a}$ can be revealed by two complementary and universal kinematic formulas. The first, and also the neatest, is *Kelvin's circulation formula* (1.1.45) for the time rate of circulation $\Gamma_{\mathcal{C}}$ along any material fluid loop $\mathcal{C}$ that spans a surface $\mathcal{S}$:

$$\frac{d\Gamma_{\mathcal{C}}}{dt} = \int_{\mathcal{S}} (\nabla \times \boldsymbol{a}) \cdot \boldsymbol{n} dS, \tag{3.2.4}$$

which reflects the *integrated* effect of $\nabla \times \boldsymbol{a}$ on the evolution of $\Gamma_{\mathcal{C}}$. The second formula concerns the *local* effect of $\nabla \times \boldsymbol{a}$ on the evolution of vorticity. Taking the inner product of (3.2.3) and the gradient of an arbitrary tensor $\mathcal{F}$ from the right yields a general relation

$$\frac{D}{Dt}\left(\frac{\boldsymbol{\omega}}{\rho} \cdot \nabla \mathcal{F}\right) = \frac{\boldsymbol{\omega}}{\rho} \cdot \nabla \left(\frac{D\mathcal{F}}{Dt}\right) + \frac{1}{\rho}(\nabla \times \boldsymbol{a}) \cdot \nabla \mathcal{F}. \tag{3.2.5}$$

Assigning $\mathcal{F}$ with different quantities may lead to a variety of results; for instance taking $\mathcal{F} = \boldsymbol{x}$ simply returns to (3.2.3). Equation (3.2.5) is mainly used for those $\mathcal{F}$ which are Lagrangian invariant, and in this case the quantity $(\boldsymbol{\omega}/\rho) \cdot \nabla \mathcal{F}$ is called the *generalized potential vorticity* (the name will be explained later), governed by

$$\frac{D}{Dt}\left(\frac{\boldsymbol{\omega}}{\rho} \cdot \nabla \mathcal{F}\right) = \frac{1}{\rho}(\nabla \times \boldsymbol{a}) \cdot \nabla \mathcal{F}, \quad \frac{D\mathcal{F}}{Dt} = 0. \tag{3.2.6}$$

For example, recall that due to (1.1.5) the label $\boldsymbol{X}$ of the fluid particles is a Lagrangian invariant, so we may set $\mathcal{F} = \boldsymbol{X}$. Denote the Lagrangian time variable by τ such that $D/Dt = \partial/\partial\tau$, and let $\mathbf{F} = \nabla_X \boldsymbol{x}$ be the *deformation gradient tensor* and $\mathbf{F}^{-1} = \nabla \boldsymbol{X}$ be its inverse such that $F^{-1}_{ij} F_{jk} = \delta_{ik}$. From (3.2.6) it follows that

$$\frac{\partial}{\partial \tau}\left(\frac{\boldsymbol{\omega}}{\rho} \cdot \mathbf{F}^{-1}\right) = \frac{1}{\rho}(\nabla \times \boldsymbol{a}) \cdot \mathbf{F}^{-1}.$$

Integrating this equation with respect to τ and taking inner product of both sides with $\mathbf{F}$ from the right, we obtain the *fundamental vorticity formula* named by Truesdell (1954):

$$\frac{\boldsymbol{\omega}}{\rho} = \left(\frac{\boldsymbol{\omega}_0}{\rho_0} + \int_0^\tau \frac{1}{\rho}(\nabla \times \boldsymbol{a}) \cdot \mathbf{F}^{-1} d\tau\right) \cdot \mathbf{F}. \tag{3.2.7}$$

Our key observation of both (3.2.4) and (3.2.6) or (3.2.7) is: *The vorticity kinetic vector $\nabla \times \boldsymbol{a}$ makes a vortical flow depend essentially on fluid particles' motion history; once the vorticity kinetic vector $\nabla \times \boldsymbol{a}$ vanishes the vortical flow will be independent* *of its history.* This is evidently a significant simplification, which happens if and only if

$$\nabla \times \boldsymbol{a} = \mathbf{0} \quad \text{or} \quad \boldsymbol{a} + \nabla\phi^* = \mathbf{0}, \tag{3.2.8}$$

where ϕ^* is a potential of acceleration. An immediate consequence of (3.2.8) is the famous **circulation theorem** (Kelvin 1869): *If and only if the acceleration is curl-free, the circulation along any material loop is time invariant*, i.e.,

$$\frac{d}{dt}\oint_C \boldsymbol{u} \cdot d\boldsymbol{x} = 0. \tag{3.2.9}$$

Equation (3.2.8) or (3.2.9) brings us to a big special class of vortical flows called *circulation-preserving flow*.

In terms of the two fundamental processes, an overall physical understanding of circulation-preserving flows can be gained. Because $\boldsymbol{a}$ and $\nabla \times \boldsymbol{a}$ are the bridges of kinematics and kinetics in the momentum equation and the vorticity transport equation, respectively, (3.2.8) implies that *in a circulation-preserving flow the evolution of shearing is a purely kinematic process*, which is *dynamically decoupled* from the compressing process. This observation suggests that the shearing process should own certain symmetries associated with a series of invariant quantities or conservation theorems.

In the rest of this section we focus on this special type of flows. We discuss these invariants first, followed by an examination of the vorticity-tube stretching and tilting as time, which is the most unique feature of vorticity-tube kinematics. Then, since now the kinetics implied by the acceleration potential ϕ^* only enters the compressing process, we can integrate the momentum equation once and obtain various Bernoulli integrals. This simplicity is of great significance and can serve as good approximation of many practical flows at least in some local regions.

3.2.2 Local Material Invariants

For circulation-preserving flow, a series of theorems about *local* vorticity invariants can be directly inferred from condition (3.2.8).

1. Potential vorticity and its generalization. The simplest specification of (3.2.6) is to assign $\mathcal{F}$ as any conservative *scalar* ϕ with $D\phi/Dt = 0$. Then $(\boldsymbol{\omega}/\rho) \cdot \nabla\phi$ is known as the *potential vorticity*, and from (3.2.6) it follows the **potential-vorticity theorem** (Ertel 1942): *The potential vorticity is Lagrangian invariant if and only if either the flow is circulation-preserving or $\nabla\phi$ is perpendicular to vector $\nabla \times \boldsymbol{a}$.* Namely,

$$\frac{DP}{Dt} = 0, \quad P \equiv \frac{\boldsymbol{\omega}}{\rho} \cdot \nabla\phi. \tag{3.2.10}$$

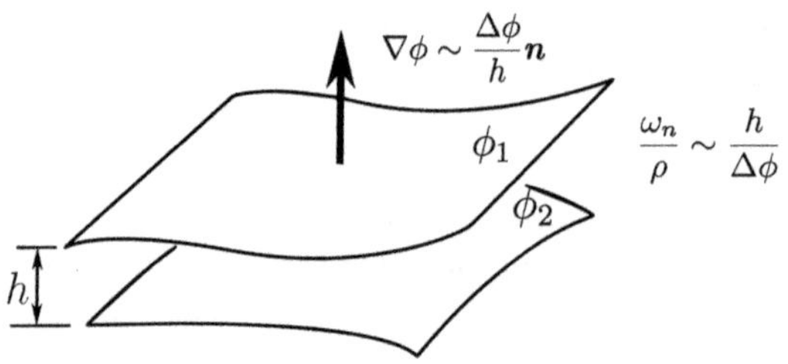

Fig. 3.9 Potential-vorticity conservation: an increase of the distance of two iso-ϕ surfaces causes vorticity enhancement

The name "potential vorticity" was first introduced by Rossby in 1936, and can be understood from the following fact (Fig. 3.9). If the distance between two neighboring iso-ϕ surfaces increases such that $|\nabla\phi|$ is reduced, then by (3.2.10) the component of $\boldsymbol{\omega}/\rho$ parallel to $\nabla\phi$ must be enhanced. If ρ varies very weakly, what is changing must be the vorticity, and the stretching of the distance between iso-ϕ surfaces has an effect similar to the vorticity-line stretching to be discussed in the next section.

The potential-vorticity theorem is most widely applied to geophysical fluid dynamics, see, e.g., Greenspan (1968) and Wu et al. (2006). Here we just give two other elementary examples.

For two-dimensional flow on the (x, y)-plane we can set $\phi = z$ with $Dz/Dt = 0$ and $\nabla z = \boldsymbol{e}_z$. Similarly, for rotationally symmetric flow we can take ϕ as the azimuthal angle θ in a cylindrical coordinates, with $r D\theta/Dt = u_\theta = 0$ and $\nabla\theta = \boldsymbol{e}_\theta/r$. Then there is

$$\frac{D}{Dt}\left(\frac{\omega}{\rho}\right) = 0 \quad \text{for two-dimensional flow,} \tag{3.2.11a}$$

$$\frac{D}{Dt}\left(\frac{\omega_\theta}{\rho r}\right) = 0 \quad \text{for rotationally symmetric flow.} \tag{3.2.11b}$$

More generally, for any conservative tensor $\mathcal{F}$ we may call $(\boldsymbol{\omega}/\rho)\cdot\nabla\mathcal{F}$ the *generalized potential vorticity* and then have the **generalized potential vorticity conservation theorem**: *For a circulation-preserving flow the generalized potential vorticity of each fluid particle is preserved*:

$$\frac{\boldsymbol{\omega}}{\rho}\cdot\nabla\mathcal{F} = \frac{\boldsymbol{\omega}_0}{\rho_0}\cdot\nabla\mathcal{F}_0. \tag{3.2.12}$$

Most of the local vorticity invariants as stated below can be derived from (3.2.12). Note that for steady flow this theorem implies the constancy of generalized potential vorticity along any streamline.

2. Cauchy vorticity theorem. (Cauchy in 1815). *In a circulation-preserving flow the vorticity field can be neither created nor destroyed. In particular, an initially irrotational flow will remain irrotational.*

Proof Set $\mathcal{F} = \boldsymbol{X}$ in (3.2.12) and noticing $\partial_i X_{0j} = \delta_{ij}$, or set $\nabla\times\boldsymbol{a} = \boldsymbol{0}$ in (3.2.7), there is

$$\frac{\boldsymbol{\omega}}{\rho} = \frac{\boldsymbol{\omega}_0}{\rho_0} \cdot \nabla_X \boldsymbol{x}, \tag{3.2.13}$$

which is called the *Cauchy vorticity formula* and by which the theorem is obvious.

3. The second and third Helmholtz vorticity-tube theorems. (Helmholtz 1858). *If and only if the flow is circulation-preserving, a material vorticity tube will move with the fluid* (the second theorem) *and its strength is time-invariant* (the third theorem).

We prove the theorems by using (3.2.13). Consider a vorticity line $\mathcal{L}_0$ at $t = 0$, on which a point $\boldsymbol{x}_0 = \boldsymbol{X}$ is its Lagrangian coordinates. By definition, along $\mathcal{L}_0$ there must be

$$\boldsymbol{\omega}_0 \times \delta\boldsymbol{x}_0 = \boldsymbol{0} \quad \text{or} \quad \boldsymbol{\omega}_0 = \lambda_0 \delta\boldsymbol{x}_0 = \lambda_0 \delta\boldsymbol{X},$$

where λ_0 is a scalar. Then at time t the fluid particles forming $\mathcal{L}_0$ have moved to line $\mathcal{L}$ on which a point from $\boldsymbol{x}_0$ has coordinates $\boldsymbol{x} = \boldsymbol{x}(\boldsymbol{X}, t)$. By (3.2.13), for the vorticity at t there is

$$\frac{\omega_i}{\rho} = \left(\frac{\omega_{0j}}{\rho_0}\right) \frac{\partial x_i}{\partial X_j} = \frac{1}{\rho_0} \lambda_0 \frac{\partial x_i}{\partial X_j} \delta X_j,$$

so that

$$\boldsymbol{\omega} = \lambda \delta\boldsymbol{x} \quad \text{or} \quad \boldsymbol{\omega} \times \delta\boldsymbol{x} = \boldsymbol{0},$$

where $\lambda = (\rho/\rho_0)\lambda_0$ is a scalar. Namely, $\mathcal{L}$ remains a material line. This result can also be stated as

$$\frac{\boldsymbol{\omega}/\rho}{|\boldsymbol{\omega}_0/\rho_0|} = \frac{\delta\boldsymbol{x}}{|\delta\boldsymbol{x}_0|}, \tag{3.2.14}$$

indicating that *the vorticity is "frozen" to material lines, and as a vorticity line is stretched the vorticity must be enhanced.*

This being the case, for circulation-preserving flow any vorticity surface consisting of vorticity lines also remains as material surface, and so must do any vorticity tube. Thus, if $\delta\boldsymbol{x}$ and $\delta\boldsymbol{S} = \boldsymbol{n}\delta S$ are the length and sectional area of a piece of vorticity tube such that $\delta\boldsymbol{x} \cdot \delta\boldsymbol{S}$ forms a material volume $\delta\mathcal{V}$, the fluid mass in $\delta\mathcal{V}$ is also invariant: $\rho\mathcal{V} = \rho_0\mathcal{V}_0$. Substituting this into (3.2.14) yields

$$\boldsymbol{\omega} \cdot \boldsymbol{n}\delta S = \boldsymbol{\omega}_0 \cdot \boldsymbol{n}_0 \delta S_0. \tag{3.2.15}$$

Then the theorems follow from collecting (3.2.14) and (3.2.15).

In his seminal paper Helmholtz (1858) established three vorticity-tube theorems. The first was stated in Sect. 3.1.1 and is unconditional. The other two concern the time variation of a vorticity tube, which hold only conditionally due to the involvement of dynamic assumption (3.2.8). They can be proved by other ways, in particular by the Kelvin circulation theorem (3.2.9) as stated in standard textbooks. The above proof based on the Cauchy theorem is a *local* argument and hence sharpens the original Helmholtz's second and third theorems. For an equivalent proof but in terms of Eulerian description see Problem 3.6.

4. Relabeling symmetry. In addition to the above conservation properties, a circulation-preserving flow enjoys yet another symmetry of significant theoretical interest, for which we just mention the result without going into the details. In Sect. 1.1.1 it is pointed out that the field and material descriptions of fluid motion are not equivalent; the latter requires the former *plus* a condition (1.1.5),

$$\frac{DX}{Dt} = \mathbf{0}, \tag{3.2.16}$$

that ensures each fluid particle will not be relabeled during the motion. But, it has been shown that *relabeling fluid particles will not change anything in the flow if and only if the flow is circulation-preserving*, so in this case the Eulerian and Lagrangian descriptions become fully equivalent without any need for (3.2.16). This property is called the *relabeling symmetry* (e.g. Salmon 1988, 1998; Wu et al. 2006).

3.2.3 Vorticity-Tube Stretching and Tilting

By the Helmholtz first vorticity theorem or (3.2.14) along with the Biot-Savart formula (3.1.38), a vorticity tube will have higher vorticity with stronger induced velocity field if a stretching makes it thinner. Opposite result would take place if the vorticity tube is fattened or shrunk. Moreover, a tilting of the tube will immediately alter the direction of its induced velocity field. These phenomena are the most important kinematic feature of a vorticity field that exists, of course, also in more general non-circulation-preserving flows. But they do not happen in two dimensions. Our question now is: what mechanisms are responsible for the stretching-thinning (or shrinking-thickening) as well as tilting of a vorticity tube.

The unique kinematic property of vorticity evolution in three-dimensional flows is reflected by the key term $-\boldsymbol{\omega}\cdot\nabla\boldsymbol{u}$ (or $-(\boldsymbol{\omega}/\rho)\cdot\nabla\boldsymbol{u}$ in compressible flow), which has no counterpart in the acceleration formula $\boldsymbol{a} = D\boldsymbol{u}/Dt$. To see its effect, we introduce again the intrinsic curvilinear $(\boldsymbol{t}, \boldsymbol{n}, \boldsymbol{b})$ triad as we did in Sect. 3.1.2, but this time along a *vorticity-line* or a vortex filament so that $\boldsymbol{\omega} = \omega\boldsymbol{t}$ with κ and τ being the curvature and torsion thereof, while the velocity $\boldsymbol{u}$ has three components (u_s, u_n, u_b). Then we have

$$\boldsymbol{\omega}\cdot\nabla\boldsymbol{u} = \omega\frac{\partial}{\partial s}(u_s\boldsymbol{t} + u_n\boldsymbol{n} + u_b\boldsymbol{b}),$$

which by the Frenet-Serret formula (3.1.5) yields

$$
\begin{aligned}
\boldsymbol{t} \cdot \nabla \boldsymbol{u} = & \left(\frac{\partial u_s}{\partial s} - u_n \kappa\right) \boldsymbol{t} + \left(\frac{\partial u_n}{\partial s} + u_s \kappa - u_b \tau\right) \boldsymbol{n} \\
& + \left(\frac{\partial u_b}{\partial s} + u_n \tau\right) \boldsymbol{b}.
\end{aligned} \tag{3.2.17}
$$

Here, the component of $\boldsymbol{t} \cdot \nabla \boldsymbol{u}$ along $\boldsymbol{t}$ makes the $\boldsymbol{\omega}$-line stretched or contracted, and those along $\boldsymbol{n}$ and $\boldsymbol{b}$ make it tilted. For small κ and τ, the $\boldsymbol{\omega}$-line stretching (or contracting) and tilting are dominated by the increments of u_s and (u_n, u_b) along s, respectively (Batchelor 1967). But a strong stretching may occur at a point of a vortex filament even if $\partial u_s / \partial s = 0$, as long as $\kappa \gg 1$ and $u_n < 0$ (outward from the curvature center).

Similar to the kinetic energy, we introduce the concept *enstrophy* $\omega^2/2$ to measure the total amount of shearing in a flow domain. Its time evolution is governed by the inner product of $\boldsymbol{\omega}$ and (3.2.2c). Since $\boldsymbol{\omega} \cdot \nabla \boldsymbol{u} \cdot \boldsymbol{\omega} = \boldsymbol{\omega} \cdot \mathbf{D} \cdot \boldsymbol{\omega}$, we obtain

$$
\frac{D}{Dt}\left(\frac{1}{2}\omega^2\right) = \alpha\omega^2 + \boldsymbol{\omega} \cdot (\nabla \times \boldsymbol{a}), \tag{3.2.18a}
$$

$$
\frac{D\omega}{Dt} = \alpha\omega + \boldsymbol{t} \cdot (\nabla \times \boldsymbol{a}), \tag{3.2.18b}
$$

where $\boldsymbol{t}$ is the unit vector along the vorticity-line direction, and

$$
\alpha \equiv \boldsymbol{t} \cdot \mathbf{D} \cdot \boldsymbol{t} - \vartheta = \frac{\partial u_s}{\partial s} - u_n \kappa - \vartheta \tag{3.2.18c}
$$

is the *stretching rate* of an infinitely thin vorticity tube or a single vorticity line. Here, as is evident, a compression ($\vartheta < 0$) of a compressible fluid will make the vorticity field more condensed, and an increase of u_s along the vorticity tube will stretch it. Either way will enhance the enstrophy and vise versa, with the latter being much more common and effective. In contrast, if a thin vorticity tube is folded with large curvature κ, then a positive u_n (pointing toward the curvature center) there will reduce α or even make $\alpha < 0$.

In Lagrangian description, (3.2.18a) or (3.2.18b) can be integrated with respect to time:

$$
\omega(\tau) = \omega(0)e^{A(\tau)} + e^{A(\tau)} \int_0^\tau \boldsymbol{t} \cdot (\nabla \times \boldsymbol{a}) e^{-A(\eta)} d\eta, \tag{3.2.19a}
$$

where

$$
A(\tau) = \int_0^\tau \alpha d\eta. \tag{3.2.19b}
$$

While the effect of $\nabla \times \boldsymbol{a}$ on the variation of ω is formally linear, that of stretching is *exponential*. Figure 3.10 shows a numerical example due to Siggia (1985), who computed the evolution of a vorticity loop which is initially elliptic on the (y, z)-

plane. Let L be its total length and σ be the cross-sectional radius, and assume that $\sigma^2 L$ is time-invariant. At time $t = 0$, σ_0 and L_0 were taken as 0.2 and about 10 respectively, and the ratio of the axes of the initial elliptical ring was 4:1. As indicated by (3.1.40), the self induction is non-uniform and causes the vortex ring to both deform quickly and stretch non-uniformly. Figure 3.10a shows a sequence of the vorticity-loop shapes, and Fig. 3.10b shows the growth of L in time, where the last three points imply an exponential growing.

The nonlinear vorticity stretching and tilting as well its *cut and reconnect* by viscosity (cf. Fig. 3.19) are crucial mechanisms in the generation of complex vortical structures. In particular, stretching is responsible for the *cascade* process in turbulence to be discussed in Sect. 11.1. In fact, turbulence may be briefly defined as *randomly stretched vortices*. The strain rate **D** that causes stretching can be either a background field induced by other vortices, or induced locally by the vortex itself. The strongest stretching and shrinking occur if ω is aligned to the stretching and shrinking principal axes of **D**, respectively. Then the stretching rate α is the maximum eigenvalue of **D**.

It should be stressed that, as seen from Fig. 3.10, when the vortex ring is stretched it is also tilted. A straight vortex cannot be unboundedly stretched: by (3.1.38), see also (3.3.30) and (3.3.31) below, this would increase the induced kinetic energy unboundedly, but the total kinetic energy is conserved (or decreasing due to dissipation). To offset the increase of kinetic energy due to stretching, therefore, there must be vortex tilting to cause a partial cancelation of the induced velocity. Contrary to the stretching, the cancelation is strongest if two neighboring vortex filaments are parallel with opposite vorticity.

3.2.4 Bernoulli Integrals

So far we have discussed the shearing process in circulation-preserving flow, for which only kinematics remains. This implies that *the kinetics of a circulation-preserving flow exists solely in the compressing process*, but as mentioned in Sect. 2.3 the shearing process can contribute to this dynamics through nonlinear terms. Let us now examine this "compressing-kinetics aspect" of circulation-preserving vortical flows.

Substituting the circulation-preserving condition (3.2.8) into the acceleration formula (3.2.1) yields

$$\frac{\partial \boldsymbol{u}}{\partial t} + \boldsymbol{\omega} \times \boldsymbol{u} = -\nabla \left(\frac{1}{2} q^2 + \phi^* \right). \tag{3.2.20}$$

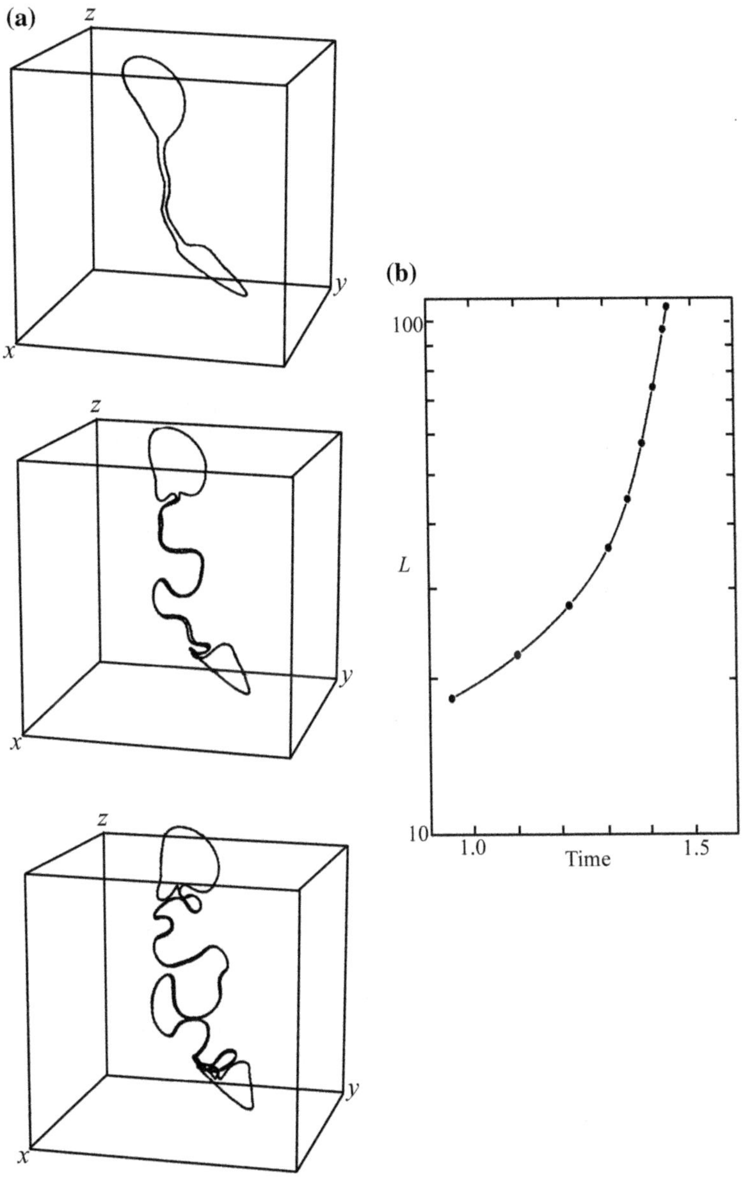

Fig. 3.10 The self-induced stretching of a vorticity loop, starting from an elliptical vortex ring on the (y, z)-plane. **a** The shape evolution of the loop, **b** the growth of the length L of the loop. From Siggia (1985)

Thus, if in a region the flow is irrotational such that $\boldsymbol{u} = \nabla\phi$, then (3.2.20) can be integrated once to yield the most commonly encountered Bernoulli integral, with (2.6.11) being a special case:

$$\frac{\partial\phi}{\partial t} + \frac{1}{2}q^2 + \phi^* = 0, \quad \text{or} \tag{3.2.21a}$$

$$\frac{D\phi}{Dt} - \left(\frac{1}{2}q^2 - \phi^*\right) = 0, \tag{3.2.21b}$$

where the integration "constant" $C(t)$ has been absorbed into ϕ. Equation (3.2.21b) is the Lagrangian form of (3.2.21a) following each material fluid elements.

Our main interest here is to seek the most general possible Bernoulli integrals. In this case, if on a line, a surface or in a volume the first two terms of (3.2.20) can be reduced to the gradient of a scalar, then (3.2.20) can be integrated once on that line, surface or in that volume, yielding a corresponding Bernoulli integral. Let us see how these can be reached.

Assume the flow is steady with nonzero Lamb vector, such that (3.2.20) becomes

$$\boldsymbol{\omega} \times \boldsymbol{u} = -\nabla\left(\frac{1}{2}q^2 + \phi^*\right). \tag{3.2.22}$$

Then by (3.1.17) the flow is generalized Beltramian, and hence there must exist a set of Lamb surfaces which are orthogonal to the Lamb vector everywhere. Denoting these surfaces by $\mathcal{S}_L$ and integrating (3.2.22) along such a surface, we obtain a *surface Bernoulli integral*

$$\frac{1}{2}q^2 + \phi^* = C(\mathcal{S}_L), \tag{3.2.23}$$

where $C(\mathcal{S}_L)$ is the integration "constant" which varies with $\mathcal{S}_L$. Note that the kinetic energy $q^2/2$ may contain the contribution of the velocity induced by a vorticity field, which and the Lamb surfaces themselves are the place where the coupling with shearing process enters the Bernoulli integral.

Alternative to (3.2.23), for steady flow (3.2.22) implies that the Lamb vector is irrotational and has a scalar potential χ, say, such that there exists a *volume Bernoulli integral*:

$$\frac{1}{2}q^2 + \phi^* + \chi = 0. \tag{3.2.24}$$

For example, in two-dimensional or rotationally axisymmetric generalized Beltrami flow (Sect. 3.1.3), in terms of the stream function or Stokes stream function ψ there is $\omega = f(\psi)$ or $\omega_\theta/r = f(\psi)$, respectively, and $\boldsymbol{\omega} \times \boldsymbol{u}$ is expressed by (3.1.30). Thus, (3.2.24) becomes

$$\frac{1}{2}q^2 + \phi^* + \int f(\psi)d\psi = 0. \tag{3.2.25}$$

The Bernoulli integrals listed in most books are various applications and simplifications of (3.2.21), (3.2.23), or (3.2.25). However, these integrals are not yet the most general Bernoulli integral for vortical flow. Recall that the circulation preserving and various vorticity-related local invariants derived therefrom are of Lagrangian nature, one may expect that the most general Bernoulli integral should be found by the Lagrangian description. This has indeed been proved true. But a big part of analysis has to be made in the reference space spanned by the particle label $\boldsymbol{X}$ and Lagrangian time $\tau = t$, in which the Cauchy vorticity formula (3.2.13) plays a key role, and therefore the resulting Bernoulli integral is essentially an inherent combination of the shearing kinematics and compressing dynamics. Here we only state the final result (for details see Wu et al. 2006): *For any circulation-preserving flow, the velocity has a decomposition*

$$\boldsymbol{u} = \nabla\Phi + f\nabla g \quad \text{with} \quad \frac{Df}{Dt} = \frac{Dg}{Dt} = 0,$$

where Φ is a function of $(\boldsymbol{X}, \tau)$, such that in Lagrangian description the desired Bernoulli integral reads

$$\frac{\partial\Phi}{\partial\tau} - \left(\frac{1}{2}q^2 - \phi^*\right) = 0, \tag{3.2.26a}$$

or in the Eulerian description

$$\frac{\partial\Phi}{\partial t} + \frac{1}{2}q^2 + f\frac{\partial g}{\partial t} + \phi^* = 0, \quad \frac{Df}{Dt} = \frac{Dg}{Dt} = 0. \tag{3.2.26b}$$

Note that we now have $\boldsymbol{\omega} = \nabla f \times \nabla g$, which appears only implicitly in this general Bernoulli integral.

3.3 Vorticity Integrals and Their Invariance

The properties of various vorticity integrals are an indispensable part of vorticity kinematics. We have seen such an integrated property of vorticity tube in Sect. 3.1.1, which leads to Helmholtz's first vorticity-tube theorem. In this section we consider various *volumetric* vorticity integrals. These include integrals of the vorticity vector (circulation in two-dimensional flow), Lamb vector $\boldsymbol{\omega} \times \boldsymbol{u}$ and its moments, vorticity moments $\boldsymbol{x} \times \boldsymbol{\omega}$ and $x^2\boldsymbol{\omega}$, helicity density $\boldsymbol{\omega} \cdot \boldsymbol{u}$, and the total kinetic energy, etc. The crucial importance of these integrals lies in the fact that, for unbounded fluid at rest at infinity, some of these integrals tend to zero and some are time invariant, either identically or conditionally when the flow is circulation preserving. These inherent invariants reflect various aspects of the conservative nature of a vortical flow field.

Our analysis will stay within kinematics but rely heavily on the exponential decay behavior of vorticity as $|\boldsymbol{x}| \to \infty$ which itself, as we found in Sect. 2.4.1, is a result of dynamics.

3.3.1 Total Vorticity and Circulation

While the Helmholtz first vorticity theorem was derived from the integral expression (3.1.1) of solenoidal nature of vorticity, (3.1.1) itself is a special case of an easily proved general *vorticity integral identity* for any tensor $\mathcal{F}$:

$$\int_V \boldsymbol{\omega} \cdot \nabla \mathcal{F} dV = \int_{\partial V} (\boldsymbol{n} \cdot \boldsymbol{\omega}) \mathcal{F} dS, \tag{3.3.1}$$

by setting $\mathcal{F} = 1$. By associating $\mathcal{F}$ with different quantities we may obtain infinitely many conserved vorticity integrals. For example, next to (3.1.1) we can set $\mathcal{F}$ as the position vector $\boldsymbol{x}$. Then since $x_{i,j} = \delta_{ij}$, (3.3.1) yields a general formula for total vorticity:

$$\int_V \boldsymbol{\omega} dV = \int_{\partial V} (\boldsymbol{n} \cdot \boldsymbol{\omega}) \boldsymbol{x} dS. \tag{3.3.2}$$

Owing to the compact nature of vorticity field, from (3.3.2) we see that for unbounded fluid or if V is externally bounded by a non-rotating boundary where the adherence condition holds, there is **total vorticity conservation theorem** found by Föppl in 1897:

$$\int_V \boldsymbol{\omega} dV = \boldsymbol{0}. \tag{3.3.3}$$

Physically, this theorem implies that the contribution of one piece of the vorticity tube to the vectorial vorticity field must be canceled by another piece with vorticity of opposite sign, which is possible if all vorticity tubes are closed. Therefore, Helmholtz's first theorem and Föppl's theorem reflect the same solenoidal nature of the vorticity field from different aspects. Equation (3.3.3) also applies to the continuous system of fluid plus solid mentioned in Sect. 3.1.1. Evidently, (3.3.3) is also consistent with the integral condition (2.4.5b) we set for an infinite space V_∞.

In two-dimensional flows on an (x, y)-plane, vorticity tubes are not closed and (3.3.2) does not apply. Rather, the integral of $\boldsymbol{\omega} = \omega \boldsymbol{e}_z$ over V becomes the total circulation along the boundary line of the flow domain, which is generically nonzero. We thus confine ourselves to the total circulation Γ_∞ in the entire free space V_∞. The integral condition (2.4.5b) reduces to requiring $\Gamma_\infty \equiv 0$, which is physically less evident than its three-dimensional counterpart. However, another general result provides a sufficient support to the validity of (2.4.5b) in two dimensions:

Total-Circulation Conservation Theorem. *For any externally unbounded two-dimensional flow, if the far-field condition* (2.4.1) *holds then the total circulation is time invariant.*

Thus, if initially $\Gamma_\infty(0) = 0$, then it will be identically zero. A famous demonstration of this theorem is the experimentally observed formation of the circulation around a two-dimensional airfoil, which is the very most important process in aerodynamics and will be addressed in Sect. 9.1.

The proof of the theorem comes directly from Kelvin's formula (1.1.45) and compactness of the vorticity field. Let C be a big material loop at and outside which the flow is irrotational, with singly-valued smooth velocity potential ϕ. Then there is $\boldsymbol{a} = \nabla(\partial_t\phi + |\nabla\phi|^2/2) = -\nabla\phi^*$, no matter if the flow is incompressible or not. Thus, by (1.4.5) there is

$$\frac{d\Gamma_C}{dt} = -\oint_C \boldsymbol{a} \cdot d\boldsymbol{x} = -\oint_C d\phi^* = 0,$$

implying that as $|\boldsymbol{x}| \to \infty$ we have

$$\frac{d\Gamma_\infty}{dt} = 0 \quad \text{or} \quad \Gamma_\infty(t) = \Gamma_\infty(0), \tag{3.3.4}$$

so the theorem is proved.

3.3.2 Lamb-Vector Integrals

To study the integrals of the Lamb vector, observe that since

$$\boldsymbol{u} \cdot \nabla\boldsymbol{u} = \boldsymbol{\omega} \times \boldsymbol{u} + \nabla\left(\frac{1}{2}q^2\right) = \nabla \cdot (\boldsymbol{u}\boldsymbol{u}) - \boldsymbol{u}\vartheta,$$

where ϑ is the dilatation, we have identity

$$\boldsymbol{\omega} \times \boldsymbol{u} + \vartheta\boldsymbol{u} = \nabla \cdot \left(\boldsymbol{u}\boldsymbol{u} - \frac{1}{2}q^2\mathbf{I}\right),$$

where $\mathbf{I}$ is the unit tensor. Thus, over any volume V there is

$$\int_V (\boldsymbol{\omega} \times \boldsymbol{u} + \vartheta\boldsymbol{u})dV = \int_{\partial V} \left(\boldsymbol{n} \cdot \boldsymbol{u}\boldsymbol{u} - \frac{1}{2}q^2\boldsymbol{n}\right) dS. \tag{3.3.5a}$$

By integration by parts, we also find

$$\int_V \boldsymbol{x} \times (\boldsymbol{\omega} \times \boldsymbol{u} + \vartheta \boldsymbol{u}) dV = \int_{\partial V} \boldsymbol{x} \times \left(\boldsymbol{n} \cdot \boldsymbol{u}\boldsymbol{u} - \frac{1}{2} q^2 \boldsymbol{n} \right) dS, \tag{3.3.5b}$$

$$\int_V \boldsymbol{x} \cdot (\boldsymbol{\omega} \times \boldsymbol{u} + \vartheta \boldsymbol{u}) dV = \int_{\partial V} \boldsymbol{x} \cdot \left(\boldsymbol{n} \cdot \boldsymbol{u}\boldsymbol{u} - \frac{1}{2} q^2 \boldsymbol{n} \right) dS + (n-2) \int_V \frac{1}{2} q^2 dV, \tag{3.3.5c}$$

where $n = 2, 3$ is the dimension of the space. Here and below $\boldsymbol{x}$ (often written as $\boldsymbol{r} = \boldsymbol{x} - \boldsymbol{x}_0$) is the distance vector from any coordinate origin, say $\boldsymbol{x}_0$, and all relevant results are independent of the choice of $\boldsymbol{x}_0$ (Problem 3.7).

For an unbounded *incompressible* fluid at rest at infinity, since as $|\boldsymbol{x}| \to \infty$ the potential velocity decays as $|\boldsymbol{x}|^{-n}$ (Sect. 2.4.2), the boundary integrals in (3.3.5a) to (3.3.5c) all vanish. Thus, we simply have

$$\int_{V_\infty} (\boldsymbol{\omega} \times \boldsymbol{u}) dV = \boldsymbol{0}, \tag{3.3.6a}$$

$$\int_{V_\infty} \boldsymbol{x} \times (\boldsymbol{\omega} \times \boldsymbol{u}) dV = \boldsymbol{0}, \tag{3.3.6b}$$

$$\int_{V_\infty} \boldsymbol{x} \cdot (\boldsymbol{\omega} \times \boldsymbol{u}) dV = (n-2) \int_{V_\infty} \frac{1}{2} q^2 dV. \tag{3.3.6c}$$

If the externally unbounded fluid has an internal boundary, say a solid surface ∂B, one may either employ the velocity adherence to cast the surface integrals over ∂B to volume integrals over B, or continue the Lamb vector into B. Both ways form a single continuous medium for which (3.3.6) still holds, although locally $\boldsymbol{\omega}$ is discontinuous across ∂B.

It should be stressed that, however, if the condition $\boldsymbol{u} = O(|\boldsymbol{x}|^{-n})$ at infinity does not hold, (3.3.6) will be violated. This happens if one only considers a subspace $V_{\text{st}} \subset V_\infty$ in which, as discussed in Sect. 2.4.3, the flow can be assumed steady and the total vorticity or circulation (in three or two dimensions, respectively) is nonzero. Then, if the flow at ∂V_{st} is *irrotational*, by (2.6.25) and (3.3.5a) we immediately obtain the total force on a body in steady flow, solely in terms of Lamb-vector integral, first given by Prandtl (1918):

$$\boldsymbol{F} = \rho \int_{V_{\text{st}}} \boldsymbol{u} \times \boldsymbol{\omega} dV, \tag{3.3.7}$$

which is the basis of classic *steady low-speed aerodynamics*.[4] In Chap. 9 we shall sharpen the derivation of (3.3.7) and extend it to more general flows.

[4] The above derivation of (3.3.7) has the same flaw as that of the Kutta-Joukowski theorem (2.6.26) as we remarked there, and will also be improved in Sect. 9.1.

3.3.3 Vortical and Potential Impulses

In an unbounded space, the integrated total momentum and angular momentum of a fluid do not converge unconditionally. For example, consider an incompressible and acyclic potential flow $\boldsymbol{u} = \nabla\phi$ in V_f bounded internally by body surface ∂B and externally by control surface Σ. The total momentum of V_f is

$$\rho \int_{V_f} \boldsymbol{u} dV = \rho \int_{\partial B} \phi \boldsymbol{n} dS + \rho \int_{\Sigma} \phi \boldsymbol{n} dS.$$

As Σ recedes to infinity, ϕ decays as $O(r^{-(n-1)})$ but the area of Σ increases as $O(r^{n-1})$ in n-dimensional space. Thus the second term remains of $O(1)$ and the integral is indeterminate, depending on the shape of Σ. Nevertheless, owing to the exponential decay of vorticity, it is preferable to use vorticity integrals to replace momentum integrals. This consideration leads to the concept of *fluid impulse* (an integrated quantity) as a close analogy of total momentum.

1. Concept of impulse. For incompressible flow, we may introduce impulse by using a vector identity given by (A.2.23), Appendix A.2.3, where a set of identities of similar function are listed and called *derivative-moment transformation* (DMT for short). We thus have, for a fluid in a finite domain V enclosed by Σ,

$$\int_V \boldsymbol{u} dV = \boldsymbol{I}_V - \frac{1}{n-1} \int_{\Sigma} \boldsymbol{x} \times (\boldsymbol{n} \times \boldsymbol{u}) dS, \quad n = 2, 3, \tag{3.3.8a}$$

where the volume integral

$$\boldsymbol{I}_V \equiv \frac{1}{n-1} \int_V \boldsymbol{x} \times \boldsymbol{\omega} dV \tag{3.3.8b}$$

is called *vortical impulse* (usually known as *hydrodynamic impulse* as introduced by Kelvin), with subscript specifying the volume over which integration is taken. When $V = V_\infty$ occupies the whole unbounded space, the subscript will be neglected. This vortical impulse is the first vectorial *vorticity moment*. In Sect. 2.4.1 we have seen that the flow impulse caused by a compact vortical system dominates the incompressible velocity far-field asymptotics. In the definition of $\boldsymbol{I}_V$ the spatial-dimension dependence of factor $1/(n-1)$ comes from the physical fact that when $n = 2$ vorticity lines never form closed loops.

Similarly, for both $n = 2$ and 3, the total moment can be cast to

$$\int_V \boldsymbol{x} \times \boldsymbol{u} dV = \boldsymbol{L}_V + \frac{1}{2} \int_{\Sigma} x^2 \boldsymbol{n} \times \boldsymbol{u} dS, \tag{3.3.9a}$$

where

$$\boldsymbol{L}_V \equiv -\frac{1}{2}\int_V x^2 \boldsymbol{\omega} dV \tag{3.3.9b}$$

is an *angular vortical impulse* or *second vorticity moment*.[5]

In an externally unbounded domain with fluid at rest at infinity, if Σ encloses the entire vorticity field in space such that the flow at and out of Σ is irrotational with $\boldsymbol{u} = \nabla\phi$, the boundary integrals in (3.3.8a) and (3.3.9a) can be greatly simplified by identities (A.2.25) and (A.2.28b):

$$-\frac{1}{n-1}\int_\Sigma \boldsymbol{x} \times (\boldsymbol{n} \times \boldsymbol{u}) dS = \int_\Sigma \phi \boldsymbol{n} dS \equiv \boldsymbol{I}_\Sigma, \tag{3.3.10a}$$

$$\begin{aligned}\frac{1}{2}\int_\Sigma x^2 \boldsymbol{n} \times \boldsymbol{u} dS &= -\frac{1}{3}\int_\Sigma \boldsymbol{x} \times [\boldsymbol{x} \times (\boldsymbol{n} \times \boldsymbol{u})] dS \\ &= \int_\Sigma \boldsymbol{x} \times \phi \boldsymbol{n} dS \equiv \boldsymbol{L}_\Sigma,\end{aligned} \tag{3.3.10b}$$

which are *potential impulse* and *angular potential impulse*, respectively, with subscript specifying the surface over which the integrals are taken.

Physically, these vortical and potential impulses can be viewed as hypothetical impulsive force and moment that brings the fluid from rest to the current motion instantaneously. To see this, suppose at $t = 0^+$ there is a velocity field $\boldsymbol{u}(\boldsymbol{x}, 0^+)$. Imagine that this velocity is suddenly generated from the fluid at rest everywhere at $t = 0^-$ by an impulsive external force density $\boldsymbol{F} = \boldsymbol{i}(\boldsymbol{x})\delta(t)$ distributed in a *finite* region. We integrate the incompressible Navier-Stokes equation over a small time interval $[0^-, \delta t]$. Since all finite terms in the equation including convection and viscous force can only have a variation of $O(\delta t)$, the finite $\boldsymbol{u}(\boldsymbol{x}, 0^+)$ must be solely generated by the infinitely large $\boldsymbol{F}$ at $t = 0$, which also causes a pressure impulse:

$$P(\boldsymbol{x}) = \int_{0^-}^{\delta t} p(\boldsymbol{x}) dt.$$

Hence, the momentum balance implies

$$\boldsymbol{u}(\boldsymbol{x}, 0^+) = -\frac{1}{\rho}\nabla P + \boldsymbol{i}(\boldsymbol{x}),$$

from which it follows that $\nabla^2 P = \rho \nabla \cdot \boldsymbol{i}$ and $\boldsymbol{\omega} = \nabla \times \boldsymbol{i}$. Then, as long as V encloses the entire $(\boldsymbol{\omega}, \boldsymbol{i})$-field such that by (A.2.23) we see at once

[5]The second vorticity moments are in general tensors, of which there are three vectorial ones: $x^2\boldsymbol{\omega}$, $\boldsymbol{x} \times (\boldsymbol{x} \times \boldsymbol{\omega})$, and $\boldsymbol{x}(\boldsymbol{x} \cdot \boldsymbol{\omega})$. Of these three only two are independent, and one may replace the integral of the first by that of the second or third. See Problem 3.11. Here, we choose $x^2\boldsymbol{\omega}$ to construct angular impulse $\boldsymbol{L}_V$.

$$\int_V \boldsymbol{i} dV = \boldsymbol{I}_V, \quad \int_V \boldsymbol{x} \times \boldsymbol{i} dV = \boldsymbol{L}_V.$$

Similarly, in a potential flow with $\boldsymbol{u} = \nabla\phi$ we simply have $P = -\rho\phi$, so there evidently is

$$-\frac{1}{\rho}\int_\Sigma P\boldsymbol{n} dS = \boldsymbol{I}_\Sigma, \quad -\frac{1}{\rho}\int_\Sigma \boldsymbol{x} \times P\boldsymbol{n} dS = \boldsymbol{L}_\Sigma.$$

Obviously, the poor convergence property of total momentum as Σ recedes to infinity is solely from these Σ-integrals; in fact $\boldsymbol{L}_\Sigma$ even diverges. But since the concern of dynamics is the rate of change of momentum rather than momentum itself, this troublesome potential integral over Σ does not appear as we shall show later.

Vortical impulses are our main concern. To gain an intuitive feeling of $\boldsymbol{I}$ and $\boldsymbol{L}$, let us look at some simple examples. In two dimensions, the simplest vortex system with $\Gamma_\infty = 0$ is a vortex couple of circulation $\mp\Gamma\boldsymbol{e}_z$ ($\Gamma < 0$) located at $x = \pm r/2$, respectively, see Fig. 3.11. Then by (3.3.8b) there is

$$\boldsymbol{I} = \boldsymbol{e}_y \Gamma r. \tag{3.3.11}$$

The fluid in between is pushed downward by the vortex couple. In three dimensions, the simplest compact vortex system with vanishing total vorticity is a closed loop C of thin vortex filament of circulation Γ. Then (3.3.8b) yields

$$\boldsymbol{I} = \frac{\Gamma}{2}\oint_C \boldsymbol{x} \times \boldsymbol{t} ds = \Gamma \int_S d\boldsymbol{S} = \Gamma \boldsymbol{S}, \tag{3.3.12}$$

where $d\boldsymbol{S} = \boldsymbol{x} \times \boldsymbol{t} ds/2$ is the vector surface element spanned by the triangle formed by $\boldsymbol{x}$ and $d\boldsymbol{x} = \boldsymbol{t} ds$, and $\boldsymbol{S}$ is the vector surface spanned by C, see Fig. 3.12. Note that $\boldsymbol{S}$ depends only on the shape of C but not the arbitrarily chosen origin O of $\boldsymbol{x}$. For example, if C is a curve on a plane of normal $\boldsymbol{n}$ then $\boldsymbol{S} = \boldsymbol{n}S$ with S being the area enclosed by C. It is very different from the area S of a cone with apex at the origin of $\boldsymbol{x}$ that depends on the arbitrarily chosen origin. Similarly, by (3.5.7) (see Problem 3.11) we have

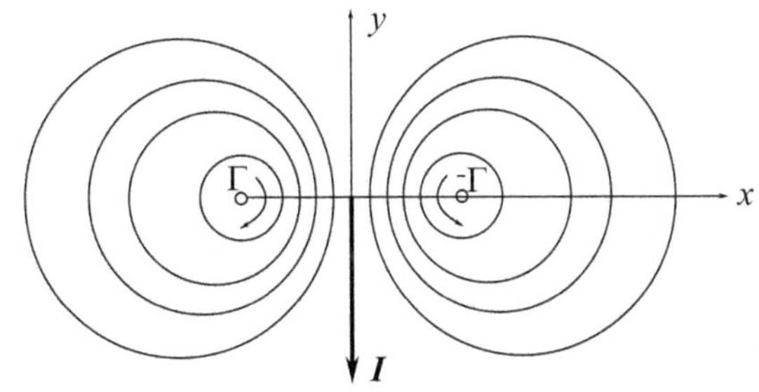

Fig. 3.11 The impulse produced by a vortex couple with $\Gamma < 0$ in two dimensions

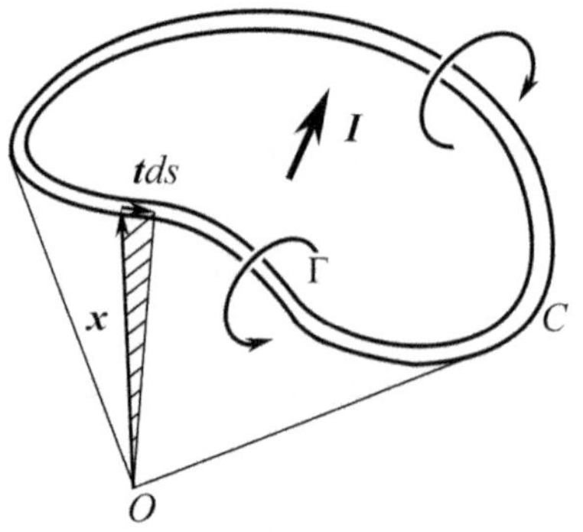

Fig. 3.12 The impulse produced by a vortex loop in three dimensions

$$\boldsymbol{L} = \frac{\Gamma}{3}\oint_C \boldsymbol{x} \times (\boldsymbol{x} \times \boldsymbol{t})ds = \frac{2\Gamma}{3}\int_S \boldsymbol{x} \times d\boldsymbol{S}. \tag{3.3.13}$$

In particular, for a thin circular vortex ring of radius r_0, in terms of cylindrical coordinates (r, θ, z) there is

$$\boldsymbol{I} = \pi r_0^2 \Gamma \boldsymbol{e}_z, \quad \boldsymbol{L} = \boldsymbol{0}, \tag{3.3.14}$$

which is a very useful result in vortex ring theory (Chap. 7).

2. Time rates of impulses. We now consider the time rates of potential and vortical impulses defined for a material fluid volume $\mathcal{V}$ in an unbounded domain. Consider the time rates of potential impulses first. Assume that $\phi(\boldsymbol{x})$ is *single-valued* and *smooth* in $\mathcal{V}$ and on $\partial\mathcal{V}$. This excludes the situation that $\partial\mathcal{V}$ has a non-zero circulation or intersects a free vortex sheet (which appears as a jump of ϕ across a surface). Such a velocity potential is referred to as *acyclic* or *non-circulatory*. Then, since $\partial\mathcal{V}$ is a material surface, there is

$$\frac{d\boldsymbol{I}_{\partial\mathcal{V}}}{dt} = \int_{\partial\mathcal{V}} \frac{D}{Dt}(\phi\boldsymbol{n}dS), \quad \frac{d\boldsymbol{L}_{\partial\mathcal{V}}}{dt} = \int_{\partial\mathcal{V}} \frac{D}{Dt}(\boldsymbol{x} \times \phi\boldsymbol{n}dS).$$

Here, we notice that (1.1.46b) simply gives

$$\frac{D}{Dt}(\boldsymbol{n}dS) = -\boldsymbol{n} \cdot \nabla\nabla\phi dS. \tag{3.3.15}$$

To proceed, it is necessary to use the Gauss theorem to cast boundary integral to volume integral and then back to boundary integral. Because (A.3.13) indicates that ϕ may have singularity in $\mathcal{V}$, we replace $\mathcal{V}$ by $\mathcal{V}_{\text{out}} = V_\infty - \mathcal{V}$, the space exterior to $\partial\mathcal{V}$ including infinity in which ϕ is regular. Then by using the Bernoulli integral, it simply follows that (Problem 3.12)

$$\frac{d\boldsymbol{I}_{\partial\mathcal{V}}}{dt} = -\frac{1}{\rho}\int_{\partial\mathcal{V}} p\boldsymbol{n}dS, \tag{3.3.16a}$$

$$\frac{d\boldsymbol{L}_{\partial\mathcal{V}}}{dt} = -\frac{1}{\rho}\int_{\partial\mathcal{V}} \boldsymbol{x}\times p\boldsymbol{n}dS. \tag{3.3.16b}$$

Note that $\partial\mathcal{V}$ does not have to recede to infinity. But if it does, since in Bernoulli integral q^2 decays sufficiently fast, the material boundary $\partial\mathcal{V}$ can be replaced by a control surface $\Sigma \to \Sigma_\infty$ on which (3.3.16) also holds asymptotically.

The time rate of vortical impulse has been extensively considered by Saffman (1992). With the help of incompressible vorticity transport equation (2.6.6),

$$\frac{D\boldsymbol{\omega}}{Dt} = \boldsymbol{\omega}\cdot\nabla\boldsymbol{u} + \nu\nabla^2\boldsymbol{\omega},$$

The time rate of vortical impulse of an arbitrary material fluid volume $\mathcal{V}$ is

$$\begin{aligned}\frac{d\boldsymbol{I}_{\mathcal{V}}}{dt} &= \frac{1}{n-1}\int_{\mathcal{V}}\left(\boldsymbol{u}\times\boldsymbol{\omega} + \boldsymbol{x}\times\frac{D\boldsymbol{\omega}}{Dt}\right)dV \\ &= \frac{1}{n-1}\int_{\mathcal{V}}[\boldsymbol{u}\times\boldsymbol{\omega} + \boldsymbol{x}\times(\boldsymbol{\omega}\cdot\nabla\boldsymbol{u})]dV + \boldsymbol{J}_\nu,\end{aligned}$$

where

$$\begin{aligned}\boldsymbol{J}_\nu &\equiv \frac{1}{n-1}\int_{\mathcal{V}}\boldsymbol{x}\times\nu\nabla^2\boldsymbol{\omega}dV = -\frac{1}{n-1}\int_{\mathcal{V}}\boldsymbol{x}\times[\nabla\times(\nabla\times\nu\boldsymbol{\omega})]dV \\ &= -\int_{\partial\mathcal{V}}\boldsymbol{n}\times\nu\boldsymbol{\omega}dS - \frac{1}{n-1}\int_{\partial\mathcal{V}}\boldsymbol{x}\times[\boldsymbol{n}\times(\nabla\times\nu\boldsymbol{\omega})]dS\end{aligned} \tag{3.3.17}$$

due to (A.2.23), which approaches zero as $\mathcal{V} \to V_\infty$ due to the compactness of vorticity field. Therefore, since for $n = 3$

$$\boldsymbol{x}\times(\boldsymbol{\omega}\cdot\nabla\boldsymbol{u}) = \nabla\cdot(\boldsymbol{\omega}\boldsymbol{x}\times\boldsymbol{u}) + \boldsymbol{u}\times\boldsymbol{\omega},$$

we obtain

$$\frac{d\boldsymbol{I}_{\mathcal{V}}}{dt} = \int_{\mathcal{V}}\boldsymbol{u}\times\boldsymbol{\omega}dV + \frac{1}{n-1}\int_{\partial\mathcal{V}}\boldsymbol{x}\times\boldsymbol{u}\omega_n dS. \tag{3.3.18a}$$

Alternatively, if V is a fixed control volume, we find

$$\frac{d\boldsymbol{I}_V}{dt} = \int_V \boldsymbol{u}\times\boldsymbol{\omega}dV + \frac{1}{n-1}\int_{\partial V}\boldsymbol{x}\times(\boldsymbol{u}\omega_n - \boldsymbol{\omega}u_n)dS. \tag{3.3.18b}$$

Now, for a *compact* vortex system surrounded by potential flow, both (3.3.18a) and (3.3.18b) are simplified. This is also true for the time rate of angular vortical impulse. Namely, for both $n = 2$ and 3 we have

$$\frac{d\boldsymbol{I}_V}{dt} = \int_V \boldsymbol{u} \times \boldsymbol{\omega} dV, \quad \frac{d\boldsymbol{L}_V}{dt} = \int_V \boldsymbol{x} \times (\boldsymbol{u} \times \boldsymbol{\omega}) dV. \tag{3.3.19}$$

Thus, the time rates of vortical impulses are solely determined by Lamb-vector integrals. Moreover, let $\boldsymbol{u} = \boldsymbol{v} + \nabla\phi_e$, where $\boldsymbol{v}$ is the velocity solely induced by $\boldsymbol{\omega}$ in V and $\nabla\phi_e$ the velocity induced by other vortices outside V. Then the integrals of $\boldsymbol{v} \times \boldsymbol{\omega}$ and $\boldsymbol{x} \times (\boldsymbol{v} \times \boldsymbol{\omega})$ over V can be extended to those over V_∞, which vanish due to (3.3.6a) and (3.3.6b). Hence, (3.3.19) is further cast to

$$\frac{d\boldsymbol{I}_V}{dt} = \int_V \nabla\phi_e \times \boldsymbol{\omega} dV, \quad \frac{d\boldsymbol{L}_V}{dt} = \int_V \boldsymbol{x} \times (\nabla\phi_e \times \boldsymbol{\omega}) dV. \tag{3.3.20}$$

Finally, the right-hand side of (3.3.20) will disappear if either the vortex system is *isolated*, namely far away from other vortices outside V such that their induced $\nabla\phi_e$ is negligible (for example, an isolated vortex ring in three dimensions and vortex couple in two dimensions); or $V \to V_\infty$ that contains all vorticity. Consequently, we arrive at

Vortical-Impulse Conservation Theorem. *Let a sufficiently smooth viscous incompressible flow occupy the entire unbounded space and satisfy far-field condition* (2.4.1). *Then the total vortical impulse and angular impulse are time invariant:*

$$\frac{d\boldsymbol{I}}{dt} = \boldsymbol{0}, \quad \frac{d\boldsymbol{L}}{dt} = \boldsymbol{0}. \tag{3.3.21}$$

The same invariance holds for isolated vortex system that occupy a finite domain but free from interaction with other vortices.

It should be stressed that the above theorem will be invalid once $\nu\boldsymbol{\omega}$ in (3.3.17) has discontinuity somewhere in V_∞, since then $\nu\nabla^2\boldsymbol{\omega}$ will have singularity. This happens when V_∞ contains a moving solid body B, where the tangent components of $\boldsymbol{\omega}$ behave as a step function across body surface ∂B (Fig. 3.2). This discontinuity makes $d\boldsymbol{I}/dt$ dependent on the unsteady flow field caused by the body motion. Remarkably, in that case the total force $\boldsymbol{F}$ on the body can just be expressed by $d\boldsymbol{I}_\infty/dt$, of which a neat proof is as follows. Referring to Fig. 1.12, we continue the fluid in V_f into the body B to make $V_f + B = V$ a single continuum without internal boundary, and let Σ retreat to infinity where the fluid is at rest. Then (1.3.4b) gives

$$\boldsymbol{F} = -\rho\frac{d\boldsymbol{I}}{dt} + \rho\frac{d}{dt}\int_B \boldsymbol{u} dV - \rho\frac{d\boldsymbol{I}_{\Sigma_\infty}}{dt} - \int_{\Sigma_\infty} p\boldsymbol{n} dS, \tag{3.3.22}$$

where the last two terms are cancelled due to (3.3.16a). Thus, the convergence problem of potential impulses is bypassed in dynamics, and we obtain an elegant general formula (Wu 1981)

$$\boldsymbol{F} = -\rho\frac{d\boldsymbol{I}}{dt} + \rho\frac{d}{dt}\int_B \boldsymbol{u} dV, \tag{3.3.23}$$

where the second term is the inertial force of the fluid displaced by the body. Parallel to (3.3.7), (3.3.23) is the basis of classic *unsteady low-speed aerodynamics* and will be further explored in Chap. 9.

3.3.4 Helicity

In this and the next subsections we consider two more vorticity integrals, the helicity and total kinetic energy. They are time-invariant if and only if the flow is circulation-preserving.

The integral of helicity density,

$$\mathcal{H} \equiv \int_V \boldsymbol{\omega} \cdot \boldsymbol{u} dV, \tag{3.3.24}$$

is known as the *helicity*, which is nontrivial only in three-dimensional flow. Moffatt (1969) found that this quantity measures the state of "*knotness*" or "*tangledness*" of vortex filaments. To see this, assume that in a domain V with $\boldsymbol{n} \cdot \boldsymbol{\omega} = 0$ on ∂V there are two thin vortex filament loops C_1 and C_2, with strengths (circulations) κ_1 and κ_2 respectively, away from which the flow is irrotational. Suppose C_1 is not self-knotted, such that it spans a piece of surface S_1 without intersecting itself, and that the circulation *along* C_1 is

$$\Gamma_1 = \oint_{C_1} \boldsymbol{u} \cdot d\boldsymbol{x} = \int_{S_1} \boldsymbol{\omega} \cdot \boldsymbol{n} dS.$$

In the present situation, Γ_1 can only come from the contribution of the filament C_2. Therefore, if C_1 and C_2 are not tangled (Fig. 3.13a) then $\Gamma_1 = 0$; but if C_2 goes through C_1 once (Fig. 3.13b) then $\Gamma_1 = \pm\kappa_2$, with the sign depending on the relative direction of the vorticity in C_1 and C_2. More generally, C_2 can go through C_1 an integer number of times (Fig. 3.13b, c), so that $\Gamma_1 = \alpha_{12}\kappa_2$, where $\alpha_{12} = \alpha_{21}$ is a positive or negative integer called the *winding number* of C_1 and C_2.

Figure 3.14 shows a trefoil knot of vortex filament loop produced experimentally (Kleckner and Irvine 2013). To find the winding number of this self-knotted vortex

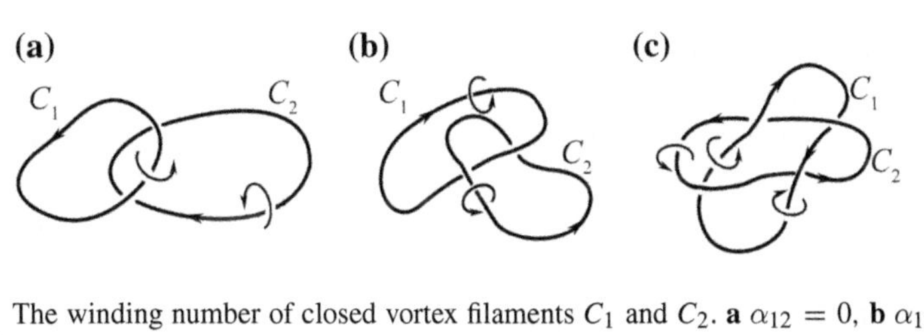

Fig. 3.13 The winding number of closed vortex filaments C_1 and C_2. **a** $\alpha_{12} = 0$, **b** $\alpha_{12} = -1$, **c** $\alpha_{12} = 2$

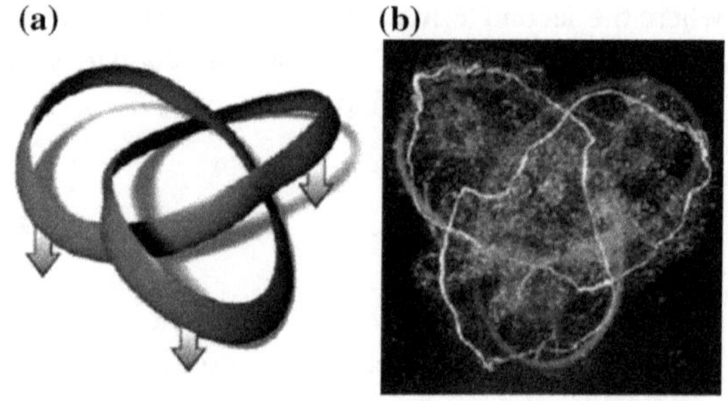

Fig. 3.14 The sudden move of a knotted "wing" with sharp trailing edge, (**a**) can generate experimentally a trefoil knot of vortex filament loop, (**b**) traced by a line of ultra-fine hydrogen bubble. From Kleckner and Irvine (2013)

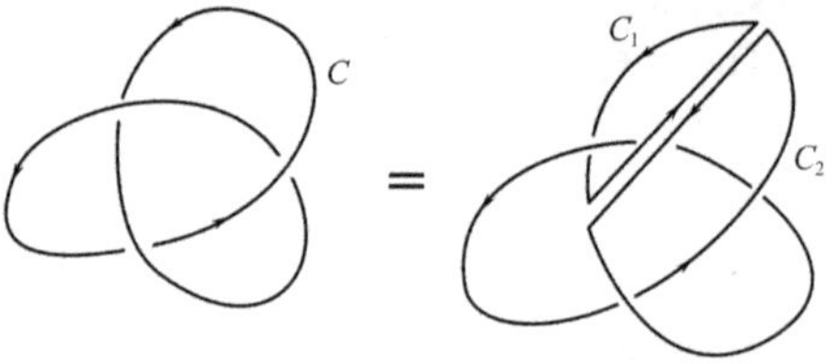

Fig. 3.15 Decomposition of a knotted vortex filament

filament, we can insert a pair of filaments of opposite κ's to decompose it into two filaments which go through each other but are not self-knotted, see Fig. 3.15. Then we have

$$\oint_C \boldsymbol{u} \cdot d\boldsymbol{x} = \oint_{C_1} \boldsymbol{u} \cdot d\boldsymbol{x} + \oint_{C_2} \boldsymbol{u} \cdot d\boldsymbol{x} = 2\kappa.$$

In general, if there are n unknotted vortex filaments, then the circulation along the ith closed filament is

$$\Gamma_i = \oint_{C_i} \boldsymbol{u} \cdot d\boldsymbol{x} = \sum_{j=1}^{n} \alpha_{ij}\kappa_j,$$

where α_{ij} is the winding number of C_i and C_j. Thus, there is

$$\Sigma_i \kappa_i \Gamma_i = \Sigma_i \oint_{C_i} \kappa_i \boldsymbol{u} \cdot d\boldsymbol{x} = \Sigma_i \Sigma_j \alpha_{ij} \kappa_i \kappa_j = \int_V \boldsymbol{\omega} \cdot \boldsymbol{u} dV, \tag{3.3.25}$$

which is precisely the helicity $\mathcal{H}$ since $\kappa_i d\boldsymbol{x}$ is nothing but $\boldsymbol{\omega} dV$ for the ith vortex filament. Therefore, the helicity measures the strengths of vortex filaments and their winding numbers.

The knotness or tangledness, characterized by the winding number, is known as the *topological property* of a curve. A topological property of a geometric configuration remains invariant under any continuous deformation. When a flow structure is a material curve like a vortex filament, the state of its knotness or tangledness is its topological property.

We now consider the time rate of helicity of a material fluid body $\mathcal{V}$, permitting variable density. For circulation-preserving flow, the Kelvin theorem (3.2.9) has ensured the time invariance of κ_i of each vortex filament; thus so must be their topology if $\mathcal{H}$ is also time invariant under the same condition. This is indeed true. By using the Beltrami equation (3.2.3) there is

$$\begin{aligned}\frac{d\mathcal{H}}{dt} &= \frac{d}{dt}\int_{\mathcal{V}} \rho\left(\frac{\boldsymbol{\omega}}{\rho}\cdot\boldsymbol{u}\right)dv = \int_{\mathcal{V}} \rho\frac{D}{Dt}\left(\frac{\boldsymbol{\omega}}{\rho}\cdot\boldsymbol{u}\right)dv \\ &= \int_{\mathcal{V}} [\boldsymbol{\omega}\cdot\nabla\boldsymbol{u}\cdot\boldsymbol{u} + (\nabla\times\boldsymbol{a})\cdot\boldsymbol{u} + \boldsymbol{\omega}\cdot\boldsymbol{a}]dv.\end{aligned}$$

Then since

$$\boldsymbol{\omega}\cdot\nabla\boldsymbol{u}\cdot\boldsymbol{u} = \nabla\cdot\left(\frac{1}{2}q^2\boldsymbol{\omega}\right), \quad \boldsymbol{\omega}\cdot\boldsymbol{a} = \nabla\cdot(\boldsymbol{u}\times\boldsymbol{a}) + (\nabla\times\boldsymbol{a})\cdot\boldsymbol{u},$$

we obtain a general kinematic formula

$$\frac{d\mathcal{H}}{dt} = 2\int_{\mathcal{V}} (\nabla\times\boldsymbol{a})\cdot\boldsymbol{u}dv + \int_{\partial\mathcal{V}} \left(\frac{1}{2}q^2\boldsymbol{\omega} + \boldsymbol{u}\times\boldsymbol{a}\right)\cdot\boldsymbol{n}dS, \tag{3.3.26}$$

which for an unbounded fluid at rest at infinity reduces to

$$\frac{d\mathcal{H}}{dt} = 2\int_{V_\infty} (\nabla\times\boldsymbol{a})\cdot\boldsymbol{u}dV. \tag{3.3.27}$$

Therefore, *the helicity of an unbounded compressible and circulation-preserving flow at rest at infinity is time-invariant* (Moffatt 1969).

3.3.5 Total Kinetic Energy

In n-dimensional free space, $n = 2, 3$, the *total kinetic energy* for incompressible flow in domain V_∞,

$$K \equiv \int_{V_\infty} \frac{1}{2}\rho q^2 dV,$$

can be expressed in terms of vorticity by two different ways. First, we write

$$q^2 = \boldsymbol{u}\cdot(\nabla\times\boldsymbol{\psi}) = \nabla\cdot(\boldsymbol{\psi}\times\boldsymbol{u}) + \boldsymbol{\omega}\cdot\boldsymbol{\psi},$$

such that

$$K = \frac{\rho}{2}\int_{V_\infty} \boldsymbol{\omega}\cdot\boldsymbol{\psi}dV + \frac{\rho}{2}\int_{\partial V_\infty} \boldsymbol{n}\cdot(\boldsymbol{\psi}\times\boldsymbol{u})dS, \tag{3.3.28}$$

which was first derived by Helmholtz (1858). Suppose that there is no total vorticity, then according the asymptotic behavior of velocity (2.4.16) the surface integral in (3.3.28) is reduced to zero, and (3.3.28) is reduced to

$$K = \frac{\rho}{2}\int_{V_\infty} \boldsymbol{\omega}\cdot\boldsymbol{\psi} dV. \tag{3.3.29}$$

In three dimensions it reads (Lamb 1932)

$$K = \frac{\rho}{8\pi}\int\int \frac{\boldsymbol{\omega}\cdot\boldsymbol{\omega}'}{r} dV dV'. \tag{3.3.30}$$

A peak of $\boldsymbol{\omega}\cdot\boldsymbol{\omega}' > 0$ in a neighborhood of a point $\boldsymbol{x}$ (after integration over $\boldsymbol{x}'$ but not yet over $\boldsymbol{x}$) must be associated with a strong alignment of neighboring vorticity lines of the same sign, which does induce a strong kinetic energy.

The second way of expressing K in terms of vorticity directly comes from (3.3.6c), also derived by Lamb:

$$K = \rho\int_{V_\infty} \boldsymbol{x}\cdot(\boldsymbol{\omega}\times\boldsymbol{u}) dV = \rho\int_{V_\infty} (\boldsymbol{x}\times\boldsymbol{\omega})\cdot\boldsymbol{u} dV, \quad n = 3 \text{ only}. \tag{3.3.31}$$

Here, the velocity $\boldsymbol{u}$ may contain contribution from a potential flow.

From its physical constituents (e.g., see (1.2.23)), it is evident that K is generically not time-invariant. But it *is* for unbounded incompressible and circulation-preserving flow. Indeed, if $\boldsymbol{a} = -\nabla\phi^*$, there is

$$\frac{D}{Dt}\left(\frac{1}{2}q^2\right) = \boldsymbol{u}\cdot\boldsymbol{a} = -\nabla\cdot(\phi^*\boldsymbol{u}).$$

Thus, for a material volume $\mathcal{V}$,

$$\frac{d}{dt}\int_{\mathcal{V}} \frac{1}{2}q^2 dv = -\int_{\partial\mathcal{V}} \phi^* u_n dS. \tag{3.3.32}$$

As $\partial\mathcal{V}$ retreats to infinity, the flow thereon becomes irrotational:

$$\boldsymbol{u} = \nabla\phi \quad \text{and} \quad \phi^* = -\left(\frac{\partial\phi}{\partial t} + \frac{1}{2}q^2\right),$$

so that the integral of the right-hand side of (3.3.32) is cast to

$$-\int_{\partial\mathcal{V}} \phi^* u_n dS = \int_{\partial\mathcal{V}} \left(\frac{\partial\phi}{\partial t} + \frac{1}{2}|\nabla\phi|^2\right)\frac{\partial\phi}{\partial n} dS,$$

which decays fast enough to zero as $|\boldsymbol{x}| \to \infty$. Therefore, *the total kinetic energy of an unbounded incompressible and circulation-preserving flow at rest at infinity is time-invariant.* This result holds for both $n = 2$ and 3.

All fluid flows are constrained in their evolution by certain conservation or sometimes the lack of conservation of relevant variables. So are vortical flows by the invariance of vorticity integrals discussed in this section, under their respective conditions. An interesting example is the fundamental difference between two- and three-dimensional flows. For the former we have total-enstrophy invariance

$$E \equiv \int_{V_\infty} \frac{1}{2}\omega^2 dV, \quad \frac{dE}{dt} = 0, \tag{3.3.33}$$

as a direct consequence of (3.2.11a), which prevents any spurious singularity to form in finite time. In contrast, for the latter E is not invariant due to vortex stretching, and it has been conjectured that unbounded vortex stretching could cause finite-time formation of singularities.

3.4 Physical Causes of Vorticity Kinetics

In preceding sections we have used the vorticity kinetic vector $\nabla \times \boldsymbol{a}$ as a symbolic notation. It is time now to examine various physical causes of this vector to close the theory of vorticity dynamics. To this end, for those physical causes taking place inside the fluid, we only need to replace $\nabla \times \boldsymbol{a}$ by the curl of various forces in momentum equation (per unit mass) that balances the vorticity kinetic vector. This is discussed in the first three subsections. However, looking at $\nabla \times \boldsymbol{a}$ alone is not enough. A material boundary is also a source of vorticity kinetics, for which we need to apply the momentum balance to the boundary without taking the curl, which is addressed in Sect. 3.4.4.

As mentioned in Sect. 3.2.1, we may use Kelvin's circulation formula (3.2.4) to examine the integrated effects of kinetics on circulation, and the generalized potential vorticity formula (3.2.6) to examine the local effect of kinetics on vorticity field. The latter can be used conveniently for steady flow, since then we can integrate (3.2.6) along any streamline C. Indeed, in terms of the intrinsic streamline coordinates (l, n, b) with $\boldsymbol{u} = (q, 0, 0)$ we simply have $D/Dt = q\partial_l$. This is especially convenient for two-dimensional steady flow on the (x, y)-plane with $\mathcal{F} = z$ and for rotationally symmetric flow on the (z, r)-plane with $\mathcal{F} = \theta$. Then we have an extension of (3.2.7):

$$\left(\frac{\omega}{\rho}\right)_l - \left(\frac{\omega}{\rho}\right)_{l_0} = \int_{l_0}^{l} \frac{1}{q\rho}(\nabla \times \boldsymbol{a}) \cdot \boldsymbol{e}_z dl, \tag{3.4.1a}$$

$$\left(\frac{\omega_\theta}{\rho r}\right)_l - \left(\frac{\omega_\theta}{\rho r}\right)_{l_0} = \int_{l_0}^{l} \frac{1}{q\rho r}(\nabla \times \boldsymbol{a}) \cdot \boldsymbol{e}_\theta dl, \tag{3.4.1b}$$

for two-dimensional and rotationally symmetric flows, respectively. In particular, if all streamlines come from far-upstream uniform flow where $\boldsymbol{\omega} = \mathbf{0}$, as the fluid moves into a region with nonzero $\nabla \times \boldsymbol{a}$, (3.4.1) indicates that $\nabla \times \boldsymbol{a}$ may *produce* an $\boldsymbol{\omega} \neq \mathbf{0}$; while on the contrary if $\boldsymbol{\omega}(l_0) \neq \mathbf{0}$ the streamline integral may eliminate it at a downstream location. As asserted in Sect. 3.2.1, the essential difference between (3.4.1) and (3.2.7) is that nonzero $\nabla \times \boldsymbol{a}$ has an accumulated effect on the vorticity at all points $\boldsymbol{x}$ through which a particle goes.

The physical constituents of vorticity kinetic vector can be seen from the momentum equation (2.2.8a) per unit mass, to which we add an external body force $\boldsymbol{f}$ for generality such that

$$\boldsymbol{a} = \frac{D\boldsymbol{u}}{Dt} = \boldsymbol{f} - \frac{1}{\rho}\nabla p + \boldsymbol{\eta} = \boldsymbol{f} - \nabla h + T\nabla s + \boldsymbol{\eta}, \tag{3.4.2a}$$

where

$$\boldsymbol{\eta} \equiv \nu_\theta \nabla\vartheta - \nu\nabla \times \boldsymbol{\omega} \tag{3.4.2b}$$

is the viscous force. Then the curl of (3.4.2) reads

$$\nabla \times \boldsymbol{a} = \nabla \times \boldsymbol{f} + \frac{1}{\rho^2}\nabla\rho \times \nabla p + \nabla \times \boldsymbol{\eta} \tag{3.4.3a}$$

$$= \nabla \times \boldsymbol{f} + \nabla T \times \nabla s + \nabla \times \boldsymbol{\eta}. \tag{3.4.3b}$$

The three terms on the right occur independently. Thus, on the one hand, *a flow will be circulation-preserving and has Bernoulli integrals if all these terms vanish*; and on the other hand, *if any of the three terms appears it will produce, dissipate, and/or annihilate vorticity.* In the following subsections we look at these terms one by one. For clarity, when we discuss one of these the other two will be assumed absent.

3.4.1 Coriolis Force in Rotating Fluid

The most familiar non-conservative body force is the *Coriolis force* observed in a rotating frame of reference, which serves as an internal source of the *relative* vorticity viewed in that frame. This effect is important in, for instance, geophysical vorticity dynamics where one observes the motion of atmosphere and ocean in the frame fixed to the rotating earth of angular velocity $\boldsymbol{\Omega}$. While the earth rotating effect on a flow of laboratory scale is negligible, it becomes strong on a geophysical flow scaled to kilometers. To study the rotating-frame effect, we first need to transform the momentum equation in an inertial frame Σ to a rotating frame Σ'. The form of scalar equations, such as continuity equation and energy equation, is not affected by the transformation.

Let $\boldsymbol{r}$ be the position vector of a fluid element from the coordinate origin. Its decompositions in the inertial frame Σ and rotating frame Σ' are $\boldsymbol{r} = x_i \boldsymbol{e}_i$ and $\boldsymbol{r} =$

$x'_i e'_i$, respectively. Since e'_i rotates with angular velocity $\boldsymbol{\Omega}$, we have $de'/dt = \boldsymbol{\Omega} \times \boldsymbol{r}$. From this basic fact follows the transformation between the velocity and acceleration measured in Σ and those in Σ'. The similar transformation of the vorticity has been given in Sect. 3.1.1. Here we just list the results (Problem 3.13):

$$\boldsymbol{u} = \frac{D\boldsymbol{r}}{Dt} = \boldsymbol{u}' + \boldsymbol{\Omega} \times \boldsymbol{r}, \tag{3.4.4a}$$

$$\boldsymbol{a} = \frac{D\boldsymbol{u}}{Dt} = \boldsymbol{a}' + 2\boldsymbol{\Omega} \times \boldsymbol{u}' + \boldsymbol{\Omega} \times (\boldsymbol{\Omega} \times \boldsymbol{r}) + \dot{\boldsymbol{\Omega}} \times \boldsymbol{r}, \tag{3.4.4b}$$

$$\boldsymbol{\omega} = \boldsymbol{\omega}' + 2\boldsymbol{\Omega}, \tag{3.4.4c}$$

where dot denotes time derivative. Assume $\dot{\boldsymbol{\Omega}} = \boldsymbol{0}$ for simplicity, then in rotating frame there are two extra forces: *Coriolis force* $\boldsymbol{f}_{\text{Cor}} = -2\boldsymbol{\Omega} \times \boldsymbol{u}'$ and *centrifugal force* $\boldsymbol{f}_{\text{cent}} = -\boldsymbol{\Omega} \times (\boldsymbol{\Omega} \times \boldsymbol{r})$. While the latter depends on only the position and has a scalar potential, the former is non-conservative and has curl:

$$\boldsymbol{f}_{\text{cent}} = -\nabla \phi_{\text{cent}}, \quad \phi_{\text{cent}} = \frac{1}{2}[(\boldsymbol{\Omega} \cdot \boldsymbol{r})^2 - \Omega^2 r^2], \tag{3.4.5a}$$

$$\nabla \times \boldsymbol{f}_{\text{Cor}} = -2\mathbf{B}' \cdot \boldsymbol{\Omega}, \tag{3.4.5b}$$

where $\mathbf{B}' = \vartheta \mathbf{I} - (\nabla \boldsymbol{u}')^T$ is the surface deformation tensor viewed in Σ'. Equation (3.4.5b) represents an internal source of relative vorticity (but not absolute vorticity). Therefore, by the momentum equation (2.6.2b), the relative momentum equation of a viscous incompressible flow in the rotating frame Σ' reads

$$\frac{D\boldsymbol{u}'}{Dt} = \boldsymbol{f}_{\text{Cor}} - \nabla \Phi - \nu \nabla \times \boldsymbol{\omega}', \tag{3.4.6}$$

where $\Phi = p/\rho + \phi_{\text{cent}}$ is the modified pressure to include the centrifugal potential. Accordingly, the relative-vorticity equation reads

$$\frac{\partial \boldsymbol{\omega}'}{\partial t} + \boldsymbol{u}' \cdot \nabla \boldsymbol{\omega}' - (\boldsymbol{\omega}' + 2\boldsymbol{\Omega}) \cdot \nabla \boldsymbol{u}' = \nu \nabla^2 \boldsymbol{\omega}', \tag{3.4.7}$$

where $-2\boldsymbol{\Omega} \cdot \nabla \boldsymbol{u}' = \mathbf{B}' \cdot \boldsymbol{\Omega}$ since now $\mathbf{B}' = -(\nabla \boldsymbol{u}')^T$.

Now, assume the absolute flow is circulation-preserving such that $\nabla \times \boldsymbol{a} = \nabla \times \boldsymbol{f}_{\text{Cor}}$. Then (3.2.4) yields

$$\frac{d\Gamma'_C}{dt} = \int_S (\nabla \times \boldsymbol{a}') \cdot \boldsymbol{n} dS = -2\boldsymbol{\Omega} \cdot \int_S d\boldsymbol{S} \cdot \mathbf{B}' = -2\boldsymbol{\Omega} \cdot \frac{D\boldsymbol{S}}{Dt} \tag{3.4.8}$$

due to (1.1.44), where $\boldsymbol{S}$ is any vector surface spanned by C. Therefore, as sketched in Fig. 3.16, if a velocity component $\boldsymbol{u}'_\perp$ normal to C makes $\boldsymbol{S}$ enlarged or reduced, and this change of $\boldsymbol{S}$ is not perpendicular to $\boldsymbol{\Omega}$, then the coupling between the frame's

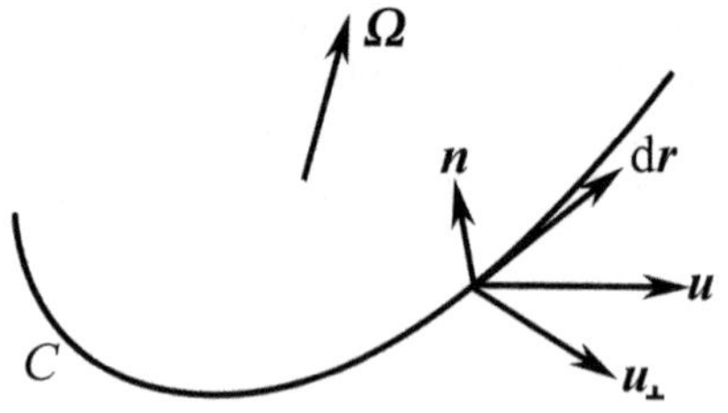

Fig. 3.16 The creation of relative vorticity inside a loop C by the Coriolis force

angular velocity $\boldsymbol{\Omega}$ and (essentially) the relative velocity gradient $\nabla\boldsymbol{u}'$ will create relative vorticity inside the loop C even if $\boldsymbol{\omega}'$ was initially zero with $\boldsymbol{u}'_0 = \nabla\phi'_0$.

We may also apply (3.2.7) to examine the creation of relative vorticity by Coriolis force. For example, in a two-dimensional steady flow with conservative absolute circulation, (3.4.1a) yields

$$\left(\frac{\omega'}{\rho}\right)_l - \left(\frac{\omega'}{\rho}\right)_{l_0} = -2\int_{l_0}^{l} \frac{1}{q'\rho}(\mathbf{B}'\cdot\boldsymbol{\Omega})\cdot\boldsymbol{e}_z dl. \tag{3.4.9}$$

3.4.2 Baroclinicity

The second terms in both (3.4.3a) and (3.4.3b) reflects the *nonlinear coupling* between shearing process and thermodynamic process. If these terms can be linearized then their contribution to $\nabla\times\boldsymbol{a}$ is gone. For compressible flow without strong external heat addition $\dot{Q}$, in smooth flow regions the entropy increment by heat conduction is generically a small disturbance of higher order than the disturbances of other thermodynamic variables. Thus in some cases (e.g., aeroacoustics) we may replace $T\nabla s$ in (3.4.2a) by $T_0\nabla s$ with T_0 being a constant reference temperature, making $\nabla T\times\nabla s \simeq 0$.

In rigorous sense, then, $\nabla\rho\times\nabla p = \mathbf{0}$ *if and only if the flow is barotropic*, i.e., in the flow only one thermodynamic variable is independent. The proof of "if" is evident, since then there must be $\nabla p = f'(\rho)\nabla\rho$ which is parallel to $\nabla\rho$. To prove the "only if", observe that the condition

$$\mathbf{0} = \nabla p\times\nabla\rho = \nabla\times(p\nabla\rho)$$

implies either $\rho = f(p)$ and hence the flow is barotropic, or $p\nabla\rho = \nabla\lambda$ has a potential. In the latter case λ can be taken as the unique independent thermodynamic variable; thus the proof is completed. Consequently, we have $\nabla p/\rho = \nabla(\int dp/\rho)$. The same argument can be applied to prove that $\nabla T\times\nabla s$ in (3.4.3b) must vanish for barotropic flow.

In contrast, for *baroclinic flow* with two independent variables, another internal vorticity source appears due to nonzero $\nabla T\times\nabla s$ or $\nabla\rho\times\nabla p$. Then (3.2.4) implies

$$\frac{d\Gamma_C}{dt} = -\oint_C \frac{1}{\rho} dp = \int_S \frac{1}{\rho^2} \boldsymbol{n} \cdot (\nabla\rho \times \nabla p) dS \tag{3.4.10a}$$

$$= \oint_C T ds = \int_S \boldsymbol{n} \cdot (\nabla T \times \nabla s) dS. \tag{3.4.10b}$$

Note that although (3.4.10b) is apparently inviscid the viscosity and heat conductivity play a key implicit role, since by entropy equation (1.2.31c) a nonuniform dissipation Φ and temperature T, as well as heat addition $\dot{Q}$, will create a nonuniform entropy distribution with $\nabla s \neq \mathbf{0}$ that can be advected downstream even into an effectively inviscid flow region.

For two-dimensional steady flow, we may use (3.4.1a) to observe the baroclinic creation of vorticity. Noticing that $\boldsymbol{n} \cdot (\nabla\rho \times \nabla p)$ and $\boldsymbol{n} \cdot (\nabla T \times \nabla s)$ are both Jacobians, see (3.1.26), where (x, y) can be any orthonormal coordinates on the flow plane. Therefore, let (x, y) be the intrinsic streamline coordinates (l, n) with $\boldsymbol{u} = (q, 0)$, by (3.4.1a) we obtain

$$\left(\frac{\omega}{\rho}\right)_l - \left(\frac{\omega}{\rho}\right)_{l_0} = \int_{l_0}^{l} \frac{1}{q\rho^3} \frac{\partial(\rho, p)}{\partial(l, n)} dl = \int_{l_0}^{l} \frac{1}{q\rho} \frac{\partial(T, s)}{\partial(l, n)} dl. \tag{3.4.11}$$

Similar result can be obtained for rotationally symmetric flow.

Example 1. Density stratification. This happens in atmosphere and sea water due to gravity. Although the flow can still be considered incompressible, density stratification makes it typically baroclinic with nonzero $\nabla\rho \times \nabla p$. In theoretical analysis, the effect of geophysical density stratification is often considered within the *Boussinesq approximation*. Assume the density variation from a reference value ρ_0 is small at each height z, and ρ_0 is related to the reference pressure p_0 by hydrostatic equation

$$\frac{dp_0}{dz} = -\rho_0 g,$$

so the effect of density stratification amounts to merely causing a *buoyancy*. Then set $(p, \rho) = (p_0 + p', \rho_0 + \rho')$ and assuming $\rho'/\rho_0 \equiv \sigma \ll 1$, the incompressible momentum equation reads

$$\frac{D\boldsymbol{u}}{Dt} = -\frac{1}{\rho_0}\nabla p' + \sigma \boldsymbol{g} + \nu\nabla^2 \boldsymbol{u}, \quad \boldsymbol{g} = -g\boldsymbol{e}_z, \quad \nu = \frac{\mu}{\rho_0}. \tag{3.4.12}$$

Thus, $\nabla\sigma \times \boldsymbol{g}$ is a source of vorticity.

Example 2. Vorticity generation in curved shock layer. Across a curved shock the entropy increases non-uniformly and a layer of ∇s behind the shock is formed. Then by (3.4.10b) or the second expression of (3.4.11), a region with $\nabla s \times \nabla T \neq \mathbf{0}$ will be a major interior vorticity source in gas dynamics. An example is shown in Fig. 3.17.

If (3.4.10b) or (3.4.11) is to be directly used to calculate the shock-generated vorticity, one has to know the detailed flow inside the thin viscous shock layer.

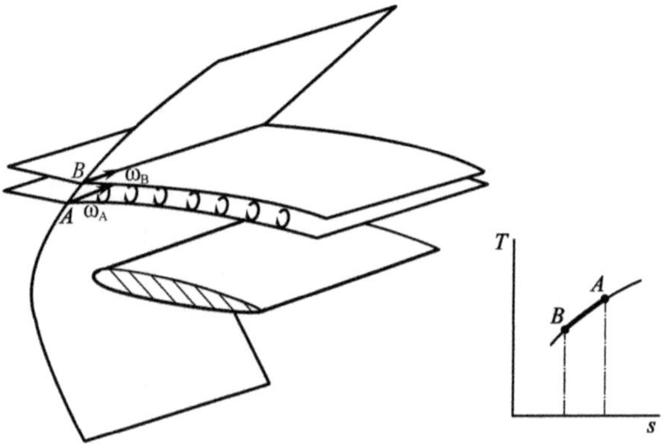

Fig. 3.17 Vortex layer behind a curved shock wave. Reproduced from Kármán (1954)

Fortunately, the generated vorticity can be solely derived from the flow quantities at both sides of the shock treated as a normal discontinuity. Here, the jump of $\rho(\boldsymbol{\omega} \times \boldsymbol{u})$ across the shock is the key information since only this vector contains vorticity in the Euler equation. For example, consider a two-dimensional uniform flow $\boldsymbol{U}$ passing a stationary curved shock S of curvature κ. Let $\boldsymbol{t}$ and $\boldsymbol{n}$ be the tangent and normal vectors varying along S, such that $\boldsymbol{t} \times \boldsymbol{n} = \boldsymbol{e}_3$. Let (s, n) be the coordinates along and normal to S. Then only the tangent component of the Lamb vector, $\rho(\boldsymbol{\omega} \times \boldsymbol{u}) \cdot \boldsymbol{t} = -\rho\omega u_n$, has a jump, which we infer from the Euler equation as

$$[\![\rho u_n \omega]\!] = [\![\rho q^2_{,s}/2 + p_{,s}]\!], \tag{3.4.13}$$

where $q^2 = u_s^2 + u_n^2$, $(\cdot)_{,s} = \partial_s(\cdot)$, and $[\![\cdot]\!] = (\cdot)_1 - (\cdot)_0$, with subscripts 0 and 1 referring to quantities ahead of and behind S, respectively. To pick up $[\![\omega]\!] = \omega_1$ from (3.4.13), we use the jump conditions $m = \rho u_n$ and ρu_n^2 due to mass and momentum conservations,

$$[\![m]\!] = 0, \quad [\![\rho u_n^2 + p]\!] = 0, \tag{3.4.14}$$

along with the identity on jumps of product

$$[\![fg]\!] = \bar{f}[\![g]\!] + [\![f]\!]\bar{g}, \tag{3.4.15}$$

where the overline means averaged value of both sides. After some algebra (e.g. Wu et al. 2006), we find

$$\omega_1 = -\frac{(1-\epsilon)^2}{\epsilon}\partial_s U_n, \quad \epsilon \equiv \frac{\rho_0}{\rho_1} \leq 1.$$

But, since

$$\partial_s U_n = \partial_s \boldsymbol{U} \cdot \boldsymbol{n} + \boldsymbol{U} \cdot \partial_s \boldsymbol{n},$$

where $\partial_s \boldsymbol{n} = -\kappa \boldsymbol{t}$ with $\kappa > 0$ being the curvature, and $\partial_s \boldsymbol{U} = \boldsymbol{0}$ because $\boldsymbol{U}$ is constant vector, we obtain

$$\omega_1 = \frac{(1-\epsilon)^2}{\epsilon} U_s \kappa. \tag{3.4.16}$$

Thus, the newly produced vorticity depends only on the velocity in front of the shock and shock curvature, as well as the density ratio across it, but completely independent of the thermodynamic process inside the shock layer. The vorticity generated by a curved shock is of $O([\![\rho]\!]^2\kappa)$. The reader may check that for the flow pattern of Fig. 3.17, (3.4.16) predicts $\omega_1 < 0$ as it should. This result can be extended to three-dimensional and unsteady flows, and the same mechanism also appears in high Mach-number turbulence (Fig. 2.5). No vorticity is generated behind a straight shock.

3.4.3 Vorticity Diffusion and Enstrophy Dissipation

The third constituent of the vorticity kinetics vector (3.4.3) is associated with fluid viscosity, which plays a double role in the interior of a flow, namely *vorticity diffusion* and *dissipation*.

Before proceed, we remark that not all viscous flows must have nonzero vorticity kinetic vector. In a viscous flow, the third term of (3.4.3) disappears if and only if there exists a potential, say σ, such that

$$\boldsymbol{\eta} = -\nabla \sigma, \tag{3.4.17}$$

which signifies a special class of viscous flows, for example the familiar two-dimensional or axisymmetric Couette flow and Poiseuille flow (Problem 3.14). Thus, circulation-preserving flow is not necessarily inviscid. Nevertheless, a generic viscous flow cannot satisfy (3.4.17).

1. Vorticity diffusion. Assume the flow is incompressible without external body force, such that $\boldsymbol{\eta} = -\nu\nabla \times \boldsymbol{\omega} = \nu\nabla^2 \boldsymbol{u}$ and

$$\nabla \times \boldsymbol{a} = -\nu\nabla \times (\nabla \times \boldsymbol{\omega}) = \nu\nabla^2 \boldsymbol{\omega}, \tag{3.4.18}$$

by which the Kelvin circulation formula (3.2.4) yields

$$\frac{d\Gamma}{dt} = -\nu \oint (\nabla \times \boldsymbol{\omega}) \cdot d\boldsymbol{x} \tag{3.4.19}$$

for any material loop. Consider now a vorticity tube enclosed by a loop C in two-dimensional viscous flow. Introduce the unit vector $\boldsymbol{t}$ tangent to C such that $d\boldsymbol{x} = \boldsymbol{t}ds$, and the unit vector $\boldsymbol{n}$ normal to C pointing out of C. Then $(\boldsymbol{t}, \boldsymbol{e}_z, \boldsymbol{n})$ form a triad moving along C with $\boldsymbol{\omega} = \omega \boldsymbol{e}_z$, see Fig. 3.18, and there is

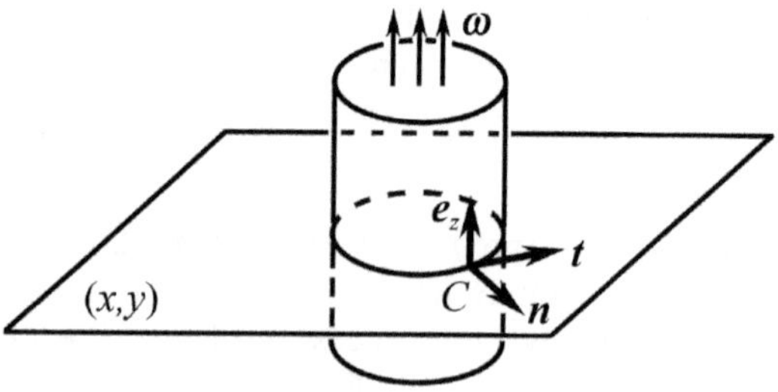

Fig. 3.18 *Cross view* of a vorticity tube, with tangential and normal vectors ***t*** and ***n*** defined at the tube boundary

$$d\boldsymbol{x} \cdot (\nabla \times \boldsymbol{\omega}) = ds(\boldsymbol{t} \times \nabla) \cdot \boldsymbol{e}_z\omega = -\boldsymbol{n} \cdot \nabla\omega.$$

Thus, we obtain

$$\frac{d\Gamma}{dt} = \oint_C \nu \frac{\partial \omega}{\partial n} ds, \qquad (3.4.20)$$

where $\nu\partial\omega/\partial n$ is the *vorticity diffusion flux*. Like the temperature gradient controls the heat conduction, ω in the tube is diffused across the tube wall and hence Γ varies as time. Diffusion spreads the existing vorticity at a point downgrade to its neighbors and thereby alters the $\boldsymbol{\omega}$-distribution. But, *diffusion never creates new vorticity in the interior of the flow.*

2. Vorticity annihilation. The second effect of viscosity on a vorticity field is to annihilate it. The mechanism is precisely analogous to the annihilation of fluid kinetic energy $q^2/2$ by viscous dissipation Φ introduced in Sect. 1.2.2. The analogy of $q^2/2$ is the *enstrophy* $\omega^2/2$ already appeared in (3.2.18a). It suffices to consider only the time rate of total enstrophy $E = \int \frac{1}{2}\omega^2 dV$ of the entire vortical flow domain, since unlike the vorticity vector once the total enstrophy vanishes there must be $\boldsymbol{\omega} = \boldsymbol{0}$ everywhere. Note that *enstrophy dissipation does not conflict but is strictly constrained by the kinematic conservation of total vorticity and circulation* stated in Sect. 3.3.1.

Assume again the flow is incompressible without external body force. By (3.4.18) and since

$$\nabla^2\boldsymbol{\omega} \cdot \boldsymbol{\omega} = \nabla \cdot (\nabla\boldsymbol{\omega} \cdot \boldsymbol{\omega}) - \nabla\boldsymbol{\omega} : (\nabla\boldsymbol{\omega})^T,$$

integrating (3.2.18a) over a material volume $\mathcal{V}$ leads to the rate of change of total enstrophy E in $\mathcal{V}$:

$$\frac{dE}{dt} = 2\overline{\alpha}(t)E + \int_{\partial\mathcal{V}} \eta dS - \int_{\mathcal{V}} \Phi_\omega dv, \qquad (3.4.21a)$$

where $\overline{\alpha}(t)$ is a mean value of the vorticity stretching rate α in $\mathcal{V}$,

$$\eta \equiv \nu\frac{\partial}{\partial n}\left(\frac{1}{2}\omega^2\right) = \nu\frac{\partial \boldsymbol{\omega}}{\partial n} \cdot \boldsymbol{\omega} \qquad (3.4.21b)$$

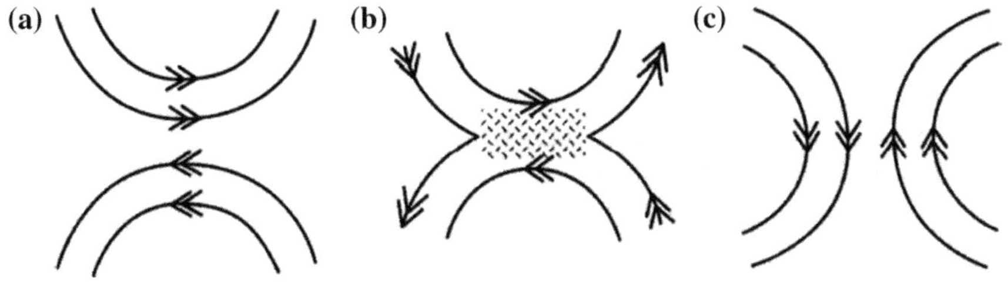

Fig. 3.19 Vorticity cancellation and enstrophy dissipation as two opposite vortices are in contact

is the *enstrophy diffusion flux*, and

$$\Phi_\omega \equiv \mu \nabla \boldsymbol{\omega} : (\nabla \boldsymbol{\omega})^T \geq 0 \tag{3.4.21c}$$

is the *enstrophy dissipation rate* analogous to the kinetic-energy dissipation rate Φ.

The peak values of enstrophy dissipation Φ_ω must occur in localized regions where the vorticity gradient is strong. In particular, if two vortices of opposite signs move toward each other and become in contact, see the sketch of Fig. 3.19, the vorticity of one vortex will be diffused into the region occupied by the other, and then Φ_ω measures how much local enstrophy $\omega^2/2$ is canceled near the common boarder of the vortices by this contact.

3.4.4 Vorticity Creation at Boundary

In addition to vorticity diffusion and dissipation, viscosity has yet another role that, along with the no-slip condition it makes the fluid boundary the most important source of vorticity. This concept has been introduced in Sect. 2.3.2 in the context of linear coupling of compressing and shearing processes on boundary, and we now further explore it. We have seen that as the on-wall signature of global compressing process, the tangent pressure gradient is transferred by the no-slip condition to the on-wall root of global shearing process, described by the *boundary vorticity flux* (BVF for short):

$$\boldsymbol{\sigma} \equiv \nu \frac{\partial \boldsymbol{\omega}}{\partial n} \quad \text{at boundary.} \tag{3.4.22}$$

The BVF exists only on the fluid side and measures the rate by which vorticity is created at the wall due to the no-slip condition, and also the rate by which the newly created vorticity is sent into the fluid by diffusion.

In addition to BVF, the boundary enstrophy flux, i.e., (3.4.21b) applied to the wall,

$$\eta = \boldsymbol{\sigma} \cdot \boldsymbol{\omega}_B \quad \text{at wall,} \tag{3.4.23}$$

may clearly reflect the relative roles of BVF and boundary vorticity ω_B. $\eta > 0$ implies that the newly produced vorticity is of the same sign of the existing one, and so enhances it; while $\eta < 0$ indicates that the newly produced vorticity weakens the existing one, and even changes the sign of ω_B.

Owing to the on-wall vorticity creation, *near-wall viscous flow can never be circulation-preserving.*[6] This boundary source of vorticity appears universally in any wall-bounded external and internal flows. It serves as the key physical root of all vortical structures therein, and also the basis for flow management by configuration design and on-wall control devices. Therefore, it is necessary to gain a clear and complete physical understanding of the boundary vorticity generation.

1. Constituents of boundary vorticity flux. The (ω, p) coupling on boundary is the most primary mechanism of the on-wall vorticity creation, but is not the only one. Consider a generic three-dimensional, viscous and incompressible flow with external body force $\boldsymbol{f}$ over a solid boundary ∂B of normal $\boldsymbol{n}$ (pointing out of fluid) that may have arbitrary shape, motion, and deformation. The momentum equation is

$$\boldsymbol{a} = \boldsymbol{f} - \frac{1}{\rho}\nabla p - \nu\nabla\times\boldsymbol{\omega}, \quad \boldsymbol{a} = \frac{D\boldsymbol{u}}{Dt}.$$

Applying this to ∂B of given acceleration $\boldsymbol{a}_B$ and using the acceleration no-slip condition (1.3.3b), we obtain a general expression for the BVF:

$$\begin{aligned}\boldsymbol{\sigma} &= \boldsymbol{n}\times\left(\boldsymbol{a} - \boldsymbol{f} + \frac{1}{\rho}\nabla p\right) + \nu(\boldsymbol{n}\times\nabla)\times\boldsymbol{\omega} \\ &= \boldsymbol{\sigma}_a + \boldsymbol{\sigma}_f + \boldsymbol{\sigma}_p + \boldsymbol{\sigma}_{\text{vis}} \quad \text{on } \partial B, \qquad (3.4.24a)\end{aligned}$$

where

$$\boldsymbol{\sigma}_a = \boldsymbol{n}\times\boldsymbol{a}_B, \qquad \boldsymbol{\sigma}_f = -\boldsymbol{n}\times\boldsymbol{f}, \qquad (3.4.24b)$$

$$\boldsymbol{\sigma}_p = \frac{1}{\rho}\boldsymbol{n}\times\nabla p, \quad \boldsymbol{\sigma}_{\text{vis}} = \nu(\boldsymbol{n}\times\nabla)\times\boldsymbol{\omega} \qquad (3.4.24c)$$

are the BVF constituents caused by the tangential components of the wall acceleration $\boldsymbol{a}_B$ (due to acceleration adherence), the external body force, the pressure gradient, and a tangent vorticity diffusion ($\boldsymbol{n}\times\nabla$ is a tangent gradient operator), respectively. The on-wall effects of $\boldsymbol{a}_B$ and $\boldsymbol{f}$ on the BVF can actually be understood in the same way as that of $\nabla p/\rho$, by looking at Fig. 2.7. In (3.4.24a), all terms on the right-

[6] In 1781, Lagrange used power-series expansion to prove that if the vorticity is an analytical function of time t, then an initially irrotational flow will remain so at later time. This Lagrange potential-flow theorem was accepted by Euler, Cauchy, and Helmholtz, among others, but cannot explain observed vortical wake behind a body that starts to move at $t = 0$ in a fluid otherwise at rest. The paradox had not been resolved till late 19th century when one realized that, while Lagrange's analyticity assumption holds true for circulation-preserving flow, it is violated by boundary vorticity creation. See Truesdell (1954) and Darrigol (2009).

hand side take their on-wall value, and $\boldsymbol{\sigma}_{\text{vis}} \equiv \mathbf{0}$ in two dimensions. Note that like pressure force, even a conservative body force may create new vorticity at boundary. By assigning various $\boldsymbol{a}_B$, (3.4.24) can cover flow over porous wall with prescribed normal blowing or suction. It can also be easily extended to compressible flow by replacing p by $\Pi = p - \mu_\theta \vartheta$.

The multiple constituents of BVF provide certain options for flow control and configuration design. As consequences of global flow behavior, $\boldsymbol{\sigma}_p$ and $\boldsymbol{\sigma}_{\text{vis}}$ are not controllable, but $\boldsymbol{\sigma}_a$ and $\boldsymbol{\sigma}_f$ are at our hand. For example, imposing a traveling wave-like body force $\boldsymbol{f}$ near a solid wall may significantly alter the local BVF and result in drag reduction. This was first confirmed by Du and Karniadakis (2000) for an electrically-conducting turbulent channel flow using the Lorentz force near the channel wall. But, so far as the on-wall vorticity creation is concerned, this $\boldsymbol{\sigma}_f$ effect may well be replaced by an equivalent $\boldsymbol{\sigma}_a$ (say, flexible wall wave), or vise versa. This was found indeed so by Zhao et al. (2004) who replaced the Lorentz force by tangent spanwise traveling wave of a flexible channel wall. On the other hand, a streamwise traveling wave can be formed by vertically oscillating flexible wall, which has shown promising application to suppress flow separation from bluff body and to result in a big drag reduction (Wu et al. 2007).

2. Strength of boundary vorticity source. The involvement of viscosity in the on-wall $(\boldsymbol{\omega}, p)$ coupling means that the coupling exists only in viscous flow. However, it is not the normal gradient of $\boldsymbol{\omega}$ but that of $\nu\boldsymbol{\omega}$ that is directly related to the tangential components of $\nabla p, \boldsymbol{f}$, and $\boldsymbol{a}_B$ which are generically of $O(1)$ after proper nondimensionalization. Namely, the BVF is *independent of* the Reynolds number. As Re increases, the newly created $\boldsymbol{\omega}$ becomes stronger but advected faster to downstream, making the vorticity confined in a thinner fluid layer adjacent to the wall to ensure $|\boldsymbol{\sigma}| = O(1)$. When $Re \gg 1$, the boundary-layer theory to be learned in Chap. 4 shows the layer thickness δ is of $O(Re^{-1/2})$, so the vorticity inside the layer is of $O(Re^{1/2})$. This picture holds true even if $Re \to \infty$ with $|\boldsymbol{\omega}| \to \infty$ but $\delta \to 0$ (the created vorticity forms a vortex sheet). On the contrary, when $Re \ll 1$ the created vorticity spreads all over the space by diffusion with much smaller normal gradient, still ensuring $|\boldsymbol{\sigma}| = O(1)$.

As a consequence of this Re-independence of BVF, it can be shown that if a material loop $C(t)$ is in part attached to the wall, then the rate of change of its circulation can also be independent of Re even if $\nu \to 0$ (Problem 3.18). This is an interesting example that the concepts of $\nu \to 0$ and $\nu \equiv 0$ are radically different.

Compared to various internal vorticity sources, the large $O(1)$ strength of boundary vorticity creation and its universal existence come from the simple fact that the adherence condition makes the fluidity completely lost for those fluid particles sticking to the wall. No adjustment of fluid motion can be done on the wall to balance externally applied tangent actions given in (3.4.24); they have to be totally transformed to the BVF. But this stiff compressing-shearing coupling on a solid wall is relaxed at an *interface* of two fluids such as water-air interface or free surface. There, a pair of normal-tangent boundary coupling relations also exists; but when $\nu \ll 1$, the fluid may freely accelerate at the interface to locally adapt the normal and tangential gradients of pressure (modified by a surface tension). A wind blowing over a sea surface, say, can drive water to move. Thus, in (3.4.24a) the dominant balance occurs between $\boldsymbol{\sigma}_a$ and $\boldsymbol{\sigma}_p$ which cancel each other, leaving a small residual $\boldsymbol{\sigma}$. Adjacent to the interface there are still vortex layers or sheets on both sides, which are however much weaker due to the smallness of BVF. For example, on a two-dimensional and steady free surface with tangent velocity U_s and curvature κ, one finds (Wu et al. 2006)

$$\sigma = O(Re^{-1/2}), \quad \omega = 2U_s\kappa. \tag{3.4.25}$$

To conclude, we list the strengths of the created vorticity by these boundary sources, along with the vorticity created by curved shock wave:

$$\begin{aligned} \text{Solid wall, } Re \gg 1: &\quad |\boldsymbol{\omega}| = O(Re^{1/2}) \\ \text{Interface, } Re \gg 1: &\quad |\boldsymbol{\omega}| = O(\kappa) \\ \text{Behind curved shock}: &\quad |\boldsymbol{\omega}| = O([\![\rho]\!]^2\kappa) \end{aligned} \tag{3.4.26}$$

If the free-surface wavelength is not very small ($\kappa \ll 1$), the vortex layer thereon is very weak and negligible. This is why the classic water wave theory can be successfully developed in the framework of potential flow. On the other hand, the curved-shock generated vortex layer could be as strong as boundary layer only in hypersonic flight, where the strong nose normal shock turns sharply to almost the direction of oncoming flow with large density jump and $\kappa \gg 1$. Generically, other internally generated vorticities not listed in (3.4.26) are of $O(1)$.

3.5 Problems for Chapter 3

3.1. Consider a bucket of water hanging on a tightly twisted rope.[7] The bucket was hold rest for $t < t_0$. Let the system rotate freely from $t = t_0$ under the action of the rope's torque. At $t = t_0^+$, the bucket rotates with angular velocity Ω, and only a thin layer of water adjacent to the wall is driven to co-rotate with the bucket due to adherence condition, see Fig. 3.20a. After sufficient time the whole water co-rotates with the bucket (so the water surface becomes concave), see Fig. 3.20b. Finally, stop the bucket suddenly so that a thin layer of water adjacent to the wall become stationary, while the rest part remains rotating due to inertia, see Fig. 3.20c. Please make a qualitative sketch of the vorticity lines in the water (in particular the vorticity lines inside the near-wall layer) for the three different cases.

3.2. Assume that the flow domain V has an external boundary ∂V and no internal boundary.

---

[7] Newton conceived this example as a mental experiment to argue that his "absolute motion" (characterized by the concave water surface) and "relative motion" (of water to the bucket, characterized by the flat water surface) have observable difference, so that they can be objectively distinguished.

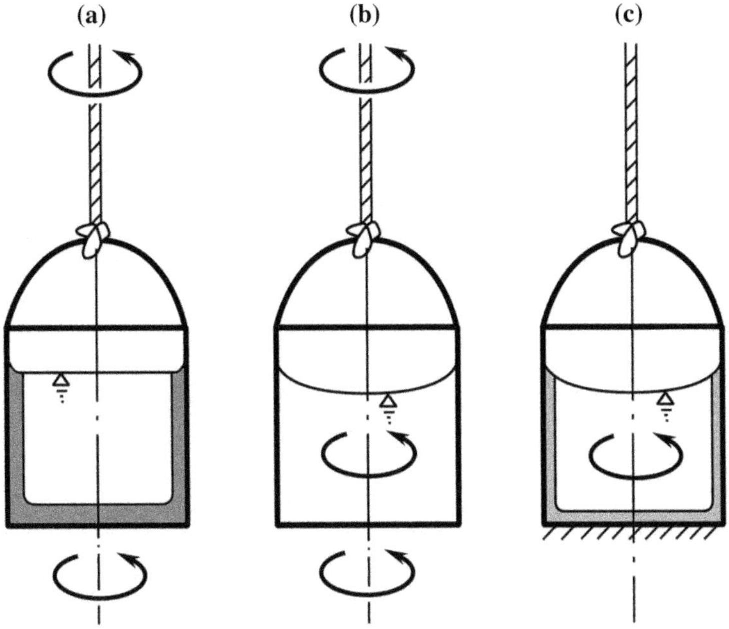

Fig. 3.20 The three stages of Newton's mental experiment of water bucket hanging on a tightly twisted rope. **a** The bucket just starts to rotate at angular velocity Ω but most part of water remains at rest. **b** The bucket and entire water rotate at angular velocity Ω. **c** Most part of the water rotates at angular velocity Ω but the bucket is at rest

(1) Prove that if either $\boldsymbol{n} \cdot \boldsymbol{u}$ or $\boldsymbol{n} \times \boldsymbol{u}$ at the domain boundary ∂V is given then the solution of problem (3.1.31a), if exists, must be unique.

(2) For viscous flow, the velocity adherence condition requires both $\boldsymbol{n} \cdot \boldsymbol{u}$ and $\boldsymbol{n} \times \boldsymbol{u}$ should be prescribed at body surface. But according to the above result, generically one would find no solution for (3.1.31a). Try to explain this apparent dilemma.

3.3. Consider a Π-shaped "horseshoe" vortex filament of circulation Γ with two "leg segments" separated by $2b$ and extending to infinity as shown in Fig. 3.21. Let $\boldsymbol{u} = (u, v, w)$ in the coordinates (x, y, z) with origin at the center of the front segment. Show that the horseshoe vortex induces a downward velocity at $(0, y, 0)$ along that segment:

$$w(y) = -\frac{\Gamma}{2\pi} \frac{b}{b^2 - y^2}. \tag{3.5.1}$$

3.4. Replace the horseshoe vortex filament in Problem 3.3 by a plat vortex sheet of width $2b$ on the (x, y) plane, which also extends to downstream infinity. Assume the sheet strength is $\boldsymbol{\gamma} = \gamma(y)\boldsymbol{e}_x$. Show that at the front edge of the sheet, $(0, y, 0)$, the vortex sheet induces a velocity

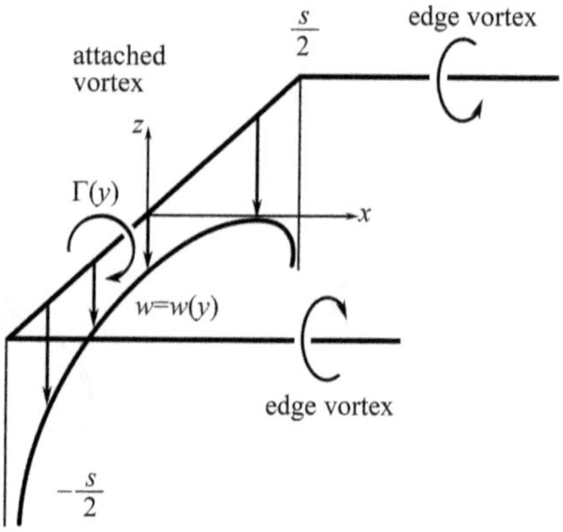

Fig. 3.21 Vertical velocity induced by a horseshoe vortex. From Oertel (2004)

$$w(y) = \frac{1}{4\pi}\int_{-b}^{b}\frac{\gamma'(y')dy'}{y-y'}. \tag{3.5.2}$$

3.5. Let $\mathbf{B} = \vartheta\mathbf{I} - (\nabla\boldsymbol{u})^T$ be the surface deformation tensor. Prove $\boldsymbol{\omega}\cdot\mathbf{B} = \mathbf{B}\cdot\boldsymbol{\omega}$ and explain the physical meaning of this term. Then show that for compressible or incompressible flow the vorticity evolution equation can be written

$$\frac{D\boldsymbol{\omega}}{Dt} + \mathbf{B}\cdot\boldsymbol{\omega} = \nabla\times\boldsymbol{a}, \tag{3.5.3}$$

and the stretching rate α defined by (3.2.18c) can also be expressed as $\alpha = -\boldsymbol{t}\cdot\mathbf{B}\cdot\boldsymbol{t}$.

3.6. We study vorticity evolution.

(1) Prove formula (3.2.5).

Hint: Use a second-rank tensor $\mathbf{S} = \{S_{ij}\}$ to carry out the algebra and then show that the result holds for any $\mathbf{S}$ of arbitrary rank.

(2) Consider the special case of (3.2.5), the Beltrami equation under the circulation-preserving condition,

$$\frac{D}{Dt}\left(\frac{\boldsymbol{\omega}}{\rho}\right) = \frac{\boldsymbol{\omega}}{\rho}\cdot\nabla\boldsymbol{u}. \tag{3.5.4}$$

Use this equation to prove that any material vorticity line remains material at any time, and then prove the Helmholtz's second and third vorticity theorems, parallel to the proof given in Sect. 3.3.

Hint: Compare (3.5.4) with the rate of change of line element, (1.1.40):

$$\frac{D}{Dt}(\delta\boldsymbol{x}) = \delta\boldsymbol{x}\cdot\nabla\boldsymbol{u},$$

and examine the rate of change

$$\frac{D}{Dt}\left(\delta\boldsymbol{x} - \lambda\frac{\boldsymbol{\omega}}{\rho}\right),$$

where λ is a scalar.

3.7. Prove that for any invariant vorticity integrals containing position vector $\boldsymbol{x}$, the result is independent of the choice of the origin from which the position is taken.

3.8. Consider Prandtl's force formula (3.3.7) for uniform incoming flow of velocity $\boldsymbol{U} = U\mathbf{e}_x$ over a stationary body. Prove that in two-dimensional flow this formula leads exactly the Kutta-Joulowski lift formula

$$\boldsymbol{F} = \rho\boldsymbol{U} \times \boldsymbol{\Gamma}, \quad \boldsymbol{\Gamma} \equiv \boldsymbol{e}_z\Gamma = \boldsymbol{e}_z\int_V \omega dV.$$

Explain why for three-dimensional flow one cannot obtain this simple result.

Hint: Write $\boldsymbol{v} = \boldsymbol{U} + \boldsymbol{u}$, where $\boldsymbol{u}$ is now the disturbance velocity induced by the vorticity field and hence the Biot-Savart formula (3.1.35) can be applied. Substitute the formula into (3.3.7) and prove the resulted double integral vanishes in two dimensions.

3.9. Consider an incompressible and viscous compact vortical structure in a finite material volume $\mathcal{V}$. By "compact" we mean that the exponentially decaying tail of the vorticity field can be neglected at $\partial\mathcal{V}$ such that we may assume $\boldsymbol{\omega} = \boldsymbol{0}$ there. Suppose that in $\mathcal{V}$ and on $\partial\mathcal{V}$ there exists a potential flow $\boldsymbol{u}_e = \nabla\phi_e$ induced by other compact vortical structures outside $\mathcal{V}$. Show that

$$\frac{d\boldsymbol{I}_{\mathcal{V}}}{dt} = \int_{\mathcal{V}} \boldsymbol{u}_e \times \boldsymbol{\omega} dV. \tag{3.5.5}$$

3.10. Let V_R be a big spherical volume of radius $R \to \infty$ with $\boldsymbol{n} = \boldsymbol{e}_R$ at ∂V_R. Then using (2.4.16) and (2.4.15b) to prove

$$\int_{V_R} \boldsymbol{u} dV = \frac{(n-1)\boldsymbol{I}}{n}, \tag{3.5.6}$$

namely $1/n$ of vortical impulse escapes from V_R for n-dimensional flow, which must be carried by $\boldsymbol{I}_{\partial V_R}$ to infinity.[8]

3.11. Show that the three vectorial second vorticity moments, $x^2\boldsymbol{\omega}, \boldsymbol{x} \times (\boldsymbol{x} \times \boldsymbol{\omega})$, and $\boldsymbol{x}(\boldsymbol{x} \cdot \boldsymbol{\omega})$, are linked by identity

$$\boldsymbol{x} \times (\boldsymbol{x} \times \boldsymbol{\omega}) = \boldsymbol{x}(\boldsymbol{x} \cdot \boldsymbol{\omega}) - x^2\boldsymbol{\omega},$$

---

[8] As pointed out by Landau and Lifshitz (1959) and Saffman (1992), the above apparent paradox is due to the assumed incompressibility of the flow where the pressure signal propagates instantly. Once we permit the flow be slightly compressible, far-field pressure wave will be governed dynamically by (2.3.11), by which (inviscid version) one can indeed confirm that the outgoing pressure wave front carries exactly $1/n$ of the vortical impulse.

and thus only two of them are independent. Then, let

$$\boldsymbol{L}'_V \equiv \frac{1}{n}\int_V \boldsymbol{x}\times(\boldsymbol{x}\times\boldsymbol{\omega})dV. \tag{3.5.7}$$

Show that for $n = 2$ $\boldsymbol{L}'_V$ simply the same as the moment $\boldsymbol{L}_V$ defined by (3.3.9b), while for $n = 3$ they are related by

$$\boldsymbol{L}'_V - \boldsymbol{L}_V = \frac{1}{6}\int_{\partial V} x^2\boldsymbol{x}(\boldsymbol{n}\cdot\boldsymbol{\omega})dS, \tag{3.5.8}$$

so $\boldsymbol{L}'_V = \boldsymbol{L}_V$ if $\boldsymbol{n}\cdot\boldsymbol{\omega} = 0$ on ∂V or $V \to V_\infty$.

3.12. Derive (3.3.16a) and (3.3.16b).

3.13. Complete the derivation of the transformation formulas (3.4.4a) and (3.4.4b) between an inertial frame of reference and a rotating frame of reference.

3.14. Consider two-dimensional steady, unidirectional, and viscous flow with $\boldsymbol{u} = (u, 0)$ in the (x, y)-plane between two parallel plates at $y = \pm b$. Prove that the flow can only take the form

$$\mu\partial_y^2 u = \partial_x p,$$

where $\partial_x p$ must be a constant, say $-G$. Then, assume (i) $G = 0$, $u(b) = U$, and $u(-b) = 0$; and (ii) $G > 0$ and $u(\pm b) = 0$, show that the corresponding solutions of the Navier-Stokes equation are

$$u_C = \frac{U}{2b}(y+b), \quad G = 0, \quad \text{(Couette flow)}, \tag{3.5.9a}$$

$$u_P = \frac{G}{2\mu}(b^2 - y^2), \quad G > 0, \quad \text{(Poiseuille flow)}. \tag{3.5.9b}$$

Show that these flows are circulation preserving. Does the Bernoulli integral $p + \rho q^2/2 = \text{const.}$ holds for these flows, and why? When and how the vorticity is generated in these flows?

3.15. Extend the relative vorticity equation (3.4.7) to include baroclinic effect due to density-stratification (within Boussinesq approximation). Show that the dimensionless form of the resulting equation reads

$$Ro\left\{\frac{\partial\boldsymbol{\omega}}{\partial t} + \nabla\times(\boldsymbol{\omega}\times\boldsymbol{u})\right\} = 2\boldsymbol{k}\cdot\nabla\boldsymbol{u} - \frac{1}{Fr}\nabla\sigma\times\boldsymbol{e}_z + Ek\nabla^2\boldsymbol{\omega}. \tag{3.5.10}$$

Here, all quantities and operators are dimensionless and viewed in the earth frame of reference, $\boldsymbol{k}$ is the unit vector along rotating axis, and

$$Ro = \frac{U}{\Omega L}, \quad Ek = \frac{\nu}{\Omega L^2} = \frac{Ro}{Re}, \quad Fr = \frac{\Omega U}{g} \tag{3.5.11}$$

with $Re = UL/\nu$, are the *Rossby number*, *Ekman number*, and *Froude number*, respectively. Explain the implication of each of these dimensionless numbers.

3.16. Consider the water motion in a rotating tank, assuming the flow is inviscid and barotropic. Observe that for fast tank rotation there is

$$\lim_{Ro \to 0} \boldsymbol{k} \cdot \nabla \boldsymbol{u} = \boldsymbol{0}, \tag{3.5.12}$$

and describe the special feature of this flow (The *Taylor-Proudman theorem*). What would happen if a ball is suspended in the water and moves horizontally?

3.17.* Consider the vortex loop associated with a finite wing as sketched in Fig. 3.3c. Assume at a point near the wingtip one may create a very strong $\nu \nabla \boldsymbol{\omega} : \nabla \boldsymbol{\omega}^T$ so that the vorticity in the loop is locally dissipated. Discuss what could happen to the entire loop and wing lift.

3.18. Consider a material loop $C(t)$ in a two-dimensional viscous flow on the (x, y)-plane, with part of the loop, say a segment AB, being attached to the wall, while the rest part remains in the fluid. Prove that as $\nu \to 0$ the rate of change of the circulation along $C(t)$ can be written as

$$\rho \frac{d\Gamma}{dt} = p(\boldsymbol{x}_B, t) - p(\boldsymbol{x}_A, t), \quad \boldsymbol{x} \in \text{points on the wall},$$

which is independent of ν or the Reynolds number.

Chapter 4
Attached and Free Vortex Layers

After the development of a general background on fundamental processes in fluid motion in Chap. 2 and basic theory of vorticity dynamics in Chap. 3, we now move onto *vortex dynamics*, i.e., the dynamics of vortices observed at large Reynolds numbers, which by definition are shear flow structures with highly concentrated vorticity. Primary vortical structures are *shear layers*, also known as *layer-like vortices* or *vortex layers*, of which those attaching to a solid surface or an interface are especially called *boundary layers*. Secondary vortical structures are *axial vortices* typically formed by the rolling up of shear layers. We discuss these two types of vortical structures in the present and next two chapters.

4.1 Parallel Shear Flows on Upper-Half Plane

Unidirectional flow is one of the simplest theoretical models of viscous shear flows for which exact solutions of the Navier-Stokes equations can be found. As detailed in all textbooks and discussed in Problem 3.14, the most famous *steady* solutions of such flows are those bounded by two parallel plates or circular pipe, i.e., the *Couette* flow (also known as the *simple shear flow*) driven by a moving boundary and the *Poiseuille flow* driven by a constant favorable pressure-gradient. In contrast, the most famous *unsteady* solutions are those on the semi-infinity domain, in particular the *Stokes first problem* (also known as the *Rayleigh problem*) driven by an impulsively-start moving flat plate and the *Stokes second problem* driven by the wall harmonic oscillation.

In this section we focus on unsteady unidirectional flows, i.e., unsteady shear flows on the (x, y) plane with $y \geq 0$. The approach can well be applied to unsteady unidirectional flows between two parallel plates or in circular pipe, see Rosenhead (1963), Batchelor (1967), Schlichting and Gersten (2000) and Panton (2013).

J.-Z. Wu et al., *Vortical Flows*, DOI 10.1007/978-3-662-47061-9_4

4.1.1 General Solution in Vorticity Formulation

In a unidirectional shear flow on the semi-infinite plane $y \geq 0$, the velocity and vorticity fields take the form

$$\boldsymbol{u} = (u(y,t), 0, 0), \quad \boldsymbol{\omega} = (0, 0, \omega(y,t)), \quad \omega(y,t) = -\frac{\partial u}{\partial y}. \tag{4.1.1}$$

Thus, the continuity equation is automatically satisfied, and since $\partial u/\partial x = 0$ the momentum equation is linearized:

$$\frac{\partial u}{\partial t} = -P(t) + \nu \frac{\partial^2 u}{\partial y^2}. \tag{4.1.2}$$

Here, $P = \partial p/\partial x$ is the streamwise pressure gradient which as is easily verified has to be independent of both x and y. We assume that the diffusion dies out at $y = \infty$, where $\omega = 0$ and $P(t)$ solely balances the local acceleration du_∞/dt.

Instead of this *momentum formulation*, in what follows we use the *vorticity formulation* as discussed in Sect. 2.6.1. By (4.1.2), the vorticity transport equation is

$$\frac{\partial \omega}{\partial t} = \nu \frac{\partial^2 \omega}{\partial y^2}. \tag{4.1.3a}$$

We use the BVF as the Neumann boundary condition at $y = 0$, obtained by applying (4.1.2) to the wall. Denote the wall velocity by $u_B = b(t)$, by (2.6.8a) the Neumann condition reads

$$\sigma = -\nu \frac{\partial \omega}{\partial y} = \frac{db}{dt} + P(t) \quad \text{at} \quad y = 0, \tag{4.1.3b}$$

which as said before is independent of the Reynolds number. Assume both the fluid and wall are at rest for $t < 0$, and at $t = 0$ there can be a tangent wall motion with speed $b(t)$. The fluid is driven to move and vorticity is generated by either $P(t)$ or db/dt, or both, see (2.6.8a).

The Neumann problem (4.1.3) can be solved by several alternative methods, say Laplace transform or Green's function. The solution is

$$\omega(y,t) = \int_{0-}^{t} \frac{\sigma(t')}{\sqrt{\pi\nu(t-t')}} \exp\left[-\frac{y^2}{4\nu(t-t')}\right] dt', \tag{4.1.4}$$

indicating that the vorticity is diffused along the y-direction without altering the total amount of the vorticity. In this flow only the BVF σ can change the total vorticity, of which the value per unit streamwise width is the difference of velocities at the wall and infinity:

$$\int_0^\infty \omega(y,t) dy = b(t) - u_\infty(t). \tag{4.1.5}$$

Owing to (4.1.3b) and the assumed condition $du/dt|_{y=\infty} = -P$, the rate of change of the total vorticity is

$$\frac{d}{dt}\int_0^\infty \omega(y,t)dy = \sigma(t), \tag{4.1.6}$$

which clearly demonstrates the physical meaning of σ.

At $t = 0$, the flow may evolve either smoothly or, ideally, impulsively. The latter happens if the boundary velocity is a step function such that $db/dt = b_0\delta(t)$ and/or P contains an impulsive part $P = U_0\delta(t)$ to generate a step-function for u_∞. For physical clarity and mathematical convenience, we set $b = b_1 + b_2$ and $u_\infty = u_{1\infty} + u_{2\infty}$, and write the initial and boundary conditions for velocity as

$$b_1(t) = \begin{cases} 0 \text{ for } t < 0, \\ b_0 \text{ for } t \ge 0, \end{cases} \quad u_{1\infty}(t) = \begin{cases} 0 \text{ for } t < 0, \\ U_0 \text{ for } t \ge 0; \end{cases} \tag{4.1.7a}$$

$$b_2(t) = f(t), \quad f(0) = 0, \quad u_{2\infty}(t) = g(t), \quad g(0) = 0, \tag{4.1.7b}$$

where $f(t)$ and $g(t)$ are smooth functions with $f = g = 0$ for $t < 0$. The corresponding BVF is split to

$$\sigma_1(t) = \gamma_0\delta(t), \quad \gamma_0 = b(0) - u_\infty(0), \tag{4.1.8a}$$

$$\sigma_2(t) = \frac{d}{dt}F(t), \quad F(t) = f(t) - g(t), \tag{4.1.8b}$$

where γ_0 is the initial vortex-sheet strength. Evidently, the roles of velocity of wall and at infinity, as well as their time rate, can well be exchanged in vorticity formulation, implying a more unified approach.

Then, substituting (4.1.8) into (4.1.4) yields general formulas for vorticity distribution

$$(\omega, u)(y,t) = (\omega_1, u_1)(y,t) + (\omega_2, u_2)(y,t),$$

where

$$\omega_1(y,t) = \frac{\gamma_0}{\sqrt{\pi\nu t}}\exp\left(-\frac{y^2}{4\nu t}\right), \tag{4.1.9a}$$

$$\omega_2(y,t) = \int_{0+}^t \frac{\sigma_2(t')}{\sqrt{\pi\nu(t-t')}}\exp\left[-\frac{y^2}{4\nu(t-t')}\right]dt'. \tag{4.1.9b}$$

In particular, the boundary vorticity at $y = 0$ reads

$$\omega_B(t) = \frac{\gamma_0}{\sqrt{\pi\nu t}} + \frac{1}{\sqrt{\pi\nu}}\int_0^t \frac{\sigma_2(t')}{\sqrt{t-t'}}dt', \tag{4.1.10}$$

indicating clearly that *boundary vorticity is a temporally accumulated effect of the boundary vorticity flux.*[1]

The velocity distribution $u(y,t)$ follows from a simple integration of (4.1.9) in y:

$$u_1(y,t) = \gamma_0 \mathrm{erfc}\left(\frac{y}{\sqrt{\nu t}}\right), \tag{4.1.11a}$$

$$u_2(x,t) = u_\infty(t) - \int_0^t \sigma_2(t')\mathrm{erf}\left(\frac{y}{\sqrt{\nu(t-t')}}\right) dt', \tag{4.1.11b}$$

where

$$\mathrm{erfc}(\eta) = \frac{2}{\sqrt{\pi}} \int_\eta^\infty e^{-\lambda^2} d\lambda = 1 - \mathrm{erf}(\eta), \quad \mathrm{erf}(\eta) = \frac{2}{\sqrt{\pi}} \int_0^\eta e^{-\lambda^2} d\lambda \tag{4.1.12}$$

are the *complementary error function* and *error function*, respectively. It is worth noticing that, as a general rule, while the velocity solution can also be derived in momentum formulation, the result of two formulations will have different but equivalent expressions and can be derived from each other. Indeed, an integration by parts on (4.1.11) will lead to the solution in momentum formulation given by, e.g. Schlichting and Gersten (2000). But the vorticity formulation often permits clearer physical understanding and neater mathematic expressions.

In what follows we present two types of similarity solutions generated by specific BVF distributions, including classic Stokes first problem (Rayleigh problem) and second problem.

4.1.2 Singular BVF: Stokes First Problem (Rayleigh Problem)

As the first special case of the general solution (4.1.4), assume in (4.1.8) $\sigma_2 = 0$ for all $t > 0$ and $\sigma = \sigma_1 = \gamma_0 \delta(t)$, so that

$$u(y,t) = \gamma_0 \mathrm{erfc}(\eta), \quad \eta = \frac{y}{2\sqrt{\nu t}}, \tag{4.1.13a}$$

$$\omega(y,t) = \frac{\gamma_0}{\sqrt{\pi \nu t}} e^{-\eta^2} = \omega_B e^{-\eta^2}, \tag{4.1.13b}$$

where η is a similarity variable and $\omega_B = \gamma_0/\sqrt{\pi \nu t}$ is the boundary vorticity. This is the classic *Stokes first problem* or *Rayleigh problem*, where the flow is solely driven by an impulsive start of wall speed b or impulsively appeared uniform external velocity u_∞ by a singular pressure gradient $P = -u_\infty \delta(t)$.

[1] In general, the spatial-temporal integral of the nonlinear term $\nabla \times (\boldsymbol{\omega} \times \boldsymbol{u})$ has an additional contribution to the $\boldsymbol{\omega}$-field and $\boldsymbol{\omega}_B$.

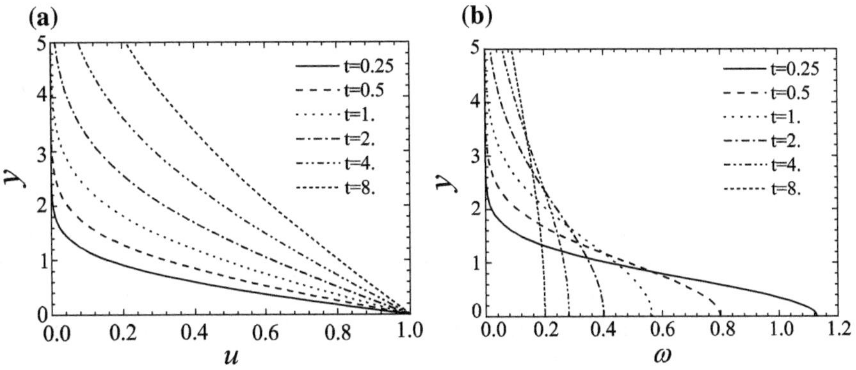

Fig. 4.1 Distributions of **a** velocity and **b** vorticity in Stokes' first problem

In momentum formulation (4.1.13a) is directly obtained by considering the similarity solution of (4.1.2) with $P = 0$. The (u, ω) distributions are shown in Fig. 4.1, where $b_0 = \nu = 1$.

Clearly, the entire vorticity is solely created at the instant $t = 0$ by the singular BVF; once in constant motion state the wall or u_∞ no longer generates any new vorticity. The sudden move of the wall or sudden appearance of pressure gradient generates a tangential discontinuity (vortex sheet) of strength γ_0 with $\omega_B = \infty$ at $t = 0^+$, which is immediately smoothed out by normal diffusion and decays as $1/\sqrt{\pi\nu t}$. This feature holds equally true for the Couette flow between parallel plates or concentric circular cylinders, see Problem 3.14.

The range of diffused vorticity in the fluid is of $O(\sqrt{\nu t})$. For $y = 4\sqrt{\nu t}$ the vorticity is already reduced to about 1 % of its wall value. Hence, for $\nu \ll 1$ the vorticity is effectively confined in a thin layer of thickness

$$\delta \sim \sqrt{\nu t}. \tag{4.1.14}$$

4.1.3 Oscillatory BVF: Stokes Second Problem

In *Stokes' second problem*, it is assumed that the wall makes tangential harmonic oscillation with $P = 0$ again and

$$b(t) = b_0 \sin nt = b_0 \mathrm{Im}\{e^{int}\}, \quad \text{such that} \quad \sigma(t) = n b_0 \cos nt. \tag{4.1.15}$$

In this case (4.1.10) yields

$$\omega_B(t) = b_0 \sqrt{\frac{2n}{\nu}} \left\{ \cos nt \; C(\sqrt{2nt/\pi}) + \sin nt \; S(\sqrt{2nt/\pi}) \right\}, \tag{4.1.16a}$$

where

$$C(x) = \int_0^x \cos\left(\frac{\pi}{2}\tau^2\right) d\tau, \quad S(x) = \int_0^x \sin\left(\frac{\pi}{2}\tau^2\right) d\tau \tag{4.1.16b}$$

are the *Fresnel integrals*.

Our main concern is so-called "stationary" or "steady-state" solution, denoted by subscript st. These adjectives do not mean the flow is time-independent, but rather a fully developed oscillatory flow after a sufficient time, so that the transient process from start (which is significant only during the first cycle of oscillation) is over and the effect of initial condition is negligible. Then (4.1.11b) yields the classic result

$$u_{st}(y,t) = b_0 e^{-ky} \sin(nt - ky), \quad k = \sqrt{\frac{n}{2\nu}}, \tag{4.1.17a}$$

$$\omega_{st}(y,t) = b_0 \sqrt{\frac{n}{\nu}} e^{-ky} \sin\left(nt - ky + \frac{\pi}{4}\right). \tag{4.1.17b}$$

The boundary vorticity ω_B follows from either (4.1.17b) or (4.1.16a) by using the fact $C(\infty) = S(\infty) = 1/2$. Therefore, the wall oscillation produces a *transverse vortical wave* that propagates along the normal and decays exponentially, see Fig. 4.2, where $b_0 = n/\nu = 1$. The effective propagation distance can be measured by the *penetration depth*

$$\delta_n = \frac{1}{k} = \sqrt{\frac{2\nu}{n}}, \tag{4.1.18}$$

and for high frequency and small ν the oscillating fluid is again confined to a thin *Stokes layer*.

If the wall oscillation is driven by a harmonic force, then as the root of the entire flow the force is in phase with the BVF but $\pi/4$ ahead of ω_{stB} (or the skin friction)

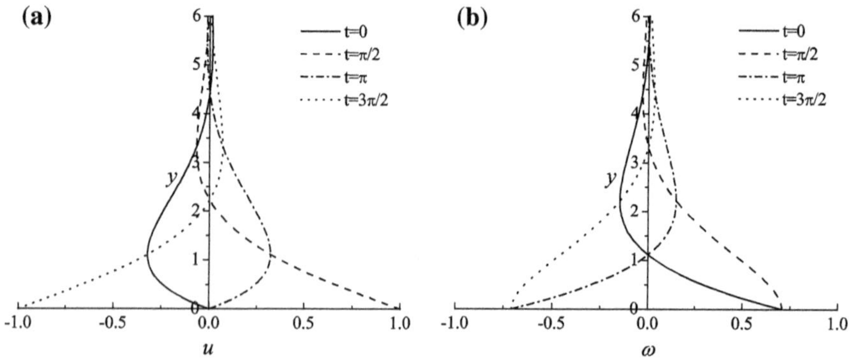

Fig. 4.2 Profiles of **a** velocity and **b** vorticity driven by oscillatory wall with $b = b_0 \sin nt$

since the latter is an accumulated effect of the former. Then, the phase of ω_{st} is in turn $\pi/4$ ahead of u_{st} since the friction between fluid layers drives the motion.[2]

If the wall oscillation (4.1.15) is replaced by a harmonic pressure gradient $P \sim e^{int}$, since the roles of db/dt and $P = -du_\infty/dt$ are exchangeable, the vorticity profile will evidently remain the same (with a sign difference). But the velocity profile is different from (4.1.17a) due to the change of its boundary condition:

$$u_{st}(y, t) = U_0[\sin nt - e^{-ky} \sin(nt - ky)]. \tag{4.1.19}$$

Compared with the original Stokes second problem, this model problem is a more realistic prototype of various compressing-caused vortical wave. It is also a critical part of the water-hammer problem (Sect. 2.3.2). Some more properties of Stokes layer will be studied later.

4.2 Boundary Layers: Formulation and Physics

Among various shear layers at large Reynolds numbers, the most primary structure is thin *boundary layers* generated by and adjacent to solid surfaces. In the whole life of a vortical flow generated by a moving body, the formation and evolution of boundary layer is the process next to the vorticity creation from solid surface (Sect. 3.4.4). In turn, boundary layers and their separation are major precedent structure and process to various vortical structures formed in later evolutional stages. In this section we discuss steady and unsteady boundary layers. Since the boundary-layer theory has become a well-defined branch of fluid dynamics as detailed in many excellent books, such as Rosenhead (1963) and Schlichting and Gersten (2000), we shall not repeat much of the standard materials but pay more attention to underlying vorticity-dynamics mechanisms.

4.2.1 From d'Alembert's Paradox to Prandtl's Theory

The emerge of the boundary layer theory was the greatest revolutionary event in the development of fluid mechanics of twentieth century. The smallness of air and water viscosities had tempted most theoreticians in 19th century to completely ignore them and focus on strictly inviscid flow model. But we have seen in Sect. 2.6.4 that this approach led to predictions in serious confliction with experimental observation, namely a body moving at uniform velocity through an inviscid fluid would experience no force at all or at least no resistive force, although d'Alembert paradox is theoretically rigorous!

[2] This may also be explained in terms of pressure gradient and acceleration, e.g., Batchelor (1967).

It is Ludwig Prandtl's (1904) ingenious conference paper on the boundary layer theory that turned the theoretical fluid dynamics to the right direction. In this paper and a series of subsequent studies, Prandtl and his students clarified:

(1) No matter how small it could be, the viscosity plays a crucial role in a thin *boundary layer* adjacent to the solid wall, in which the flow must be highly rotational and causes a skin friction drag;

(2) If a body is *streamlined*, i.e., the boundary layer remains attached on its surface, outside the layer the potential-flow model can well be applied;

(3) The boundary layer may separate from a smooth body surface under an adverse pressure gradient and form *free shear layer* that in turn rolls into concentrated vortices and can significantly alter the global flow behavior;

(4) It is the vorticity in these vortical structures that is responsible (and solely responsible if the flow is incompressible) for the entire force exerted to the body by the fluid, including both lift and drag.

These physical findings of Prandtl formed the very basis of modern fluid dynamics and aerodynamics at large Reynolds numbers. The above item (1) implies that all solid bodies moving in air or water must be enclosed by boundary layers as their "clothes", and there is no "naked body" in reality. Item (2) implies that, for a streamlined body, whenever the thickness of its "cloth" is negligible, the flow will be very close to the potential flow. But item (3) warns that big difference between real flow and potential flow will appear whenever boundary layer separates from the body surface before reaching the trailing edge—an important issue to be discussed in Chap. 8. Item (4) will be explored in Chap. 9.

Remarkably, the above discoveries are not yet the full story of Prandtl's 1904 conference paper. He also invented a new powerful mathematical method to find the solution of the boundary layer flow and match it with the outer potential flow as a whole. This method has later been systematically formulated, known as the *matched asymptotic expansion*. It inspired quite a number of *singular perturbation methods*, which form one of the major achievements of twentieth-century applied mathematics.

Although Prandtl's method can well be applied to fully nonlinear field equations, its essential idea can be illustrated by the following simple example of linear ordinary differential equation [Lighthill (1995), whose presentation is followed here]:

$$\epsilon y'' + y' + ky = 0, \quad 0 < x < 1, \tag{4.2.1a}$$

$$y = 0 \text{ at } x = 0, \quad y = 1 \text{ at } x = 1, \tag{4.2.1b}$$

where $\epsilon \ll 1$ is an extremely small parameter like the viscosity of air or water. The exact solution of this problem can be easily deduced (Problem 4.1), but we now seek its approximate solution. According to regular perturbation method, we would first consider the "inviscid" solution with $\epsilon = 0$ and then improve it by a process of successive approximation, say an expansion in powers of ϵ. Then we obtain

$$y' + ky = 0, \tag{4.2.2}$$

of which a solution satisfying $y(1) = 1$ is

$$y = e^{k(1-x)} \quad \text{with} \quad y(0) = e^k, \tag{4.2.3}$$

which is a good approximation in a large part of the domain where $\epsilon y'' \ll 1$, but cannot satisfy the required $y(0) = 0$ (like the no-slip condition at a wall). Since (4.2.2) is only a first-order equation, the "extra" condition at $x = 0$ makes it impossible to solve (4.2.1) by any regular perturbation even with higher powers of ϵ. The fact is, as can be justified by exact solution of (4.2.1), that in a narrow region near $x = 0$ the solution must have very rapid change from 0 to e^k, with very large y' and enormously large y'', so that $\epsilon y'' + y'$ is much larger than ky. Thus, in the rapid-change region (4.2.1a) can be approximated by

$$\epsilon y'' + y' = 0, \tag{4.2.4}$$

whose solution is

$$y = c(1 - e^{-x/\epsilon}) \quad \text{with} \quad y(0) = 0, \tag{4.2.5}$$

where the constant c is the asymptotic value of y for $x/\epsilon \gg 1$. As shown in Fig. 4.3, this rapid-change region is indeed very narrow and soon disappears as y tends to c. Therefore, if we choose $c = e^k$ to make the value of (4.2.5) just outside the rapid-change region coincide with the limit of (4.2.3) as $x \to 0$, then as the figure shows the two approximate solutions are matched. This selection of c is the so-called *limiting match rule*.

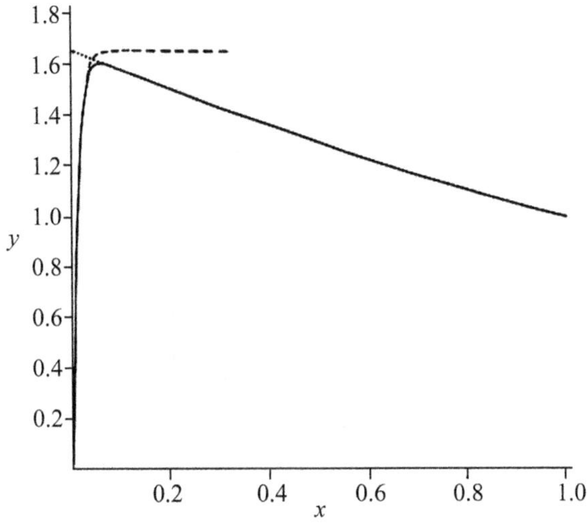

Fig. 4.3 Approximate solution of (4.2.1). *Solid line* exact solution. *Dotted line* "outer" solution (4.2.3). *Broken line* "Boundary-layer" solution (4.2.5). From Lighthill (1995)

4.2.2 From Rayleigh Problem to Boundary Layer Equations

The incompressible unidirectional shear flows with exact analytical solutions of Sect. 4.1 represent the simplest large-Re unsteady shear flows. In particular, in the Rayleigh problem there is a vorticity layer adjacent to the wall of uniform thickness $\delta \sim \sqrt{\nu t}$. After vorticity is impulsively created at $t = 0$, two processes take place at $t > 0$. One is the normal diffusion of ω that brings the vorticity to a distance δ above the wall. The other is the streamwise advection of ω by the flow with a velocity $u \sim U$ to a downstream distance about $x \sim Ut$, with U being a characteristic velocity at the outer edge of the layer. This yields $\delta \sim \sqrt{\nu x/U}$ so that the ratio of the length scales of diffusion and advection is

$$\frac{\delta}{x} \sim \sqrt{\frac{\nu}{Ux}} = Re_x^{-1/2}, \tag{4.2.6}$$

where Re_x ($\gg 1$) is the local Reynolds number at streamwise location x. If we introduce a transformation $x = Ut$ so that $t = x/U$, then in vorticity formulation we obtain the steady-flow counterpart of (4.1.3a):

$$U\frac{\partial \omega}{\partial x} = \nu \frac{\partial^2 \omega}{\partial y^2}, \quad y \in [0, \infty). \tag{4.2.7}$$

However, this equation is of course only a crude approximation of real boundary-layer flow. An obvious drawback is that, owing to the simplified assumption $dx/dt = U$ instead of $dx/dt = u(y) \leq U$, (4.2.7) must overestimate the vorticity (and hence also the velocity) inside the thin layer. After all, for an infinite flat plate the origin of x is indeterminate. Therefore, a natural step forward is to consider a shear flow over a *semi-infinite* flat plate aligned to the incoming flow at large Re, of which the sharp leading edge defines a natural distance x along the plate. The thin shear layer adjacent to the wall and starting from the leading edge is a steady *boundary layer* sketched in Fig. 4.4. But then mathematical complexity appears immediately. Unlike the thin parallel shear layer in the Rayleigh problem, since now δ increases as $\sqrt{x}$, there must be fluid particles continuously entering the boundary layer at different x as shown in Fig. 4.4. In so doing the particles' velocity decreases from U, so $u_x < 0$ and hence $v_y > 0$ (we often use subscripts to denote partial derivatives), indicating that there must be $v \neq 0$ and (u, v) must depend on x as well as y and t.

In order to estimate the order of magnitude of various quantities more carefully, consider two-dimensional shear flow with $\rho = 1$. Let L be the characteristic streamwise length scale and δ the thickness of the shear layer, then the analysis of Sect. 4.1 has revealed the basic scalings[3]

[3]It is of interest to compare this estimate for shear layer with that for a shock layer of thickness δ, where the advection and diffusion *both along the streamwise direction* are balanced, so there is $\rho u^2/\delta \sim \mu u/\delta^2$, yielding $\delta/L \sim Re_L^{-1}$, much smaller than $Re^{-1/2}$.

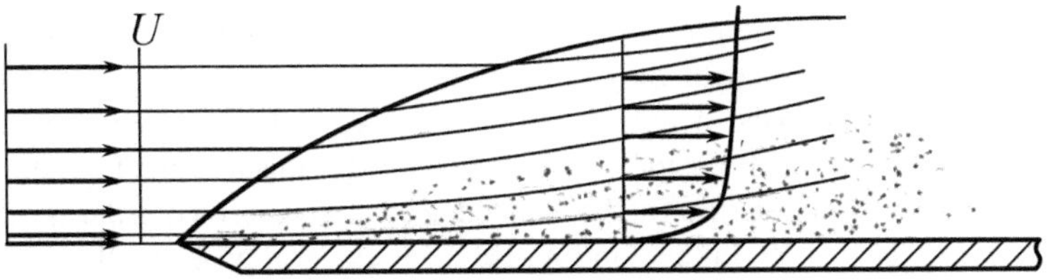

Fig. 4.4 Boundary layer over a semi-infinite flat plate aligned to the incoming flow

$$|\partial/\partial x| = O(1), \quad |\partial/\partial y| = O(1/\delta), \quad \delta/L = Re^{-1/2} \ll 1. \tag{4.2.8a}$$

Hence, in the incompressible continuity and momentum equations (2.6.2) there is

$$u \sim u\frac{\partial u}{\partial x} \sim v\frac{\partial u}{\partial y} \sim 1, \quad \frac{\partial^2 u}{\partial y^2} \sim \delta^{-2}, \quad v \sim \frac{\partial v}{\partial t} \sim u\frac{\partial v}{\partial x} \sim \frac{\partial p}{\partial y} \sim \delta. \tag{4.2.8b}$$

In particular,

$$\omega = -\frac{\partial u}{\partial y} + O(\delta) = O(\delta^{-1}), \quad p_B = p_e + O(\delta^2), \tag{4.2.8c}$$

where B and e denote quantities on the wall and the outer edge of the boundary layer, respectively. Therefore, although physically one expects that at large Re_x the flow is still nearly parallel or *quasi-parallel*, (4.2.8b) shows that once $v \neq 0$ the nonlinear advection cannot be ignored, with uu_x and vu_y being of the same order.

Let us continue with two-dimensional unsteady flow with $\rho = 1$. The estimate (4.2.8c) indicates that *pressure variation along the normal is negligible*, and hence at any $y \in [0, \delta)$ we can drop the y-component of (2.6.2b) and simply replace p by p_e, which is governed by the Bernoulli equation at the outer edge of the boundary layer:

$$\frac{\partial \phi}{\partial t} + p_e(x,t) + \frac{1}{2}U_e^2(x,t) = p_\infty + \frac{1}{2}U_\infty^2. \tag{4.2.9}$$

Consequently, by (4.2.8b) what remains in (2.6.2) for the boundary-layer flow is

$$\frac{\partial u}{\partial t} + u\frac{\partial u}{\partial x} + v\frac{\partial u}{\partial y} = \frac{\partial U_e}{\partial t} + U_e\frac{\partial U_e}{\partial x} + \nu\frac{\partial^2 u}{\partial y^2}, \tag{4.2.10a}$$

$$\frac{\partial u}{\partial x} + \frac{\partial v}{\partial y} = 0, \tag{4.2.10b}$$

along with boundary conditions

$$y = 0: \quad u = v = 0; \quad y/\delta \to \infty: \quad u = U_e. \tag{4.2.10c}$$

These are the famous incompressible *boundary layer equations* for flow over a semi-infinite flat plate, first discovered by Prandtl (1904).

Note that in (4.2.10c) the location δ of the outer edge of the boundary layer itself is unknown before really solving the equations. To bypass this trouble, Prandtl (1904) points out that the flow condition at the outer edge can be prescribed in the asymptotic limit $Re \to \infty$ or $\delta \to 0$. In this limiting case, for external potential flow the outer edge of the boundary layer is seen at $y = \delta \to 0$, completely neglecting the layer's thickness; but inside the layer we introduce a local scale $Y = y/\delta = O(1)$ such that the outer edge retreats to $Y = y/\delta \to \infty$ (the normal velocity is accordingly scaled to $V = v/\delta = O(1)$). This is the *limiting match rule*, which now requires

$$\lim_{Y\to\infty} u_{\text{in}}(x, Y) = \lim_{y\to 0} u_e(x, y). \tag{4.2.11}$$

Return to vorticity formulation, (4.2.8c) indicates that the vorticity equation is simply the y-derivative of (4.2.10a):

$$\frac{\partial \omega}{\partial t} + u\frac{\partial \omega}{\partial x} + v\frac{\partial \omega}{\partial y} = \nu\frac{\partial^2 \omega}{\partial y^2}, \tag{4.2.12a}$$

where (u, v) are related by (4.2.10b). Except initial condition, the boundary conditions are

$$\begin{cases} y = 0: & -\sigma(x,t) \equiv \nu\frac{\partial \omega}{\partial y} = \frac{\partial U_e}{\partial t} + U_e\frac{\partial U_e}{\partial x}, \\ y/\delta \to \infty: & \omega \to 0. \end{cases} \tag{4.2.12b}$$

Once again, the inhomogeneous term in (4.2.10a) is shifted to the Neumann boundary condition at the wall in (4.2.12b), indicating that no BVF, no boundary layer.

Among various ways, (4.2.12) can be solved by an iteration procedure. We write

$$u_{n-1}\frac{\partial \omega_n}{\partial x} + v_{n-1}\frac{\partial \omega_n}{\partial y} = \nu\frac{\partial^2 \omega_n}{\partial y^2}, \quad n = 1, 2, \ldots \tag{4.2.13a}$$

under constraint (4.2.10b) and boundary conditions

$$\begin{cases} y/\delta \to \infty: & \omega_n = 0; \\ y = 0: & \nu\frac{\partial \omega_n}{\partial y} = -\frac{\partial p_e}{\partial x}, \quad n = 1, 2, \ldots. \end{cases} \tag{4.2.13b}$$

For $n = 0$, we assume a uniform flow $(U, 0)$. For $n = 1$ the equation for ω_1 is just (4.2.7). We set $(u_0, v_0) = (1, 0)$. Then ω_1 satisfies (4.2.7) and the general solution can be directly inferred from (4.1.4):

$$\omega_1(x, y) = \int_0^x \frac{\sigma(x')}{\sqrt{\pi\nu U(x - x')}} \exp\left[-\frac{Uy^2}{4\nu(x - x')}\right] dx'. \tag{4.2.14}$$

In principle, the iteration procedure could be carried out to $n > 1$ by numerical integrations. Things will become much simpler if similarity variable exists, which we discuss below.

4.2.3 Blasius Boundary Layers

We now illustrate the above iteration procedure by considering a self-similarity solution of boundary-layer equations, with singular $\sigma = \delta(x)$. Unless otherwise stated, below we work on dimensionless variables scaled by characteristic velocity U and length L, and ν, which define a global Reynolds number $Re = UL/\nu \equiv \epsilon^{-1}$ with $\epsilon \ll 1$.

Assumes $\partial_x p_e = 0$ and hence $U_e = U = \text{const}$. Then (4.2.10) is reduced to

$$u\frac{\partial u}{\partial x} + v\frac{\partial u}{\partial y} = \epsilon\frac{\partial^2 u}{\partial y^2}, \tag{4.2.15a}$$

$$\frac{\partial u}{\partial x} + \frac{\partial v}{\partial y} = 0. \tag{4.2.15b}$$

This problem was first solved by Blasius (1908) for the stream function ψ by similarity variable $\eta = y/\sqrt{\epsilon x}$ using a series expansion. But the approach has some limitation. Mathematically, no single series converges over the full range of y. Physically, since $\sigma = 0$, in momentum formulation the root of vorticity in the boundary layer is unclear. To remove these shortcomings, we now revisit this problem by using the forgoing iteration procedure.

First, by introducing a similarity variable

$$\eta = \frac{y}{2\sqrt{\epsilon x}}, \tag{4.2.16}$$

we can cast (4.2.10a) to an ordinary differential equation

$$\frac{d^2u}{d\eta^2} + 2(\eta u - w)\frac{du}{d\eta} = 0, \quad w(\eta) \equiv \sqrt{\frac{x}{\nu}}v, \quad \eta \in [0, \infty), \tag{4.2.17}$$

by which we construct an iteration procedure

$$\frac{d^2u_n}{d\eta^2} + 2(\eta u_{n-1} - w_{n-1})\frac{du_n}{d\eta} = 0, \quad n = 1, 2, 3, \dots . \tag{4.2.18}$$

The error of each equation is of $O(u_n - u_{n-1})$ and $O(w_n - w_{n-1})$.

Next, let the BVF be a delta function $\sigma = \delta(x)$. Hence, like (4.1.13) and (4.2.14) yields

$$\omega_1(x, y) = \frac{1}{\sqrt{\pi \epsilon x}} e^{-\eta^2}, \quad u_1(x, y) = \mathrm{erf}(\eta), \quad v_1(x, y) = \sqrt{\frac{\epsilon}{\pi}}(1 - e^{-\eta^2}). \tag{4.2.19}$$

Therefore, *all vorticity is created right at the leading edge* and then simply diffused into the fluid and advected downstream.

Having known (4.2.19), we may switch to momentum formulation, in which the solutions of (4.2.18) for $n \geq 2$ amount to a sequence of iterative numerical integrals. Denote

$$F_{n-1}(\eta) \equiv -2 \int_0^\eta (\xi u_{n-1} - w_{n-1}) d\xi, \tag{4.2.20a}$$

then there is

$$u_n = (C_{n-1})^{-1} \int_0^\eta e^{F_{n-1}(\xi)} d\xi, \quad C_{n-1} \equiv \int_0^\infty e^{F_{n-1}(\xi)} d\xi, \tag{4.2.20b}$$

$$w_n = (C_{n-1})^{-1} \int_0^\eta e^{F_{n-1}(\xi)} \xi d\xi, \tag{4.2.20c}$$

$$\omega_n = -(C_{n-1})^{-1} \frac{1}{2\sqrt{\epsilon x}} e^{F_{n-1}(\eta)}. \tag{4.2.20d}$$

The profiles of (u_n, w_n, ω_n) for $n = 1, 2, 3$ are shown in Fig. 4.5. Also shown is the exact (numerical) solution obtained by iterative integrations up to $n = 7$.[4]

4.2.4 Further Issues

1. Skin friction. The key information obtained from boundary-layer solutions is the vorticity ω at $y = 0$ or skin friction coefficient $c_f = 2\tau_w/\rho U^2 = -2\epsilon\omega/\rho U^2$, and the total friction drag coefficient C_f over the plate of length L. For the present case these are

$$c_f(x) = \frac{0.664}{\sqrt{Re_x}}, \tag{4.2.21a}$$

$$C_f = \frac{1}{L}\int_0^L c_f dx = \frac{1.328}{\sqrt{Re_L}}. \tag{4.2.21b}$$

Actually, boundary-layer approximation holds only for $Re_x = Ux/\nu \gg 1$ and is not applicable to the leading edge. In fact, instead of $\sigma = \delta(x)$ as we have set, the

[4] Mao and Xuan (2010), private communications.

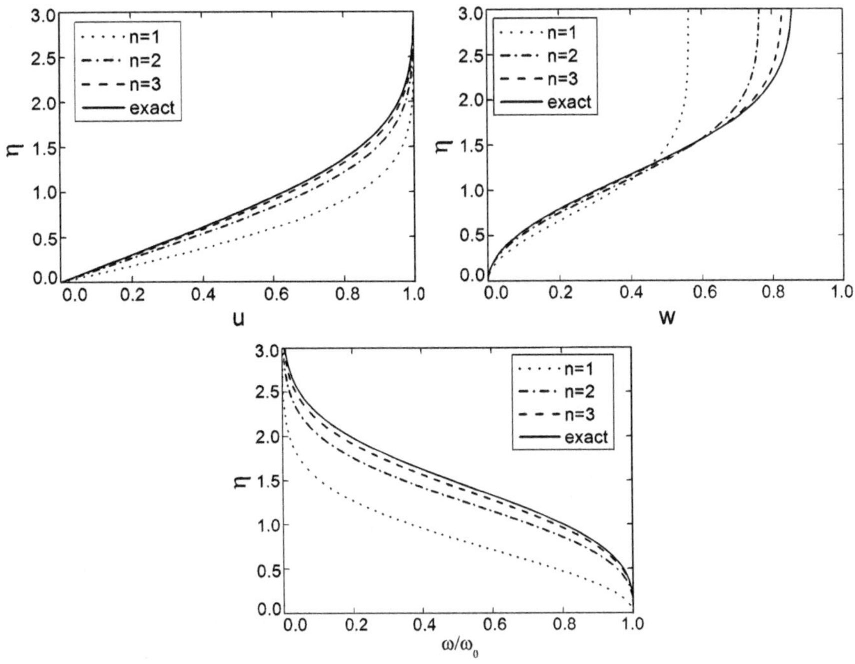

Fig. 4.5 Iterative solution sequence of (4.2.18): Variation of u, $v\sqrt{x/\nu}$, and ω/ω_0

flow up to a distance x where $Re_x \sim 1$ has to be described by the full Navier-Stokes equation, and it has been found that (4.2.21b) should be revised to

$$C_f = 1.328 Re_L^{-1/2} + 2.326 Re_L^{-1} + O(Re_L^{-3/2}). \tag{4.2.22}$$

2. Boundary-layer thickness. At the outer edge of boundary layer, although the limiting match rule (4.2.11) requires u matches u_e, v does not match $v_e = 0$ there. Rather, from (4.2.10b) and (4.2.11) it generally follows that

$$\frac{1}{\sqrt{\epsilon}}(v - V_e)\Big|_{Y\to\infty} = \frac{d}{dx}\int_0^\infty (U_e - u)dY, \tag{4.2.23}$$

indicating a *displacement* effect to the outer flow associated with the growing of the boundary-layer thickness as x. For the present problem there is

$$v(x,\delta) = \sqrt{\frac{\epsilon}{x}}w(\infty) = 0.860\sqrt{\frac{\epsilon}{x}}, \tag{4.2.24}$$

see Fig. 4.5.

Conventionally, the boundary layer thickness is quantified as the y location where $u = 0.99u_e$ since away for this location the flow can well be viewed as irrotational.

In addition to this thickness, two more thicknesses with clear dynamic implication are important in applications. The slowdown of the flow inside the boundary layer causes deficiency of mass flux and momentum flux compared to external inviscid flow (in dimensional form):

$$\int_0^\delta \rho(u_e - u)dy \quad \text{and} \quad \int_0^\delta \rho u(u_e - u)dy.$$

Dividing these expressions by ρu_e and ρu_e^2, respectively, yields two normal length scales to measure these decrements:

$$\delta^* \equiv \int_0^\delta \left(1 - \frac{u}{u_e}\right) dy, \quad \theta \equiv \int_0^\delta \frac{u}{u_e}\left(1 - \frac{u}{u_e}\right) dy, \tag{4.2.25}$$

which are called the *displacement thickness* and *momentum thickness*, respectively. δ^* is just the distance from the wall through which the outer inviscid flow is displaced due to the boundary layer. It is easily found that the three thicknesses, δ, δ^*, and θ are proportional to $x/\sqrt{Re_x}$, with coefficients varying as $\partial p/\partial x$; see common textbooks.

3. Boundary layer on curved wall. The boundary-layer theory can be extended to curved wall by using a *boundary-layer coordinate system*, where (x, y) are defined as the arc length along the wall and normal distance from the wall, respectively. Since the layer's thickness is small, as long as the wall curvature radius $R(x)$ remains of $O(1) \gg \delta$, the streamwise boundary-layer equation (4.2.10a) remains true. Meanwhile, the curvature of all coordinate lines $y=$constant across δ can be approximated by $1/R(x)$, and the y-component of the momentum equation (2.3.6b) is simplified to

$$\frac{1}{\rho}\frac{\partial p}{\partial y} = \frac{u^2}{R(x)}, \tag{4.2.26}$$

indicating that for balancing the centrifugal effect of the flow around a curved wall a normal pressure gradient is necessary. The y-integral of (4.2.26), i.e., the difference between $\partial p_e/\partial x$ at the outer edge of the layer and $\partial p/\partial x$ inside the layer, is of $O(\delta)$ and still negligible.

4. Boundary layer in three-dimensional flow. In a three-dimensional boundary layer there are two coordinates (x, y) with velocity components (u, v) tangent to the wall, and one normal coordinate z with velocity component w. The choice of the orientation of (x, y) depends on convenience. In addition, two physical effects make the problem more involved than two-dimensional flows. First, the streamlines of the outer flow $\boldsymbol{u}_e(x, y, t)$ are no longer parallel but may converge or diverge. Second, these streamlines have curvature. It is this latter effect that gives rise to some new flow pattern. To understand this, consider a steady flow and let x be aligned to the outer-flow streamline direction so that $\boldsymbol{U}_e = (U_e, 0)$. Recall that in terms of the intrinsic streamline coordinates the two-dimensional and steady Navier-Stokes equation takes the form of (2.3.6). We now apply them on a *stream surface* S at the outer edge of

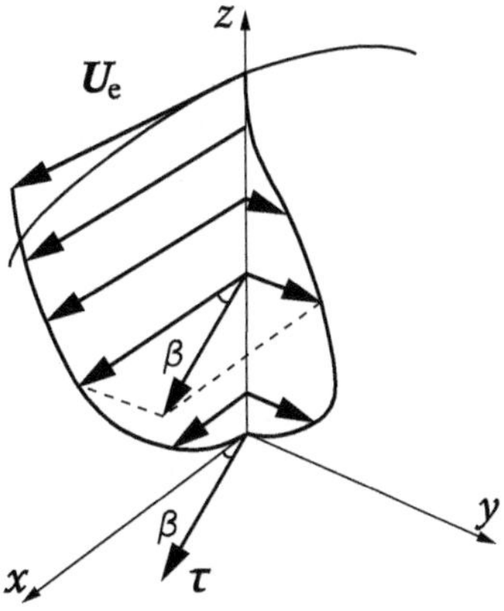

Fig. 4.6 Secondary flow inside a three-dimensional boundary layer, caused by the curvature of outer streamlines

the boundary layer by neglecting the viscous terms. In dimensionless form these equations now read

$$U_e \frac{\partial U_e}{\partial x} = -\frac{1}{\rho}\frac{\partial p}{\partial x}, \quad \kappa U_e^2 = -\frac{1}{\rho}\frac{\partial p}{\partial y}, \tag{4.2.27}$$

where κ is the curvature of the streamlines on S. If $\kappa = 0$, there will be no pressure gradient in the y-direction at any z inside the boundary layer, and hence the component momentum equation in that direction is homogeneous in v. Due to the homogeneous boundary conditions $v_e = v_B = 0$, there can only be trivial solution $v = 0$ for all z. But the curvature of the outer streamlines causes a nonzero lateral pressure gradient $\partial p/\partial y$, which can generate a lateral flow in the boundary layer that vanishes at the outer edge and bottom of the layer, see Fig. 4.6. This phenomenon is called the *secondary flow*, whose appearance is often associated with that of normal vorticity component ω_3 inside the boundary (Wu et al. 2006).

Both parallel shear-flow and boundary layer enjoy an important common feature: to the leading-order approximation, the compressing-shearing coupling at boundary is degenerated from "two-way interaction" such as (2.6.8a) to "one-way interaction", by which *the pressure field dominates the development of the vorticity inside boundary layer, but is not affected by the appearance of the layer.* This feature along with the smallness of the displacement thickness δ^* enables one to compute the outer potential flow over a streamlined body first, i.e., assuming $\delta^* = 0$, and then embed a boundary layer to the body surface. In this way, the boundary layer theory has been successfully employed to innumerable engineering flow problems for many decades in pre-computer era. Today, while it is more convenient and efficient to directly solve the full Navier-Stokes equations numerically rather than finding the potential-flow solution first and then embedding a boundary layer, the boundary layer theory and its major solutions still shed brilliant light onto the physical picture of large-Re flow over a streamlined body surface as the most basic flow pattern of modern fluid dynamics.

4.2.5 Vorticity Dynamics in Boundary Layer

A boundary layer is an *attached vortex layer* characterized by the very strong peaks of vorticity and its normal derivatives. Owing to the simplification from two-way to one-way interaction, the boundary-layer approximation provides an easy platform for exercising the main concepts of vorticity dynamics. We have seen the great convenience of this simplification in the use of vorticity formulation to solve boundary-layer equations and reveal the key role of BVF. We now turn to other issues of vorticity dynamics in the boundary-layer flow.

First, we may estimate the order of magnitude of boundary enstrophy flux η (not to be confused with the similarity variable used in preceding subsections) and enstrophy dissipation rate defined in Sect. 3.4.3:

$$\eta = \omega_B \sigma = O(Re^{1/2}), \quad \Phi_\omega = \epsilon \left(\frac{\partial \omega}{\partial y} \right)^2 = O(Re), \tag{4.2.28}$$

both are very strong. In comparison, the dissipation rate of the kinetic energy is $\Phi \sim \epsilon \omega^2 = O(1)$ only.

Another important quantity in vorticity dynamics, the Lamb vector $\boldsymbol{l} \equiv \boldsymbol{\omega} \times \boldsymbol{u}$, also has strong peak behavior and deserves an analysis. Its formation is associated with that of the vorticity, but it is zero on a stationary boundary due to the no-slip condition. Only after the vorticity is produced from the boundary and diffused into the fluid with local velocity $\boldsymbol{u} \neq \boldsymbol{0}$, can the flow gain a Lamb vector in the near-wall sublayer where the viscous force still has strong influence. To see the relevant mechanisms, we make a Taylor expansion of the near-wall Lamb vector as follows, where all normal derivatives are represented by tangent ones on the boundary (Yang et al. 2007). In a three-dimensional flow and in terms of aforementioned boundary-layer coordinates (x, y, z) with z varying along the normal, the Lamb vector at a small normal distance h from the wall can be written as

$$\boldsymbol{l} = \boldsymbol{l}_B + \left. \frac{\partial \boldsymbol{l}}{\partial z} \right|_B h + \frac{1}{2} \left. \frac{\partial^2 \boldsymbol{l}}{\partial z^2} \right|_B h^2 + O(h^3), \tag{4.2.29}$$

where subscript B denotes taking the value on the boundary. Then one finds that

$$\boldsymbol{l} = -\boldsymbol{e}_z \omega_B^2 h - \frac{1}{\nu} \sigma_n \boldsymbol{\omega}_B h^2 + \frac{3}{2\nu} (\boldsymbol{\omega}_B \cdot \boldsymbol{\sigma}) \boldsymbol{e}_z h^2 + O(h^3). \tag{4.2.30}$$

Here, $\sigma_n = -\nu \nabla_\pi \cdot \boldsymbol{\omega}_B$ is the normal component of BVF with the subscript π denoting the tangent components of any vector, and $\boldsymbol{\omega}_B \cdot \boldsymbol{\sigma}$ is boundary enstrophy flux. The normal and tangent components of $\boldsymbol{l}$ near the boundary are

$$l_z = -\omega_B^2 h + \frac{3}{2\nu} (\boldsymbol{\omega}_B \cdot \boldsymbol{\sigma}) h^2 + O(h^3), \tag{4.2.31a}$$

$$\boldsymbol{l}_\pi = (\nabla_\pi \cdot \boldsymbol{\omega}_B) \boldsymbol{\omega}_B h^2 + O(h^3). \tag{4.2.31b}$$

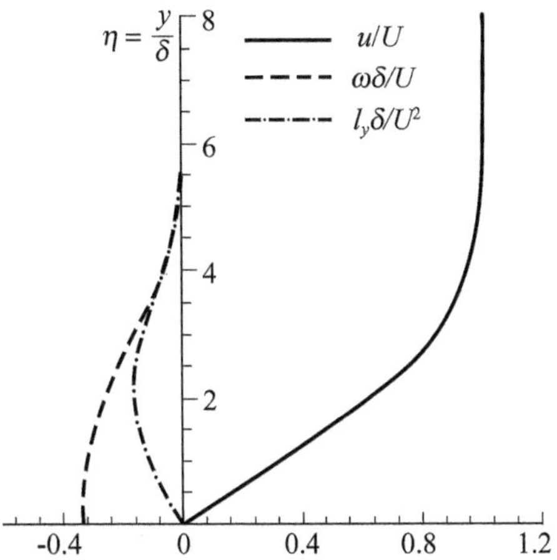

Fig. 4.7 Profiles of the velocity, vorticity, and normal component of the Lamb vector in the Blasius boundary layer

Thus, in a region sufficiently close to the boundary the Lamb vector is solely determined by the boundary vorticity and its flux. The normal Lamb vector is formed by the enstrophy ω^2 at the first order and its flux divided by ν at the second order, while the tangent Lamb vector appears at the second order if both boundary vorticity and normal BVF are nonzero. Since the boundary layer has thickness $\delta = O(Re^{-1/2})$, the vorticity and BVF are of $O(Re^{1/2})$ and $O(1)$, respectively. A term-by-term magnitude estimate in (4.2.30) shows that for the series to converge there should be $h \sim \delta$. Therefore, because in a boundary layer tangent derivatives are usually smaller than normal ones by a factor of $O(Re^{-1/2})$, so is $|\mathbf{l}_\pi| \ll |l_z|$.

Figure 4.7 shows the profiles of vorticity and Lamb vector in a Blasius boundary layer on (x, y)-plane. The Lamb vector is strong only in a thin off-wall layer, with a maximum l_y at $\eta \simeq 2.16$. It will then be advected downstream.

A realistic Lamb-vector distribution in a turbulent boundary layer over an elliptic wing, computed by Marongiu et al. (2013) using Reynolds-averaged Navier-Stokes equations, is shown in Fig. 4.8 that exemplifies the above estimate. In this three-dimensional flow the coordinates (x, y, z) are along streamwise, spanwide, and vertical directions, respectively. The peak values of $-l_z$ and $-l_x$ are located in the *buffer layer* and of $O(Re^{1/2})$ and $O(1)$, respectively.[5] Downstream the trailing edge these components drops quickly to $O(10^{-4})$ and $O(10^{-3})$, respectively.

We now look at the relation of the Lamb vector and the velocity field in the whole boundary layer. Consider two-dimensional steady flow as example. Equation (2.6.3) yields a natural Helmholtz decomposition of the Lamb vector:

$$\boldsymbol{l} = -\nabla H - \epsilon \nabla \times \boldsymbol{\omega}, \quad H = p + \frac{1}{2}q^2. \tag{4.2.32}$$

[5]See Sect. 11.1.2 for the concept of buffer layer in turbulent boundary layers.

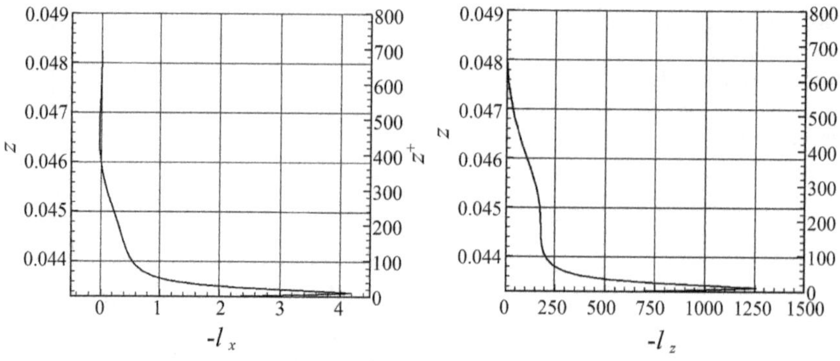

Fig. 4.8 Profiles of Lamb-vector components $-l_x(z)$ and $-l_z(z)$ on central sectional plane at $y = 0$ and $x = 0.191$ of an elliptic wing of aspect ratio $AR = 7$. The sectional airfoil is NACA-0012 and the coordinates are scaled by root-chord length c_r. $Re = 3 \times 10^6$, $M_\infty = 0.01$, $\alpha = 4° \cdot \boldsymbol{l} \equiv \boldsymbol{\omega} \times \boldsymbol{u}$ is scaled by a reference value $l_{\text{ref}} = U^2 S_w/(2c_r^3)$. The coordinate on the right is in turbulent wall unit z^+ (for its definition see Sect. 11.1.2). From Marongiu et al. (2013)

Since

$$\boldsymbol{l} = -\boldsymbol{e}_x \omega v + \boldsymbol{e}_y \omega u, \quad -\nabla \times \boldsymbol{\omega} = -\boldsymbol{e}_x \frac{\partial \omega}{\partial y} + \boldsymbol{e}_y \frac{\partial \omega}{\partial x},$$

we have

$$l_x = -\omega v = \frac{1}{2}\frac{\partial}{\partial x}(U_e^2 - u^2) - \epsilon \frac{\partial \omega}{\partial y}, \tag{4.2.33a}$$

$$l_y = \omega u = -\frac{1}{2}\frac{\partial u^2}{\partial y} + O(\delta). \tag{4.2.33b}$$

While (4.2.33a) is nothing but the boundary layer equation (4.2.15a), (4.2.33b) is merely a trivial identity. However, it is this normal component of the Lamb vector that has strong peak of $O(\delta^{-1})$ compared to $l_x = O(1)$, yielding

$$-\int_0^\delta l_y dy = \frac{1}{2}U_e^2 = O(1), \quad -\int_0^\delta l_x dy = O(\delta). \tag{4.2.34}$$

Note that Prandtl's formula (3.3.7) suggests, and as will be further shown in Chap. 9, that the strong l_y-integral over the boundary layers of an airfoil in steady flow contributes to the entire lift to the airfoil. In contrast, the skin-friction is $\epsilon \omega_B = O(\delta)$, much smaller than the lift.

The Lamb vector also controls the viscous loss of the total head H inside the boundary layer. By (4.2.32) we have

$$\boldsymbol{u} \cdot \nabla H = -\epsilon \boldsymbol{u} \cdot (\nabla \times \boldsymbol{\omega}) = -\epsilon[\nabla \cdot (\boldsymbol{\omega} \times \boldsymbol{u}) + \omega^2]. \tag{4.2.35}$$

Thus, let $\boldsymbol{u} = q\boldsymbol{t}$ and $d\boldsymbol{x} = \boldsymbol{t}ds$ be the arc element along any streamline entering the boundary layer from upstream incoming flow, there is

$$H_\infty - H(\boldsymbol{x}) \equiv \Delta H = \epsilon \int_{-\infty}^{P} \frac{1}{q}[\nabla \cdot (\boldsymbol{\omega} \times \boldsymbol{u}) + \omega^2] ds, \tag{4.2.36}$$

where ΔH denotes the viscous loss of the total head at point $P(\boldsymbol{x})$ as fluid particles move into the boundary layer along that streamline.

4.3 High-Frequency Oscillatory Boundary Layer

This section considers some aspects of unsteady boundary layer. In laminar regime without considering instability waves (Chap. 10), unsteady shear flows mainly fall into two basic types: transient flow from one steady state to another, and time-periodic flow, with the Rayleigh flow and Stokes layer treated in Sect. 4.1 being their simplest prototypes, respectively. While in Sect. 4.2 we extended the Rayleigh (Stokes first) problem to boundary layers, we now do the similar extension, from Stokes layer to oscillating boundary layers.

Consider two-dimensional and incompressible flow. We use the vorticity formulation to analyze a general two-dimensional and unsteady large-Re flow above a flat wall at $y = 0$. In Sect. 4.1 we have seen that this formulation is equivalent to the velocity formulation, as long as at $t = 0$ the velocity satisfies the adherence condition. The exact vorticity transport equation reads

$$\omega_t + u\omega_x + v\omega_y = \nu(\omega_{xx} + \omega_{yy}), \tag{4.3.1a}$$

where subscripts denote partial derivatives. Assume the wall is *flexible*, being able to perform arbitrary periodic motion and/or deformation $b(x, t)$ in its own plane. Then the boundary conditions for solving (4.3.1a) are

$$-\nu\omega_y = b_t + bb_x + \frac{1}{\rho}p_x \quad \text{at } y = 0, \quad \omega = 0 \quad \text{at } y = \infty. \tag{4.3.1b}$$

We split the flow quantities including the wall velocity $b(x, t)$ into mean and fluctuating parts, and focus on the latter. The fluctuating vorticity equation and its boundary condition are:

$$\begin{aligned} &\underline{\omega'_t} + (\overline{u}\omega'_x + \overline{v}\omega'_y) + (u'\overline{\omega}_x + v'\overline{\omega}_y) + (u'\omega'_x + v'\omega'_y) \\ &\quad - (\overline{u'\omega'_x} + \overline{v'\omega'_y}) = \nu(\omega'_{xx} + \underline{\omega'_{yy}}), \end{aligned} \tag{4.3.2a}$$

$$-\underline{\nu\omega'_y} = \sigma' = \underline{b'_t} + b'b'_x - \overline{b'b'_x} + \frac{1}{\rho}\underline{p'_x} \quad \text{at } y = 0, \tag{4.3.2b}$$

$$\omega' = 0 \quad \text{at } y = \infty. \tag{4.3.2c}$$

Here, a few terms are underlined of which the implication will soon be made clear. The equations for (u', v', ω') are underdetermined due to the existence of terms like $\overline{u'\omega'_x}$ and $\overline{v'\omega'_y}$. But now a key observation is: if the characteristic frequency n of the fluctuation is very large, one may approximate the solution by taking $n \to \infty$ such that the Stokes layer has thickness $\delta_n = \sqrt{2\nu/n} \to 0$. Then, similar to what we observed in the steady boundary layer theory, in (4.3.2) y and v need to be rescaled. Consequently, Lin (1957) points out that, to the leading order only the underlined terms are retained (where p'_x must be a high-frequency fluctuation as well to balance the oscillating flow) and all other terms can be dropped:

$$\omega'_t = \nu\omega'_{yy}, \tag{4.3.3a}$$

$$\sigma' = -\nu\omega'_y = b'_t + \frac{1}{\rho}p'_x \quad \text{at } y = 0, \quad \omega' = 0 \text{ at } y = \infty. \tag{4.3.3b}$$

This is precisely our problem of general parallel flow over infinite flat plate (Sect. 4.1.1), of which the integral solution is known:

$$\omega'(x, y, t) = \int_{0^-}^{t} \frac{\sigma'(x, \tau)}{\sqrt{\pi\nu(t-\tau)}} \exp\left[-\frac{y^2}{4\nu(t-\tau)}\right] d\tau, \tag{4.3.4a}$$

$$u'(x, y, t) - b'(x, t) = -\int_{0}^{t} \sigma'(x, \tau)\text{erf}\left(\frac{y}{\sqrt{\nu(t-\tau)}}\right) d\tau. \tag{4.3.4b}$$

This result is an asymptotically exact Navier-Stokes solution in the limit $n \to \infty$, *no matter what the background flow and the fluctuating amplitude could be*. Unlike strictly parallel flows, now the appearance of arbitrary background flow makes the solution x-dependent with a small normal velocity $v'(x, y, t)$. For example, if we introduce a harmonic component to the velocity at the outer edge of a boundary layer over a stationary wall with x-dependent amplitude $U_0(x)$, such that

$$U(x, t) = \overline{U}(x) + U'(x, t), \quad U'(x, t) = U_0(x)\cos nt,$$

then we have a double-layer structure, a steady boundary layer of thickness $\delta = \sqrt{\nu L/\overline{U}}$ and a Stokes layer of $\delta_n \sim \sqrt{\nu/n}$ as defined by (4.1.18). If moreover the latter has very large frequency or Strouhal number,

$$\left(\frac{\delta_n}{\delta}\right)^2 \sim \frac{\overline{U}}{nL} = St^{-1} \equiv \epsilon \ll 1, \tag{4.3.5}$$

then the Stokes layer exists only as a sublayer of the boundary layer, with thickness about $5\sqrt{\nu/n}$, which for $n = 100$ and 10^4 Hz in air is only 0.8 and 0.08 mm, respectively (Lighthill 1978).

Now the fluctuating boundary vorticity flux is

$$\sigma' = -\frac{1}{\rho}\frac{\partial p'}{\partial x} \simeq -U_t' = nU_0 \sin nt,$$

which by (4.3.4b) yields the Stokes wave solution studied in Sect. 4.1.3. In the stationary state, the streamwise velocity $u'(x, y, t)$ is given by (4.1.19), from which and the continuity equation also follows the small normal velocity $v'(x, y, t) = O(\delta_n)$:

$$u' = U_0(x)[\cos nt - e^{-\eta}\cos(nt - \eta)], \tag{4.3.6a}$$

$$v' = -\delta_n \frac{dU_0}{dx}\left\{\eta \cos nt + \frac{1}{\sqrt{2}}\cos\left(nt + \frac{3\pi}{4}\right) + \frac{e^{-\eta}}{\sqrt{2}}\cos\left(nt - \eta - \frac{\pi}{4}\right)\right\}, \tag{4.3.6b}$$

where

$$\eta = \sqrt{\frac{n}{2\nu}}\,y = \frac{y}{\delta_n}. \tag{4.3.7}$$

In (4.3.6b), the first term comes from the continuity of the main flow, the second term is the displacement effect of the boundary layer on the external flow, and the third term tends to zero outside the boundary layer. Notice the phase shifts in u' and v' relative to the external flow $U_0(x)\cos nt$ are y-dependent. The vorticity wave is

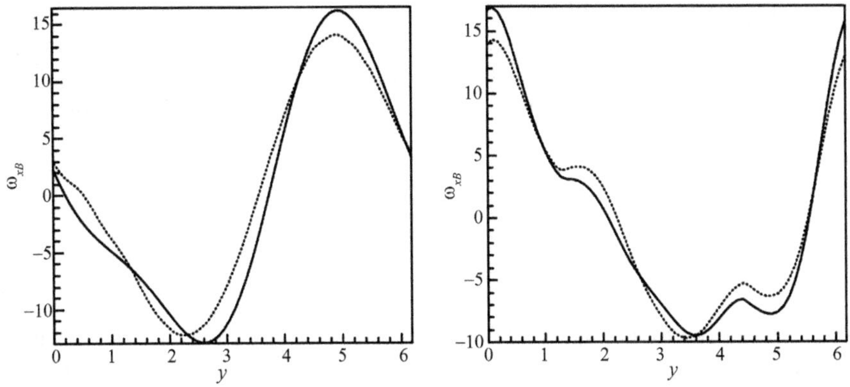

Fig. 4.9 Channel turbulence control by transverse traveling waves of flexible lower wall able to deform and oscillate on its own plane with spanwise velocity $b(y, t)$. Comparison of streamwise boundary vorticity ω_{xB} calculated from Lin's theory (*solid lines*) and direct numerical simulation (*dashed lines*) for **a** sinusoidal wave $b(y, t) = I_w \sin(ky - nt)$, and **b** "Gaussian-distribution wave" $b(y, t) = \pm I_w e^{(ky-nt)^2}$, where the $\pm$ signs are for first and second half of the cycle, respectively. From Yang (2004)

$$\omega' = U_0(x)\sqrt{\frac{n}{\nu}}e^{-\eta}\cos\left(nt - \eta + \frac{\pi}{4}\right), \tag{4.3.8}$$

which has a phase lead of $\pi/4$ over the velocity fluctuations.

The above standing harmonic wave e^{int} in Lin's high-frequency theory can be generalized to a traveling wave with arbitrary periodic function of $k_x x - nt$ with k_x being the wave number, for which the solution (4.3.4a) and (4.3.4b) still holds. For example, conceive a channel turbulence between two parallel walls. Assume the lower wall at $z = 0$ is ideally flexible such that it can produce a high-frequency spanwise traveling wave of arbitrary wave pattern. Then one can use Lin's theory to predict the Stokes wave on a cross-flow (y, z)-plane. The predicted streamwise boundary vorticity ω_{xB} at the wall is compared with direct numerical simulation in Fig. 4.9 for a sinusoidal and a non-sinusoidal traveling wave (Yang 2004). The good agreement signifies how the Stokes layer is independent of the strong background turbulence.

4.4 Free Steady Vortex Layers

We now turn from boundary layers to free vortex layers, such as shear layers, jet, and wake, etc. Like boundary layers, the small thickness of free vortex layers also implies the same estimates given by (4.2.8), and hence Prandtl's boundary-layer equations (4.2.10a) and (4.2.10b) are still effective, but under different boundary condition. Being no longer bounded by a rigid wall, unsteady free vortex layers are more vulnerable to various disturbances than boundary layers. They easily become unstable and evolve to different geometric patterns. This instability problem will be discussed in Chaps. 10 and 11. Here we are confined to the simplest part of the subject, the steady similarity solutions of (4.2.15) of typical free vortex layers.

4.4.1 Free Shear Layer

Consider two uniform streams of identical fluid moving in the same x-direction, with different velocity U_1 and $U_2 = \lambda U_1$ $(\lambda < 1)$, separated by a thin flat plate with trailing edge at $(x, y) = (0, 0)$. Then for $x > 0$ the streams come into contact to form a *free shear layer*, of which the thickness is gradually growing as x increases as shown in Fig. 4.10. Since the pressure outside the streams is constant, (4.2.15) is applicable if $Re = U_1 L/\nu \gg 1$. The boundary conditions are

$$u \to U_1 \quad \text{as} \quad y \to \infty, \qquad u \to U_2 \quad \text{as} \quad y \to -\infty. \tag{4.4.1a}$$

Besides, if $v \to v(\pm\infty)$ as $y \to \pm\infty$, by the global momentum integral Ting (1959) proves that there must be

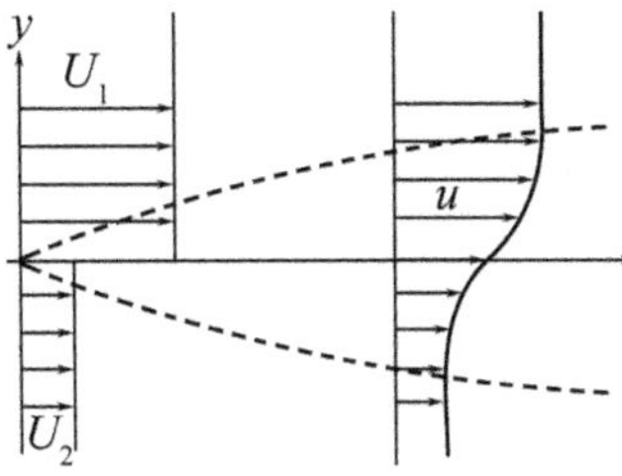

Fig. 4.10 Steady free shear layer formed between two parallel streams

$$v(\infty) = -\lambda v(-\infty), \tag{4.4.1b}$$

i.e., the faster stream expands slower, indicating that the central line of the free shear layer will turn to the slower-stream side.

Once again we may cast the problem to the similarity Eq. (4.2.17) and solve it by iterative equation (4.2.18), with only a corresponding change in the boundary condition for iterative integrals (4.2.20), which converge to the exact solution at about $n = 4$. The profiles for iterative (u, w) and the exact sectional vorticity profiles are shown in Fig. 4.11.[6] The last plot clearly shows the locations of vorticity peaks indeed shift toward $y < 0$.

4.4.2 *Jet*

Consider a steady two-dimensional *jet* discharged from an orifice (a long slit perpendicular to the flow plane), which drives the surrounding fluid to move along. The surrounding pressure is again uniform, and hence (4.2.15) applies. Due to the symmetry with respect to the x-axis, it suffices to observe the flow at upper part. Then the boundary conditions are

$$y = 0: \quad u = U_m, \quad v = 0, \quad \frac{\partial u}{\partial y} = 0, \tag{4.4.2a}$$

$$y = \infty: \quad u = 0, \quad v = -v_e, \quad \frac{\partial u}{\partial y} = 0. \tag{4.4.2b}$$

Unlike boundary layer or free shear layer, now at $y = \infty$ there is no streamwise flow but rather a nonzero normal transverse $-v_e$. To understand the overall pattern of this jet flow, we first integrate (4.2.15b) to obtain

$$y = \infty: \quad v = -v_e = \int_0^\infty \frac{\partial v}{\partial y} dy = -\frac{d}{dx}\int_0^\infty u dy = -\frac{1}{2}\frac{dQ}{dx}, \tag{4.4.3}$$

[6]Mao and Xuan (2010), private communications.

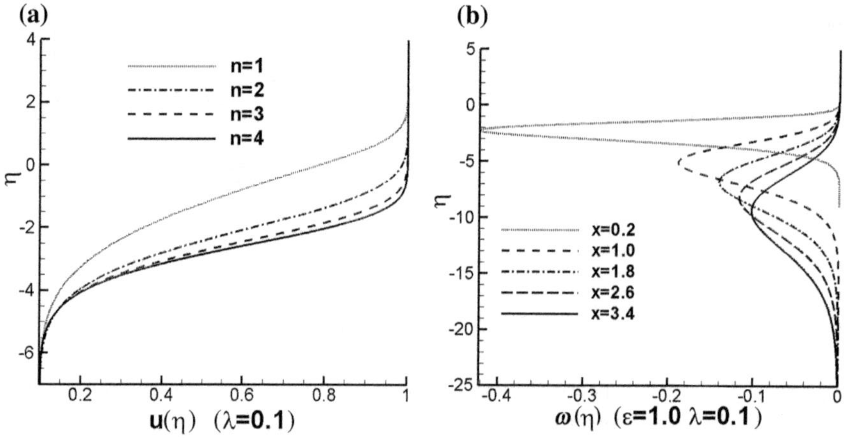

Fig. 4.11 Profiles in steady free shear layer. **a** Iterative and exact profiles of similarity velocities u with $n = 1, 2, 3$ and 4 (exact). **b** Sectional profiles of exact vorticity solution

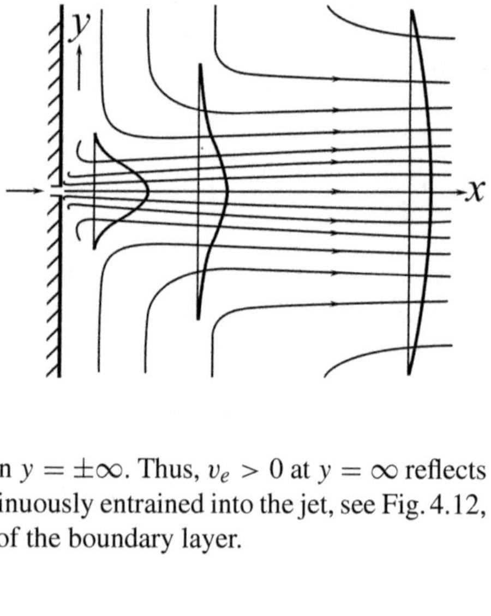

Fig. 4.12 Two-dimensional jet from orifice. From Schlichting and Gersten (2000)

where Q is the total volume flux between $y = \pm\infty$. Thus, $v_e > 0$ at $y = \infty$ reflects the fact that the surrounding fluid is continuously entrained into the jet, see Fig. 4.12, just opposite to the displacement effect of the boundary layer.

Moreover, since by (4.4.2) there is

$$\int_0^\infty v\frac{\partial u}{\partial y}dy = -\int_0^\infty u\frac{\partial v}{\partial y}dy = \int_0^\infty u\frac{\partial u}{\partial x}dy,$$

the integral of (4.2.15a) (in dimensional form) yields

$$2\int_0^\infty u\frac{\partial u}{\partial x}dy = \frac{d}{dx}\int_{-\infty}^\infty u^2 dy = \nu\frac{\partial u}{\partial y}\bigg|_0^\infty = 0,$$

which confirms the total momentum conservation:

$$K = \int_{-\infty}^{\infty} u^2 dy = \text{const.} \tag{4.4.5}$$

Note that at $x = 0$ there must be $u = \infty$.

The above differences of the jet from boundary layer and free shear layer make the similarity variable $\eta = y/\sqrt{2\epsilon x}$ no longer appropriate, and we have to try a different one. Let

$$u(x, y) \sim x^m \bar{u}(\eta), \quad \eta = \frac{y}{x^n},$$

where m, n are unknown numbers. Then from (4.4.5) we find $2m + n = 0$, while from the balance between the advection and diffusion,

$$u\frac{\partial u}{\partial x} \sim x^{2m-1}(m\bar{u} - n\eta\bar{u}'), \quad \nu\frac{\partial^2 u}{\partial y^2} \sim \nu x^{m-2n}\bar{u}'',$$

we find $m + 2n = 1$. Hence we set (the chosen coefficients are for convenience)

$$u = 6\nu x^{-1/3}\bar{u}(\eta), \quad v = 2\nu x^{-2/3}\bar{v}(\eta), \quad \eta = \frac{y}{x^{2/3}}, \tag{4.4.7}$$

where the exponent of v comes from (4.2.15b). The two equations in (4.2.15) are now cast to

$$-\bar{u}^2 + (\bar{v} - 2\eta\bar{u})\bar{u}' = \frac{1}{2}\bar{u}'', \tag{4.4.8a}$$

$$2\eta\bar{u}' + \bar{u} = \bar{v}'. \tag{4.4.8b}$$

Integrating (4.4.8b) yields

$$\bar{v} = 2\eta\bar{u} - f, \quad f(\eta) \equiv \int_0^{\eta} \bar{u}\, d\eta,$$

by which (4.2.15a) is cast to an ordinary integral-differential equation

$$\bar{u}'' + 2(f\bar{u}' + \bar{u}^2) = 0.$$

It is more convenient to work on the equation for $f(\eta)$:

$$f''' + 2(ff'' + f'^2) = (f'' + 2ff')' = 0, \tag{4.4.10}$$

which is just the conventionally used similarity equation for the stream function $\psi = 6\nu x^{1/3} f(\eta)$. By using $f(0) = f''(0) = 0$, (4.4.10) can be integrated twice to yield analytical solution (Schlichting and Gersten 2000)

$$f(\eta) = \alpha \tanh \alpha\eta,$$

where α is determined by the total momentum flux K:

$$K = 36\nu^2\alpha^4 \int_{-\infty}^{\infty} \operatorname{sech}^4 \alpha\eta d\eta = 48\nu^2\alpha^3.$$

The final form of the jet solution is

$$u = 0.454 \left(\frac{K^2}{\nu x}\right)^{1/3} (1 - \tanh^2 \tilde{\eta}), \quad \tilde{\eta} = \alpha\eta, \tag{4.4.11a}$$

$$v = 0.550 \left(\frac{K\nu}{x^2}\right)^{1/3} [2\tilde{\eta}(1 - \tanh^2 \tilde{\eta}) - \tanh \tilde{\eta}], \tag{4.4.11b}$$

$$\tilde{\eta} = 0.275 \left(\frac{K}{\nu^2}\right)^{1/3} \frac{y}{x^{2/3}}. \tag{4.4.11c}$$

The profiles predicted by (4.4.11a) and (4.4.11b) are plotted in Fig. 4.13. The solution $u(\eta)$ has been well verified by experiments at sufficiently downstream x-stations.

So far we have used dimensional variables because of the lack of a constant external velocity to define the Reynolds number. Now we may define a Reynolds number by the maximum velocity U_m, which is given by (4.4.11a) with $\eta = 0$. Then the approximate boundary-layer equation (4.2.15) holds for the jet flow if

$$Re_x = \frac{U_m x}{\nu} \sim \left(\frac{Kx}{\nu^2}\right)^{2/3} \gg 1. \tag{4.4.12}$$

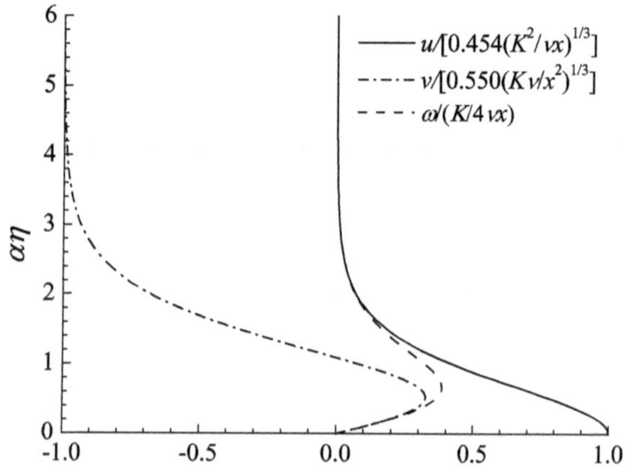

Fig. 4.13 The profiles of velocity (u, v) and vorticity ω

Besides, we have

$$v_e = -0.550 \left(\frac{K\nu}{x^2}\right)^{1/3}, \quad Q = 3.302(K\nu x)^{1/3}, \tag{4.4.13}$$

thus

$$\frac{v_e}{U_m} \sim Re_x^{-1} \ll 1. \tag{4.4.14}$$

Finally, under the present boundary-layer approximation we find

$$\omega \simeq -\frac{\partial u}{\partial y} = \frac{K}{4\nu x} \tanh \tilde{\eta}(1 - \tanh^2 \tilde{\eta}), \tag{4.4.15}$$

see also Fig. 4.13. The ratio of $\partial v/\partial x$ and $\partial u/\partial y$ is

$$\frac{|\partial v/\partial x|}{|\partial u/\partial y|} \sim \frac{|v_e|/x}{K/4\nu x} \sim Re_x^{-2} \ll 1,$$

so the former is indeed well negligible.

4.4.3 Far Wakes

Consider a symmetric airfoil aligned to the free stream. The boundary layers at the upper and lower surfaces of the airfoil merge at its trailing edge, to form a vortical region known as the *wake* of the airfoil. Since the fluid from boundary layers carries lower kinetic energy than the external flow, the wake profile has a velocity defect as shown in Fig. 4.14, just opposite to the jet profile, and (4.2.15) is again applicable.

For an arbitrary stationary body in an otherwise uniform stream, the specific wake-flow pattern depends on the body shape and can be very complicated and unsteady. But at far downstream as $x \to \infty$, the viscous diffusion has smeared out the direct effect of the body presence and continuously spread the wake vorticity. Consequently, the far wake flow becomes quasi-parallel, in which the advection dominates over

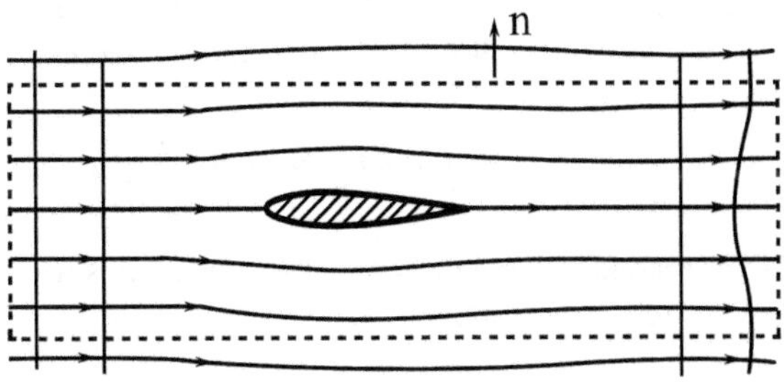

Fig. 4.14 Two-dimensional wake behind a symmetric airfoil. After Batchelor (1967)

streamwise diffusion, and transverse gradient dominates over streamwise one. These features make the far wake flow again governed by the same Eq. (4.2.15).

Moreover, the continual spreading of the wake also tends to reduce the friction forces to make the velocity more uniform. Specifically, consider the steady far wake of a body that experiences only a drag force along the flow direction. Then not only (4.2.15) is applicable but also the local velocity differs from the free-stream velocity $\boldsymbol{U} = U\boldsymbol{e}_x$ by only a small amount at large x, namely $|U - u| \ll U$ such that (u, v) can be approximated by $(U, 0)$. This brings us at once to the familiar parallel shear-flow model. In dimensional form it reads

$$U\frac{\partial u'}{\partial x} = \nu\frac{\partial^2 u'}{\partial y^2}, \quad u' = U - u(x, y). \tag{4.4.16a}$$

But now the boundary conditions are homogeneous:

$$y = 0: \quad \frac{\partial u'}{\partial y} = 0; \quad y \to \infty: \quad u' = 0. \tag{4.4.16b}$$

Thus, $u' \equiv 0$ is a trivial solution. We seek nontrivial similarity solution by setting (Schlichting and Gersten 2000)

$$u' = CU\left(\frac{x}{L}\right)^{-m} f(\eta), \quad \eta = \frac{y}{2}\sqrt{\frac{U}{\nu x}}, \tag{4.4.17}$$

where L is scaled to the body length and C is a constant which can be, remarkably, linked to the drag. To see this we first need a drag formula in terms of u'.

Return to the force formula (1.3.6) for steady flow in terms of a far-field control surface Σ on which the viscous stress is negligible. Now let Σ be a rectangular loop as shown by the dash line in Fig. 4.14. Denote the line elements of upstream face 1 and downstream face 2 of Σ by $\mp dy$, and those of side faces by ds. For face 1 locating sufficiently far upstream we may take $u = U$, and on the entire Σ we may take $p = p_\infty$. Thus by (1.3.6) the drag reads

$$D = \rho\int (U^2 - u^2)dy - \rho\int u\boldsymbol{u}\cdot\boldsymbol{n}ds.$$

No matter how large the side faces are apart from each other there must be $\int \boldsymbol{u}\cdot\boldsymbol{n}ds \neq 0$, since as shown in Fig. 4.14 the continuity requires

$$\int \boldsymbol{u}\cdot\boldsymbol{n}ds = \int (U - u_2)dy = \int u'dy.$$

In the second integral of D over side boundaries we may approximate u by U. Thus the desired drag formula is

$$D = \rho \int_{-\infty}^{\infty} u(U-u)dy \simeq \rho U \int_{-\infty}^{\infty} u' dy. \tag{4.4.18}$$

Since D is independent of x, so must be the integral of (4.4.17):

$$\int_{-\infty}^{\infty} u' dy = \frac{D}{\rho U} = 2CU \left(\frac{x}{L}\right)^{-m} \sqrt{\frac{\nu x}{U}} \int_{-\infty}^{\infty} f(\eta) d\eta, \tag{4.4.19}$$

which requires $m = 1/2$. By (4.4.16) the equation for $f(\eta)$ is

$$f'' + 2\eta f' + 2f = f'' + 2(\eta f)' = 0,$$

which has solution $f(\eta) = e^{-\eta^2}$. Then by (4.4.19) there is

$$C_D \equiv \frac{2D}{\rho U^2 L} = 4\sqrt{\pi} Re^{-1/2} C, \quad Re = \frac{UL}{\nu},$$

and hence finally the wake solution reads

$$\frac{u'(x,y)}{U} = \frac{C_D}{4\sqrt{\pi}} Re^{1/2} \left(\frac{x}{L}\right)^{-1/2} e^{-\eta^2}, \tag{4.4.20}$$

indicating that the velocity defect is reduced as $x^{-1/2}$ and the wake width increases as $x^{1/2}$. Experiments have shown that for x greater than about 3 times of body size L the pressure indeed reaches a uniform value and this *laminar wake law* indeed holds true.

4.5 Problems for Chapter 4

4.1. Find the exact solution of problem (4.2.1) and examine the error distribution of its boundary-layer approximation.

4.2. Consider the transient Couette flow between two parallel plates of distance h, driven by the suddenly started motion of the lower plate at $y = 0$. The vorticity field is still governed by the homogeneous heat equation (4.1.3a), but now the Green's function is known as (e.g., Barton 1989)

$$\hat{G}(y,t;y',t') = H(t-t')\frac{2}{h}\left\{\frac{1}{2} + \sum_{n=1}^{\infty} \cos ky \cos ky' e^{-\nu k^2 (t-t')}\right\}, \quad k = \frac{n\pi}{h}. \tag{4.5.1}$$

Write down the Neumann condition prescribed at parallel boundaries, and show that the vorticity and velocity solutions are

$$\omega(y,t) = b_0 \frac{2}{h}\left\{\frac{1}{2} + \sum_{n=1}^{\infty} \cos ky e^{-\nu k^2 t}\right\}, \tag{4.5.2a}$$

$$u(y,t) = b_0 \left\{1 - \frac{y}{h} - \frac{2}{\pi}\sum_{n=1}^{\infty} \frac{1}{n} \sin ky e^{-k^2 \nu t}\right\}. \tag{4.5.2b}$$

where the series term represents the transient effect. Discuss the vorticity production mechanism and the role of upper plate at $y = h$.

4.3. The Blasius boundary-later equation (4.2.15a) can be expressed in terms of stream function ψ so that (4.2.15b) is automatically satisfied. It is well known that one can introduce a dimensionless similarity variable η and set $\psi = \sqrt{\nu U x} f(\eta)$, so that the boundary-layer equation is reduced to an ordinary differential equation

$$ff'' + 2f''' = 0. \tag{4.5.3}$$

Write down the boundary conditions for $f(\eta)$, compose a code to find the numerical solution of (4.5.3), and compute the profiles of velocity u, vorticity ω, and Lamb-vector component $\boldsymbol{e}_y \cdot (\boldsymbol{\omega} \times \boldsymbol{u})$. Compare your plot with Fig. 4.7.

4.4*. It is often convenient to express oscillating quantities in a Stokes layer by complex variable. Confirm now that the oscillating velocities (4.1.19) and (4.3.6a) have a common normal distribution due to diffusion, as a complex function of y:

$$\frac{u}{U_0} e^{-int} = 1 - \exp(y\sqrt{in/\nu}). \tag{4.5.4}$$

Plot the curves of real and imaginary parts of the function on the right-hand side versus $\eta = y\sqrt{n/2\nu}$. Show the asymptotic behavior of the curves as $\eta \to \infty$. Discuss the phase difference between the externally imposed flow $U_0 e^{int}$ and inside the Stokes layer. Infer the variation of the phase lag of vorticity as η and explain its physical implication.

Chapter 5
Vortex Sheet Dynamics

Although well-developed theories have been available for various attached boundary layers, the motion of free shear layers as well as their effect on global flow and reaction to solid bodies is far more complicated. Only some simplest solutions with prescribed shear-layer locations are known, like those in Sect. 4.4. To have a tractable theory for arbitrary free shear layers, we now proceed from the boundary-layer approximation at $Re \gg 1$ to the asymptotic state of a viscous flow as $Re \to \infty$ so that the shear-layer thickness $\delta \to 0$. Then wall boundary layer and separated free shear layer are simplified to *attached vortex sheet* and *free vortex sheet*, respectively. In this model only the vorticity integral along the normal, the *vortex sheet strength* $\boldsymbol{\gamma}(\boldsymbol{x}, t)$, enters the theory. This flow model is especially useful at very large Re and if one's main concern is the global vortical-flow performance rather than detailed local processes inside thin shear layers.

Throughout this chapter the flow is assumed to be incompressible.

5.1 Basic Properties of Free Vortex Sheet

Let $\boldsymbol{n}$ be the unit normal vector of a boundary layer or free shear layer of thickness $\delta \ll 1$ at $Re \gg 1$, in which the vorticity is dominated by the normal velocity gradient:

$$\boldsymbol{\omega} = \nabla \times \boldsymbol{u} \simeq \boldsymbol{n} \times \frac{\partial \boldsymbol{u}}{\partial n} = \frac{\partial}{\partial n}(\boldsymbol{n} \times \boldsymbol{u}).$$

This relation becomes exact as $\delta \to 0$. Its normal integral yields

$$\boldsymbol{\gamma} = \lim_{\delta \to 0} \int_0^\delta \boldsymbol{\omega} dn = \boldsymbol{n} \times [\![\boldsymbol{u}]\!], \tag{5.1.1}$$

which defines the vortex sheet strength. The jump notation $[\![\boldsymbol{u}]\!] = \boldsymbol{u}^+ - \boldsymbol{u}^-$ is used here as the difference of the velocities at two sides of the sheet. Unless otherwise

J.-Z. Wu et al., *Vortical Flows*, DOI 10.1007/978-3-662-47061-9_5

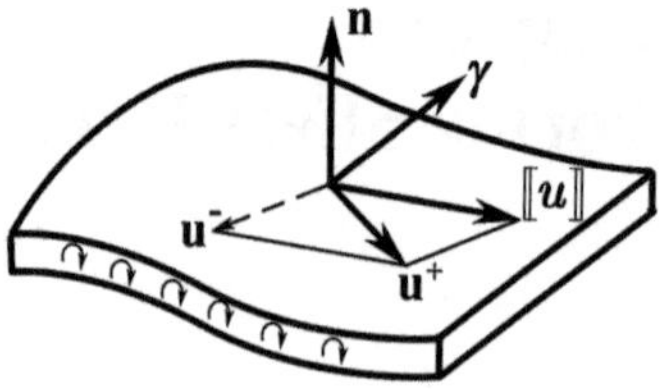

Fig. 5.1 Vortex sheet and its relevant quantities

stated, we follow the convention that $\boldsymbol{n}$ points from side $-$ to side $+$. The geometric relation among $\boldsymbol{u}^+$, $\boldsymbol{u}^-$, $[\![\boldsymbol{u}]\!]$, and $\boldsymbol{\gamma}$ is shown in Fig. 5.1. In two-dimensional flow on the (x, y)-plane, it is convenient to describe a vortex sheet C (a curve of tangent discontinuity) in terms of the intrinsic curvilinear coordinates (s, n) along C with unit vectors $(\boldsymbol{t}, \boldsymbol{n})$ and velocity $\boldsymbol{u}^\pm = (u_s^\pm, u_n)$. Then by (5.1.1) the sheet strength is

$$\boldsymbol{\gamma} = \gamma(s)\boldsymbol{e}_z = \boldsymbol{n} \times [\![u_s]\!]\boldsymbol{t} = -[\![u_s]\!]\boldsymbol{e}_z. \tag{5.1.2}$$

5.1.1 *Strength and Velocity of Free Vortex Sheet*

We start from free vortex sheet. Across such a sheet S the jumps of flow properties have to satisfy the continuity equation and inviscid momentum equation. Let the sheet move at velocity $\boldsymbol{u}_\gamma$, since $[\![\boldsymbol{u}_\gamma]\!] = \boldsymbol{0}$ the continuity requires

$$\rho[\![\boldsymbol{u} - \boldsymbol{u}_\gamma]\!] \cdot \boldsymbol{n} = \rho[\![\boldsymbol{u}]\!] \cdot \boldsymbol{n} = 0. \tag{5.1.3}$$

Thus, $[\![\boldsymbol{u}]\!]$ must be along a tangential direction, and (5.1.1) implies

$$[\![\boldsymbol{u}]\!] = \boldsymbol{\gamma} \times \boldsymbol{n}. \tag{5.1.4}$$

Then the inviscid momentum balance requires that the jump of $\rho\boldsymbol{u}(u_n - u_{\gamma n})$ must be balanced by that of p, which by (5.1.3) can be written as

$$\rho(u_n - u_{\gamma n})[\![\boldsymbol{u}]\!] = -[\![p]\!]\boldsymbol{n}, \tag{5.1.5}$$

of which the left- and right-hand sides are along a tangent and the normal directions, respectively, so both sides must vanish simultaneously:

$$[\![p]\!] = 0, \tag{5.1.6a}$$

$$u_{\gamma n} = u_n. \tag{5.1.6b}$$

Equation (5.1.6a) indicates that *a vortex sheet cannot stand any pressure jump*, in consistency with the fact that across a shear layer at large Re the normal pressure gradient is negligible. Equation (5.1.6b) determines the normal component of the sheet velocity $\boldsymbol{u}_\gamma$. The determination of the tangent components of $\boldsymbol{u}_\gamma$ is more complicated; it has been found that (Wu 1995)

$$\boldsymbol{u}_\gamma = \overline{\boldsymbol{u}} \equiv \frac{1}{2}(\boldsymbol{u}^+ + \boldsymbol{u}^-) \quad \text{if and only if} \quad \boldsymbol{\omega} \cdot \boldsymbol{n} = 0 \ \text{ on } S. \tag{5.1.7}$$

The condition of this result holds for any two-dimensional vortex sheets and those three-dimensional ones away from which the flow is irrotational. Unless otherwise stated, in what follows we assume this is so.

5.1.2 Circulation, Lamb Vector, and Bernoulli Equation

Let P be a point on a vortex sheet S, and take a curve C to connect P^+ and P^-. If the sheet has an edge, the curve can be made go across S only once, see Fig. 5.2. The circulation along C is

$$\oint_C \nabla\phi \cdot d\boldsymbol{x} = \oint_C d\phi = [\![\phi_P]\!] \equiv \Gamma. \tag{5.1.8}$$

Unlike the conventionally defined circulation, now this Γ is defined only for points consisting the material vortex sheet, and is a *point function* independent of the shape of C. $[\![\phi]\!]$ or Γ is nothing but the strength of *doublet* or *dipole* distributed on S. Then by (5.1.1) there is

$$\boldsymbol{\gamma} = \boldsymbol{n} \times \nabla_\pi [\![\phi]\!] = \boldsymbol{n} \times \nabla_\pi \Gamma, \quad \nabla_\pi \Gamma = [\![\boldsymbol{u}]\!] = \boldsymbol{\gamma} \times \boldsymbol{n}, \tag{5.1.9}$$

where and below the suffix π denotes tangent components of a vector on a surface, i.e.,

$$\boldsymbol{f}_\pi = \boldsymbol{f} - \boldsymbol{n} f_n = \boldsymbol{n} \times (\boldsymbol{f} \times \boldsymbol{n}).$$

Thus there always is $\boldsymbol{n} \times \nabla = \boldsymbol{n} \times \nabla_\pi$. In two dimensions (5.1.9) becomes

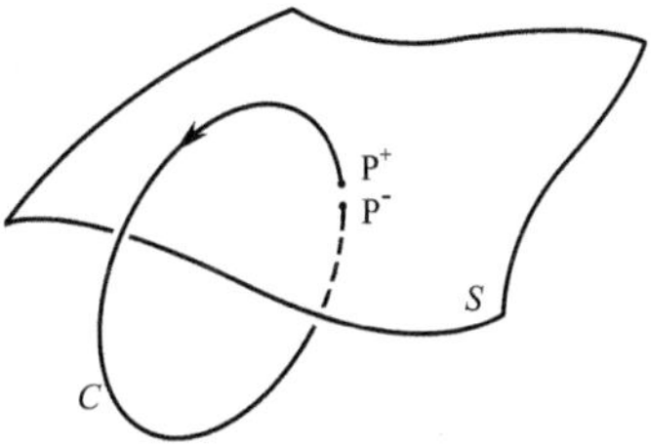

Fig. 5.2 A loop across the vortex sheet once

$$\gamma = -[\![u_s]\!] = -\frac{\partial \Gamma}{\partial s}. \tag{5.1.10}$$

Moreover, we can define the *vortex-sheet Lamb vector* $\gamma \times \overline{\boldsymbol{u}}$. Like the conventional Lamb vector $\boldsymbol{\omega} \times \boldsymbol{u}$, $\gamma \times \overline{\boldsymbol{u}}$ represents a vortex force that causes a local rate of change of the fluid momentum. If we make decomposition $\overline{\boldsymbol{u}} = \overline{\boldsymbol{u}}_\pi + u_n \boldsymbol{n}$, then by (5.1.4) there is

$$\gamma \times \overline{\boldsymbol{u}} = \gamma \times \overline{\boldsymbol{u}}_\pi + u_n [\![\boldsymbol{u}]\!], \tag{5.1.11}$$

where the first and second terms are along the normal and tangent directions, respectively, which are in turn from the tangent and normal components of $\overline{\boldsymbol{u}}$, respectively. For example, in two-dimensional flow with $\overline{\boldsymbol{u}} = \overline{u}_s \boldsymbol{t} + u_n \boldsymbol{n}$ we have

$$(\gamma \times \overline{\boldsymbol{u}})_{2D} = \gamma(\boldsymbol{n}\overline{u}_s - \boldsymbol{t}u_n), \quad \gamma = |\gamma|. \tag{5.1.12}$$

Next, if the kinetic energy $q^2/2$ at both sides of a vortex sheet has a jump, then this jump is solely expressible by the normal component of the vortex-sheet Lamb vector, because

$$\begin{aligned}\frac{1}{2}[\![q^2]\!] &= \frac{1}{2}(\boldsymbol{u}^+ + \boldsymbol{u}^-) \cdot (\boldsymbol{u}^+ - \boldsymbol{u}^-) = \overline{\boldsymbol{u}} \cdot [\![\boldsymbol{u}]\!] \\ &= \overline{\boldsymbol{u}}_\pi \cdot \nabla \Gamma = -\boldsymbol{n} \cdot (\gamma \times \overline{\boldsymbol{u}}_\pi) = -\gamma |\overline{\boldsymbol{u}}| \sin \beta,\end{aligned} \tag{5.1.13}$$

where β is the angle between γ and $\overline{\boldsymbol{u}}$. The geometric relation involved in the last expression of (5.1.13) is shown in Fig. 5.3 on the tangential plane at a point of a vortex sheet. The continuous normal velocity $u_n^\pm = \overline{u}_n = u_n$ does not affect the geometry and is not shown. Note that $\boldsymbol{u}_\pi^\pm - \overline{\boldsymbol{u}}_\pi$ are the velocity increments induced by γ and along the direction of $\pm[\![\boldsymbol{u}]\!]$, both having magnitude $\gamma/2$. For given nonzero γ and $|\overline{\boldsymbol{u}}|$, when $\beta = \pi/2$ (which is always so in two-dimensional flow) the jump of kinetic energy reaches maximum.

We can now relate Γ, $\gamma \times \overline{\boldsymbol{u}}$, and $[\![q^2]\!]/2$ by the jump of a Bernoulli integral at both sides of any material surface S of tangent discontinuity. Since the flow outside the sheet is assumed from the same upstream fluid and irrotational, and since the

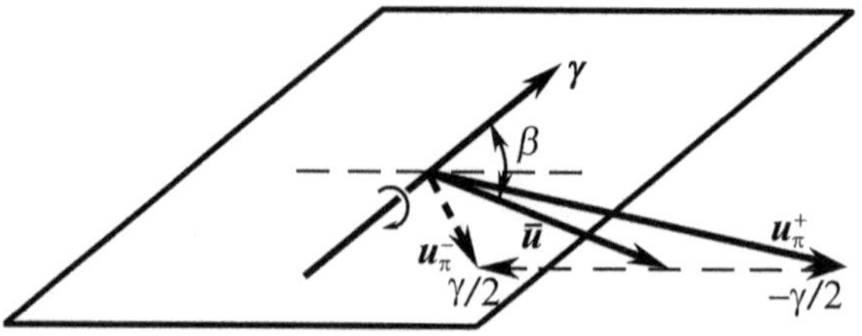

Fig. 5.3 Vector geometry on the tangential plane of a vortex sheet at a point. For two-dimensional flow the vortex sheet is perpendicular to flow plane, with $\beta = \pm\pi/2$ and $\boldsymbol{u}_\pi^+ \| \boldsymbol{u}_\pi^-$ along the intersection of the plane and sheet. After von Kármán and Burgers (1935)

jump of advection acceleration of ϕ is $[\![\boldsymbol{u}\cdot\nabla\phi]\!] = [\![q^2]\!]$ and for on-sheet variables $D/Dt = \partial/\partial t + \overline{\boldsymbol{u}}\cdot\nabla$, by (5.1.13) both (3.2.21a) and (3.2.21b) lead to

$$-\frac{1}{\rho}[\![p]\!] = \frac{\partial\Gamma}{\partial t} + \overline{\boldsymbol{u}}_\pi\cdot\nabla\Gamma = \frac{D\Gamma}{Dt} \tag{5.1.14a}$$

$$= \frac{\partial\Gamma}{\partial t} - \boldsymbol{n}\cdot(\gamma\times\overline{\boldsymbol{u}}) = \frac{\partial\Gamma}{\partial t} - \overline{u}\gamma\sin\beta. \tag{5.1.14b}$$

Specifically, for free vortex sheet there is

$$\frac{D\Gamma}{Dt} = \frac{\partial\Gamma}{\partial t} - \boldsymbol{n}\cdot(\gamma\times\overline{\boldsymbol{u}}) = 0. \tag{5.1.15}$$

For unsteady flow $\partial\Gamma/\partial t$ is the local time rate at a field point that sits on the sheet instantaneously.

In particular, for a *steady* free vortex sheet, the second term of (5.1.15) vanishes, namely $\gamma\times\overline{\boldsymbol{u}}$ can only have tangent component. However, the geometry and location of steady vortex sheets are time-independent, so a nonzero u_n is impossible; but then (5.1.11) indicates that $\gamma\times\overline{\boldsymbol{u}}$ can only have normal component. Consequently, the vortex-sheet Lamb vector has to be zero. In three-dimensional flow, this implies that $\gamma\|\overline{\boldsymbol{u}}$, and by (5.1.13) $\boldsymbol{u}^+$ and $\boldsymbol{u}^-$ have the same magnitude and make the same angle with γ at both sides, see Fig. 5.3. On the other hand, in two-dimensional flow and rotationally symmetric flow with $\overline{u}_s = \overline{u}$ but $\beta = \pm\pi/2$, there is $\gamma\overline{u} = 0$. Namely, either $\gamma \neq 0$ and $u^+ = -u^-$, which could happen only in rare cases; or generically $\gamma = 0$ and $u^+ = u^- = \overline{u}$.

To summarize, we may state

Theorem *Suppose the flows at both sides of a free vortex sheet are from the same upstream fluid and irrotational. Then:*

(1) For the circulation Γ defined as an on-sheet point function, the Kelvin circulation theorem equally applies to free vortex sheet.

(2) The vortex sheet has nonzero Lamb vector only if its motion is unsteady. When the sheet is steady, in three dimensions the flows at both sides have the same kinetic energy but different tangent-velocity directions, with $\gamma\|\overline{\boldsymbol{u}}$; while in two-dimensional and rotationally symmetric flows there is $\gamma\overline{u} = 0$.

5.2 Attached Vortex Sheet and Its Separation

5.2.1 Attached and Bound Vortex Sheet

Attached vortex sheet is the simplified model of boundary layer. We first observe that, the rate of change of the sheet strength γ is closely related to boundary vorticity flux (BVF), now defined by $\boldsymbol{\sigma} = -\nu \partial \boldsymbol{\omega}/\partial n$.[1] Suppose we are solving the time-dependent Navier-Stokes equation along with continuity equation in a small time step $\Delta t \ll 1$. The solution procedure can be split into two substeps, solving the Euler equation under no-through condition and the *Stokes equation* [an approximate theory for low-*Re* flow, see Wu et al. (2006)] under no-slip condition, which takes care of fluid advection and diffusion, respectively (assuming $\rho = 1$):

$$\frac{\partial \boldsymbol{u}_e}{\partial t} + \boldsymbol{u}_e \cdot \nabla \boldsymbol{u}_e = -\nabla p_e, \quad \boldsymbol{n} \cdot (\boldsymbol{u}_e - \boldsymbol{b}) = 0, \tag{5.2.1a}$$

$$\frac{\partial \boldsymbol{u}_s}{\partial t} + \nabla p_s = -\nu \nabla \times \boldsymbol{\omega}_s, \quad \boldsymbol{n} \times (\boldsymbol{u}_e - \boldsymbol{b}) = \boldsymbol{0}, \tag{5.2.1b}$$

where $\boldsymbol{b}(\boldsymbol{x}, t)$ is the wall velocity. The error of this *fractional-step* approach has been proved to be $O(\nu \Delta t)$. Assume at $t = 0$ the flow satisfies the Navier-Stokes equations. After Δt, the Euler substep produces a slip velocity $\boldsymbol{u}_s$; then once the Stokes substep starts, the no-slip condition implies that an attached vortex sheet $\boldsymbol{\gamma} = (\boldsymbol{u}_s - \boldsymbol{b}) \times \boldsymbol{n}$ is produced. It has then been rationally shown and numerically confirmed that, to bring $\boldsymbol{u}_s$ to zero at the end of Δt there must be (for details see Wu et al. (2006), Sect. 4.5)

$$\boldsymbol{\sigma} = -\nu \frac{\partial \boldsymbol{\omega}}{\partial n} = \frac{\boldsymbol{\gamma}}{\Delta t} + O(Re^{-1/2}) \tag{5.2.2}$$

with $\boldsymbol{n}$ pointing out of the body, and hence the production rate of $\boldsymbol{\gamma}$ is nothing but the boundary vorticity flux. Note that physically the boundary $(p, \boldsymbol{\omega})$ coupling always exists, which however is bypassed by the fractional-step method.

Next, attached vortex sheet does not satisfy the preceding theorem for free vortex sheet, since the fluid at its inner side (the bottom of the boundary layer) has smaller total pressure $p + \rho q^2/2$; but for a translational body with velocity $\boldsymbol{b}$ its property can be quickly determined. For example, a thin wing S flying at velocity $\boldsymbol{b} = \boldsymbol{U}_B$ at a small angle of attack is always dressed in a pair of attached vortex sheets on its upper $(+)$ and lower $(-)$ surfaces, forming a sandwich structure. The location and shape of this pair are known, and they still satisfy (5.1.6a) and (5.1.7) so that

$$\boldsymbol{\gamma}^{\pm} = \boldsymbol{n}^{\pm} \times (\boldsymbol{u}^{\pm} - \boldsymbol{U}_B), \quad \overline{\boldsymbol{u}}^{\pm} = \frac{1}{2}(\boldsymbol{u}^{\pm} + \boldsymbol{U}_B), \tag{5.2.3}$$

where $\boldsymbol{n}^{\pm}$ are the unit normals of the upper and lower attached vortex sheets pointing from inner to outer sides, and $\boldsymbol{u}^{\pm} = \nabla \phi^{\pm}$ denote the potential velocities at the outer sides of these sheets. Since the outer flow can be quickly solved by inviscid potential-flow theory, so will be $\boldsymbol{\gamma}^{\pm}$. This is a considerable advantage in using vortex sheet models.

---

[1] In this chapter the unit normal vector of a vortex sheet is defined to point from $-$ side to $+$ side, see Fig. 5.1, so for attached vortex sheet with $-$ side adjacent to the wall $\boldsymbol{n}$ points out of the wall. This adds a minus sign to the BVF definition.

We now further assume the wing thickness is negligible, such that the sandwich structure can be viewed as a single net "vortex sheet" known as a *bound vortex sheet* (denoted by suffix b) which, unlike free vortex sheet, *can stand pressure jump*. Take the upward unit vector $\hat{\boldsymbol{n}} = \boldsymbol{n}^+ = -\boldsymbol{n}^-$ as the normal of the wing, from (5.2.3) it follows that

$$\gamma_b = \gamma^+ + \gamma^- = \hat{\boldsymbol{n}} \times \nabla[\![\phi]\!] = \hat{\boldsymbol{n}} \times \nabla_\pi \Gamma_b, \quad \overline{\boldsymbol{u}}_b = \boldsymbol{U}_B + \frac{1}{2}(\boldsymbol{u}^+ + \boldsymbol{u}^-). \quad (5.2.4)$$

Thus, once ϕ is solved, all properties of a bound vortex sheet are known. The momentum jump equation (5.1.5) is not applicable to bound vortex sheet. Rather, for *steady flow* over a bound sheet, (5.1.14b) yields

$$[\![p]\!] = \rho \overline{u}_b \gamma_b \sin\beta \quad \text{or} \quad -[\![p]\!]\boldsymbol{n} = -\rho \overline{\boldsymbol{u}}_b \times \gamma_b, \quad (5.2.5)$$

which for two-dimensional flow over a thin airfoil reaches the maximum value since $\beta = \pi/2$:

$$[\![p]\!] = \rho \overline{u}_b \gamma_b. \quad (5.2.6)$$

This is a local analogy to the Kutta-Joukowski lift formula (2.6.26), but now the tangent component of $\overline{\boldsymbol{u}}_b$ plays the role of the flow velocity $\boldsymbol{U} = -\boldsymbol{U}_B$ at infinity. In fact, it can be shown (Problem 5.1) that the integral of (5.2.6) over the airfoil chord recovers exactly (2.6.26).

Bound vortex sheets have been a model used extensively in classic aerodynamics, where they mimic a solid thin wing along with its attached boundary layers as a whole at $Re \to \infty$. Then the flow becomes internally unbounded. This is a big simplification in theoretical analysis, for example permitting the use of Biot-Savart formula (3.1.37).

Example: "Coffee-spoon experiment". Consider a two-dimensional flow in a (y, z)-plane of an inviscid fluid at rest at infinity, caused by a flat plate at $|y| \le b$ moving downward along its normal with uniform speed $-\boldsymbol{W} = -\boldsymbol{e}_z W$ (Fig. 5.4a). A Galilean transformation casts the problem to a vertical incoming flow $\boldsymbol{W}$ over the stationary plate (Fig. 5.4b), for which we can easily find the velocity potential and stream function, and thereby the tangential and normal velocities (v, w) as well as the distributions of bound circulation and vortex-sheet strength:

$$\phi(y, \pm 0) = \mp W\sqrt{b^2 - y^2}, \quad (5.2.7a)$$

$$v(y, \pm 0) = \mp \frac{Wy}{\sqrt{b^2 - y^2}}, \quad w(y, \pm 0) = 0, \quad (5.2.7b)$$

$$\Gamma(y) = 2W\sqrt{b^2 - y^2}, \quad (5.2.7c)$$

$$\gamma(y) = \boldsymbol{e}_x \frac{2Wy}{\sqrt{b^2 - y^2}}. \quad (5.2.7d)$$

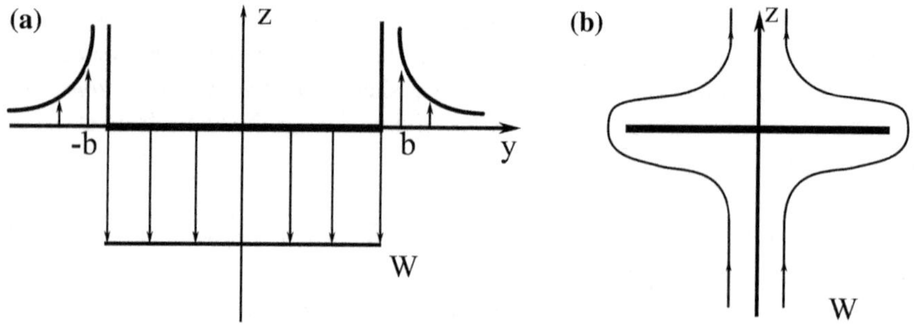

Fig. 5.4 Instantaneous flow caused by a flat plate (a "coffee spoon") moving downward with normal velocity $-W$. **a** Viewed in laboratory frame of reference. **b** Viewed in co-moving frame of reference. The normal velocity profile $w(y)$ at $|y| > b$ outside the plate is calculated from the complex velocity potential

This flow may serve as a simplified model of the so-called "*coffee spoon experiment*", like what happens if we disturb a cup of coffee by a spoon or row a boat.

5.2.2 Kutta Condition and Vortex-Sheet Separation

Equation (5.2.7d) shows that as $y \to \pm b$, γ has a singularity of $|b^2 - y^2|^{-1/2}$ type, and hence the attached flow pattern of Fig. 5.4b is unrealistic. Instead, the true flow pattern caused by the same plate motion in viscous flow is shown in Fig. 5.5. The difference lies in the fact that viscous flow must satisfy an important regularity condition known as the **Kutta condition** or **Joukowski hypothesis**: *Any steady flow cannot turn around a sharp corner with a nonzero velocity.*[2] This is easily understood, for otherwise a nonzero velocity would have more than one directions or the acceleration would be infinity at the corner. The difference between Figs. 5.4b and 5.5 indicates that the control effect of this condition on global flow behavior over a body with sharp edge is very critical.

Originally, the Kutta condition was introduced at the sharp trailing edge of an airfoil in two-dimensional potential flow. It is well known that as the flow is transformed to that over a circular cylinder by conformal mapping, one can add arbitrary circulation to the cylinder flow and hence to the airfoil flow as well. The airfoil flow with $\Gamma = 0$ and unrealistic singular trailing-edge flow is only one of infinitely many possible solutions with indeterminate Γ, owing to the double connectivity of the flow domain. The Kutta condition assigns a unique Γ to the airfoil, which is the sum

[2]First proposed by Kutta in 1902 and then by Joukowski independently in 1906. This condition has been stated in different ways. For example, *the flow must leave the trailing edge smoothly* (Glauert 1926, p. 119); *in the final steady flow, the rear stagnation point shall coincide with the trailing edge of the airfoil* (Kármán and Burgers 1935, p. 137; Batchelor 1967, p. 437). These statements are equivalent for steady flow, but not for unsteady flow (see Sect. 8.5.2).

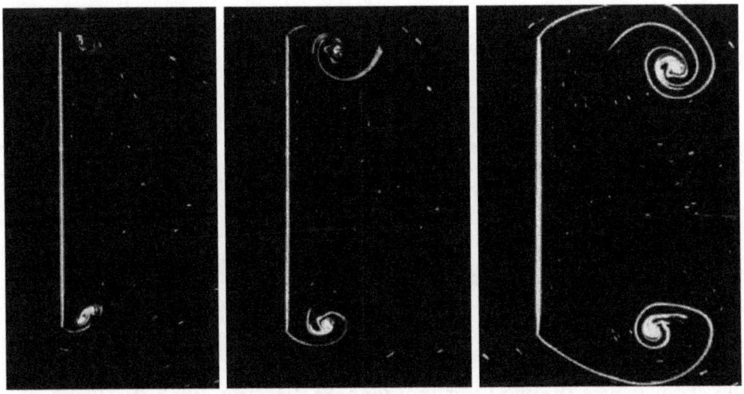

Fig. 5.5 Impulsive motion of a plate normal to itself from right to left at Reynolds number 88 based on the plate breadth L. The *photos* were taken as the plate had moved 0.079, 0.26, and 0.93 L. From Van Dyke (1982)

of the strengths of the physically realistic upper and lower attached vortex sheets (bound vortex sheet). In three dimensions the flow domain is singly-connected, so there would be no room to add the condition.

In reality, however, the Kutta condition exists for both two- and three-dimensional flows, which should be a natural result of fluid viscosity $\mu \neq 0$ (Glauert 1926; van Kármán and Burgers 1935; Batchelor 1967). However, although the condition has been routinely applied to inviscid steady and attached wing flow with sharp edge, its real formation involves highly transient viscous processes, of which the central issue is *unsteady flow separation*. This complicated issue will be introduced later in Sect. 8.5, where one can make detailed examination of the formation process of the Kutta condition. Here we just use the condition to analyze the behavior of vortex-sheet separation.

In a viscous and steady flow at $Re \to \infty$, the Kutta condition can do more than requiring an attached vortex sheet to leave (separate from) sharp edge. It also holds at a *separation line* in three-dimensional flow even on smooth surface (for more on flow separation see Chap. 8). Besides, the Kutta condition can be used to determine the initial strength and orientation of a free vortex sheet as it just leaves the body. A general formulation for steady flow is as follows.

In a three-dimensional flow, assume a free vortex sheet γ leaves the body surface along a *separation line* L thereon, which can be a sharp edge or on a smooth surface due to early separation (for the latter case the location of L has to be determined by viscous theory). The flows at both sides of the sheet are assumed from the same upstream fluid. Let $\boldsymbol{e}_1$ and $\boldsymbol{e}_2$ be the unit tangent vector along L and that tangent to the body surface and orthogonal to $\boldsymbol{e}_1$, respectively, such that $\boldsymbol{u}^{\pm} = u_1^{\pm}\boldsymbol{e}_1 + u_2^{\pm}\boldsymbol{e}_2$, see Fig. 5.6a. We examine how the Kutta condition along with the Bernoulli equation determines the initial behavior of the free vortex sheet.

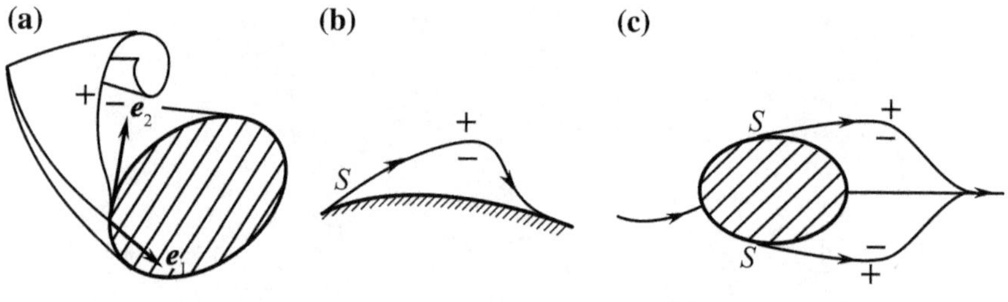

Fig. 5.6 Vortex sheet separating from a smooth surface. **a** Three-dimensional flow; **b** and **c** two-dimensional flow

If the sheet leaves the wall not in a tangent direction, then the Kutta condition implies that there must be $u_2^{\pm} = 0$ and hence the flow can only be along the $\pm \boldsymbol{e}_1$-directions. If the flow is steady, by (5.1.15) there must be $\boldsymbol{u}^+ = \pm \boldsymbol{u}^-$. If $\boldsymbol{u}^+ = \boldsymbol{u}^-$, we have $[\![\boldsymbol{u}]\!] = \boldsymbol{0}$, implying no free vortex sheet at all; while if $\boldsymbol{u}^+ = -\boldsymbol{u}^-$, we have $\overline{\boldsymbol{u}} = \boldsymbol{0}$ and hence even a free vortex sheet exists it cannot shed off from the body. Thus, *a steady vortex sheet can leave the body surface only along a tangential direction.*

Here we digress to make an important remark. It should be born in mind that the above assertion and many others in this chapter on vortex-sheet dynamics are valid only for the vortex-sheet model at $Re \to \infty$, but not directly applicable to viscous flow at finite Re. For example, Chap. 8 will discuss boundary-layer separation at finite Re, where the separated shear layer will leave the body surface at a nonzero angle. This is also the case of Fig. 5.5. But this angle approaches zero as $Re \to \infty$, in consistence with the foregoing analysis. Moreover, the Kutta condition itself holds firmly only for steady flow. In Sect. 8.5.2 we shall see that, in real viscous flow the condition takes time to be established through a set of transient processes, in which some fluid particles may even move around a sharp edge in very short time and to a very short distance.

Return to vortex-sheet separation problem. In two dimensions (Fig. 5.6b, c) there is no velocity component along the $\boldsymbol{e}_1$ direction. Hence, for steady flow $\boldsymbol{u}^+ = \pm \boldsymbol{u}^- = \pm u_2 \boldsymbol{e}_2$, and by the same reasoning, *if all streamlines are from the same upstream flow, then no steady vortex sheet can shed off from a two-dimensional body*. Besides, an attached vortex bubble can be included here as a portion of the body, see Fig. 5.6b, c; but inside the bubble the flow has lower total pressure. It can then be shown that the outer boundary sheet of the bubble must still leave the wall along a tangential direction, with the inside fluid at rest at the cusp.

The above general analysis agrees the classic *free-streamline theory* for two-dimensional and steady flow (e.g. Batchelor 1967, Sect. 6.13) constructed in late 1870s by Helmholtz, Kirchhoff and others, for modeling separated flow behind a bluff body. There, not only the body contour is represented by a "zero-streamline" (stream function $\psi = 0$), but also so is the separating vortex sheet that divides the main flow and a "dead-water region" behind the body where the fluid is assumed at rest. The pressure p is continuous across the sheet, and is constant along the sheet (hence so is the kinetic energy $q^2/2$ on the flow side). Although the theory's predicted drag is not in agreement with experiments, it does give a correct description of the *local* behavior near the separation point as we found before: the vortex sheet leaves the surface tangentially, and in the cusp region between the sheet and wall the fluid is locally at rest.

Note that so far all of our discussion on vortex sheet separation only tells *how* the sheet leaves a body surface, but it does not tell *where* the separation takes place if the body surface is smooth. The non-uniqueness of the inviscid solution implies that there can be many solutions with the sheet leaving the surface tangentially but at different locations. This also happens in the free-streamline theory. It is of interest that the attempt of using the free-streamline theory to fix the separation point ended with a no-solution dilemma.

Let x be the coordinate along a curved body surface, and the separating vortex sheet leave the surface at point $x = x_s$. Denote $\xi = x - x_s$. By the free-streamline theory, as $\xi \to 0^{\pm}$, the sheet curvature κ and relative pressure p are given by (e.g. Sychev et al. 1998)

$$\kappa(x) = -k\xi^{-1/2} + \kappa_0 + O(\xi^{1/2}), \quad \xi \to 0^+; \tag{5.2.8a}$$

$$p(x) = k(-\xi)^{-1/2} + \frac{16}{3}k^2 + O[(-\xi)^{1/2}], \quad \xi \to 0^-; \tag{5.2.8b}$$

$$p(x) = 0, \quad \xi > 0, \tag{5.2.8c}$$

where κ_0 is the wall curvature and k is an arbitrary constant related to the position of x_s.

Now, if $k < 0$, (5.2.8a) implies that the separating vortex sheet would intersect the body surface, which is physically impossible. If $k > 0$, (5.2.8b) implies an infinitely large adverse pressure gradient at x_s, so the separation would have happened before x_s, which should also be rejected. What left is the choice $k = 0$, so by (5.2.8a) $\kappa = \kappa_0$ at x_s, known as a *smooth separation* (for flow over circular cylinder, this condition predicts that separation occurs at polar angle $\theta = 55°$). However, by (5.2.8b) the pressure gradient would vanish, and there should be no separation at all!

Remarkably, it was this dilemma that finally led to the right solution a century after Kirchhoff, which has to be in viscous-flow regime and will be discussed in Sect. 8.3.2.

5.3 Motion of Free Vortex Sheet

5.3.1 Rolling up and Kaden's Similarity Law

Free vortex sheet is also the simplest model for understanding the formation of concentrated vortices. If it is desired to produce a vortex by solid body moving relative to fluid, one might first consider to rotate a circular cylindrical water container, of which for a discussion see Problem 3.1 (Fig. 3.20). However, for fluid of small viscosity, the required time for the water to become fully rotating would be infinitely long as $\mu \to 0$, but what we see when we disturb a cup of coffee by a spoon or row a boat, say, is that an axial vortex is formed right after the spoon or paddle starts to move. Moreover, for given circulation Γ and vortex core radius a with velocity $v = \Gamma/2\pi a$ there, reducing a to a point would require infinite kinetic energy. The smaller the core is, the more energy is necessary to produce it. This however conflicts the common existence of many thin-core axial vortices (vortex filaments) of finite energy.

To resolve these apparent paradoxes, notice that in Fig. 5.5 what appears first is a vortex sheet shedding from the moving plate's edge, which then rolls up quickly into a spiral. This phenomenon has been studied theoretically by Kaden (1931), who shows that a vortex sheet with singular strength at its end, once becomes free, must roll into tight spirals. Based on this finding, Betz (1950) asserts that *the only known mechanism for the quick formation of axial vortices in a fluid of small viscosity is the rolling up of thin shear layers*, as sketched in Fig. 5.7a. The centrifugal force of the vortex-sheet turns is balanced by the suction force at the spoon's edge, and these turns must eventually be smeared out by viscosity to form a thin smooth vortex core with a size that matches the correct energy level.

In general, the aforementioned singular point around which the vortex-sheet rolls up is not necessarily at the end of the sheet. It has also been found that, under small regular periodic disturbance, an originally flat and endless infinite vortex sheet may evolve to spontaneous curvature discontinuities, which develop to an array of *double-branch spirals*, i.e., one branch runs into the core and the other runs out of it; see Fig. 5.7b.

Let us now look at the essence of Kaden's (1931) discovery. Consider for simplicity the rolling-up process of a semi-infinite sheet located initially at $x \in [0, \infty)$,

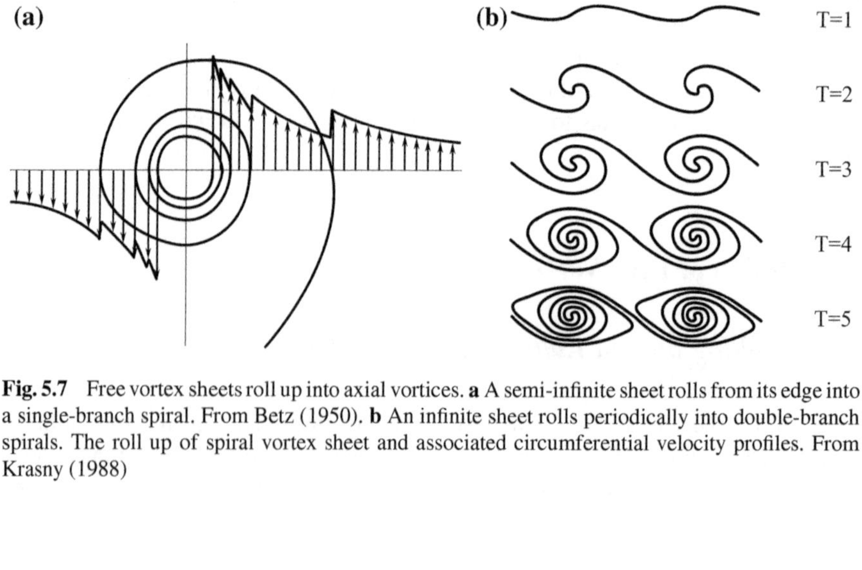

Fig. 5.7 Free vortex sheets roll up into axial vortices. **a** A semi-infinite sheet rolls from its edge into a single-branch spiral. From Betz (1950). **b** An infinite sheet rolls periodically into double-branch spirals. The roll up of spiral vortex sheet and associated circumferential velocity profiles. From Krasny (1988)

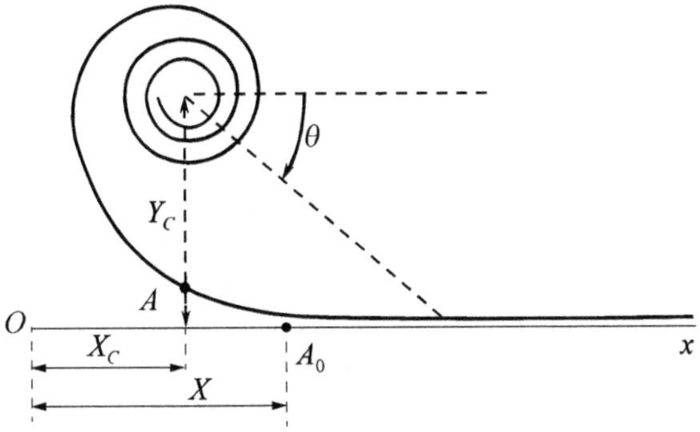

Fig. 5.8 Sketch of the roll-up of a semi-infinitely extended vortex sheet. Adapted from von Kármán and Burgers (1935)

see Fig. 5.8, of which the circulation and strength at $t = 0$ are specified as

$$\Gamma(x)_{t=0} = 2kx^{1-n}, \quad \gamma(x)_{t=0} = \frac{d\Gamma}{dx} = 2(1-n)kx^{-n}, \tag{5.3.1}$$

where k is a dimensional constant and $n \in [0, 1]$ is a number that in finite-wing aerodynamics characterizes spanwise wing-loading distribution. For example, $n = 1/2$ and $n = 0$ correspond to the wing loadings on elliptical and delta wings, respectively (for the implication of wing-loading see Sect. 5.4.1 and Chap. 9). Figure 5.7 suggests that as the rolling-up proceeds for $t > 0$, there appears an inner core region where the spiral is almost axisymmetric, so in polar coordinates (r, θ) the Kelvin circulation theorem yields

$$\frac{D\Gamma}{Dt} = \frac{\partial \Gamma}{\partial t} + \frac{\Gamma}{r^2}\frac{\partial \Gamma}{\partial \theta} = \frac{\partial \Gamma}{\partial t} = 0, \tag{5.3.2}$$

indicating that inside each circle the total vorticity remains time-invariant and we may write $\Gamma = \Gamma(r)$ even if the global flow is still unsteady and non-axisymmetric. Now Γ and k have dimensions L^2/T and L^{1+n}/T, respectively; so a dimensional analysis leads at once to

$$\Gamma(r) = 2k(\lambda r)^{1-n}, \quad 0 < n < 1, \tag{5.3.3}$$

where λ is a constant to be determined shortly. Consequently, the circumferential velocity must take the form

$$v = \frac{\Gamma}{2\pi r} = \frac{\beta}{r^n}, \quad \beta \equiv \frac{k\lambda^{1-n}}{\pi}. \tag{5.3.4}$$

Since for a material point at r_P on the vortex sheet there is $r_P d\theta/dt = v$, by (5.3.3) and (5.3.4) we have $\theta_P - \theta_0 = \beta t r^{-(1+n)}$, from which follows a *similarity law* first derived by Kaden (1931) for $n = 1/2$ (Moore and Saffman 1973):

$$r = \left(\frac{\beta t}{\theta - \theta_0}\right)^{\frac{1}{1+n}} \quad \text{as} \quad \theta \to \infty. \tag{5.3.5}$$

As $\theta \to \infty$, the arc length of the sheet is also infinity:

$$s = \int^{\infty} \sqrt{1 + \frac{1}{r^2}\left(\frac{dr}{d\theta}\right)^2}\, r d\theta \sim \int^{\infty} \frac{d\theta}{\theta^{2/3}} = \infty, \tag{5.3.6}$$

indicating that the vortex sheet is stretched, making its strength $\gamma = -\partial\Gamma/\partial s$ [see (5.1.10)] decreases as θ increases. Thus the tangential velocity jump across each turn also decreases as shown in Fig. 5.7. On the other hand, associated with spiral roll, the vortex sheet feeds vorticity into the core at a rate

$$\lim_{\delta\to 0} \int_0^{\delta} u_s \omega dn = \lim_{\delta\to 0} \int_0^{\delta} (\boldsymbol{\omega} \times \boldsymbol{u}) \cdot \boldsymbol{n} dn. \tag{5.3.7}$$

Because Γ and v are time-independent inside $r = a_0 \sim (\beta t)^{1/(1+n)}$, the vorticity feeding from the sheet must cause a core expanding. But when t is so large that the tightly rolled-up spiral has entrained vorticity from the sheet's segments far from $s = 0$, the growth of a_0 will stop to reach a saturated steady core. By then the feeding vortex sheet will be disconnected from the core.

Suppose that the vorticity within a distance X of the initial flat sheet enters a circular core of radius $R(X)$, see Fig. 5.8. We can then obtain an R-X relation, and thereby obtain an estimate of the unknown constant λ introduced in (5.3.3). To this end we use the time-invariance of the total circulation Γ and the second vorticity moment (angular impulse) L (Sect. 3.3.3) as the sheet within X evolves to a circular core in R. The invariance of the former is ensured by the total circulation conservation theorem (Sect. 3.3.1), and that of the latter is ensured by the moment-free status of the vortex about its center. Namely, we have[3]

$$\Gamma = 2\pi \int_0^R r\omega dr = \int_0^X \gamma(x) dx, \tag{5.3.8a}$$

$$L = -\pi \int_0^R r^3 \omega dr = 2(1-n)k \int_0^X (x - X_c)^2 x^{-n} dx, \tag{5.3.8b}$$

where

$$\boldsymbol{X}_c = \frac{\int \boldsymbol{x}\omega dS}{\int \omega dS} \tag{5.3.9}$$

defines the *vorticity centroid* and by (5.3.4)

[3] The sign difference from (5.1.10) comes from the choice of a clockwise direction of Γ in Fig. 5.8.

$$\omega = \frac{1}{r}\frac{\partial(rv)}{\partial r} = \beta(1-n)r^{-(1+n)} \tag{5.3.10}$$

is the core vorticity. Then by (5.3.1), (5.3.3), and (5.3.10), from (5.3.9) and (5.3.8) follows that

$$X_c = \frac{1-n}{2-n}X, \quad R = \frac{X}{\lambda}, \quad \lambda = 2-n. \tag{5.3.11}$$

Thus, λ may be viewed as a "stretching factor": as the vorticity of the sheet within X rolls into the circle R it is stretched. For more detailed discussion see Tong et al. (2009).

5.3.2 Methods of Computing Vortex Sheet Motion

The above discussion on vortex-sheet roll up along with Kaden's similarity law provides only a qualitative physical picture. If one considers a thin wing S_b moving through an "inviscid" incompressible flow (viscous flow at $Re = \infty$) and represents the wing by a bound vortex sheet, the sheet strength $\gamma_b(\boldsymbol{x}, t)$ is determined by requiring its induced normal velocity equal to the wing's prescribed normal velocity $U_n(\boldsymbol{x}, t)$ for all $\boldsymbol{x} \in S_b$. If there is also a wake vortex sheet S_w shedding off and rolling up, the shape and strength $\gamma_w(\boldsymbol{x}, t)$ of S_w, as well as its location, motion, and deformation are all to be solved. This amounts to solving the disturbance potential flow equation $\nabla^2\phi = 0$ with S_w being a tangent discontinuity, which itself is a material surface (stream surface for steady flow) and on which $[\![p]\!] = 0$ and $D\Gamma/Dt = 0$. Besides, the total vorticity (or total circulation in two dimensions) must be conserved. Note that a free wake vortex sheet is under the induction of both itself and the bound sheet γ_b, and γ_w may also induce a velocity affecting the normal velocity on S_b, and hence alters γ_b.

Except certain linearized approximations by which the above bound-free vortex sheet problem can be solved analytically, see Chap. 9, in general this problem has to be solved numerically. A numerical method commonly applied to steady and unsteady aerodynamics is the *panel method*, in which the bound and free vortex sheets are discretized to small panels, and on each panel one assign a fundamental solution of $\nabla^2\phi = 0$ such as point source/sink, doublet, or vortex, of which the strengths are to be solved. For a systematic presentation of the method the reader is referred to Katz and Plotkin (2001).

Instead of solving a Laplace equation with vortex sheet as tangent discontinuity, one may use the Biot-Savart formula (3.1.37),

$$\boldsymbol{u}(\boldsymbol{x}) = \frac{1}{2(n-1)\pi}\int_S \frac{\gamma(\boldsymbol{x}') \times \boldsymbol{r}}{r^n} dS(\boldsymbol{x}'), \tag{5.3.12}$$

to formulate an integral-differential equation to solve γ under specific initial-boundary conditions. In two-dimensional flow on the (x, y)-plane and in terms of complex variable $Z = x + iy$, (5.3.12) is a line integral

$$\frac{dZ^*}{dt} = \frac{1}{2\pi i}\int \frac{\gamma(s)ds}{Z - Z(s)}, \tag{5.3.13}$$

where $Z(s)$ defines the shape of the vortex sheet, s is the arc length or other proper coordinate, and the asterisk denotes complex conjugate. These formulas are necessary for solving vortex sheet strength and its motion in various specific problems. Generically, this leads to an *integral equation*, of which the solution will yield the unknown tangent velocity $\boldsymbol{u}_\pm$ at both surfaces of the wing, and then $\gamma_b = \boldsymbol{n} \times [\![\boldsymbol{u}]\!]$. Later in Chap. 9 we shall meet this kind of integral equations. Here we only discuss the *self-induced evolution* of a free vortex sheet in two-dimensional flow on (x, y)-plane, from a prescribed initial condition. Namely, we apply (5.3.13) to all points at the sheet and solve it. Since $\gamma(s)ds = -d\Gamma(s)$ is the infinitesimal circulation increment, we may reversely view s as functions of the Lagrangian invariant Γ; thus (5.3.13) leads to the well-known *Birkhoff-Rott equation*:

$$\frac{\partial}{\partial t}Z^*(\Gamma, t) = -\frac{1}{2\pi i}\mathrm{pv}\int_{\Gamma_0}^{\Gamma_e} \frac{d\Gamma'}{Z(\Gamma, t) - Z(\Gamma', t)}, \tag{5.3.14}$$

where pv means taking Cauchy principal value to handle the singularity when $\Gamma' \to \Gamma$, and Γ_0 and Γ_e are the circulations at the two end points of the sheet.

Equation (5.3.14) is a highly nonlinear and singular integral-differential equation. A numerical method and example for solve (5.3.14) will be mentioned shortly.

5.4 Formation of Wing Vortices

This section exemplifies the initial formation process of two important aerodynamic vortices via vortex-sheet rolling up. The formation process will be completed after viscous effect is taken into account, which we discuss in the next chapter.

5.4.1 Formation of Wingtip Vortices

Consider a steady three-dimensional incoming flow $\boldsymbol{U} = (U, 0, 0)$ in Cartesian coordinates (x, y, z) over a finite wing located at $x \in [0, c(y)]$, $y \in [-b, b]$ at an angle of attack $\alpha \ll 1$. Here, $c(y)$ is the chord length at spanwise y-station with the root-chord being $c(0) = c_0$. Let S_b be the wing area and $\Lambda = S_b/c_0^2$ define the wing *aspect ratio*. When $\Lambda \gg 1$, at each y-station the sectional airfoil flow is approximately two-dimensional, so that the Kutta-Joukowski formula (2.6.26) gives

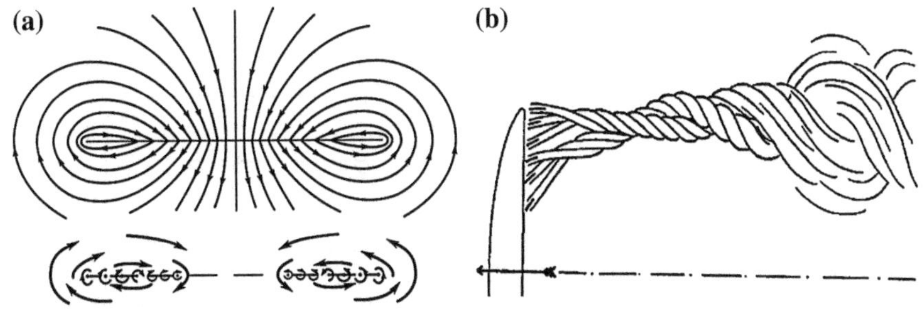

Fig. 5.9 Sketch of wingtip vortices formed by rolling-up of wake vortex sheet behind a finite-span wing. **a** Velocity induced by streamwise vortex sheet viewed from *downstream*. **b** *Top view* After Lanchester (1907)

a spanwise *wing-loading distribution* $l(y) = \rho U \Gamma(y)$ measured by $\Gamma(y)$, which can be conceived as the circulation of a bound vortex that plays the role of the wing. But by the Helmholtz first vorticity-tube theorem, the bound vortex cannot be a single vorticity tube. Rather, it has to be a bundle of thin tubes, of which each turns to x-direction at a y-location, and hence a free vortex sheet of strength $\gamma(y) = -\boldsymbol{e}_x d\Gamma/dy$ is formed at the trailing edge that sheds into wake according to the Kutta condition. This initially flat wake vortex sheet will roll up to form a pair of *wingtip vortices*, of which the existence and formation were first observed by Lanchester, see his sketch shown in Fig. 5.9.

Let us construct a simple vortex-sheet model for this process. Because the variation of flow quantities in the x-direction is much milder than that on the transverse (y, z)-plane, we may set $x = Ut$ to cast this steady three-dimensional problem approximately to an unsteady two-dimensional problem on the (y, z)-plane at varying t (varying downstream location in original problem), with upward normal velocity $W = U \sin\alpha \simeq U\alpha$. For a wing of elliptic load distribution, the circulation and strength of the bound vortex sheet are given by (5.2.7), see Fig. 5.4:

$$\Gamma(y) = 2U\alpha\sqrt{b^2 - y^2}, \quad \gamma(y) = \boldsymbol{e}_x \frac{2U\alpha y}{\sqrt{b^2 - y^2}}. \tag{5.4.1}$$

Now, let an observer pass the trailing edge at $t = 0$ (or the wing is suddenly dissolved). From Figs. 5.4a and 5.5 it is clear that strong circulations must be quickly formed at $y = \pm b$ and $t = 0^+$, which in viscous flow come from the natural development of the boundary layers at both sides of the plate. Set $y = y' - b$ in (5.4.1), we may focus on the vortex sheet rolling near $y' = 0$ by letting $b \to \infty$ and $y'/b \to 0$ so that the influence of the flow at the opposite wingtip on the flow near $y' = 0$ is negligible. Then the initial circulation and the strength of the vortex sheet near $y' = 0$ is precisely the preceding semi-infinite vortex-sheet model that rolls up to tight concentric circles as predicted by the similarity law (5.3.5) with $n = 1/2$.

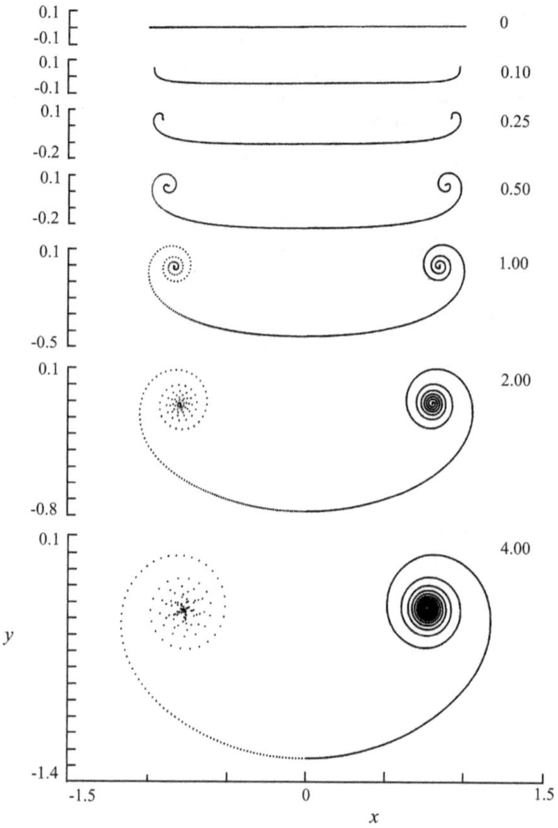

Fig. 5.10 Early stage of the trailing vortex sheet rolling up calculated by vortex-blob method (Krasny 1987). The *left half* of the plot shows the blob positions, noticing the stretching of the rolled-up sheet as predicted by (5.3.6); and the *right half* is their interpolation curve

With this simplified model, one may apply the kinematic Birkhoff-Rott equation (5.3.14) to calculate the formation of a wingtip vortex. This was first done by Krasny (1987) numerically, using a discrete vortex-blob method. The key to success is to smooth out the kernel function so that the singularity in (5.3.14) as $\Gamma' \to \Gamma$ is removed. Figure 5.10 shows Krasny's result, where the tightly rolled part forms a set of nearly concentric circles and agrees the prediction of Kaden's similarity law.

5.4.2 Formation of Leading-Edge Vortex

The steady flow over a slender delta wing is a typical situation where the vortex sheets separate from sharp leading edges due also to the Kutta condition, and roll into axial vortices above the wing, see the sketch in Fig. 5.11. The leading-edge vortices can enhance the lift at larger angles of attack. While in the downstream portion of the flow the leading and trailing edge vortices merge to a complex system, in the front portion where the upstream influence of the trailing edge is negligible, the flow has

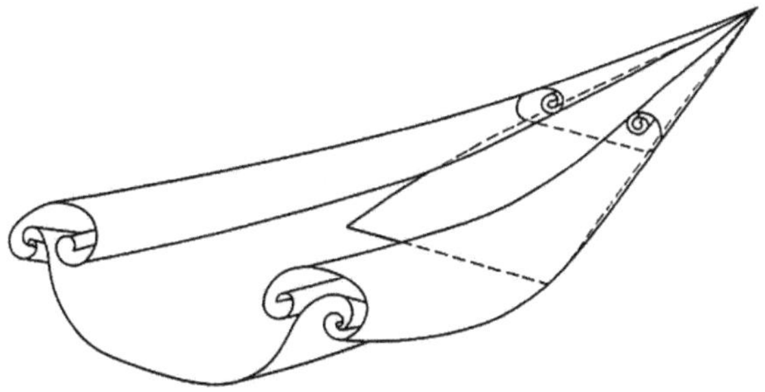

Fig. 5.11 Vortex sheets around a slender delta wing with sharp leading edges at large angles of attack. From Küchemann (1978)

a spatial similarity that it depends linearly on x. Such a flow is called *conical flow*, for which independent variables (r, θ, x) can be reduced to $(r/x, \theta)$.

For a conical flow the Birkhoff-Rott equation (5.3.14) cannot be applied, but the aforementioned potential-flow solution with tangent discontinuity at vortex sheet can still be found analytically by series expansion (Küchemann and Weber 1965). Like in the trailing-vortex formation problem, the solution is invalid near the wing's leading edge where the Kutta condition should be satisfied. Thus the starting point of the sheet's trace has to be artificially chosen at a point away from the wing. The calculated trace of the vortex sheet and profiles of velocity $\boldsymbol{u} = (u_r, u_\theta, u_x)$ for a slender delta wing are plotted in Fig. 5.12, compared with a carefully measured result by Earnshaw (1961). The agreement is remarkable.

Figure 5.12 also indicates that, however, as $r/z \to 0$ the measured u_θ increases first and then abruptly drops to zero, but by invisicd theory $u_\theta \to \infty$; and, the measured axial velocity u_x has an abrupt and strong finite growth with maximum at $r/z = 0$, but by inviscid theory u_x is even more singular than u_θ at $r/z = 0$. This fact exemplifies that the physical processes in axial-vortex formation cannot be vortex-sheet rolling up alone. Viscous effect has to be included to smooth out those singularities.

We notice a basic difference between the leading-edge and trailing-edge spirals. In the leading-edge spiral, the fluid entering the core between two turns of the spiral can flow away axially, but this mechanism is absent in two-dimensional unsteady flow. For the latter the core has to sufficiently widen out as t to accommodate all the entrained fluid. In the original steady and three-dimensional problem this implies that the trailing-edge spirals expand as x increases.

5.5 On the Role of Vortex-Sheet Dynamics

We conclude this chapter by a bird's-eye view of the overall role of vortex-sheet dynamics in fluid dynamics.

1. From inviscid theory with $\mu \equiv 0$ to viscous theory with $\mu \to 0$. As seen in Kirchhoff's free-streamline theory mentioned in Sect. 5.2.2, the early use of the

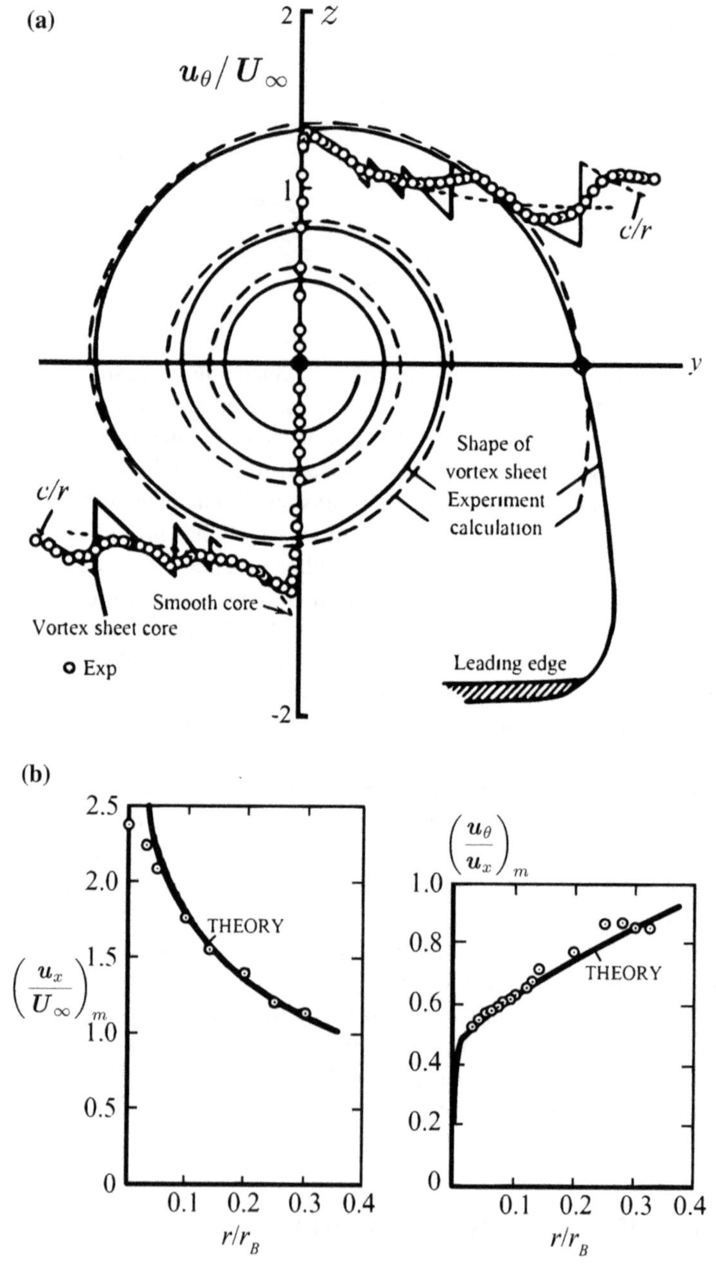

Fig. 5.12 Theoretical and experimental results for a delta wing of sweeping angle $\chi = 76°$ at angle of attack $\alpha = 20°$ and $Re = U_\infty x/\nu = 2 \times 10^6$. **a** Trace of vortex sheet and circumferential velocity u_θ in a transverse plane. **b** Mean velocity components u_x/U_∞, u_θ/U_∞, and u_θ/u_x in vortex core. r_B is the reference radius of the intersection of the experimental trace and the y-axis of plot (**a**), which was chosen to start the theoretical calculation of the sheet. From Küchemann (1978)

concept of vortex sheet was of inviscid nature. Indeed, vortex sheet is mathematically admitted by the Euler equation, which is consistent with the theory of characteristics of the Euler equation and inviscid integral conservation laws. An Euler solution with vortex sheet is known as a *weak Euler solution*. However, after Prandtl's boundary-layer theory, one realized that even an attached vortex sheet must exist between the outer potential flow and the wall where the no-slip condition should be satisfied. Free vortex sheets are the result of separation of attached sheets. Then the oversimplified "dead water" in Kirchhoff's theory has been replaced by various separated flows. In other words, owing to Prandtl, the inviscid flow model with $\mu \equiv 0$ and tangent discontinuity has evolved to a *viscous theory* with $\mu \neq 0$ but $\mu \to 0$ or $Re \to \infty$. The latter has remarkably high flexibility. On the one hand, it can fully utilize the efficient methods of inviscid theory to solve global flows (mostly irrotational).[4] On the other hand, a vortex sheet satisfies the general vorticity theorems and hence can mimic the skeletons of attached and free shear layers associated with rich motion patterns, including vortex-sheet rolling up to form concentrated vortices as seen in Sect. 5.4. Consequently, vortex-sheet dynamics has successfully served as the very basis of the entire classic (low-speed) aerodynamics to be outlined in Sects. 9.2 and 9.3.

From the classic inviscid potential-flow theory to the boundary-layer theory at large and finite Reynolds numbers, and then to viscous theory with $Re \to \infty$, one sees a spiral-up development. This evolution reveals that behind the puzzling d'Alembert's Paradox is the truth that the large-Re flow over a streamlined body dressed in attached boundary layers is indeed very close to the potential flow. Thus, Lighthill (1963, 1995) claims that, due to this evolution, d'Alembert's Paradox has become **"d'Alembert's theorem"**: *if ever the flows around a steadily moving body could be made quite close to a potential flow, then the resistive force should likewise become quite close to that zero force which an exactly potential flow would exert.*

2. Vortex sheet and singular perturbation. The above essential feature of vortex-sheet dynamics can be further understood in terms of the concept of singular perturbation (Lagerstrom 1975), of which three general aspects are worth noticing:

Firstly, For incompressible flow at large Reynolds number, the Navier-Stokes (N-S) equation has a small parameter $\epsilon = Re^{-1} \ll 1$. Taking $\epsilon \to 0$ leads to the Euler equation as its less accurate approximation.[5] Accordingly, under the *same adherence boundary condition*, by taking $\epsilon \to 0$, the N-S solution should degenerate to an Euler solution. The limiting case of N-S solution at $\epsilon \to 0$ is called the *Euler limit* of the N-S solution. Reversely, by adding perturbation terms with small but nonzero ϵ to the Euler solution (e.g., inserting a boundary layer between inviscid main stream and the wall) one may obtain the asymptotic N-S solutions. Figure 5.13

---

[4]A flow region with vortex sheet at its boundary is not necessarily irrotational. In Sect. 8.4.1 we shall see that a steady flow enclosed by cyclic vortex sheet can have constant vorticity, which was also found by Prandtl (1904).

[5]Many other pairs of equations have the same correspondence, for example theory of compressible flow and incompressible flow (Chap. 2), Newton's gravitational theory and Einstein's gravitational theory, etc.

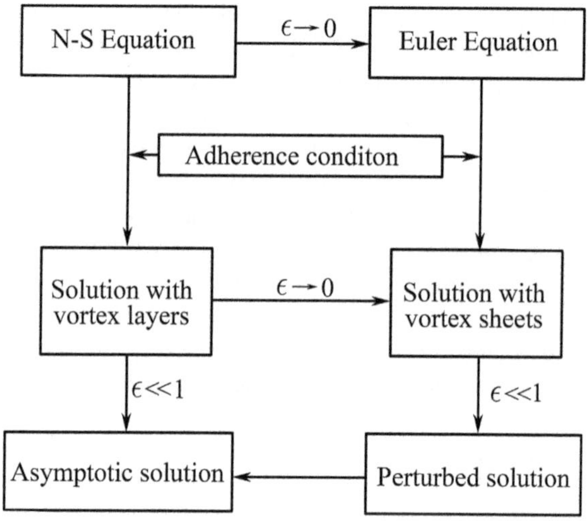

Fig. 5.13 Relations between the N-S equation and Euler equation as well as their solutions. Adapted from Wu et al. (1993)

demonstrates these relations between the two equations and their solutions (Wu et al. 1993), in which vortex-sheet dynamics stands at the upper and middle rows.

Secondly, while surfaces of tangent discontinuity are admitted by the Euler equation, they are not by the N-S equation. Associated with this difference is the Euler equation being one order lower than the N-S equation. Thus, the perturbation problem must be singular; a uniformly effective Euler solution cannot be an Euler limit of the N-S solution. In other words, *the Euler limit of an N-S solution is not an Euler solution satisfying only the no-through normal velocity condition, but rather a combination of one of such solutions with proper vortex sheets.*

Thirdly, the solutions of the Euler equation for given boundary conditions are not unique. Under the same boundary conditions and as $\epsilon \to 0$, the N-S solution approaches only one of possible Euler solutions as its *relevant Euler limit*. It is this relevant Euler limit that can be perturbed to an N-S solution for finite $\epsilon \ll 1$, see the bottom row of Fig. 5.13. There, not only an attached vortex sheet of $\epsilon \to 0$ can be perturbed to a boundary layer of $\epsilon \ll 1$, but also a free vortex sheet can be perturbed to a viscous free vortex layer.

5.6 Problems for Chapter 5

5.1. Prove that, for two-dimensional steady flow with $\mu \to 0$, from (5.2.6) one may derive the Kutta-Joukowski theorem (2.6.26).

5.2. In Sect. 5.2.2 we concluded that a free vortex sheet must leave a solid body tangentially. Thus we may write $\boldsymbol{u}^+ = u_1^+\boldsymbol{e}_1 + u_2^+\boldsymbol{e}_2$ but $\boldsymbol{u}^- = u_1^-\boldsymbol{e}_1$ for the velocities on the outer and inner sides of the sheet, respectively; for the definition of $\boldsymbol{e}_1$ and $\boldsymbol{e}_2$ see Fig. 5.6a. Given a fixed separation line L and $\boldsymbol{u}^+$, Show that for steady flow there is (Wu et al. 2006)

$$[\![\boldsymbol{u}]\!] = (u_1^+ - u_1^-)\boldsymbol{e}_1 + u_2^+\boldsymbol{e}_2, \tag{5.6.1a}$$

$$\overline{\boldsymbol{u}} = \frac{1}{2}[(u_1^+ + u_1^-)\boldsymbol{e}_1 + u_2^+\boldsymbol{e}_2]. \tag{5.6.1b}$$

Then, let $\boldsymbol{n} = \boldsymbol{e}_1 \times \boldsymbol{e}_2$ point from the $-$ side to the $+$ side, show that

$$\gamma = u_2^+\boldsymbol{e}_1 + (u_1^- - u_1^+)\boldsymbol{e}_2, \tag{5.6.2a}$$

$$q^{+2} = u_1^{+2} + u_2^{+2} = u_1^{-2}. \tag{5.6.2b}$$

Therefore, the initial free vortex is completely determined.

Chapter 6
Axisymmetric Columnar Vortices

In the beginning of Chap. 4 we remarked that, in a flow with small viscosity or at large Reynolds numbers, primary vortical structures are shear layers and secondary (but stronger) vortical structures are *axial vortices* (often simply called vortices henceforth) in which fluid particles rotate around common axes. Then in Chap. 5 we used vortex-sheet model to explain the initial formation process of axial vortices as the rolling up of free shear layers. It is now appropriate to discuss certain basic theoretical models of axial vortices.

Vortices are typically either unboundedly extended (columnar vortices) to be discussed below, or closed loops (vortex rings) to be discussed in the next chapter. Of various axial vortices the most fundamental ones are those having axisymmetry, which this book is confined to. Furthermore, only incompressible laminar vortices will be considered. More advanced non-axisymmetric vortex models, such as vortex dipoles, vortices in strained field, in array, or in a multiple-vortex system, are treated in Wu et al. (2006). For a comprehensive monograph on axial vortices, including various columnar vortex models, vortex-filament dynamics, two-dimensional point-vortex systems, as well as their stability and experimental methods, see Alekseenko et al. (2007).

Unless otherwise stated, throughout this and next chapters we assume the flow domain is unbounded and each vortex model is an isolated entity. But it should be born in mind that mathematically exact vortex solutions may not be physically realistic. The artificial isolation of vortex solutions for mathematical simplicity often leads to some unrealistic feature of the solutions, which may violate certain global conservative properties that true physical vortices must satisfy. In particular, the total-circulation conservation theorem implies that columnar vortices should appear in pair with opposite circulations (often called vortex dipoles), such as wingtip vortices. Only when the distance of two vortices is large enough such that their interaction is negligible, can theories on isolated columnar vortex be meaningful.

J.-Z. Wu et al., *Vortical Flows*, DOI 10.1007/978-3-662-47061-9_6

6.1 General Background

6.1.1 Governing Equations and Their Simplifications

1. General equations in cylindrical coordinates. In this chapter we use cylindrical coordinates (r, θ, z) with velocity components (u, v, w). The continuity and momentum equations read

$$\frac{\partial(ru)}{\partial r} + \frac{\partial v}{\partial \theta} + \frac{\partial(rw)}{\partial z} = 0, \tag{6.1.1a}$$

$$\frac{\partial u}{\partial t} + u\frac{\partial u}{\partial r} + \frac{v}{r}\frac{\partial u}{\partial \theta} + w\frac{\partial u}{\partial z} - \frac{v^2}{r} = -\frac{1}{\rho}\frac{\partial p}{\partial r} + \nu\left(\nabla^2 u - \frac{u}{r^2} - \frac{2}{r^2}\frac{\partial v}{\partial \theta}\right), \tag{6.1.1b}$$

$$\frac{\partial v}{\partial t} + u\frac{\partial v}{\partial r} + \frac{v}{r}\frac{\partial v}{\partial \theta} + w\frac{\partial v}{\partial z} + \frac{uv}{r} = -\frac{1}{\rho r}\frac{\partial p}{\partial \theta} + \nu\left(\nabla^2 v - \frac{v}{r^2} + \frac{2}{r^2}\frac{\partial u}{\partial \theta}\right), \tag{6.1.1c}$$

$$\frac{\partial w}{\partial t} + u\frac{\partial w}{\partial r} + \frac{v}{r}\frac{\partial w}{\partial \theta} + w\frac{\partial w}{\partial z} = -\frac{1}{\rho}\frac{\partial p}{\partial z} + \nu\nabla^2 w, \tag{6.1.1d}$$

where

$$\nabla^2 = \frac{\partial^2}{\partial r^2} + \frac{1}{r}\frac{\partial}{\partial r} + \frac{1}{r^2}\frac{\partial^2}{\partial \theta^2} + \frac{\partial^2}{\partial z^2}. \tag{6.1.1e}$$

The corresponding vorticity components $(\omega_r, \omega_\theta, \omega_z)$ are given by

$$\omega_r = \frac{1}{r}\frac{\partial w}{\partial \theta} - \frac{\partial v}{\partial z}, \quad \omega_\theta = \frac{\partial u}{\partial z} - \frac{\partial w}{\partial r}, \quad \omega_z = \frac{1}{r}\frac{\partial(rv)}{\partial r} - \frac{1}{r}\frac{\partial u}{\partial \theta}. \tag{6.1.2}$$

2. Kinematics of axisymmetric flow. For axisymmetric flow we are now concerned, (6.1.1a) is ensured by introducing the *Stokes stream function* ψ such that

$$u = -\frac{1}{r}\frac{\partial \psi}{\partial z}, \quad w = \frac{1}{r}\frac{\partial \psi}{\partial r}. \tag{6.1.3}$$

The remaining velocity component v does not enter the axisymmetric continuity equation and can be expressed by $C = rv$ (differing from the conventional circulation around a circle centered at $r = 0$ by a factor $1/2\pi$). Then the axisymmetric version of (6.1.2) is cast to

$$\omega_r = -\frac{1}{r}\frac{\partial C}{\partial z}, \quad \omega_z = \frac{1}{r}\frac{\partial C}{\partial r}, \tag{6.1.4a}$$

$$\omega_\theta = -\left[\frac{\partial}{\partial r}\left(\frac{1}{r}\frac{\partial \psi}{\partial r}\right) + \frac{1}{r}\frac{\partial^2 \psi}{\partial z^2}\right]. \tag{6.1.4b}$$

The role of C for ω_r and ω_z is exactly the same as that of ψ for u and w. Contours of C and ψ on an (r, z)-plane are the intersections of vorticity surfaces and stream surfaces with the plane, respectively.

3. Dynamics of axisymmetric flow. Dynamically, (6.1.1b)–(6.1.1d) are reduced to

$$\frac{Du}{Dt} - \frac{v^2}{r} = -\frac{1}{\rho}\frac{\partial p}{\partial r} + \nu\left(\nabla^2 u - \frac{u}{r^2}\right), \tag{6.1.5a}$$

$$\frac{Dv}{Dt} + \frac{uv}{r} = \nu\left(\nabla^2 v - \frac{v}{r^2}\right), \tag{6.1.5b}$$

$$\frac{Dw}{Dt} = -\frac{1}{\rho}\frac{\partial p}{\partial z} + \nu\nabla^2 w, \tag{6.1.5c}$$

where

$$\frac{D}{Dt} = \frac{\partial}{\partial t} + u\frac{\partial}{\partial r} + w\frac{\partial}{\partial z}, \tag{6.1.6a}$$

$$\nabla^2 = \frac{1}{r}\frac{\partial}{\partial r}\left(r\frac{\partial}{\partial r}\right) + \frac{\partial^2}{\partial z^2}. \tag{6.1.6b}$$

The curl of (6.1.5) gives the component vorticity equations:

$$\frac{D\omega_r}{Dt} = \boldsymbol{\omega}\cdot\nabla u + \nu\left(\nabla^2\omega_r - \frac{\omega_r}{r^2}\right), \tag{6.1.7a}$$

$$\frac{D\omega_\theta}{Dt} = \boldsymbol{\omega}\cdot\nabla v + \frac{\omega_\theta u - \omega_r v}{r} + \nu\left(\nabla^2\omega_\theta - \frac{\omega_\theta}{r^2}\right), \tag{6.1.7b}$$

$$\frac{D\omega_z}{Dt} = \boldsymbol{\omega}\cdot\nabla w + \nu\nabla^2\omega_z. \tag{6.1.7c}$$

In the above, (6.1.5b) and (6.1.7b) can be simplified by using C to replace v and ω_θ/r to replace ω_θ, respectively:

$$\frac{DC}{Dt} = \nu\left[r\frac{\partial}{\partial r}\left(\frac{1}{r}\frac{\partial C}{\partial r}\right) + \frac{\partial^2 C}{\partial z^2}\right], \tag{6.1.8a}$$

$$\frac{D}{Dt}\left(\frac{\omega_\theta}{r}\right) = \nu\left(\nabla^2 + \frac{2}{r}\frac{\partial}{\partial r}\right)\left(\frac{\omega_\theta}{r}\right) + \frac{1}{r^4}\frac{\partial C^2}{\partial z}, \tag{6.1.8b}$$

which govern the azimuthal and meridional motions, respectively. C and ω_θ/r may serve as the basic variables to be solved, and other quantities can be inferred therefrom. These two equations are coupled solely through the z-dependence of v.

6.1.2 Simplified Axisymmetric Model Equations

In a columnar vortex, observe that if w is nonzero and r-dependent, ω_θ will coexist with ω_z such that velocity and vorticity lines are *helical*. This kind of vortices are called *swirling vortices*, having nonzero helicity density $\boldsymbol{\omega} \cdot \boldsymbol{u}$. On the other hand, in an axisymmetric vortex ring ω_θ is dominating; a vortex ring is swirling only when ω_r and/or ω_z appears.

A special class of axisymmetric vortical flows is worth noticing. In Sect. 3.1.2 we saw that the flow will be generalized Beltramian if the Lamb vector is curl-free, $\boldsymbol{u} \cdot \nabla \boldsymbol{\omega} = \boldsymbol{\omega} \cdot \nabla \boldsymbol{u}$, then the vorticity transport equation linearized and hence easier for finding analytical solutions. In cylindrical coordinates this condition can be written, by using (6.1.3) and (6.1.4),

$$(u\partial_r + w\partial_z)\boldsymbol{\omega}^* = (\omega_r\partial_r + \omega_z\partial_z)\boldsymbol{u}^* \quad \text{or} \quad \frac{\partial(\psi, \omega_i^*)}{\partial(r,z)} = \frac{\partial(C, u_i^*)}{\partial(r,z)}, \tag{6.1.9}$$

where we use notations $\boldsymbol{u}^* = (u, v/r, w)$, $\boldsymbol{\omega}^* = (\omega_r, \omega_\theta/r, \omega_z)$ and $i = 1, 2, 3$. This advantage will be used below to construct three special types of vortex models.

1. Stretch-free columnar vortices. When the flow field is z-independent, $\omega_r = 0$ and (6.1.9) becomes trivially $0 = 0$. Then by the continuity equation there must be $u = K(t)/r$, but except singular line-vortex solution (3.1.39) the regularity of u at $r = 0$ implies $K = 0$. Consequently, the viscous term in (6.1.5a) disappears, while in (6.1.8) the viscous terms are solely balanced by the unsteady terms:

$$\frac{v^2}{r} = \frac{1}{\rho}\frac{\partial p}{\partial r}, \tag{6.1.10a}$$

$$\frac{\partial C}{\partial t} = \nu r \frac{\partial}{\partial r}\left(\frac{1}{r}\frac{\partial C}{\partial r}\right), \tag{6.1.10b}$$

$$\frac{\partial}{\partial t}\left(\frac{\omega_\theta}{r}\right) = \nu\left(\nabla^2 + \frac{2}{r}\frac{\partial}{\partial r}\right)\left(\frac{\omega_\theta}{r}\right). \tag{6.1.10c}$$

Note that (6.1.10b) and (6.1.10c) are decoupled and can be solved one by one; then pressure follows from (6.1.10a).

One special case of stretch-free vortices has $w = 0$ and $\omega_\theta = 0$. Then the flow appears only on a two-dimensional (r, θ)-plane governed by (6.1.10a) and (6.1.10b) (Sect. 6.2.1). On the other hand, viscous stretch-free vortex rings are governed by (6.1.10c) (Chap. 7).

2. Rotationally symmetric vortices. It is characterized by $v = C = 0$ such that $\boldsymbol{u}^* = (u, 0, w)$ and $\boldsymbol{\omega}^* = (0, \omega_\theta/r, 0)$. Condition (6.1.9) is satisfied, implying that kinematically there is

$$-r\omega_\theta = r\frac{\partial}{\partial r}\left(\frac{1}{r}\frac{\partial \psi}{\partial r}\right) + \frac{\partial^2 \psi}{\partial z^2} = -r^2 f(\psi, t). \tag{6.1.11}$$

Dynamically, by (6.1.8b) the flow is governed by

$$\frac{Df}{Dt} = \nu\left(\nabla^2 + \frac{2}{r}\frac{\partial}{\partial r}\right) f, \quad f \equiv \frac{\omega_\theta}{r}. \tag{6.1.12}$$

This model is also a basis of circular vortex-ring theory of Chap. 7.

3. Inviscid and steady swirling flows. For this flow $\nu = 0$ and $\partial/\partial t = 0$. Then there is

$$\boldsymbol{\omega} \times \boldsymbol{u} = -\nabla H, \quad \boldsymbol{u} \cdot \nabla H = 0, \tag{6.1.13}$$

and (6.1.8) is simply reduced to

$$\boldsymbol{u} \cdot \nabla C = 0, \quad \boldsymbol{u} \cdot \nabla\left(\frac{\omega_\theta}{r}\right) = \frac{1}{r^4}\frac{\partial C^2}{\partial z}. \tag{6.1.14}$$

Thus, both H and C are invariant along a streamline characterized by $\psi =$ constant, or $H = H(\psi)$ and $C = C(\psi)$. It can then be shown that (Problem 6.2)

$$\omega_r = u\frac{dC}{d\psi}, \quad \omega_z = w\frac{dC}{d\psi}, \quad \frac{\omega_\theta}{r} = \frac{C}{r^2}\frac{dC}{d\psi} - \frac{dH}{d\psi}. \tag{6.1.15}$$

Combine the expression for ω_θ with (6.1.4b), we obtain a single differential equation to be solved for steady inviscid axisymmetric flows:

$$r\frac{\partial}{\partial r}\left(\frac{1}{r}\frac{\partial \psi}{\partial r}\right) + \frac{\partial^2 \psi}{\partial z^2} = r^2\frac{dH}{d\psi} - C\frac{dC}{d\psi}. \tag{6.1.16}$$

This equation is called the *Bragg-Hawthorne equation* or the *Squire equation*. Note that $\boldsymbol{u}$ can be z-dependent.

6.2 Two-Dimensional Stretch-Free Vortices

Two-dimensional stretch-free vortices are the simplest models, governed by (6.1.10a) and (6.1.10b). Evidently, if ν is assumed zero, such vortices must be steady; while if $\nu \neq 0$ they must be unsteady.

6.2.1 Steady and Inviscid Pure Vortices

For steady flow, the right-hand side of (6.1.10b) must vanish, thus it follows that $\Gamma = 2\pi C = Ar^2 + B$ or

$$v(r) = Ar + \frac{B}{r}, \tag{6.2.1}$$

where A and B are arbitrary constants. If the flow occurs between two rotating coaxial circular cylinders, these constants can be determined by boundary conditions and (6.2.1) represents an exact and smooth Navier-Stokes solution known as the *Taylor-Couette flow*. But we are now considering an unbounded domain where A must vanish to conserve a finite total circulation as $r \to \infty$. Thus $\Gamma = \Gamma_0$ is a constant and we return to the most elementary *point-vortex* solution (3.1.39) on an (r, θ)-plane with singular velocity at $r = 0$, or a straight *line vortex* in three-dimensional space. Correspondingly, $\omega(r) = \Gamma_0 \delta(r)$ is a delta-function. One's interest in this simplest vortex solution lies in the fact that it is a fundamental solution of the Laplace equation and can be linearly superimposed to mimic multi-vortex system (Wu et al. 2006; Alekseenko et al. 2007), but only to a limited extend.

To avoid the singularity of (3.1.39), Rankine proposed in 1882 to use step function for ω with finite v:

$$\omega = 2\Omega, \quad v = \Omega r, \quad r \le a, \tag{6.2.2a}$$

$$\omega = 0, \quad v = \Omega \frac{a^2}{r}, \quad r > a, \tag{6.2.2b}$$

where a is the radius of the vortex where $v = v_{\max}$. This radius can be similarly defined for other axial vortices, and the flow region within a is called the *vortex core*. The velocity field is continuous but not smooth across the core boundary, implying a viscous shear-stress jump there which would cause an infinite acceleration. Thus this *Rankine vortex* is not a global Navier-Stokes solution.

Both point vortex and Rankine vortex are associated with a pressure drop from the far-field value p_∞ (Problem 6.3). These vortices are the solutions of the Euler equations and Navier-Stokes equations, but at the expense of having singular behaviors that cannot exhibit the real role of viscosity.

6.2.2 Unsteady and Viscous Pure Vortices

Since in two-dimensional flow any smooth and viscous vortex solutions must be *unsteady*, we now use them to ask how a viscous line vortex $v = \Gamma_0/2\pi r$ evolves as time. Namely, we consider the initial-value problem of (6.1.10b) or equivalently

$$\frac{\partial \omega_z}{\partial t} = \frac{\nu}{r}\frac{\partial}{\partial r}\left(r\frac{\partial \omega_z}{\partial r}\right), \tag{6.2.3}$$

under the condition

$$\omega(r, t) = \Gamma_0 \delta(r) \quad \text{at} \quad t = 0. \tag{6.2.4}$$

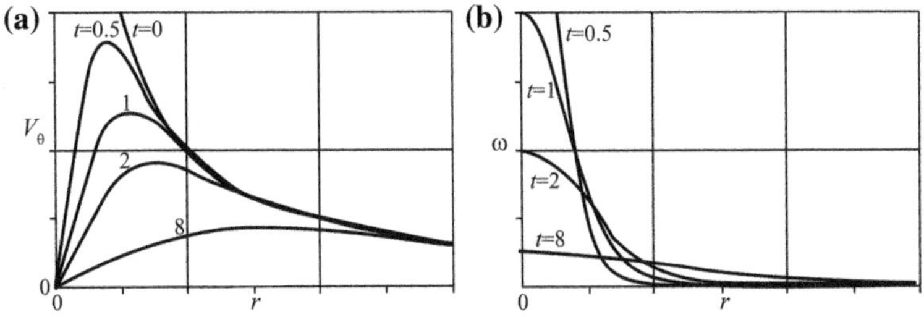

Fig. 6.1 The decay of, **a** circumferential velocity v and, **b** vorticity in the Oseen vortex

This problem permits a similarity solution, known as the *Oseen-Lamb vortex*:

$$v(r,t) = \frac{\Gamma_0}{2\pi r}\left[1 - \exp\left(-\frac{r^2}{4\nu t}\right)\right], \tag{6.2.5a}$$

$$\omega_z(r,t) = \frac{\Gamma_0}{4\pi\nu t}\exp\left(-\frac{r^2}{4\nu t}\right). \tag{6.2.5b}$$

The vortex has a finite circulation $\Gamma(r,t) = 2\pi r v(r,t)$ that has property

$$\Gamma(0,0) = \Gamma_0, \quad \Gamma(0,t) = 0, \quad \Gamma(\infty,t) = \Gamma_0, t > 0.$$

The vortex-core radius a is defined as the location where $v = v_{\max}$, which by (6.2.5) gives $a = 2.24\sqrt{\nu t}$, so the vortex core is expanding as time due to diffusion, see Fig. 6.1. The behavior of (6.2.5) for $r \gg a$ approaches that of line vortex. For small r, there is $v \simeq \Gamma_0 r/(8\pi\nu t)$, similar to a solid rotation. These two regions merge smoothly around $r = a$.

For a viscous axial vortex, one may introduce a Reynolds number $R_\Gamma = \Gamma_0/\nu \sim v_{\max}a/\nu$ based on the total circulation Γ_0. Larger R_Γ implies thinner core within which the vorticity is highly concentrated. But, unlike the Rankine vortex, (6.2.5) indicates an exponentially decaying "tail" of vorticity outside the core. Just like the thickness δ of attached or free shear layers cannot be sharply identified, neither can the core radius a of viscous axial vortices. In this sense the Oseen-Lamb vortex is an axisymmetric counterpart of the Stokes first problem (Rayleigh problem) discussed in Sect. 4.1.2. However, while as $\delta \to 0$ a shear layer becomes a vortex sheet across which the tangential velocity has a *finite* jump, now as $a \to 0$ an axial vortex becomes a singular line around which the circumferential velocity approaches *infinity*. This fact implies that axial vortices of large R_Γ are associated with very big kinetic energy peaks in the flow field and are the strongest type of vortical structures.

The Oseen-Lamb vortex is actually the simplest model of a sequence of solutions (Wu et al. 2006). Next to it in the same sequence is the *Taylor vortex*:

$$v(r,t) = \frac{Mr}{8\pi\nu t^2}\exp\left(-\frac{r^2}{4\nu t}\right), \tag{6.2.6a}$$

$$\omega_z(r,t) = \frac{M}{2\pi\nu t^2}\left(1-\frac{r^2}{4\nu t}\right)\exp\left(-\frac{r^2}{4\nu t}\right), \tag{6.2.6b}$$

where M represents the total angular momentum about the axis:

$$M = \int_0^\infty 2\pi r^2 v dr. \tag{6.2.7}$$

This solution has *zero total circulation* (because ω changes sign once) and finite M. Note that (6.2.6) is nothing but the time derivative of (6.2.5).

6.3 Radial-Axial Flow Coupling and Stretched Vortices

When an axial vortex has nonzero axial velocity w, the fluid path is no longer closed circles but becomes spirals. Such a vortical flow is known as *swirling flow* very often encountered in practice. Swirling jets and wakes can be found in external flows around flying vehicles and in internal flows of many industrial facilities. In the design of high-efficiency combustion chambers or control of airplane wakes, it is very important to understand the dynamic characteristics of swirling flows, in particular their stability behavior. Besides, swirling flows exist in many atmospheric phenomena, such as tornadoes and dust devils.

A vortex will subject to axial stretching or shrinking if axial velocity w is z-dependent. A positive stretching rate implies the enhancement of the vortex, which for incompressible flow is associated with the reduction of the core radius. The z-dependence of w may come from different mechanisms. One is a kinematic coupling between *radial* and axial velocity components via the continuity equation (6.1.1a), by which if ru depends on r then w must depend on z. Another is a dynamic coupling between *azimuthal* and axial velocity components via the pressure. We discuss these two mechanisms in this and next sections, respectively.

Consider steady flow and assume u and v are independent of z. Then by continuity the axial velocity can depend on z only linearly. Namely,

$$u = u(r), \quad C = rv = \Gamma(r), \quad w = zf(r). \tag{6.3.1}$$

Then (6.1.8a) for $\Gamma = 2\pi C$ is reduced to

$$\frac{d^2\Gamma}{dr^2} = \left(\frac{1}{r}+\frac{u}{\nu}\right)\frac{d\Gamma}{dr}, \tag{6.3.2}$$

which under the boundary conditions $\Gamma(0) = 0$ and $\Gamma(\infty) = \Gamma_0$ has solution

$$\Gamma(r) = \Gamma_0 \frac{H(r)}{H(\infty)}, \tag{6.3.3a}$$

where

$$H(r) = \int_0^r x \exp\left(\int_0^x \frac{u(s)}{\nu} ds \right) dx. \tag{6.3.3b}$$

Thus, once $u(r)$ or $f(r)$ in (6.3.1) is given the solution will be determined. Let us construct two exact but only locally effective Navier-Stokes solutions with arbitrary stretching rate, within this framework.

6.3.1 Burgers Vortex

As the simplest model under condition (6.3.1), assumes f is a constant γ. By continuity, this yields

$$w(z) = \gamma z, \quad u(r) = -\frac{1}{2}\gamma r, \tag{6.3.4}$$

so that $\gamma > 0$ implies an axial stretching. Then from (6.3.3a) follows the famous *Burgers vortex*:

$$v(r) = \frac{\Gamma_0}{2\pi r}\left(1 - e^{-\beta r^2}\right), \tag{6.3.5a}$$

$$\omega(r) = \frac{\gamma \Gamma_0}{4\pi\nu} e^{-\beta r^2}, \tag{6.3.5b}$$

where $\beta = \gamma/4\nu$ measures the ratio of stretching rate and viscosity. The core radius is estimated by $a \sim (\nu/\gamma)^{\frac{1}{2}}$, which exists only if $\gamma > 0$. If $\gamma = 0$ then $\omega = v = 0$; while if $\gamma < 0$ the vorticity will run away from the axis. Note that if we freeze the time in (6.2.5) at $t = 1/\gamma$ then the velocity distribution of the Oseen-Lamb vortex is the same as that of Burgers vortex. This is because the radial flow $-\gamma r/2$ brings the far-field vorticity to the vortex core, which exactly compensates the viscous diffusion.

The Burgers vortex is the first stretched vortex solution to model turbulent eddies. As a remarkable feature, its total dissipation per unit axial length is finite but *independent of* ν, even if $\nu \to 0$:

$$\rho\nu \int_0^\infty 2\pi r \omega^2(r) dr = \frac{\rho\gamma\Gamma_0^2}{4\pi}.$$

A finite and ν-independent total dissipation is a fundamental assumption in turbulence theory. Thus the Burgers vortex has served as a building block of some vortex models for fine-scale turbulent structures.

6.3.2 Sullivan Vortex

We now generalize the Burgers vortex by permitting an r-dependent axial velocity w. A natural choice is to assume $f(r)$ has the same exponential term as $v(r)$ in (6.3.5):

$$f(r) = \gamma\left(1 - b\mathrm{e}^{-\beta r^2}\right), \quad \beta = \frac{\gamma}{4\nu},$$

where b is an undetermined constant. Substituting this into (6.3.1) yields

$$u(r) = -\frac{\gamma}{2}r + \frac{2b\nu}{r}(1 - \mathrm{e}^{-\beta r^2}),$$

and then $v(r)$ comes from (6.3.3a). But only $b = 3$ can ensure the steady axial momentum equation (6.1.5c). Therefore, the three velocity components are found to be

$$u(r) = -\frac{1}{2}\gamma r + \frac{6\nu}{r}\left(1 - \mathrm{e}^{-\beta r^2}\right), \tag{6.3.7}$$

$$v(r) = \frac{\Gamma_0 H\left(\beta r^2\right)}{2\pi r H(\infty)}, \tag{6.3.8}$$

$$w(r, z) = \gamma z\left(1 - 3\mathrm{e}^{-\beta r^2}\right), \tag{6.3.9}$$

where

$$H(x) \equiv \int_0^x \exp\left(-\eta + 3\int_0^\eta \frac{1 - \mathrm{e}^{-\zeta}}{\zeta} d\zeta\right) d\eta, \quad H(\infty) = 37.905. \tag{6.3.10}$$

This solution is known as the *Sullivan vortex*. Its vorticity components are

$$\omega_r = 0, \tag{6.3.11a}$$

$$\omega_\theta = -\frac{3\gamma^2}{2\nu} rz\mathrm{e}^{-\beta r^2}, \tag{6.3.11b}$$

$$\omega_z = \frac{\gamma\Gamma_0}{2\nu H(\infty)} \exp\left(-\beta r^2 + 3\int_0^{\beta r^2} \frac{1 - \mathrm{e}^{-\zeta}}{\zeta} d\zeta\right). \tag{6.3.11c}$$

Unlike Burgers vortex, in Sullivan vortex and all vortices studied in Sects. 6.4 and 6.5 the axial flow w is r-dependent. This implies that inside the vortices there appears a shear $\partial w/\partial r$ that coexists and interacts with swirl. In Chap. 10 we shall see that the shear-swirl interaction has very important influence on the stability character of these vortices.

The most striking property of the Sullivan vortex is that it permits a *two-cell structure*. From (6.3.7), there will be $u(r_0) = 0$ at $r_0 = 3.36\sqrt{\nu/\gamma}$ that satisfies

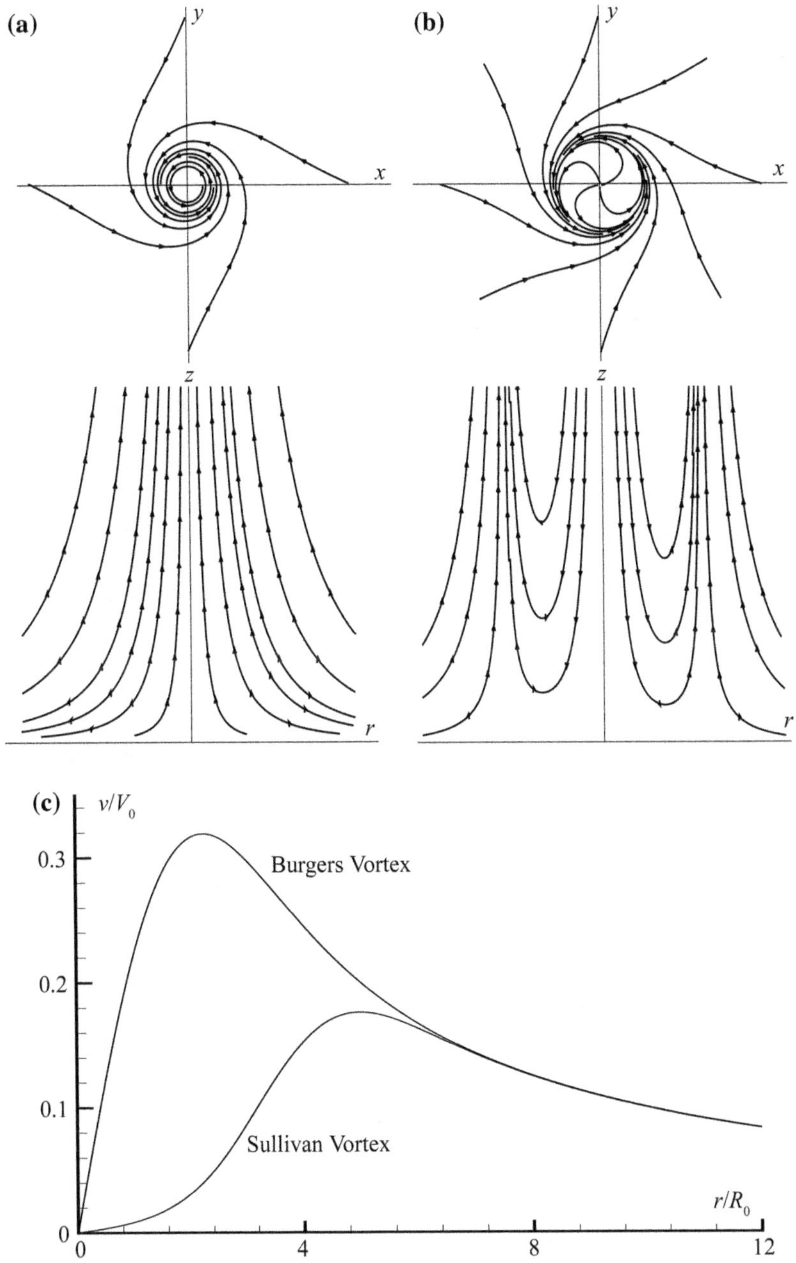

Fig. 6.2 The velocity profiles of one- and two-cell vortices. **a** One cell (*Burgers vortex*). **b** Two cells (*Sullivian vortex*). **c**. The circumferential velocity, where $V_0 = \Gamma/2\pi r_0$, $R_0 = (\nu/\gamma)^{1/2}$

$\beta r_0^2 = 3(1 - e^{-\beta r_0^2})$, and u changes sign across r_0. Thus, $r = r_0$ is a *limit circle*: fluid outside r_0 will move inward to r_0, while that inside r_0 will move outward to r_0. Therefore, near r_0 there must be a strong axial flow, which in turn requires an axial flow of opposite direction near $r = 0$. So $w(r, z)$ has to change sign somewhere between r_0 and the axis. The one-cell Burgers vortex and two-cell Sullivan vortex are compared in Fig. 6.2. The two-cell structure has been observed in some hurricanes, so the Sullivan vortex is of special interest in meteorology.

6.4 Azimuthal-Axial Flow Coupling and Batchelor Vortex

As exemplified by the Burgers and Sullivan vortices, radially variable radial velocity $u(r)$ can cause axially variable axial velocity $w(z)$ and lead to vortex stretching through the continuity equation. This is an inviscid and kinematic mechanism. We now turn to a viscous and dynamic coupling between azimuthal velocity v or $C = rv$ and axial velocity w. It makes a generic axial vortex *spontaneously* three-dimensional with variable $w(r, z)$, even if initially we start from a two-dimensional model.

Such a dynamic azimuthal-axial velocity coupling is a critical feature of aircraft *trailing vortices* (or *wingtip vortices*), see the sketch of Fig. 3.3c, which are of great importance in aeronautics. As an aircraft takes off or lands at relatively large angles of attack, its trailing vortex couple will produce a strong downwash and upwash wake flow in between and outside the couple, respectively, see Fig. 6.3. This strong swirling wake threatens the safety of following aircrafts. The safety distance between two successively aircrafts determines the permissible frequency of aircraft taking-off and landing on a runway, and how to reduce the hazard of trailing vortices is very important for a busy big airport. One has to investigate the stability of these vortices, and accordingly design some devices to promote their instability and decay. Then the safety distance between two successively aircrafts can be shortened.

The instability of trailing vortices will be discussed in Chap. 10. In this and next sections we first study the physical mechanism of azimuthal-axial flow coupling in a generic vortex core, and then present two theoretical models of trailing vortex. To reduce mathematical complexity, our discussion will be based on further simplified equations.

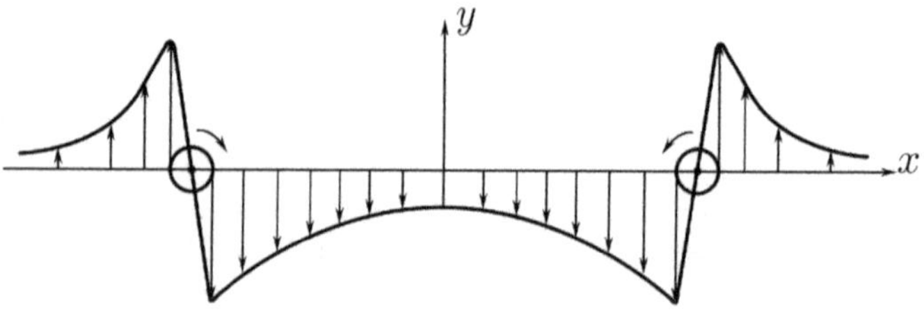

Fig. 6.3 Sketch of velocity profiles induced by a pair of aircraft trailing vortices

6.4.1 Slender and Light-Loading Approximation

Axial vortices usually appear as slender fluid bodies, in which the radial scale is much smaller than the axial scale and hence the radial variation of flow quantities is much larger than the axial variation. For a slender vortex with straight axis, the following *slender-flow approximation* is assumed:

$$r \ll 1, \quad \frac{\partial}{\partial z} \ll \frac{\partial}{\partial r}, \quad u \ll w. \tag{6.4.1}$$

This situation is evidently a counterpart of boundary layers and free shear layers as thin layer-like fluid bodies. Thus, by setting $C = rv$, to the leading order (6.1.5) is reduced to

$$\frac{C^2}{r^3} = \frac{1}{\rho}\frac{\partial p}{\partial r}, \tag{6.4.2a}$$

$$\frac{DC}{Dt} = \nu r \frac{\partial}{\partial r}\left(\frac{1}{r}\frac{\partial C}{\partial r}\right), \tag{6.4.2b}$$

$$\frac{Dw}{Dt} = -\frac{1}{\rho}\frac{\partial p}{\partial z} + \frac{\nu}{r}\frac{\partial}{\partial r}\left(r\frac{\partial w}{\partial r}\right), \tag{6.4.2c}$$

which is also called *quasi-cylindrical approximation*. It has two significant simplifications. First, (6.4.2a) is independent of viscosity, indicating that the centrifugal force is solely balanced by the radial pressure gradient, and hence can be computed after the flow field is solved:

$$\frac{p_\infty - p}{\rho} = \int_r^\infty \frac{C^2}{r^3} dr. \tag{6.4.3}$$

Second, the Laplace operator ∇^2 in the w-equation is degenerated to a parabolic operator, requiring only three boundary conditions at vortex axis, outer edge of the core, and upstream flow.

It is worth mentioning that at the opposite end of slender vortices there also exist flattened disk-like axial vortices in nature and technology, such as hurricanes and typhoons. Their axial length is much smaller than radial scale. For the relevant basic theory see Chap. 12 of Wu et al. (2006).

Moreover, for a large-span wing the interaction between the two *trailing vortices* is weak, and each vortex of the pair can be described by (6.4.2). Assume the wing moves to the z direction with constant speed $-W$. In the frame of reference fixed to the wing (*wing frame*) with coordinate axis coinciding with a trailing vortex, there is $\boldsymbol{u} = (u, v, w + W)$ with $\boldsymbol{u} = W\boldsymbol{e}_z$ at $r = \infty$. If the wing load is not very heavy, we may set $|w - W| \ll W$ and hence $u \simeq 0$ by continuity. This is called the *light-loading approximation*. Thus, while (6.4.2a) and (6.4.3) remain the same, (6.4.2b) and (6.4.2c) can be further linearized:

$$W\frac{\partial C}{\partial z} = \nu r\frac{\partial}{\partial r}\left(\frac{1}{r}\frac{\partial C}{\partial r}\right), \tag{6.4.4a}$$

$$W\frac{\partial w}{\partial z} = -\frac{1}{\rho}\frac{\partial p}{\partial z} + \frac{\nu}{r}\frac{\partial}{\partial r}\left(r\frac{\partial w}{\partial r}\right). \tag{6.4.4b}$$

Notice that C has a one-way influence on p, and p has a one-way influence on w, so (6.4.4a) and (6.4.4b) can be solved sequentially.

6.4.2 Azimuthal-Axial Flow Coupling

Let W be constant axial velocity at $r = \infty$ (the wing flight velocity is $-W$). Assume the flow satisfies slender and light-loading approximations, so that it is governed by (6.4.3) and (6.4.4), the latter having been linearized and parabolized. The boundary conditions for $C(r, z)$ at $r = 0$ and ∞ are

$$C(z, 0) = 0, \qquad C(z, \infty) = C_0, \qquad w(z, \infty) = W. \tag{6.4.5}$$

Another boundary condition at upstream end $z = 0$ should also be imposed.

By (6.4.3), at a point $P(r, z)$ in the vortex core we have

$$\frac{p_\infty - p}{\rho} = \frac{1}{2}\int_r^\infty \frac{1}{r^2}\frac{\partial C^2}{\partial r}dr + \frac{C^2}{2r^2}. \tag{6.4.6}$$

If the vorticity in the core is one-signed as is normally the case in trailing vortices, by (6.4.2a) there is $\partial C^2/\partial r > 0$. Hence the right-hand side of (6.4.6) must be positive, implying that *the pressure has a minimum inside the vortex core* (This property is not universally true).

On the other hand, recall that for inviscid steady flow the Bernoulli equation holds along a stream surface or a vorticity surface (Sect. 3.2.4), and an inviscid axial vortex can be viewed as a stream tube bounded externally by irrotational flow. In Sect. 4.2.5 we have seen that as streamlines pass through a steady and viscous vortical-flow region the total-head must have a loss ΔH. Thus, assume all fluid comes from the far-upstream uniform flow $\boldsymbol{W} = (0, 0, W)$, we may formally write down a viscous Bernoulli equation and obtain the pressure at a point P inside the viscous sub-core:

$$\frac{p_\infty - p}{\rho} = \frac{1}{2}\left[\frac{C^2}{r^2} + (w^2 - W^2)\right] + \Delta H. \tag{6.4.7}$$

Combining this and (6.4.3) yields

$$w^2 = W^2 + \int_r^\infty \frac{1}{r^2}\frac{\partial C^2}{\partial r}dr - 2\Delta H. \tag{6.4.8}$$

Therefore, *inside the vortex core the viscous effect always causes an axial-flow deficit, while a radial variation of circumferential velocity always causes an axial-flow increment* (Batchelor 1964). By (6.4.6) and (6.4.7), this axial-flow increment comes from the radial pressure gradient necessary for balancing the centrifugal force, which leads to a low pressure at the axis and larger axial velocity. This effect is usually dominating in the upstream portion of a trailing vortex. But the viscous effect may become dominant at far downstream to make the core axial flow deficit. The disturbance axial flow vanishes only when the two opposite effects just cancel each other.

While in (6.4.8) the viscous effect on the C-dependence of w is merely denoted symbolically by ΔH, it can be quantitatively revealed by a combination of (6.4.2a) and (6.4.4), yielding (Batchelor 1964)

$$W\frac{\partial w}{\partial z} - \frac{\nu}{r}\frac{\partial}{\partial r}\left(r\frac{\partial w}{\partial r}\right) = \frac{2\nu}{W}\int_r^\infty \frac{C}{r^2}\frac{\partial}{\partial r}\left(\frac{1}{r}\frac{\partial C}{\partial r}\right)dr. \tag{6.4.9}$$

6.4.3 Batchelor Vortex

The first and simplest theoretical model of trailing vortex, which exhibits the azimuthal-axial flow coupling in its core structure, is the steady *Batchelor vortex*. Introducing similarity variable

$$\eta = \frac{Wr^2}{4\nu z}. \tag{6.4.10a}$$

The boundary conditions (6.4.5) should now be imposed at $\eta = 0$ and ∞. The upstream condition at $z = 0$ has to be the same as $\eta = \infty$. Now (6.4.4a) is cast to $C''(\eta) + C'(\eta) = 0$ (prime denotes derivative), and hence

$$C(\eta) = C_0(1 - e^{-\eta}). \tag{6.4.10b}$$

This solution will be the same as the Oseen-Lamb vortex if we set $z = Wt$ such that η becomes $r^2/(4\nu t)$ or solve (6.1.10b) instead. Then (6.4.2a) is cast to

$$\frac{p_\infty - p}{\rho} = \frac{C_0^2 W}{8\nu z}P(\eta), \tag{6.4.11a}$$

with

$$P(\eta) = \int_\eta^\infty \frac{(1 - e^{-\xi})^2}{\xi^2}d\xi = \frac{(1 - e^{-\eta})^2}{\eta} + 2\mathrm{ei}(\eta) - 2\mathrm{ei}(2\eta), \tag{6.4.11b}$$

where $\mathrm{ei}(\eta) = \int_\eta^\infty (e^{-\xi}/\xi)d\xi$ is the exponential integral. Finally, denote

$$w = \frac{C_0^2}{8\nu z}\bar{w}(\eta), \tag{6.4.12a}$$

from (6.4.4b) follows an inhomogeneous similarity equation for $\bar{w}(\eta)$

$$\eta\frac{d^2\bar{w}}{d\eta^2} + (1+\eta)\frac{d\bar{w}}{d\eta} + \bar{w} = P + \eta\frac{dP}{d\eta}. \tag{6.4.12b}$$

After some algebra, Batchelor obtains the solution

$$w = W - \frac{C_0^2}{8\nu z}\left(\ln\frac{Wz}{\nu}\right)e^{-\eta} + \frac{C_0^2}{8\nu z}Q(\eta) - \frac{BW^2}{8\nu z}e^{-\eta}, \tag{6.4.13a}$$

where B is a constant determined by upstream condition and

$$Q(\eta) = e^{-\eta}[\ln\eta + \mathrm{ei}(\eta) - 0.807] + 2\mathrm{ei}(\eta) - 2\mathrm{ei}(2\eta). \tag{6.4.13b}$$

The solution (6.4.10b) and (6.4.13) is called the *Batchelor vortex*. This model is effective when $\eta = O(1)$, which implies the core radius $a \sim O(\sqrt{\nu z/W})$, comparable to the boundary-layer thickness. For a rectangular wing of span $2b$ and chord length c with aspect ratio $A = 2b/c$, the core size of trailing vortices is $a \sim 0.1b$; thus Batchelor's theory can be applied when $z \sim 0.01cRe_cA^2$, where $Re_c = Wc/\nu$ is the Reynolds number based on chord. Typically, this estimate indicates that z is about $O(10^3c)$, i.e., at far downstream of the trailing vortex.

At further downstream with very large z, we may ignore the relatively small term $Q(\eta)$ in (6.4.13a), and nondimensionalize the Batchelor vortex model by the following characteristic velocity and length scales

$$W_s(z) \equiv \frac{C_0^2}{8\nu z}\ln\frac{Wz}{\nu} + \frac{BW^2}{8\nu z}, \quad r_s(z) \equiv \sqrt{\frac{4\nu z}{W}}. \tag{6.4.14a}$$

Indeed, define dimensionless variables (denoted by asterisk)

$$q^* = \frac{C_0}{r_s W_s}, \quad r^* = \frac{r}{r_s}, \quad w^* = \frac{w}{W_s}, \tag{6.4.14b}$$

where q^* is the *swirling ratio* that measures the relative importance of the circumferential and axial velocities. Then the velocity and vorticity profiles of the Batchelor vortex take the form

$$u^*(r^*) = 0, \quad v^*(r^*) = \frac{q^*}{r^*}(1 - e^{-r^{*2}}), \quad w^*(r^*) = W^* - e^{-r^{*2}}, \tag{6.4.15a}$$

$$\omega_z^*(r^*) = 2q^*e^{-r^{*2}}, \quad \omega_\theta^* = -2r^*e^{-r^{*2}}. \tag{6.4.15b}$$

This vortex model is often called the *q-vortex*. Note that both r_s and W_s depend on not only the wing flight speed W and viscosity ν but also the z-station of the trailing vortex. While the z-dependence of the core size is simply via η defined by (6.4.10a), that of the axial stretching rate $\partial w/\partial z$ may change sign as z increases. In particular, even if there is an axial-flow excess ($w > W$) at upstream, the axial velocity must eventually become a deficit due to viscous dissipation. However, in real flows the transition from $w > W$ to $w < W$ in a trailing vortex occurs much more upstream than Batchelor's model predicts; an improved theory is needed which will be addressed in the next section.

Regardless the original motivation of the Batchelor vortex, the q-vortex (6.4.15) can be viewed as an *inviscid* and steady solution for a family of slender swirling vortices. Actually, from the inviscid and steady version of (6.4.2) we see at once that $v(r, z)$ and $w(r, z)$ can be chosen independently and quite arbitrarily. The q-vortex is just a reasonable choice, which has been found to serve as a good approximation to various realistic slender vortices, both in free space and in a pipe (of course the boundary layer at the pipe wall should be excluded). In this case, the scales r_s and W_s can be inferred from fitting experimental data (Leibovich 1984) or numerical solution (e.g., Zhang et al. 2009) at different z-stations.

In this direction, a widely used scaling is to set r_s as the vortex-core radius $r_0(z)$ and W_s as the velocity difference $\Delta W = w(0) - W$, with $w(0)$ being the axial velocity at the axis $r = 0$. This scaling defines two dimensionless parameters

$$q^* = \frac{\Omega(0)r_0}{\Delta W}, \quad a^* = \frac{W}{\Delta W}, \tag{6.4.16}$$

where $\Omega(0)$ is the rotation rate at the axis. Then in (6.4.15) the profiles of v^* and ω_z^* remain the same but w^* and ω_θ^* are generalized to

$$w^*(r^*) = W^* + a^* e^{-r^{*2}}, \quad \omega_\theta^* = 2a^* r^* e^{-r^{*2}}. \tag{6.4.17}$$

In this model, the axial stretching or shrinking of the vortex is reflected only implicitly by the z-dependence of parameters (q^*, a^*). This generalized version of (6.4.15) is still called the q-vortex or Batchelor vortex. The new parameter a^* can be both positive and negative to distinguish the jet-like and wake-like axial-velocity profiles (cf. Sects. 4.4.2 and 4.4.3), which is critical in vortex instability and breakdown (Chap. 10).

6.5 Trailing Vortex with Composite Core Structure

We have seen that the Batchelor vortex can describe wing-tip vortices only at very far downstream with $z = O(10^3 c)$. Evidently, it cannot describe the wake flow near the wing trailing edge, where the tip vortex is being formed from the rolling up of trailing vortex sheet. To improve the prediction capability of theoretical trailing-vortex model, therefore, one should take into account of the tightly wound turns of the trailing vortex sheet as well as the viscous effect necessary for a smooth vortex core. This is done by Moore and Saffman (1973), which we discuss now. Similar theory has been developed by Hall (1966) for steady conical leading-edge vortex with viscous core.

6.5.1 Composite Core Structure

As one moves toward a tip vortex near the trailing edge, the first structure that can be identified as the axisymmetric vortex core is the tightly wound vortex sheet given by (5.3.5), of which the outer boundary can therefore be crudely defined as the *first* core radius:

$$a_0(t) \sim (\beta t)^{\frac{1}{1+n}}. \tag{6.5.1}$$

The existence of this inviscid core differs from the core of the Batchelor vortex, where the flow outside the viscous core is assumed to directly become uniform as $r \to \infty$.

Then, as r reduces from a_0, the distance between two neighboring turns is eventually as small as the diffusion thickness of spiral vortex layers, where the zigzag behavior of the v-profile is smeared out. As r further reduces, the viscous effect becomes dominant and forms a viscous sub-core that can smooth out various singularities inherent in inviscid theory. Recall that in the Rayleigh problem (Sect. 4.1.2) the natural viscous length scale is $\sqrt{\nu t}$. This is also true for the present problem and we can define a *viscous core*

$$a_{\rm vis}(t) \sim \sqrt{\nu t}, \tag{6.5.2}$$

which should be effectively infinitesimal compared to $a_0(t)$. Thus,

$$(\nu t)^{\frac{1}{2}} \ll (\beta t)^{\frac{1}{1+n}} \quad \text{or} \quad t \gg \left(\frac{\nu^{1+n}}{\beta^2}\right)^{\frac{1}{1-n}}. \tag{6.5.3}$$

For elliptical wing load distribution with $n = 1/2$, (6.5.3) implies $t \gg \nu^3$, which is achievable very quickly (Problem 6.7) and therefore represents an improvement of Batchelor's trailing vortex model.

Actually, in between $a_{\rm vis}$ and a_0 there is yet another core structure, because not only inside $a_{\rm vis}$ there is no more spiral structure, but also the flow pattern between $a_{\rm vis}$ and a_0 may lose the spirals because as t increases the turns are thickened by diffusion. In fact, after the core is already smoothed out by viscosity, the viscous term $-\nu\nabla\times\boldsymbol{\omega}$ in the momentum equation becomes small accordingly. For example, by (5.3.5) the radial distance between successive turns is $r^{n+2}/[(1+n)\beta t]$, which will be smaller than $a_{\rm vis}$ if

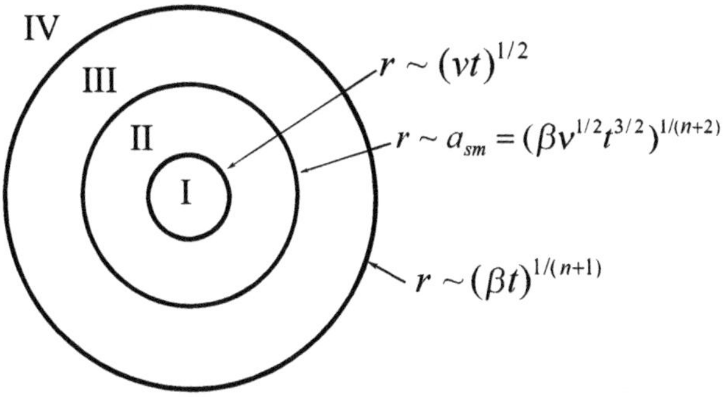

Fig. 6.4 The core structure of a typical axial vortex. I. Viscous inner region of radius $O(\sqrt{\nu t})$. II. Smoothed-out spiral with essentially inviscid velocity distribution. III. Tightly wound inviscid spiral. IV. External region containing unrolled-up part of vortex sheet. After Saffman (1992)

$$r \ll (\beta \nu^{\frac{1}{2}} t^{\frac{3}{2}})^{\frac{1}{2+n}} = a_{\rm sm},$$

say, thus $a_{\rm sm}$ defines a *smooth* vortex core inside which the flow is rotational but effectively inviscid before entering the viscous core. This region has no counterpart in boundary layer theory. Therefore, a vortex formed by vortex-sheet rolling up consists of three regions as sketched in Fig. 6.4.

6.5.2 Moore-Saffman Trailing Vortex

We now use the composite core structure of Fig. 6.4 to construct an improved trailing-vortex model, still based on slender and light-loading approximation (6.4.4) along with (6.4.3) used for Batchelor vortex. These linearized equations can be alternatively formulated in a frame of reference fixed to the ground by a Galilean transformation $z = Wt$, in which the flow becomes two-dimensional and unsteady. The only difference of the Moore-Saffman approach from Batchelor's lies in boundary conditions. Since the viscous core is enclosed by a much larger inviscid core, instead of letting the flow directly become uniform as $r \to \infty$, in the viscous subcore the radial variable r should be rescaled to $r^* = r/\sqrt{\nu t}$ and the solution of (6.4.4) with $r^* \to \infty$ should match the corresponding inviscid core solution as $r \to 0$. This asymptotic matching leads to improved outer boundary conditions

$$r^* \to \infty \text{ and } r \to 0: \quad v \sim \frac{\beta}{r^n}, \quad \frac{p - p_\infty}{\rho} \sim -\frac{1}{2n}\frac{\beta^2}{r^{2n}}, \quad w' \sim \frac{\beta^2}{2W}\left(\frac{1}{n} - 1\right) r^{-2n}, \tag{6.5.4}$$

where $w' = w - W$ and the light-loading approximation $|w'| \ll W$ is assumed. Here, the estimate of v comes from (5.3.4), that of p from substituting $\Gamma = \beta r^{1-n}$

into (6.4.3), and that of w' from substituting the p-estimate into (6.4.4b). Besides, just like the Batchelor vortex is assumed to evolve from a line-vortex at $z = 0$, now in terms of $t = z/W$ we take the values of the inviscid solution at $t = 0$ as the initial conditions of viscous solution of (6.4.4):

$$v(r, 0) = \frac{\beta}{r^n}, \quad w'(r, 0) = \frac{\beta^2}{2W}\left(\frac{1}{n} - 1\right) r^{-2n}. \tag{6.5.5}$$

Thus, $w' > 0$ (jet-like) when $n < 1$, and $w' = 0$ (the Lamb-Oseen vortex) only when $n = 1$, for which the second and third terms of (6.4.8) are canceled.

We now seek similarity solution of this viscous subcore problem. Once again we introduce similarity variable

$$\eta = -\frac{Wr^2}{4\nu z} = -\frac{r^2}{4\nu t}, \quad -\infty < \eta < 0, \tag{6.5.6a}$$

and write

$$v = \frac{\beta}{(\nu t)^{n/2}} V_n(\eta), \quad \frac{p}{\rho} = -\frac{\beta^2}{(\nu t)^n} P_n(\eta), \quad w' = \frac{\beta^2}{W(\nu t)^n} W_n(\eta). \tag{6.5.6b}$$

Then a substitution of (6.5.6) into (6.4.4) yields a set of similarity equations and relation:

$$\eta \frac{d^2 V_n}{d\eta^2} - (1 - \eta)\frac{dV_n}{d\eta} - \frac{n}{2}\left(1 + \frac{1}{\eta}\right) V_n = 0, \tag{6.5.7a}$$

$$P_n(\eta) = -\frac{1}{2}\int_{-\infty}^{\eta} \frac{V_n^2(\eta)}{\eta} d\eta, \tag{6.5.7b}$$

$$\eta \frac{d^2 W_n}{d\eta^2} + (1 - \eta)\frac{dW_n}{d\eta} - nW_n = -nP_n - \eta \frac{dP_n}{d\eta}. \tag{6.5.7c}$$

The homogeneous equation (6.5.7a) along with the above initial-boundary conditions for v can be solved analytically:

$$V_n(\eta) = 2^{-n}\Gamma\left(\frac{3}{2} - \frac{1}{2}n\right)(-\eta)^{1/2} M\left(\frac{1}{2} + \frac{1}{2}n; 2; \eta\right), \tag{6.5.8}$$

where Γ is the Gamma function and M is the confluent hypergeometric function of the first kind. Then, P_n follows from (6.5.7b) with known V_n. The remaining inhomogeneous equation (6.5.7c) can be solved numerically. The value of n should be larger as z increases. Moore and Saffman (1973) noticed that $W_n(0)$ becomes negative for $n > 0.44$, implying the transition of axial flow at vortex core from jet-like to wake-like. Figure 6.5 shows the velocity profiles $V_n(r)$ and $W_n(r)$ for a

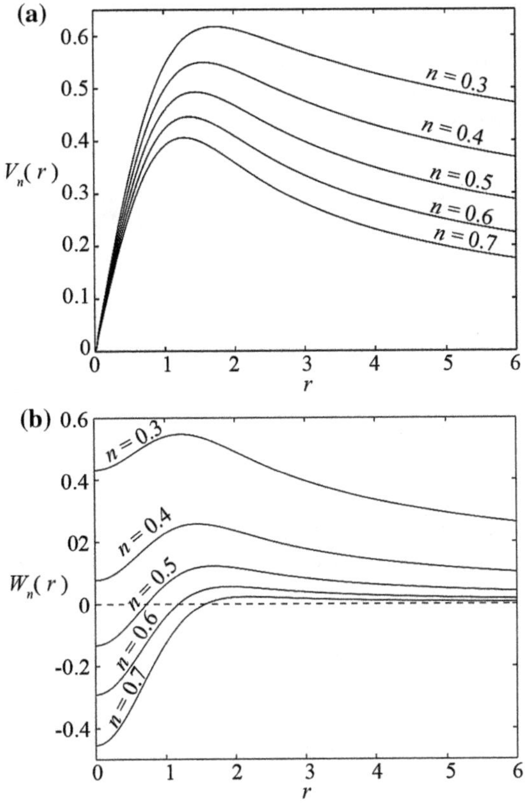

Fig. 6.5 Velocity profiles in viscous subcore of laminar trailing vortex, which match the inviscid temporal similarity solution and axial velocity at the core boundary. **a** Circumferential velocity $V_n(r)$. **b** Axial velocity $W_n(r)$. From Feys and Maslowe (2014)

few values of n calculated by Feys and Maslowe (2014). Indeed, for $n < 0.44$ the axial velocity is jet-like; while for larger n the vortex core becomes wake-like, with jet-like velocity only at larger r.

Recent experimental studies by Lee and Pereira (2010) and del Pino et al. (2011) have shown that both jet-like and wake-like profiles can appear depending on not only downstream distance but also the wingtip geometry and angle of attack, see Fig. 6.6, and the predicted velocity-vorticity profiles by Moore-Saffman model are in better agreement with experimental data than that by Batchelor model. These experimental findings have motivated Feys and Maslowe (2014) to perform a stability analysis of the Moore-Saffman vortex, which will be outlined in Sect. 10.2.2.

6.6 Problems for Chapter 6

6.1. Explain why (6.1.1) to (6.1.8) have some extra terms compared to their partners in Cartesian coordinates. Give a clear interpretation of the physics behind the four extra terms in (6.1.5a) and (6.1.5b).

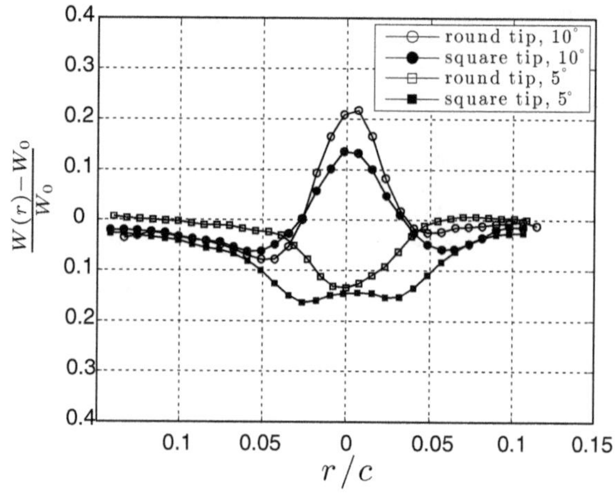

Fig. 6.6 Axial velocity profiles from Lee and Pereira (2010). The chord length is $c = 28$ cm and the measurements were made at a distance $x/c = 5$ downstream of the wingtip. From Feys and Maslowe (2014)

6.2. Derive (6.1.15).

6.3. Derive the expressions of $p_\infty - p(r)$ for the point vortex and Rankine vortex.

6.4. For a steady Burgers vortex (6.3.5), analyze how the advection, stretching, and diffusion terms in the vorticity transport equation are balanced. Moreover, prove the pressure distribution in a Burgers vortex is

$$p(r, z) = p_0 - \frac{\rho\gamma^2}{2}\left(\frac{r^2}{4} + z^2\right) + \rho\int_0^r \frac{v^2}{r}dr, \tag{6.6.1}$$

where p_0 is the stagnation pressure.

6.5. Find the Lamb vector $\boldsymbol{l} = \boldsymbol{\omega} \times \boldsymbol{u}$ of the Burgers vortex, derive its Helmholtz-decomposed two constituents, $\boldsymbol{l}_\parallel$ and $\boldsymbol{l}_\perp$. Show that they are geometrically orthogonal and the transverse part shrinks at rate $\gamma/2$.

6.6. Consider an inviscid and steady axial vortex bounded externally by a stream tube with variable external boundary $r = a(z)$. Assume the flow for $r > a$ is irrotational. Denote the velocity components in cylindrical coordinates (r, θ, z) by (u, v, w) with $C(r, z) = rv$ and flow quantities at $r = 0$ and $r = a$ by subscripts 0 and a. Due to variable $a(z)$, the axial velocity w also depends on z, with dw/dz being its axial variation rate. An interesting observation is that dw/dz depends on $\partial_z C$ or $\partial_r C$. To see this, prove

$$\frac{dw_0^2}{dz} - \frac{dw_a^2}{dz} = 2\int_0^a \frac{1}{r^3}\frac{\partial C^2}{\partial z}dr = -2\int_0^a \frac{1}{r^3}\frac{u}{w}\frac{\partial C^2}{\partial r}dr. \tag{6.6.2}$$

Namely, *an axial-variation of axial velocity at the outer edge of the core always causes a magnification effect at the vortex axis* (Hall 1966). Give a detailed physical interpretation of this result.

Hint. The derivation of (6.6.2) is similar to that of (6.4.8). For the physical discussion you may consult Sect. 10.3.1 for a similar but different analysis.

6.7. Assume a rectangular wing of span $2b$ and chord c may produce approximately elliptic loading near the wingtip. Prove that (6.5.3) can be cast to

$$z \gg \frac{4Wb^2\nu^3}{\Gamma_0^4} = \frac{16cA^2}{C_l^4 Re_c^3},$$

where $C_l = 2\Gamma_0/(Wc)$ is the wing root lift coefficient, $A = 2b/c$ is the aspect ratio, and $Re_c = Wc/\nu$. Then estimate the range of z for which the composite core structure shown in Fig. 6.4 should be used.

Chapter 7
Vortex Rings

Compared to isolated columnar vortex models with infinitely extended axes, some of which have certain inherently unrealistic properties, vortex rings of finite extent are a more perfect form of realistic vortical structures. They can be easily produced in experiments (e.g., by puffing a mass of fluid from a cylinder by a piston) and have frequent appearance in nature and technology. Ever since Helmholtz (1858), vortex rings have been actively investigated for more than 150 years, as reviewed recently by Meleshko et al. (2012). For various aspects of vortex-ring physics and theory see Shariff and Leonard (1992), Lim and Nickels (1995), and Akhmetov (2009), among others. In recent decades, vortex rings have been identified as the key vortical structures for producing thrust by insects, fishes and animals (Dabiri 2009). Accordingly, one's knowledge of vortex-ring evolution has been significantly enriched through extensive experimental, numerical, and theoretical studies. In this chapter we discuss incompressible, axisymmetric, and circular vortex rings, assuming fluid density $\rho = 1$.

The initial formation of vortex rings is also an inviscid process of vortex-sheet rolling up. This is clearly demonstrated by the elegant experimental visualization by Didden (1979) and corresponding numerical simulation by Nitsche and Krasny (1994). Their results are comparatively shown in Fig. 7.1. But the similarity law for the tightly rolled vortex sheet has thus far not been fully clarified.

7.1 General Formulation and Properties

7.1.1 Governing Equations

Consider an isolated and inviscid circular vortex ring moving in an unbounded fluid at rest at infinity. If along the ring's circular axis $r = R$ the circulatory velocity $v = rd\theta/dt$ is nonzero, the ring is said *swirling*. Otherwise it is *swirl-free*. This chapter is confined to swirl-free vortex rings.

J.-Z. Wu et al., *Vortical Flows*, DOI 10.1007/978-3-662-47061-9_7

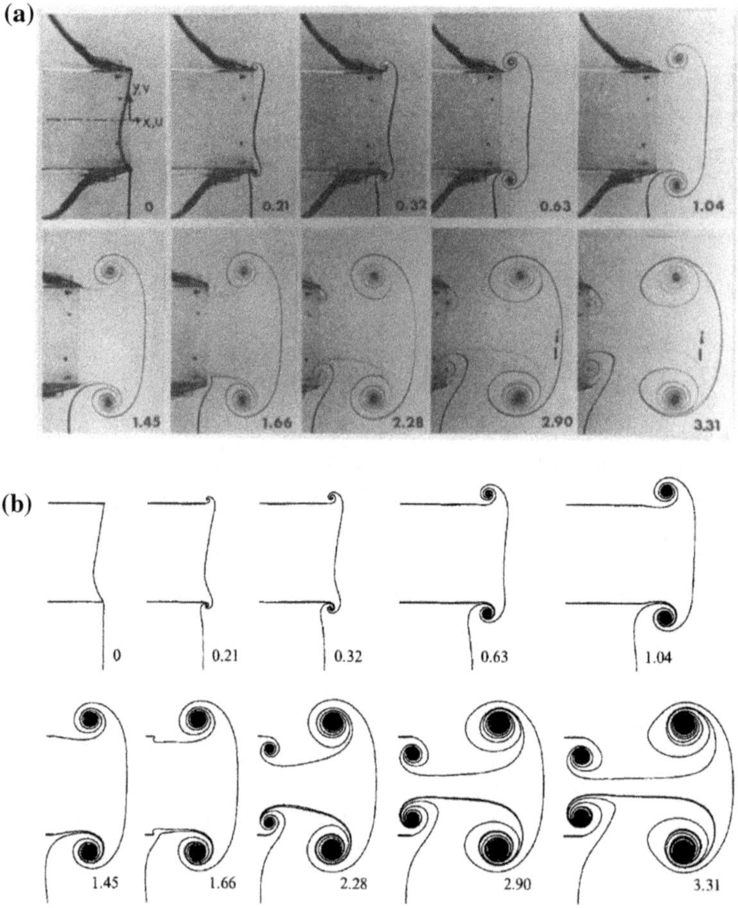

Fig. 7.1 Visualization of inviscid formation of vortex ring via vortex-sheet rolling up, generated by a cylinder-piston system. **a** Experiment by Didden (1979). **b** Simulation by Nitsche and Krasny (1994). From Nitsche and Krasny (1994)

In the study of axisymmetric vortex rings two sets of coordinates are needed. We use cylindrical coordinates (r, θ, z) to define the ring axis and location as sketched in Fig. 7.2, by which the components of velocity $\boldsymbol{u} = \nabla \times \boldsymbol{A}$, vorticity $\boldsymbol{\omega} = \nabla \times \boldsymbol{u} = -\nabla^2 \boldsymbol{A}$, and Stokes stream function ψ are given by

$$\boldsymbol{u}(r, z) = (u, v, w), \quad \boldsymbol{\omega}(r, z) = (0, \omega, 0), \tag{7.1.1a}$$

$$\boldsymbol{A}(r, z) = (0, A_\theta, 0), \quad \psi(r, z) = r A_\theta, \tag{7.1.1b}$$

respectively. $\boldsymbol{\omega}$ and $\boldsymbol{A}$ have only toroidal components. Besides, to describe the flow field inside the vortex core, we use local polar coordinates (σ, ϕ) or Cartesian coor-

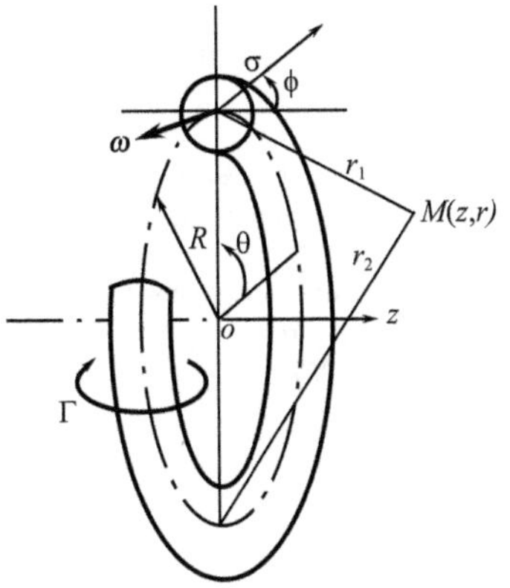

Fig. 7.2 Vortex-ring geometry and cylindrical coordinates. r_1 and r_2 will be defined in Sect. 7.1.3

dinates on the meridional plane of the ring (Fig. 7.2),

$$(r - R, z - Z) = (\sigma \sin \phi, \sigma \cos \phi), \tag{7.1.2}$$

with origin at the toroidal centroid $(R(t), Z(t))$ to be defined later. The local radial and circulatory velocities relative to the toroidal center circle are denoted by u_σ and u_ϕ, respectively.

For this kind of rotationally symmetric flow, the continuity equation reads

$$\frac{1}{r}\frac{\partial(ru)}{\partial r} + \frac{\partial w}{\partial z} = 0. \tag{7.1.3}$$

Moreover, since the flow is *generalized Beltramian*, we have (Sect. 3.1.3)

$$\frac{\omega}{r} = f(\psi, t), \tag{7.1.4}$$

so the flow is governed by (6.1.11) kinematically and (6.1.12) dynamically:

$$-r^2 f(\psi, t) = r\frac{\partial}{\partial r}\left(\frac{1}{r}\frac{\partial \psi}{\partial r}\right) + \frac{\partial^2 \psi}{\partial z^2}, \tag{7.1.5a}$$

$$\frac{Df}{Dt} = \nu\left(\nabla^2 + \frac{2}{r}\frac{\partial}{\partial r}\right) f. \tag{7.1.5b}$$

This pair of equations implies a forth-order equation for ψ. Unlike two-dimensional generalized Beltrami flows for which the differential operator on the right-hand side of both equations of (7.1.5) would be ∇^2, now since (7.1.5a) and (7.1.5b) have different operators, the solution procedure is quite complicated. But (7.1.5b) will be trivially satisfied if we consider inviscid vortex rings and assume the flow relative to the ring is *steady*, or assume $f(\psi, t) = \omega/r = \text{constant}$. It then suffices to solve the

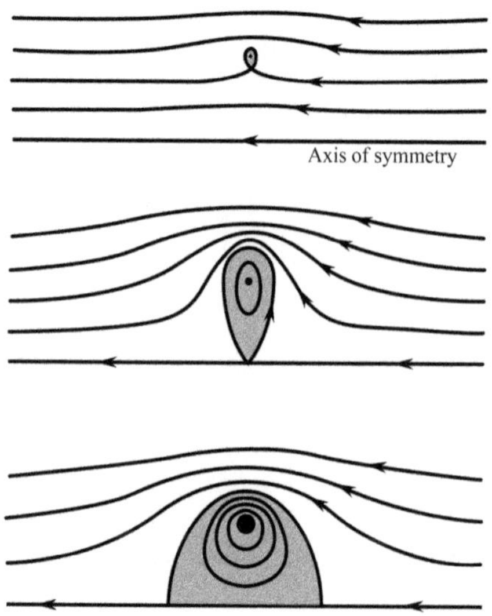

Fig. 7.3 Sketch of streamlines of steady flow relative to a vortex ring for various values of the ratio a/R. The *inner black area* marks the core of vorticity, and the *shaded area* represents fluid carried along with the ring. From Batchelor (1967)

kinematic equation (7.1.5a) alone, which for *given* distribution of $\omega = rf(\psi)$ is a Poisson equation.[1]

A vortex ring may be conceived to be formed by bending a straight columnar vortex to a loop. A thin vortex ring is a loop of vortex filament. As the simplest prototype of various toroidal axial vortices, a circular vortex ring may induce a body of fluid to move with it, which can be determined via the Biot-Savart formulas (Sect. 3.1.4) from given ring geometry and circulation. Depending on the ratio of core size a and ring radius R, the ring-carried fluid may exhibit different topologies as shown in Fig. 7.3.

In the frame of reference fixed to stationary fluid at infinity, a free isolated vortex ring cannot be also stationary due to its *self-induced motion* (Sect. 3.1.4). The ring curvature enables its different segments to induce each other to move, a kinematic mechanism not seen for straight axial vortices. While this self induction is nonuniform for non-axisymmetric rings, e.g., Fig. 3.10, for an axisymmetric ring all elements of unit arc-length have the same contribution, leading to a uniform motion in the z-direction. For an isolated vortex ring moving with constant velocity $\boldsymbol{U} = U\boldsymbol{e}_z$, the flow relative to the ring observed in a comoving frame, say $\boldsymbol{u}' = \boldsymbol{u} - \boldsymbol{U} = (u, 0, w')$, may appear steady and is convenient to work on. The Stokes streamfunction in the comoving frame is given by

$$\Psi = -\frac{1}{2}Ur^2 + \psi, \tag{7.1.6}$$

[1]For more general inviscid and steady axisymmetric vortex ring with $v \neq 0$, the flow is still generalized Beltramian since $\boldsymbol{u} \times \boldsymbol{\omega} = \nabla H$. But in this case the flow is governed by the more general Bragg-Hawthorne equation (6.1.16), along with the three component expressions for $\boldsymbol{\omega}$ given by (6.1.4).

by which one obtains the streamline pattern of Fig. 7.3.

The calculation of vortex ring's self-induced translational velocity $\boldsymbol{U}$ is a non-trivial and actually a central task of vortex-ring theory. Note that the local-induction approximation (3.1.40) of the Biot-Savart formula has indicated that in the limiting case with $a \to 0$, a line vortex ring must induce a singular axial velocity at the ring location. Therefore, line vortex ring can never be a realistic model.

7.1.2 Integral Invariants

The motion and evolution of an isolated vortex ring in an unbounded fluid at rest at infinity are constrained by the invariance of certain vorticity integrals discussed in Sect. 3.3. These include vortical impulse $\boldsymbol{I}$ given by (3.3.8b) that is invariant even for viscous flow, as well as total circulation Γ and kinetic energy K that are invariant for circulation-preserving flow. In terms of cylindrical coordinates and for axisymmetric flow, the volume integral of any scalar f, $\int f dV$, is reduced to a surface integral $2\pi \int r f dS$ on a meridional plane where $\omega \neq 0$, with $dS = drdz = \sigma d\sigma d\phi$ at the vortex core. Therefore, the total circulation of the ring is

$$\Gamma = \int_S \omega dS, \tag{7.1.7}$$

and by (3.3.8b) the vortical impulse has only a z-component (its r-component is cancelled due to axisymmetry):

$$\boldsymbol{I} = \frac{\boldsymbol{e}_z}{2} \int r\omega dV = \boldsymbol{e}_z \pi \int_S r^2 \omega dS. \tag{7.1.8}$$

On the other hand, the total kinetic energy K has two alternative expressions in terms of vorticity, (3.3.29) and (3.3.31), which now read

$$K = \frac{1}{2} \int A_\theta \omega dV = \pi \int_S \psi \omega dS \quad \text{or} \tag{7.1.9a}$$

$$= \int (rw - zu)\omega dV = 2\pi \int_S (r^2 w - rzu)\omega dS, \tag{7.1.9b}$$

respectively. Recall that both $\boldsymbol{I}$ and K are finite due to the fast decay of $\boldsymbol{\omega}$ as shown in Sect. 2.4. Also notice that for isolated vortex ring moving in a fluid at rest at infinity two vectorial Lamb-vector integrals, that of $\boldsymbol{\omega} \times \boldsymbol{u}$ and $\boldsymbol{x} \times (\boldsymbol{\omega} \times \boldsymbol{u})$, vanish as shown by (3.3.6a) and (3.3.6b).

Then, we may use the invariance of I to define the radial and axial locations of the *centroid* (R, Z) of the toroid:

$$R = \frac{\int r^3 \omega dS}{\int r^2 \omega dS} = \frac{\pi}{I} \int r^3 \omega dS, \quad (7.1.10a)$$

$$Z = \frac{\int r^2 z \omega dS}{\int r^2 \omega dS} = \frac{\pi}{I} \int r^2 z \omega dS. \quad (7.1.10b)$$

Here, the time derivative of Z gives the uniform velocity $\boldsymbol{U} = U\boldsymbol{e}_z$ of the ring due to self induction:

$$U = \frac{dZ}{dt} = \frac{\pi}{I} \int (wr^2 + 2urz)\omega dS. \quad (7.1.11)$$

Since the vorticity flux ωdS (circulation about a vortex filament of cross area dS) is also invariant, a use of (7.1.9b) and (7.1.11) leads to an expression of U in terms of I and K, known as the *Helmholtz-Lamb identity* (see also Problem 7.2):

$$2IU = K + 6\pi \int rzu\omega dS, \quad K = \pi \int \psi \omega dS. \quad (7.1.12)$$

Therefore, once the Stokes streamfunction ψ and the velocity are solved from a given vorticity distribution in the core (and hence given $\boldsymbol{I}$), the ring's self-induced motion can be determined.

Alternative to and neater than (7.1.12), Benjamin (1976) proved the following variational principle:

Theorem *In a circulation-preserving flow, a steadily moving axisymmetric vortex ring is realizable as the maximum state of the kinetic energy K subjected to the constraint of constant vortical impulse $\boldsymbol{I}$.*

This principle establishes a *variational relation* between the variations of K and $\boldsymbol{I}$ through the ring's translational velocity $\boldsymbol{U}$:

$$\delta K = \boldsymbol{U} \cdot \delta \boldsymbol{I}, \quad (7.1.13)$$

of which a clear proof can be found in Fukomoto and Moffatt (2008) and will not be presented here.

7.1.3 Stokes Streamfunction

When $\omega(\psi)$ is prescribed within a compact region in an unbounded fluid at rest at infinity, the integral solution of (7.1.5a) for vector potential $\boldsymbol{A}$ has been given by the second expression of (3.1.33), which by $\boldsymbol{A} = (0, \psi/r, 0)$ and the notations of Fig. 7.2 leads to

$$\psi(r, z) = \frac{1}{4\pi} \int \frac{r\omega' \boldsymbol{e}'_\theta \cdot \boldsymbol{e}_\theta}{|\boldsymbol{x} - \boldsymbol{x}'|} dV' = \int \omega' F(r, r', z - z') dS', \quad (7.1.14)$$

where $dS' = dr'dz'$. The kernel function there,

$$F(r, r', z - z') \equiv \frac{rr'}{4\pi} \int_0^{2\pi} \frac{\cos(\theta - \theta')d\theta'}{\sqrt{(z - z')^2 + r^2 + r'^2 - 2rr'\cos(\theta - \theta')}}, \quad (7.1.15)$$

is the Stokes streamfunction generated by a line vortex ring of unit circulation with $\omega = \delta(z - z')\delta(r - r')$. This integral can be carried out and expressed by the first and second kinds of complete elliptic integrals,

$$K(\lambda) = \int_0^{\pi/2} \frac{d\varphi}{\sqrt{1 - \lambda^2 \sin^2 \varphi}}, \quad E(\lambda) = \int_0^{\pi/2} \sqrt{1 - \lambda^2 \sin^2 \varphi} d\varphi, \quad (7.1.16)$$

respectively, with λ being their modulus. The result has two alternative forms (Lamb 1932), and the neater one (but its derivation is more involved) is as follows. Let the moving point $\boldsymbol{x}'$ be along a circle with varying θ' and fixed (r', z') and define

$$r_1 = \sqrt{(z - z')^2 + (r - r')^2}, \quad r_2 = \sqrt{(z - z')^2 + (r + r')^2} \quad (7.1.17)$$

as marked in Fig. 7.2, which are just r_{max} and r_{min} on this circle, respectively. Then set $\varphi = (\theta - \theta')/2 - \pi/2$ and choose the modulus as

$$\lambda = \frac{r_2 - r_1}{r_2 + r_1}, \quad (7.1.18a)$$

one finds

$$F(r, r', z - z') = \frac{r_1 + r_2}{2\pi}[K(\lambda) - E(\lambda)]. \quad (7.1.18b)$$

Therefore, the Stokes streamfunction finally reads

$$\psi = \frac{1}{2\pi} \int (r_1 + r_2)[K(\lambda) - E(\lambda)]\omega' dS'. \quad (7.1.19)$$

The streamlines of line vortex ring, viewed in a comoving frame, are shown in Fig. 7.4.

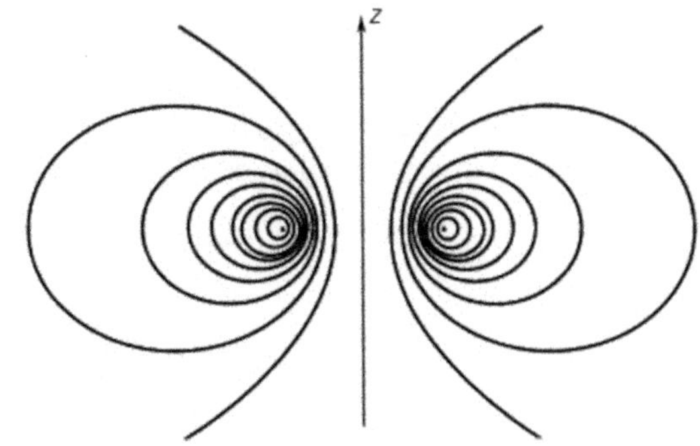

Fig. 7.4 Streamlines of a circular line vortex. From Lamb (1932)

7.2 Inviscid Vortex Rings

This section considers a few relatively simple inviscid vortex ring solutions. We first assume the vortex core is very thin and seek its approximate solution. We then go to the opposite extreme to consider the fattest vortex ring, the Hill spherical vortex. Finally, we show that Hill's vortex is just a member of a whole family of vortex rings from thin to fat, of which the solutions can be numerically obtained.

7.2.1 Thin-Core Vortex Ring

We start from circular vortex rings with very thin torus core of radius $a \ll R$ or $\epsilon = a/R \ll 1$ (the upper plot of Fig. 7.3), where R is the radius of ring circle. For this ring-like filament of circulation Γ and constant curvature $\kappa = 1/R$, by the local-induction approximation (3.1.40) the self-induced velocity is a constant $\boldsymbol{U} = U\boldsymbol{e}_z$ with $U \sim (\Gamma/4\pi R)\log(R/a)$, where a represents a small radial scale of the core area. But the quantitative determination of U relies on detailed vorticity distribution and vortex-core geometry.

The simplest model of thin-core vortex ring is a torus of circular cross-section, to which we apply local coordinates (σ, ϕ) defined by (7.1.2) with $\sigma \leq a$. Then we have approximation

$$\omega(\sigma, \phi) \simeq \omega_0(\sigma)\frac{r}{R} = \omega_0(\sigma)\left(1 + \frac{\sigma}{R}\sin\phi\right) \tag{7.2.1a}$$

$$= \omega_0(\sigma)[1 + O(\epsilon)], \quad \epsilon = \frac{a}{R} \ll 1. \tag{7.2.1b}$$

Thus, to the leading order the vorticity distribution in the core is axisymmetric. Thus, similar to (6.1.4a), there is

$$\omega_0(\sigma) = \frac{1}{\sigma}\frac{\partial}{\partial\sigma}(\sigma u_\phi) = \frac{1}{2\pi\sigma}\frac{d\gamma}{d\sigma},$$

where

$$\gamma(\sigma) \equiv 2\pi\sigma u_\phi(\sigma) = 2\pi\int_0^\sigma \omega_0'\sigma' d\sigma' \tag{7.2.2}$$

is the circulation of toroidal vorticity tube of cross-sectional radius σ, with $\gamma(a) = \Gamma$. Let us now calculate the leading-order asymptotic form of the streamfunction given by (7.1.19).

In this case there is $r_1 \ll r_2 \simeq 2R$, $\lambda \simeq 1$, and $r_1 + r_2 \simeq 2R$. Notice that

$$1 - \lambda^2 = \frac{4r_1r_2}{(r_1 + r_2)^2} \simeq \frac{2r_1}{R},$$

the elliptic integrals are approximated by

$$E(\lambda) \simeq 1, \quad K(\lambda) \simeq \frac{1}{2}\log\frac{16}{\lambda'^2} = \frac{1}{2}\log\frac{8R}{r_1}.$$

Thus, to the leading order we obtain

$$\begin{aligned}\psi &= \frac{R}{2\pi}\int\left(\log\frac{8R}{r_1} - 2\right)\omega_0'(\sigma')dS' \\ &= \frac{R}{2\pi}\int_0^a\int_0^{2\pi}(\log 8R - 2 - \log r_1)\omega_0'(\sigma')\sigma' d\sigma' d\phi',\end{aligned} \tag{7.2.3}$$

where only $r_1 = \sqrt{\sigma^2 + \sigma'^2 - 2\sigma\sigma'\cos(\phi - \phi')}$ depends on $\phi - \phi'$ but the integration of $\log r_1$ over $\phi' \in [0, 2\pi)$ is known:

$$\frac{1}{2\pi}\int_0^{2\pi}\log[\sigma^2 + \sigma'^2 - 2\sigma\sigma'\cos(\phi - \phi')]^{1/2}d\phi = \begin{cases}\log\sigma' \text{ if } \sigma' > \sigma, \\ \log\sigma \text{ if } \sigma > \sigma'.\end{cases} \tag{7.2.4}$$

Thus, by using (7.2.2) and integration by parts, in (7.2.3) there is

$$\frac{R}{2\pi}\int_0^a\int_0^{2\pi}\log r_1\omega_0'\sigma' d\sigma' d\phi' = \frac{R\Gamma}{2\pi}\left(\log a - \frac{1}{\Gamma}\int_\sigma^a\frac{\gamma'}{\sigma'}d\sigma'\right).$$

Therefore, to the leading order, the thin-core Stokes streamfunction reads

$$\psi(\sigma) = \frac{R\Gamma}{2\pi}\left(\log\frac{8R}{a} - 2 + \frac{1}{\Gamma}\int_\sigma^a\frac{\gamma(\sigma')}{\sigma'}d\sigma'\right), \tag{7.2.5}$$

which holds for *arbitrary* vorticity distribution $\omega(\sigma)$ in thin toroidal core.

Having obtained ψ, one may proceed to derive the induced velocity near the core, and then to calculate the ring's translational speed U and streamlines in a comoving frame. This approach is known as *direct method*, of which the details are given by, e.g., Tong et al. (2009) and Wu et al. (2006), where ψ is calculated to $O(\epsilon)$ and has ϕ-dependence. Here we take a simpler approach, just focusing on the determination of U by using the Helmholtz-Lamb identity (7.1.12) (and dropping terms of $O(\epsilon) \ll 1$ as before). This is known as *energy method*. For impulse, by (7.1.8) we simply have

$$I \simeq \pi R^2\Gamma. \tag{7.2.6}$$

For kinetic energy, by (7.1.9a) we find

$$K = 2\pi^2 \int_0^a \psi\omega\sigma d\sigma = \pi \int_0^a \psi(\sigma) d\gamma(\sigma) = \pi \left(\psi\gamma \Big|_0^a - \int_0^a \gamma d\psi \right)$$
$$= \frac{R\Gamma^2}{2} \left(\log \frac{8R}{a} - 2 + A \right), \tag{7.2.7a}$$

where

$$A \equiv \frac{1}{\Gamma^2} \int_0^a \frac{\gamma^2(\sigma)}{\sigma} d\sigma = \frac{4\pi^2}{\Gamma^2} \int_0^a \sigma u_\phi^2(\sigma) d\sigma. \tag{7.2.7b}$$

Moreover, for the second term of $2IU$ in (7.1.12), we have

$$u \simeq \frac{\gamma(\sigma)}{2\pi\sigma} \cos\phi, \quad r = R + \sigma \sin\phi, \quad z = \sigma \cos\phi,$$

so there is

$$6\pi \int_S rzu\omega dS \simeq 3 \int_0^{2\pi} d\phi \int_0^a (R + \sigma \sin\phi) \cos^2\phi \gamma(\sigma)\omega(\sigma)\sigma d\sigma$$
$$= \frac{3R}{2\pi} \int_0^{2\pi} \cos^2\phi d\phi \int_0^a \gamma(\sigma) d\gamma(\sigma) = \frac{3}{4} R\Gamma^2. \tag{7.2.8}$$

Therefore, a substitution of (7.2.6)–(7.2.8) into (7.1.12) yields (Fraenkel 1970; Saffman 1970)

$$U = \frac{\Gamma}{4\pi R} \left(\log \frac{8R}{a} - \frac{1}{2} + A \right). \tag{7.2.9}$$

In particular, if $\omega = \omega_0$ is uniform in the core, there is $\gamma^2/\Gamma^2 = \sigma^4/a^4$, implying that $A = 1/4$. Thus (7.2.9) is reduced to a formula first obtained by Kelvin in 1867:

$$U = \frac{\Gamma}{4\pi R} \left(\log \frac{8R}{a} - \frac{1}{4} \right). \tag{7.2.10}$$

It is of interest to see how one infers (7.2.9) from Benjamin's variational principle (7.1.13). Consider an infinitesimal virtual change of ring size and coordinate,

$$(R, \sigma) \to (\hat{R}, \hat{\sigma}) = (R, \sigma) + (\delta R, \delta\sigma).$$

The variation is subjected to the constraint that the flow preserves circulation and volume. The volume invariance implies

$$2\pi^2\sigma^2 R = 2\pi^2\hat{\sigma}^2\hat{R}, \quad \text{so that} \quad 2\frac{\delta\sigma}{\sigma} = -\frac{\delta R}{R},$$

which also holds if $\sigma = a$. It can be found that the circulation invariance implies $\delta\gamma(\sigma) = \delta\Gamma = 0$, and also $\delta A = 0$ (Fukumoto and Moffatt 2008). Thus, the variation of (7.2.7a) reads

$$\delta K = \frac{\Gamma^2}{2}\left[\delta R\left(\log\frac{8R}{a} + A - 2\right) + R\left(\frac{\delta R}{R} - \frac{\delta a}{a}\right)\right]$$
$$= \frac{\Gamma^2}{2}\left(\log\frac{8R}{a} - \frac{1}{2} + A\right)\delta R.$$

On the other hand, by (7.2.6), the invariance of impulse implies $\delta I = 2\pi\Gamma R\delta R$. Substituting these δK and δI into (7.1.13) recovers (7.2.9) at once, without any need for the calculation of (7.2.8). That calculation could be very tedious as one proceeds to higher-order perturbations.

The fact that, with the same circulation smaller vortex ring runs faster, explains an interesting phenomenon. If two coaxial vortex rings 1 and 2 are produced sequentially one after another under the same condition, their mutual induction will increase R_1 but reduce R_2, so ring 2 will run through ring 1 and the process may be repeated a few times. This phenomenon is known as *leap frogging*, of which an experimental visualization is shown in Fig. 7.5.

7.2.2 Hill's Spherical Vortex

We now turn to the opposite extreme of thin-core vortex ring and assume a very fat vortex core of constant vorticity, filling a sphere:

$$\frac{\omega}{r} = f(\psi) = \begin{cases} -\Omega & \text{if } r^2 + z^2 < a^2, \\ 0 & \text{if } r^2 + z^2 > a^2. \end{cases} \tag{7.2.11}$$

Thus, the vorticity has no diffusion, and its advection is balanced by tilting (Problem 7.1). Hill in 1894 found that (7.1.5a) has a solution in the sphere

$$\psi = \frac{1}{10}\Omega r^2(r^2 + z^2 - a^2),$$
$$u = -\frac{1}{5}\Omega rz, \quad w = \frac{1}{5}\Omega(2r^2 + z^2 - a^2). \tag{7.2.12}$$

On the sphere surface $\psi = 0$, the tangent velocity is

$$q = \sqrt{u^2 + w^2} = \frac{1}{5}\Omega ar, \quad r^2 + z^2 = a^2. \tag{7.2.13}$$

The Hill spherical vortex is the "fattest" circular vortex ring, and Hill's solution is the only exact analytical model of all vortex rings. The streamlines and vorticity

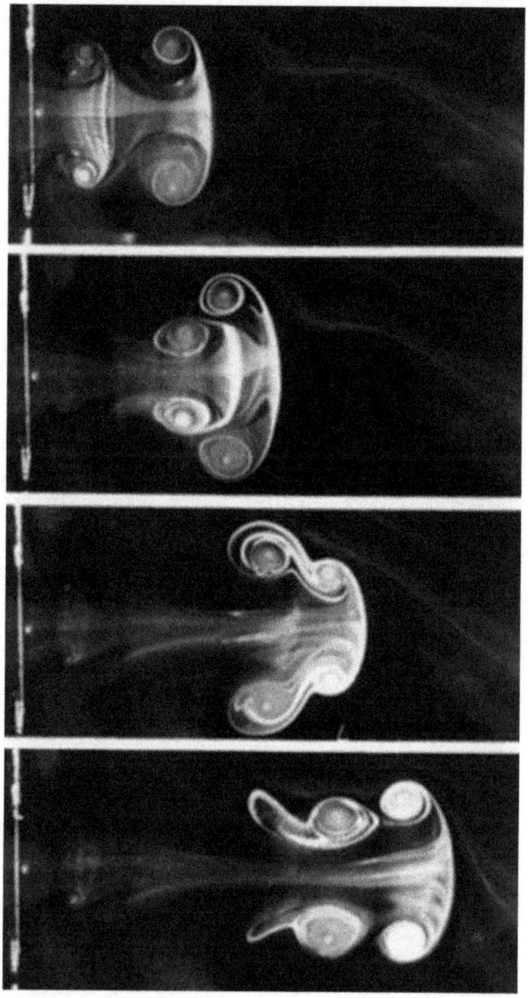

Fig. 7.5 Leap frogging of two vortex rings. From Van Dyke (1982)

lines of the Hill vortex are plotted in Fig. 7.6. On the plane $z = 0$ we have $u = 0$ while w has the same magnitude but opposite signs at $r = 0$ and $r = a$. In the upper and lower hemispheres divided by the plane $z = 0$, the radial velocity u has opposite signs. At the intersectional circle of the plane $z = h$ and the sphere, there is $w = \Omega(2r^2 + h^2 - a^2)/5$, which has opposite signs inside and outside the circle $r = \sqrt{(a^2 - h^2)/2}$. Moreover, by the Bernoulli integral (3.2.25) for generalized Beltrami flow, the pressure is determined up to a constant $p_0 = p(0, 0)$:

$$p - p_0 = \frac{1}{50} r \Omega^2 [r^2(r^2 - a^2) + 2z^2 a^2 - z^4] + 2r\Omega \nu z. \tag{7.2.14}$$

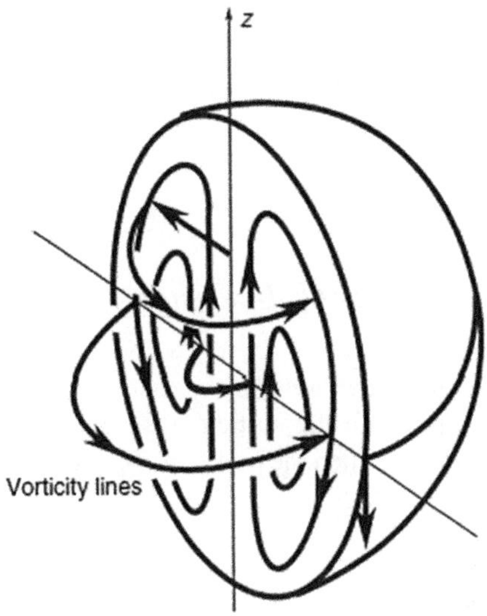

Fig. 7.6 The streamlines and vorticity lines in Hill's spherical vortex. Based on Panton (2013)

In the above analysis our coordinate system is fixed to the spherical vortex, outside which the flow is irrotational. By (7.2.12), at the vortex center $(r, z) = (0, 0)$ the induced velocity is $\boldsymbol{u}_0 = -\Omega a^2 \boldsymbol{e}_z/5$, which is balanced by a potential flow with free-stream velocity $U\boldsymbol{e}_z$ while keeping its shape unchanged. Then the constants Ω and p_0 are determined by the continuity of velocity and pressure at the sphere. For potential flow over sphere, the velocity and pressure at the surface $r^2 + z^2 = a^2$ are given by (e.g., Milne-Thomson 1968)

$$q = \frac{3}{2a} U r, \tag{7.2.15a}$$

$$p = \frac{1}{8} U^2 \left(\frac{9z^2}{a^2} - 5 \right) + p_\infty. \tag{7.2.15b}$$

Comparing (7.2.15a) and (7.2.13) gives $\Omega = 15U/(2a^2)$, which also ensures the continuity of normal vorticity $\boldsymbol{\omega} \cdot \boldsymbol{n}$. Then (7.2.14) yields

$$p - p_0 = \frac{9}{8a^2} U^2 z^2 + 2\Omega \nu z \quad \text{on the sphere,}$$

of which the comparison with (7.2.15b) gives, in the Euler limit $\nu \to 0$,

$$p_0 = p_\infty - \frac{5}{8} U^2. \tag{7.2.16}$$

Thus, p_0 is smaller than p_∞ by an amount proportional to ρU^2.

Like the Rankine vortex, the vorticity inside the sphere jumps from $-\Omega a$ to zero across the boundary. Hence, although inside a spherical boundary the Hill vortex is both exact generalized Beltrami Navier-Stokes solution and Euler solution, after combining with the external potential flow the global flow is no longer a smooth viscous solution.

7.2.3 Fraenkel-Norbury Vortex Ring Family

We have mentioned in the context of (7.1.5) that the solution procedure will be greatly simplified if the vorticity distribution in the core satisfies $\omega = \Omega r$ with Ω being a constant. Then the problem stays within kinematics, governed by a Poisson equation

$$\frac{\partial^2 \psi}{\partial r^2} - \frac{1}{r}\frac{\partial \psi}{\partial r} + \frac{\partial^2 \psi}{\partial z^2} = -\Omega r^2. \tag{7.2.17}$$

Evidently, Hill's spherical vortex (7.2.11) is just a special solution of (7.2.17). At the other extreme of thin-core vortex rings (Sect. 7.2.1), (7.2.17) simply specifies the core-vorticity distribution to (7.2.1a), so the constant ω_0 is its leading-order approximation, to which Kelvin's formula (7.2.10) applies.

Having learned two special solutions of (7.2.17) at opposite extremes, it is natural to explore all possible solutions of the same kind with various core radius, from thin to fat. This leads to a family of steady Euler solutions known as the *Fraenkel-Norbury family*, which also satisfy the Navier-Stokes equation inside the core region A with its cross-sectional shape to be solved. This family have been found in remarkable agreement with experimental measurement and viscous numerical solutions.

The problem can be stated as (Norbury 1973): given the free-stream axial velocity $(0, 0, U)$, the vorticity constant Ω, and a positive constant $k > 0$, find the Stokes stream function Ψ and boundary ∂A of the core cross section A such that

$$\left(\frac{\partial^2}{\partial r^2} - \frac{1}{r}\frac{\partial}{\partial r} + \frac{\partial^2}{\partial z^2}\right)\Psi(r, z) = \begin{cases} -\Omega r^2 & \text{in } A, \\ 0 & \text{outside } A, \end{cases} \tag{7.2.18a}$$

$$\Psi \text{ and } \nabla\Psi \text{ are continuous across } \partial A, \tag{7.2.18b}$$

$$\Psi = k \quad \text{on } \partial A, \tag{7.2.18c}$$

$$\Psi + \frac{1}{2}Ur^2 \to 0 \quad \text{as } r^2 + z^2 \to \infty. \tag{7.2.18d}$$

Observe that if ∂A is assumed known, then the solution for Ψ is obtained:

$$\Psi(r, z) = -\frac{1}{2}Ur^2 + \Omega \int_A F(r, r', z - z')dr'dz', \tag{7.2.19}$$

where F is given by (7.1.18b). So the problem focuses on the determination of the boundary shape ∂A. We do this by combining (7.2.18c) and (7.2.19), which leads to a nonlinear integral equation for the core boundary ∂A. To simplify this problem, Norbury used the radial distance L of the midpoint of the core at $z = 0$ as length scale (see Fig. 7.7), introduced a single parameter $\alpha > 0$ by defining the

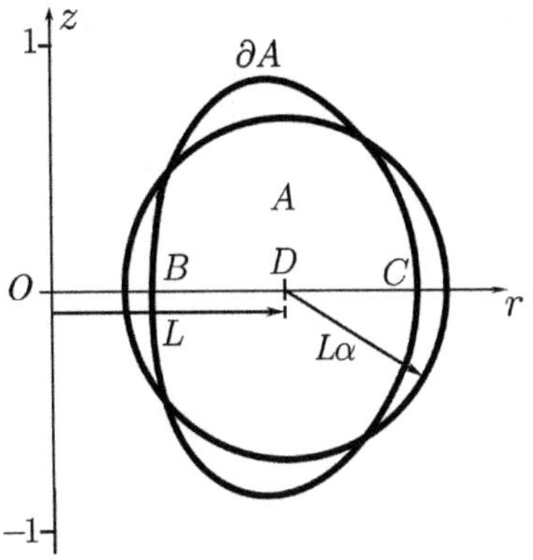

Fig. 7.7 Meridional cross-section A of Fraenkel-Norbury vortex rings, specified by parameter α (=0.7). The ring radius is $L = (OB + OC)/2$. After Norbury (1973)

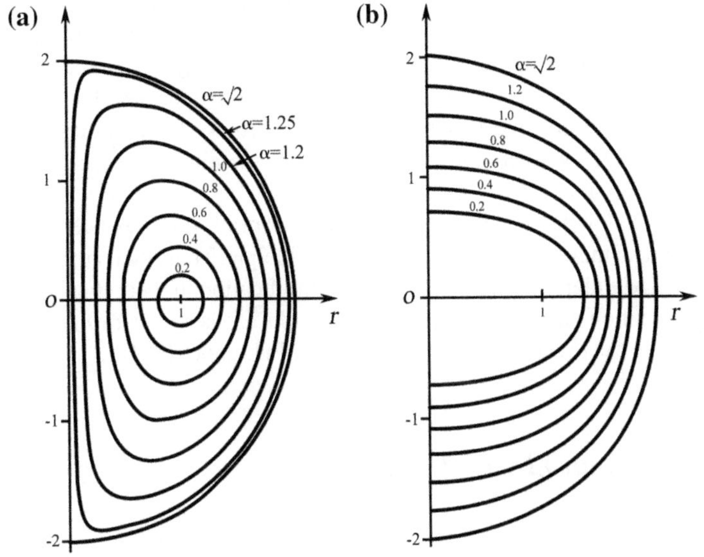

Fig. 7.8 Fraenkel-Norbury vortex rings. **a** Core shapes. **b** Dividing streamlines. After Norbury (1973)

area $A = \pi L^2\alpha^2$, and chose a velocity scale $U^* = \Omega L^2\alpha^2$. Then the dimensionless form of the problem (7.2.19), scaled by L and U^*, amounts to solving a nonlinear integral equation for the core boundary ∂A, with single parameter α:

$$k(\alpha) = -\frac{1}{2}U(\alpha)r^2 + \frac{1}{\alpha^2}\int_{A(\alpha)} F(r, r', z - z')dr'dz', \quad (r, z) \in \partial A(\alpha). \quad (7.2.20)$$

As found numerically by Norbury (1973) and proved mathematically by Fraenkel and Burgers (1974), for each $\alpha \in (0, \sqrt{2}]$ only one solution set $\{k, U, \partial A\}$ of (7.2.20) exists. The case $\alpha = \sqrt{2}$ is precisely the Hill spherical vortex, while the case $\alpha \ll 1$ is the thin-core vortex ring addressed in Sect. 7.2.1. Figure 7.8 shows the computed core boundary ∂A and dividing streamline $\Psi = 0$ of the family for different α. The latter encloses a fluid body carried along by the vortex ring.

7.3 Evolution of Viscous Vortex Rings

Real vortex rings are mostly in between of very thin and very fat extremal cases. Yet the viscosity may cause the expansion of core size and slow down the ring travel speed, making an initially thin ring at large Reynolds number $R_\Gamma = \Gamma/\nu$ become a fat ring and finally stop. Viscous vortex rings evolve from being generated to being dissipated. Recently, careful experiments and numerical simulations have provided rich information on this time evolution process at its different stages. Accordingly, theoretical analyses have been developed to cover these stages. It has been found that, although the total kinetic energy K is conservative only for circulation-preserving flow, if the viscosity is turned on it only affects the vorticity distribution inside the core to $O(\epsilon^3)$, $\epsilon = a/R$, so that both Helmholtz-Lamb identity (7.1.12) and Benjamin's variational principle (7.1.13) are still approximately applicable. Even the unsteady (quasi-steady) slowing down motion of viscous vortex ring has been found to be compatible with these identity and variational principle. Thus, their application range is significantly widened, leading to a great simplification in theoretical studies.

This subsection outlines briefly the main results of viscous vortex ring evolution. For details see the reviews of Dabiri (2009) and Fukumoto (2010).

7.3.1 Early Stage at $\nu T/R_0^2 \ll 1$

The initial formation process of viscous vortex rings has been demonstrated by Fig. 7.1. It can also be seen from Figs. 7.5 and 7.9 below. Typically, a viscous vortex ring may be produced by a given cylinder-piston system. At an initial $R_{\Gamma_0} = \Gamma_0/\nu \gg 1$, the piston motion generates a thin cylinder-like vortex layer that rolls into ring-like spirals. Vorticity is continuously fed into the spirals to increase the ring circulation Γ, associated with entraining surrounding fluid into the vortex core.

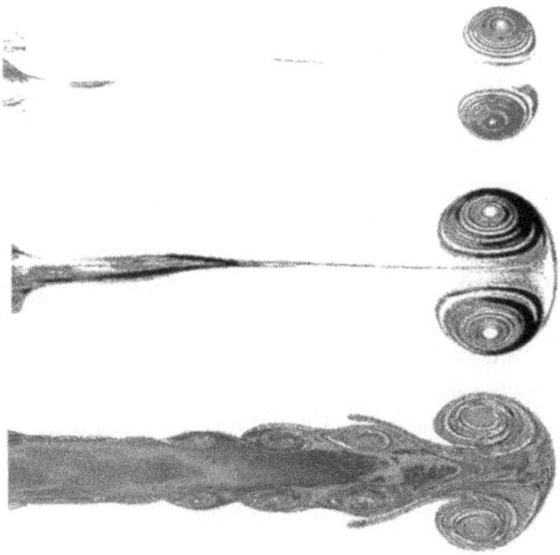

Fig. 7.9 Visualization of vortex rings generated by a cylinder-piston system. The ratio L/D is called the "formation number" of the rings. **a** $L/D = 2$, $R_\Gamma \simeq 2800$; **b** $L/D = 3.8$, $R_\Gamma \simeq 6000$; and **c** $L/D = 14.5$. From Gharib et al. (1998)

Having said this, we now begin with an already formed thin viscous vortex ring with $\epsilon = a/R \ll 1$ and $R_\Gamma \gg 1$, for which the travel speed formula (7.2.9) can be applied with arbitrary axisymmetric distributed vorticity $\omega_0 = \omega_0(\sigma)$ in the core. In most cases $\omega(\sigma)$ has been experimentally found to have Gaussian distribution, for which a good model without swirl is the Oseen-Lamb vortex given by (6.2.5), diffused from a singular line vortex ring of radius R_0 at $t = 0$ (which is a natural time origin). Namely, in the local polar coordinates (σ, ϕ) we assume

$$\omega_0 = \frac{\Gamma}{4\pi\nu t} e^{-\sigma^2/4\nu t}, \quad u_\phi(\sigma) = \frac{\Gamma}{2\pi\sigma}\left(1 - e^{-\sigma^2/4\nu t}\right). \tag{7.3.1}$$

Substituting this into (7.2.9) yields (Tung and Ting 1967; Saffman 1970)

$$U \simeq \frac{\Gamma}{4\pi R_0}\left[\log\frac{8R_0}{2\sqrt{\nu t}} - \frac{1}{2}(1 - \gamma + \log 2)\right], \tag{7.3.2}$$

where $\gamma = 0.57721566\ldots$ is Euler's constant and $(1 - \gamma + \log 2)/2 \simeq 0.558$. This result indicates that the core radius of thin viscous ring is $a \simeq 2\sqrt{\nu t}$. Thus, in the following evolution analysis we can introduce a dimensionless time-dependent parameter

$$\eta(t) \equiv \frac{\sqrt{\nu t}}{R_0}, \quad 0 \le \eta < \infty, \tag{7.3.3}$$

with η^2 being a natural dimensionless time scale.

A second-order matched asymptotic expansion to $O(\eta^2)$ has been found by Fukumoto and Moffatt (2008), and it improves (7.3.2) to

$$U \simeq \frac{\Gamma}{4\pi R_0}\left(\log\frac{4}{\eta} - 0.558 - 3.672\eta^2\right), \tag{7.3.4}$$

which is valid to $O(\eta^3)$. The use of Benjamin's variational principle (7.1.13) had considerably simplified their analysis. The extra term in (7.3.4) reflects the physics that, as one bends a straight columnar vortex to a torus, the vorticity lines in the convex side is stretched with enhanced ω, while those in the concave side is contracted with reduced ω. This is an $O(\eta)$ effect and appears on cross-sectional plane (σ, α) as an *effective dipole*.

In addition to the travel speed U, by the invariance of $\boldsymbol{I}$ for viscous flow, the radial motion of a viscous vortex ring can also be estimated. In terms of the peak-vorticity circle of radius R_{pv} and radial vorticity centroid R_{c} defined by (7.1.10), it has been found

$$R_{\text{pv}} \simeq R_0 + 4.590\eta^2(t), \quad R_{\text{c}} = \frac{\pi}{I}\int r^3\omega r dr d\theta \simeq R_0 + 3\eta^2(t). \tag{7.3.5}$$

Equations (7.3.2) or (7.3.4) and (7.3.5) are valid for $\eta \ll 1$. They characterize the early-stage evolution of the vortex ring: diffusing at constant Γ and R_Γ. The diffusion slows down the travel speed and enlarges ring radius.

7.3.2 Matured Stage at $\nu T/R_0^2 = O(1)$

Once a vortex ring is formed, the entrainment of surrounding fluid into the vortex core cannot be indefinitely continued. For example, in a cylinder-piston system, if D is the diameter of cylinder exit and $L = \overline{U}_p t$ is the piston stroke length with $\overline{U}_p$ being the mean piston speed, there is a dimensionless critical time L/D of value about 4, after which the discharged fluid mass cannot be all enrolled into the ring but partially trails the ring as a jet that takes over the continuous increase of total circulation in the flow till the piston stops (Gharib et al. 1998). Namely, at this critical time the vortex ring is pinched off as an "*optimal vortex ring*" that carries the maximum Γ. This situation is shown in Fig. 7.9. At this matured stage there is $\eta^2(t) = \nu t/R_0^2 = O(1)$, and the vortex core becomes so fat that vorticity cancelation takes place at the symmetry axis $r = 0$, and hence the total circulation Γ is no longer constant.

A simple theoretical model was proposed by Saffman (1970) for the travel speed $U(t)$ and ring radius $R(t)$ of this matured stage of viscous vortex rings. Assume $\Gamma(t) \sim U(t)R(t)$ and $I \sim R^2\Gamma$ to be consistent with (7.2.6), so we can write

$$U(t) = \frac{I}{kR^3(t)}, \tag{7.3.6}$$

with k being a constant (for Hill's spherical vortex $k = 16\pi$). On the other hand, the vorticity Eq. (6.1.12) can be rewritten as

$$\frac{\partial \omega}{\partial t} + \frac{\partial}{\partial r}(u\omega) + \frac{\partial}{\partial z}(w\omega) = \nu \left[\frac{\partial^2 \omega}{\partial r^2} + \frac{\partial}{\partial r}\left(\frac{\omega}{r}\right) + \frac{\partial^2 \omega}{\partial z^2} \right]. \tag{7.3.7}$$

Since $\omega = 0$ at infinity and $\partial\omega/\partial r = 0$ at $r = 0$ due to symmetry, on the meridional plane the integration of (7.3.7) yields

$$\frac{d\Gamma}{dt} = \int_{-\infty}^{\infty} dz \int_0^{\infty} \frac{\partial \omega}{\partial t} dr = -\nu \int_{-\infty}^{\infty} \left(\frac{\omega}{r}\right)_{r=0} dz. \tag{7.3.8}$$

When the value range of z in which $\omega \neq 0$ is about the same as R, an argument of order of magnitude indicates that the integral is proportional to $-\nu U/R$. Thus we may write

$$\frac{d}{dt}(UR) = -k' \frac{\nu U}{R}, \tag{7.3.9}$$

with k' being another constant. Combining (7.3.6) and (7.3.9) then yields

$$R^2 \simeq R_0^2 + k'\nu t, \tag{7.3.10a}$$

$$U \simeq \frac{I}{k}(R_0^2 + k'\nu t)^{-3/2} = \frac{I}{k(k'\nu t)^{3/2}} \left(1 - \frac{3}{2k'\eta} + \cdots \right), \tag{7.3.10b}$$

where $\eta = O(1)$ is defined by (7.3.3). This model exhibits a good fit to some experimental measurements, the best fit of (7.3.10) (where R was taken as R_{pv}, see (7.3.5)) to the data being $k = 14.4$ and $k' = 7.8$.

7.3.3 Late Stage at $\nu T/R_0^2 \gg 1$

After the matured stage, a viscous vortex ring evolves to the stage $\eta = \sqrt{\nu t}/R_0 \gg 1$ and $R_\Gamma \ll 1$, as Γ is weakened by vorticity cancelation at $r = 0$. In this case nonlinear advection terms in (7.3.7) can be dropped and the vorticity is governed by unsteady low-R_Γ *Stokes equation*

$$\frac{\partial \omega}{\partial t} = \nu \left[\frac{\partial^2 \omega}{\partial r^2} + \frac{\partial}{\partial r}\left(\frac{\omega}{r}\right) + \frac{\partial^2 \omega}{\partial z^2} \right], \tag{7.3.11}$$

of which the general vector form is (2.3.8a). Again with line vortex ring as initial condition, Fukumoto and Kaplanski (2008) found that the solution of (7.3.11) and circulation are

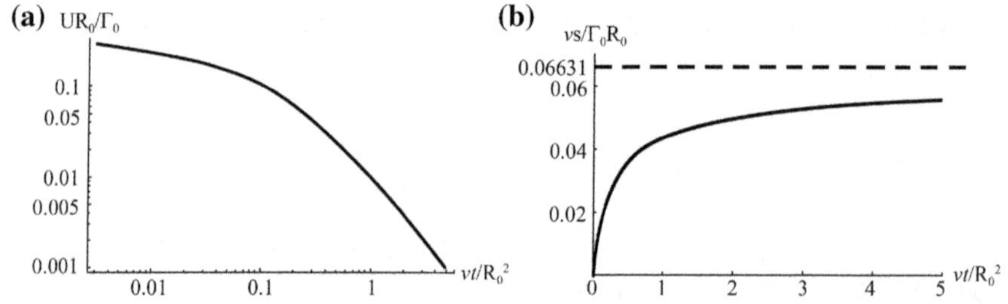

Fig. 7.10 Late stage of viscous vortex ring at $R_\Gamma \ll 1$ predicted by the uniformly effective solution U_{unif}. **a** Temporal variation of dimensionless speed. **b** Distance traveled as function of time. The *upper horizontal line* is the upper bound $\nu s_{\max}/\Gamma_0 R_0$. Adapted from Fukumoto and Kaplanski (2008)

$$\omega = \frac{\Gamma_0 R_0}{4\sqrt{\pi}(\nu t)^{3/2}} \exp\left(-\frac{r^2 + R_0^2 + z^2}{4\nu t}\right) I_1\left(\frac{R_0 r}{2\nu t}\right), \tag{7.3.12}$$

$$\Gamma = \Gamma_0\left[1 - \exp\left(-\frac{1}{4\eta}\right)\right], \tag{7.3.13}$$

where $\eta \gg 1$ is again defined by (7.3.3) and $I_1 = O(1)$ is the first kind modified Bessel function. The predicted vorticity distribution (7.3.12) is in good agreement with available experimental data, not only for $R_\Gamma \ll 1$ but also even for $R_\Gamma \gg 1$.

Then, by using the Helmholtz-Lamb identity (7.1.12), the analytic solutions for the kinetic energy K and vortex speed U can be found in terms of generalized hypergeometric function of η^{-2}, valid for all $t \geq 0$ and hence the whole η range, although the formulation came from the assumption of $\eta \gg 1$. This uniformly effective solution for travel speed, denoted by U_{unif}, is plotted in Fig. 7.10a. The asymptotic behavior of U_{unif} at $\eta \gg 1$ yields a *decay law*

$$U \simeq \frac{7I}{240\sqrt{2}(\pi\nu t)^{3/2}}\left(1 - \frac{33}{196}\frac{1}{\eta^2}\right) \simeq 0.0037\frac{I}{(\nu t)^{3/2}}(1 - 0.168\eta^{-2}). \tag{7.3.14}$$

Comparing this with (7.3.10b) leads to an estimate

$$k \simeq 10.15, \quad k' \simeq 8.91, \tag{7.3.15}$$

in reasonable agreement with the aforementioned values inferred from experiment.

The viscous vortex ring cannot exist indefinitely long nor travel indefinitely far. It will reach asymptotically a maximum travel distance (Fig. 7.10b)

$$s_{\max} = \int_0^\infty U_{\text{unif}}(\tau)d\tau = \frac{5\Gamma_0 R_0}{24\pi\nu} = \frac{5I}{24\pi^2 R_0 \nu} = 0.06631\frac{\Gamma_0 R_0}{\nu}. \tag{7.3.16}$$

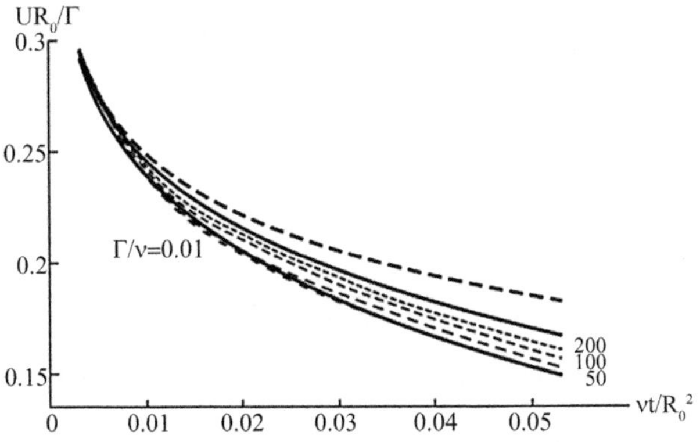

Fig. 7.11 Variation of dimensionless speed of viscous vortex ring with dimensionless time for $\eta^2 \ll 1$. The *upper* and *lower solid lines* are predicted by (7.3.4) for $R_\Gamma \gg 1$ and (7.3.17) for $R_\Gamma \ll 1$, respectively. The *thick dash line* is Saffman's (1970) result (7.3.2) without the last term of (7.3.4), i.e., the dipole effect. *Dashed lines* are values read off from numerical simulations. From Fukumoto (2010).

Remarkably, the opposite asymptotic approximation of $U_{\rm unif}$ at $\eta \ll 1$,

$$U \simeq \frac{\Gamma_0}{4\pi R_0}\{\log(4\eta^{-1}) - 0.5580 - 4.5[\log(4\eta^{-1}) - 1.0580]\eta^2\}, \qquad (7.3.17)$$

differs only slightly from (7.3.4) for $R_\Gamma \gg 1$. These distinct solutions are found to serve as upper and lower bounds for the evolution of viscous-ring speed in early stage at $\nu t \ll R_0^2$, see Fig. 7.11, where curves are not very sensitive to R_Γ. It has also been found that the low-R_Γ theory may well cover the matured stage in a wide range of Reynolds number.

7.4 Problems for Chapter 7

7.1. Prove that in any steady and swirl-free vortex rings the advection of ω_θ is always balanced by the vorticity-tube tilting.

7.2. Consider an isolated and inviscid circular vortex ring moving with constant velocity $\boldsymbol{U}$, such that in the frame comoving with the ring the flow is steady with velocity field $\boldsymbol{u}' = \boldsymbol{u} - U\boldsymbol{e}_z = (u, 0, w')$.

(1) Using (7.1.9b) to show that

$$K = 2\boldsymbol{U}\cdot\boldsymbol{I} + \int \boldsymbol{u}'\cdot(\boldsymbol{x}\times\boldsymbol{\omega})dV. \qquad (7.4.1)$$

(2) Prove that (7.4.1) is identical to (7.1.12).

Hint. For steady and inviscid flow under assumption (7.1.1a), the transport equation (6.1.7b) reads

$$u\frac{\partial \omega}{\partial r} + w'\frac{\partial \omega}{\partial z} = \frac{\omega u}{r}.$$

Multiplying both sides by $r^2 z$ to obtain identity

$$\omega(r^2 w' - rzu) = -3rzu\omega + \frac{\partial}{\partial z}(r^2 z w' \omega) + \frac{\partial}{\partial r}(r^2 z u \omega),$$

so the integral of the left-hand side over unbounded domain simply yields

$$2\pi \int_S \omega(r^2 w' - rzu) dS = -6\pi \int_S rzu\omega dS.$$

(3)* Discuss the physical implication of (7.4.1).

7.3. For rotationally symmetric viscous flow, prove that in an unbounded fluid at rest at infinity the vortical impulse $\boldsymbol{I}$ given by (7.1.8) is time-invariant.

Hint. Multiply (7.3.11) by r^2 and integrate the result over the meridional plane, noticing the fact that the integral of $\boldsymbol{e}_z \cdot (\boldsymbol{\omega} \times \boldsymbol{u})$ vanishes.

Chapter 8
Flow Separation and Separated Flows

8.1 Orientation

As stated at the end of Chap. 5, if at large Reynolds numbers a moving body is dressed in fully attached boundary layers, the flow can be solved by potential-flow theory with good accuracy, and then a boundary-layer correction yields the friction force to the body. This kind of attached flow is known as *simple flow*. Unfortunately, in most practical situations with flows of $Re \gg 1$ over either a bluff body (e.g., a circular cylinder) or a streamlined body with large incidence, owing to the appearance of strong adverse pressure gradient, it is very often that a thin boundary layer may break away from a smooth surface and form a *free shear layer* before reaching the trailing edge. This process is known as the *boundary-layer separation*. The separated free shear layer enters into the interior of the main flow and may alter the global flow pattern, from simple attached flow to *complex separated flow*.[1] Figure 8.1 illustrates an example, where the separation causes a big loss of the lift (stall). A clear physical understanding of boundary-layer separation and separated flow is therefore of critical importance for good configuration design and flow control in various engineering problems.

Flow separation can be quantitatively studied at two levels. The first level deals with *generic separation* (*fluid-particle separation*, or simply separation), in which a set of near-wall fluid particles no longer moves along the wall but turns into the interior of the flow. The relevant separation mechanism can be analyzed in an *infinitesimal* neighborhood of the separation point (two-dimensional) or line (three-dimensional). However, this approach cannot fully cover the above-described strong *boundary-layer separation*, where the whole boundary layer is ejected into the interior of the flow. To understand the relevant mechanism one has to make a thorough exploration of

[1]An attached boundary layer will eventually leave the trailing edge of a finite body, which in a broad sense is also a separation.

J.-Z. Wu et al., *Vortical Flows*, DOI 10.1007/978-3-662-47061-9_8

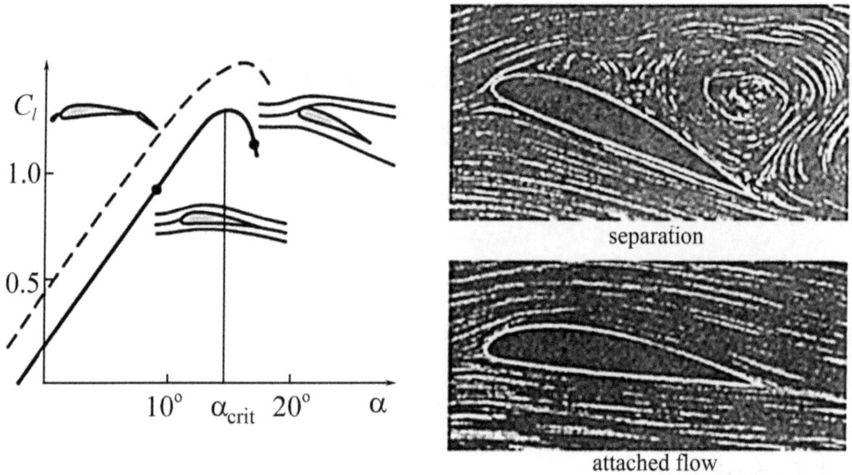

Fig. 8.1 The lift coefficient C_l versus angle of attack α of an airfoil (*real line*). The two dots on the curve correspond to the attached flow at small α and separated flow at large α as shown on the *right*. The lift is lost owing to severe boundary-layer separation at large α. The *dashed curve* shows the effect on the lift of flaps at leading and trailing edges used in taking off and landing. From Oertel (2004)

the dynamic processes of separating shear flow and its interaction with surroundings in a *small but finite* separation zone. The effort at this second level has led to a remarkable extension of Prandtl's original boundary-layer theory.

There are two types of global separated flow: *closed bubble type* and *free vortex-layer type*. The former happens when the flow has certain special symmetry, for which a simple asymptotic theory has been developed. The latter occurs in more general situation, where the free shear layer separated from either a sharp edge or smooth surface rolls into axial vortices that dominate the flow behavior, and for which no analytical theory is available. What one has is a qualitative theory mainly developed for three-dimensional flow, which combines one's knowledge on local generic separation theory and global analysis of *flow topology*.

Section 8.2 considers steady generic separation in two and three dimensions. In Sect. 8.3 we introduce steady boundary-layer separation theory. In Sect. 8.4 we discuss steady separated flows of both bubble type and free vortex-layer type, with the latter being our main concern. Unsteady separation and separated flows involve more difficult issues. The local theories developed so far on generic unsteady separation and boundary-layer separation will be outlined in Sect. 8.5, followed by three examples of complex unsteady separated flows of practical interest.

8.2 Generic Steady Flow Separation

At the first level of analysis focusing on an infinitesimal region of generic separation, the basic method is Taylor expansion around an on-wall separation point or line. The expansion coefficients are constrained by the Navier-Stokes equation and continuity equation. In so doing some elements of qualitative theory of dynamic system, in particular the concept of fixed points of vector fields, play a key role.

8.2.1 Separation Criteria

1. Two-dimensional separation criterion. Consider a two-dimensional steady flow separation on an (x, y)-plane along a curved wall $y = 0$ in terms of boundary-layer coordinates. Since $\partial_y p = 0$ in boundary-layer approximation, under the same adverse streamwise pressure gradient the fluid at smaller y is more easily retarded due to their lower forward momentum. This streamwise slowdown is associated with $\partial_x u < 0$, which by the continuity causes an *upwelling* associated with growth of v away from the wall. For steady flow, because $u = v = 0$ for all fluid particles on the wall, slightly above the wall u will be retarded to zero at a point $x = x_s$ only if $(\partial_y u)_s = 0$ or $\tau_w(x_s) = -\mu\omega_B = 0$, where ω_B is the vorticity at the wall. This point is the *separation point*. Moreover, downstream x_s there must be a reversed stream at the lower part of the boundary layer that also sends fluid into the upwelling stream, implying $\partial_x \tau_w(x_s) < 0$. Namely, at the separation point x_s there must be (Prandtl 1904)

$$\tau_w(x_s) = \mu(\partial_y u)_s = -\mu\omega_B(x_s) = 0, \tag{8.2.1a}$$

$$(\partial_x \tau_w)_s = \mu(\partial_x \partial_y u)_s = -\mu(\partial_x \omega_B)_s < 0. \tag{8.2.1b}$$

This feature signifies the formation of a *separated shear layer*. From x_s initiates a special streamline called *separation streamline* that divides the incoming and reversed upwelling streams, see Fig. 8.2a. It serves as the skeleton of the separated shear layer. The associated adverse pressure gradient is shown in Fig. 8.2b. Note that (8.2.1) can also be applied to rotationally symmetric flow, where one may examine the separation of the sectional flow on any meridian plane.

Alternative to the above momentum consideration, we may explain the separation process in terms of vorticity dynamics, in which the local dynamic equation (2.3.7a) on the boundary vorticity flux (BVF) driven by tangent pressure gradient,

$$\sigma \equiv -\nu\frac{\partial \omega}{\partial y} = \frac{1}{\rho}\frac{\partial p}{\partial x}, \tag{8.2.2}$$

adds a powerful means for quantitative analysis. Figure 8.2b shows that at certain distance upstream x_s we have favorable pressure gradient $\partial_x p = \rho\sigma < 0$, which

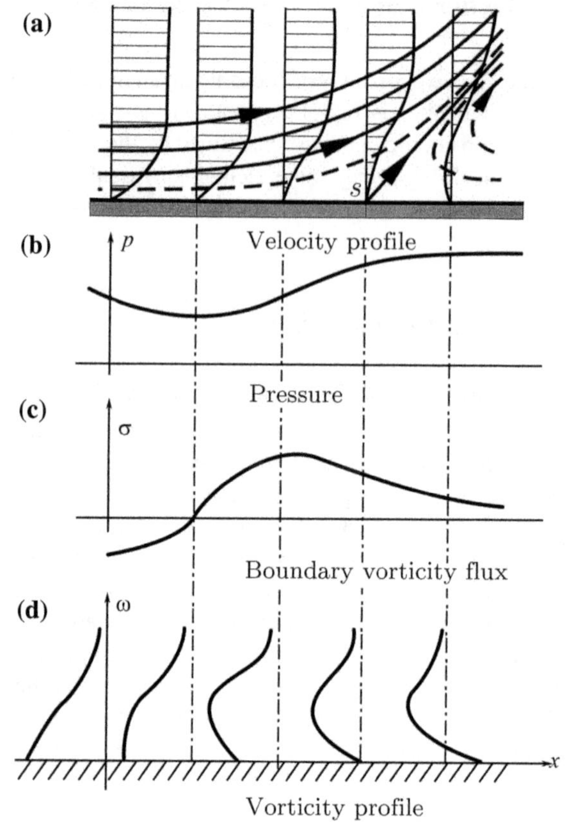

Fig. 8.2 Sketch of **a** velocity profiles and streamlines, **b** pressure, **c** boundary vorticity flux, and **d** vorticity profiles at different x-locations of a flat-plate flow as the pressure gradient changes from favorable to adverse. **a** and **b** are adapted from Oertel (2004)

generates $\omega < 0$. The accumulation of this negative vorticity in space and time forms a fully attached boundary layer with $\tau_w > 0$. Then $\partial_x p$ increases from negative to positive. When σ crosses zero, we still have $\omega < 0$ in the whole layer as is the case of the Blasius solution shown in Fig. 4.7. But when $\sigma > 0$, it will create new $\omega > 0$ to cancel a part of the existing negative ω_B. This cancellation continues as σ keeps positive, till the existing negative ω_B is completely canceled at the separation point, after which we have $\omega_B > 0$. This process is sketched in Fig. 8.2c, d. As an accumulative effect of the sign change in σ, the sign change in ω_B always happens more downstream (and at a later time if the flow is unsteady). Therefore, while separation is signified by $\omega_B = 0$, its physical cause is adverse pressure and associated BVF. According to the discussion of Sect. 2.6.1 on momentum formulation and vorticity formulation, we see that Fig. 8.2a, b belong to momentum formulation, while (c) and (d) belong to vorticity formulation. Armed by the unique equation (8.2.2), the latter is physically more informative for understanding the whole separation process upstream x_s rather than only focusing on the separation point.

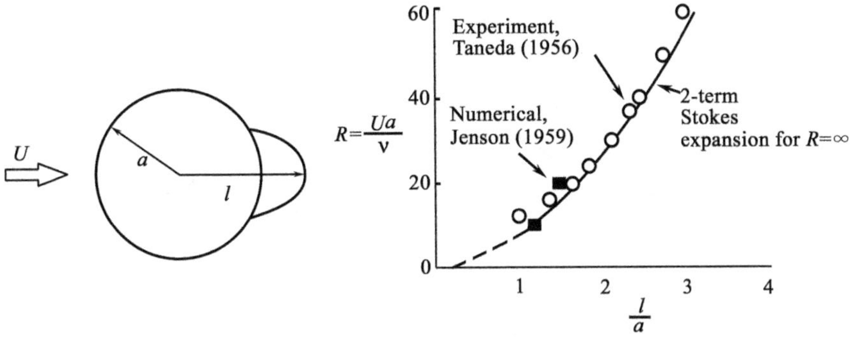

Fig. 8.3 The length of separated bubble from a sphere versus Reynolds number at small $Re = Ua/\nu$. Reproduced from Van Dyke (1975)

It should be recognized that (8.2.1) is nothing but the leading term of the Taylor expansion of the flow behavior at an infinitesimal neighborhood of the separation point, which is only a first-level interpretation of boundary-layer separation. It is necessary but insufficient for identifying the separation. A separation line determined by (8.2.1) may either go to the interior of global flow and hence be associated with the boundary-layer breaking away, or just remain very close to the wall at the bottom of the boundary layer to enclose a thin separating bubble but without altering the attached feature of the whole boundary layer. Actually, (8.2.1) holds independent of the Reynolds number. It is even applicable to separation at small Re, for which there is no layer-like flow structure but widely distributed vorticity field. Figure 8.3 shows an example for a rotationally symmetric flow past a sphere at low Re, to which (8.2.1) applies. It has been found that, if Re based on sphere's radius is greater than 8, then the flow can no longer keep fully attached but a separated bubble appears at the lee side of the sphere. Problem 8.1 gives a quantitative prediction of the behavior of this separated bubble.

2. Three-dimensional separation criterion. Unless truly two-dimensional flow (e.g., a flow on a soap-film surface), a plane flow is actually a three-dimensional flow with all vorticity lines along the $\boldsymbol{e}_z$ direction, of which the one with $\tau_w = -\mu\omega_B = 0$ is a *separation line*, which is *perpendicular* to the skin-friction lines (will often be called $\boldsymbol{\tau}$-lines), see Fig. 8.4a. This pattern may occur approximately in real flows, say on the mid-portion of a straight wing of large aspect ratio. The sectional flow there does not change much at different spanwise location, and hence the separation behavior will be quasi two-dimensional.

In three-dimensional space, however, the $\boldsymbol{\tau}$-lines no longer keep parallel. They either diverge from or converge to each other, associated with some complex nature of three-dimensional boundary layers mentioned in Sect. 4.2.4. When the $\boldsymbol{\tau}$-lines converge as sketched in Fig. 8.4b, separation may happen in a zone with nowhere $\boldsymbol{\tau} = \mathbf{0}$ but *along* one of the converging $\boldsymbol{\tau}$-lines. To see why this happens, consider two $\boldsymbol{\tau}$-lines of distance l shown in Fig. 8.4b, above which we take a flow tube of

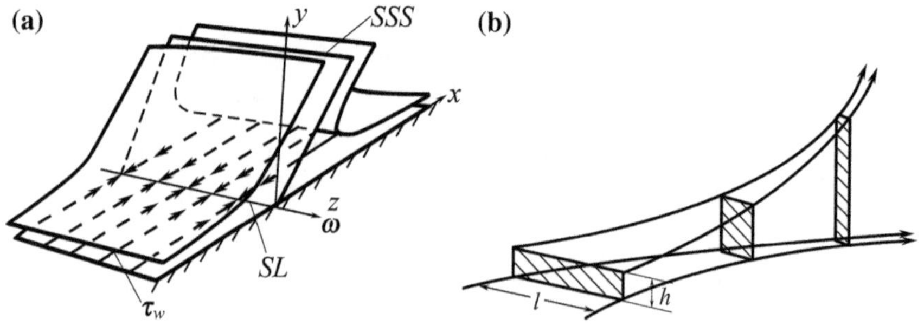

Fig. 8.4 Flow separation in three dimensions. **a** Quasi two-dimensional separation, where the separation point in two dimensions becomes a separation line (SL) perpendicular to the quasi-parallel skin-friction lines before and after separation. Separation streamline in two dimensions becomes a separation stream surface (SSS). **b** More often, three-dimensional flow separation may happen with nowhere $\boldsymbol{\tau}_w = \mathbf{0}$ but due to $\boldsymbol{\tau}_w$-lines converging

rectangular cross-section $l \times h$ with mean mass flux $\rho h l \bar{u}$. For sufficiently small h the velocity profile is linear, which gives $\partial u/\partial x_3 = 2\bar{u}/h$ and hence $\tau = 2\mu\bar{u}/h$ or $\bar{u} = \tau h/(2\mu)$. Thus, since the mass flux $\rho h l \bar{u} = h^2 l\tau/2\nu$ must be constant along the tube, one obtains (Lighthill 1963)

$$h = C\left(\frac{\nu}{l\tau}\right)^{\frac{1}{2}}. \tag{8.2.3}$$

Therefore, h will grow unboundedly not only when $\tau \to 0$ as is in two-dimensional case, but also when $l \to 0$, i.e., $\boldsymbol{\tau}$*-lines converging* toward each other which causes the fluid upwelling shown in Fig. 8.4b. As a three-dimensional counterpart of the Prandtl's two-dimensional criterion (8.2.1), (8.2.3) is a local criterion which is necessary and sufficient for generic flow separation.

Figure 8.5 exemplifies this three-dimensional separation pattern by a numerical result of laminar flow over a prolate spheroid with long-short axis ratio equal to 2, at incidence α to incoming flow and large Re. In this quasi-steady flow two separated free vortex layers roll into concentrated vortices of opposite rotating directions. The

Fig. 8.5 Fluid particle traces in a laminar flow over an 1/2 prolate spheroid at incidence $\alpha = 30°$ and $Re = 3.5 \times 10^4$ based on the length of short axis. From Wu et al. (2000)

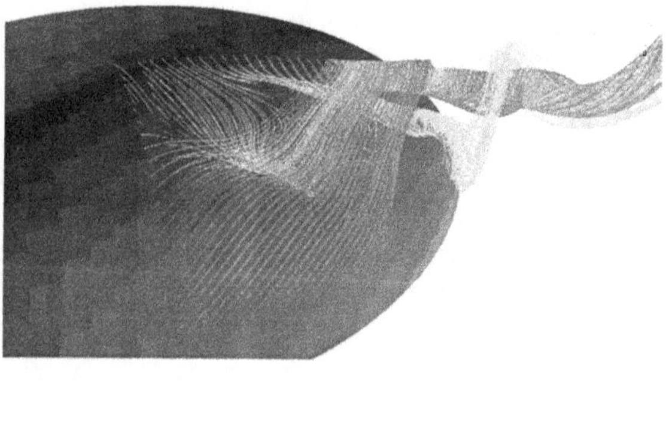

upper vortex is from the primary separation, which induces a secondary separation that leads to the lower vortex. From this figure we may identify two narrow separation zones on the wall, which behave very differently from two-dimensional separation. The $\boldsymbol{\tau}$-lines are visualized by particle tracers, which in each zone converge toward the respective separation line as predicted by (8.2.3), causing strong upwelling of neighboring fluid.

8.2.2 Dynamic System and Fixed Points

To put the above Prandtl's and Lighthill's intuitive separation criteria in a rational theoretical framework and thereby to further examine various aspects of separation patterns, we need to construct near-wall dynamic system that describes the evolution of fluid particles. The prototype of this system is nothing but (1.1.2):

$$\frac{d\boldsymbol{x}}{dt} = \boldsymbol{u}(\boldsymbol{x}, t). \tag{8.2.4}$$

For steady flow $\boldsymbol{u}$ is independent of t, and (8.2.4) describes an *autonomous* vector field. An isolated point $\bar{\boldsymbol{x}}$ with $\boldsymbol{u}(\bar{\boldsymbol{x}}) = \boldsymbol{0}$, say, is called a *fixed point* of (8.2.4), since once a particle is at $\bar{\boldsymbol{x}}$ it will no longer move. Unlike ordinary point where $\boldsymbol{u} \neq \boldsymbol{0}$, there can be more than one vector lines entering or leaving a fixed point, of which the pattern defines the behavior of particles very close to $\bar{\boldsymbol{x}}$. This is precisely the characteristics of the separation point s in Fig. 8.2a, where in addition to the on-wall vector lines $\boldsymbol{\tau}_w$ going toward and leaving s there also initiates a separation streamline from s. However, the vector field (8.2.4) cannot be directly applied, since by adherence all points on the wall are its non-isolated fixed points. We thus need to modify (8.2.4) to a new near-wall vector field able to describe both streamlines inside the flow and $\boldsymbol{\tau}$-lines on the wall, and to permit only s as its isolated on-wall fixed point.

In three-dimensional flow, the near-wall behavior of a vector field can be described by an orthonormal curvilinear frame $(\boldsymbol{e}_1, \boldsymbol{e}_2, \boldsymbol{e}_3)$ with $\boldsymbol{e}_1$ and $\boldsymbol{e}_2$ being two tangent unit vectors and $\boldsymbol{e}_3 = \hat{\boldsymbol{n}}$ the unit normal out of the wall. Since for autonomous system the time scale does not alter the vector-field pattern, we may introduce a new variable t' to replace t, defined by $\mu dt' = x_3 dt$, such that

$$\dot{\boldsymbol{x}} = \frac{\mu}{x_3}\boldsymbol{u}(\boldsymbol{x}) \equiv \tilde{\boldsymbol{u}}(\boldsymbol{x}) \tag{8.2.5}$$

with dot denoting d/dt', which defines a new vector field $\tilde{\boldsymbol{u}}(\boldsymbol{x}) = (\tilde{u}_1, \tilde{u}_2, \tilde{u}_3)$. The vector lines of $\tilde{\boldsymbol{u}}$ for $x_3 > 0$ coincide with that of $\boldsymbol{u}$, but those for $x_3 = 0$ coincide with that of the skin-friction lines since as $x_3 \to 0$ there is $\tilde{\boldsymbol{u}} \to \mu\partial_3\boldsymbol{u} = \boldsymbol{\tau}_w$. In this way, the separation point is an isolated fixed point of system (8.2.5).

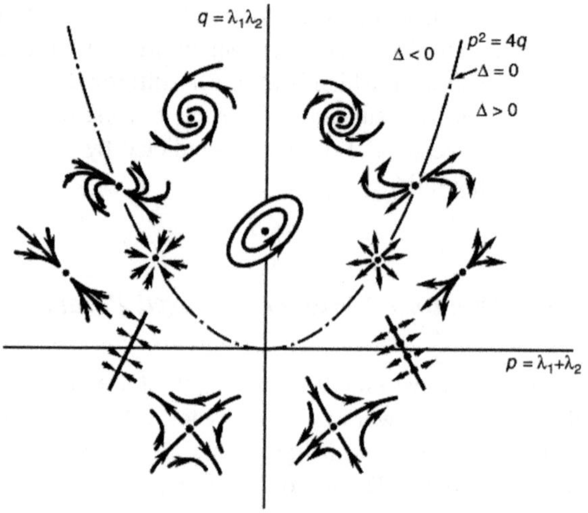

Fig. 8.6 The fixed points of two-dimensional linear systems

A three-dimensional system (8.2.5) can be decomposed into three two-dimensional subsystems on three orthogonal planes, of which the behavior near fixed points is easier to examine due to the well-known classification of fixed points on a two-dimensional plane. Let us consider such a flow plane, re-denoted here as an (x, y)-plane. Near a fixed point of (8.2.5), we may write the system as

$$\dot{x} = ax + by + P_2(x, y), \quad \dot{y} = cx + dy + Q_2(x, y), \tag{8.2.6}$$

where $P_2, Q_2 = O(r^2)$ with $r = \sqrt{x^2 + y^2}$ are analytic near the fixed point. The locally linearized system of (8.2.6) is familiar:

$$\dot{x} = ax + by, \quad \dot{y} = cx + dy, \tag{8.2.7}$$

of which the eigenvalues λ_1, λ_2 are determined by characteristic equation $\lambda^2 - p\lambda + q = 0$, where $p = \lambda_1 + \lambda_2 = a + d$ and $q = \lambda_1\lambda_2 = ad - bc$. According to the sign of the discriminant $\Delta = p^2 - 4q$, the feature of λ_1 and λ_2, and hence the behavior of various fixed points of (8.2.7), can be inferred. The result is summarized in Fig. 8.6, the *p-q diagram.*

From the figure one sees the following types of fixed points[2]:

1. *Hyperbolic fixed points.* $q \neq 0$, so both eigenvalues λ_1 and λ_2 have nonzero real part.

[2]Here, we say a fixed point is "stable" if all neighboring vector lines point toward it, and "unstable" if some of neighboring vector lines point out of it. This stability concept is not to be confused with either the stability of topological structures outlined in Sect. 8.4.4, or the flow stability addressed in Chap. 10.

(1) $q < 0$: λ_1, λ_2 are real roots of opposite signs, and the fixed points are unstable *saddles*.

(2) $0 < q < (p/2)^2$: λ_1, λ_2 are different real roots of the same sign, and the fixed points are *nodes*, which are stable if $p < 0$ and unstable if $p > 0$.

(3) $q > (p/2)^2$: λ_1, λ_2 are conjugate complex roots with nonzero real part, and the fixed points are *spirals*. They are stable if $p < 0$ and unstable if $p > 0$.

(4) $q = (p/2)^2$ and $p \neq 0$: λ_1, λ_2 are the same nonzero real root. In this case, we have *stars* (*critical nodes*) or *inflected nodes* depending on the form of the coefficient matrix being

$$\text{either} \quad \begin{bmatrix} \lambda & 0 \\ 0 & \lambda \end{bmatrix} \quad \text{or} \quad \begin{bmatrix} \lambda & 1 \\ 0 & \lambda \end{bmatrix},$$

respectively. In the latter case, the trajectories approach the fixed point along a single direction, and tend to be parallel to the same direction again at infinity.

It is easily seen that node, spiral, and star or inflected node are topologically the same, known as *topological nodes*. An unstable node is called a *source*, while an asymptotically stable node is a *sink*.

2. *Nonhyperbolic fixed points*.

(1) $q > 0$ and $p = 0$: λ_1, λ_2 are pure imaginary roots, and the fixed points are *centers* which are neutrally stable.

(2) $q = 0$ and $p \neq 0$: One of λ_1 and λ_2 is zero and the fixed points form a straight line rather than being isolated. The trajectories are a set of parallel straight lines.

If (8.2.7) has no eigenvalue with real part zero, it is called a *hyperbolic linear system*, and its fixed points are all hyperbolic (or non-degenerate) fixed points. *Hyperbolic fixed points* are robust, and we have

Hartman-Grobman theorem: *If* $\bar{\mathbf{x}}$ *is an isolated hyperbolic fixed point of* (8.2.5), *then there is a neighborhood of* $\bar{\mathbf{x}}$ *in which* $\mathbf{u}$ *is topologically equivalent to the linear vector field* (8.2.7).

By the Hartman-Grobman theorem, non-hyperbolic fixed points cannot retain their topological classes in the nonlinear system (8.2.6), which on the other hand may have more kinds of non-hyperbolic fixed points, called higher-order or multiple fixed points. One such example is the *saddle-node*, consisting half saddle on one side and half node on the other side.

For example, consider a nominally two-dimensional steady flow over a cylinder of arbitrary sectional shape in three-dimensional space, either fully attached or with separated bubble. The zero-$\boldsymbol{\tau}$ points in a flow plane is actually degenerated fixed "points" of the $\boldsymbol{\tau}$-field, i.e., the p-axis of Fig. 8.6 in the span direction. They cannot remain in real nonlinear flow. Thus, the separation pattern of Fig. 8.4a is oversimplified; rather, as can be confirmed by careful numerical simulation, a realistic pattern should be a spanwise periodic saddle-node combination shown in Fig. 8.7. Of course the near-wall flow must have corresponding periodic structures and become more complicated. This also explains why in three-dimensional space any nominally two-dimensional or rotationally symmetric steady flows and vortices always have a tendency to become fully three-dimensional.

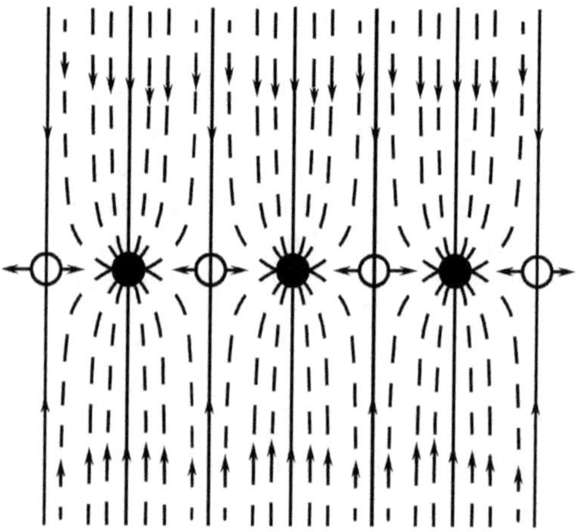

Fig. 8.7 Saddle-node combination along a nominally two-dimensional *separation line* or *reattachment line*

8.2.3 Near-Wall Dynamic System for Flow Separation

We now apply the dynamic system (8.2.5) and fixed-point classification to Navier-Stokes flow near separation zone. Since generic flow separation involves only the flow behavior in an infinitesimal neighborhood of the separation point, it suffices to consider only linearized version of (8.2.5) by Taylor expansion. Without causing confusion, we shall drop subscripts w and B below from skin-friction $\boldsymbol{\tau}_w$ and boundary vorticity $\boldsymbol{\omega}_B$, respectively.

1. Two-dimensional separation point as a semi-saddle. In two dimensions, let the origin of (x, y) coordinates be at the separation point x_s, and $P(x, y)$ be a point near x_s with $y \geq 0$ and $|\boldsymbol{x}| \ll 1$. Then the Taylor-expansion for $\tilde{\boldsymbol{u}}_P$ takes the form

$$\dot{\boldsymbol{x}}_P = (1 + x\partial_x + \cdots)\left(1 + y\partial_y + \frac{1}{2}y^2\partial_y^2 + \cdots\right)\left(\frac{\mu \boldsymbol{u}}{y}\right), \tag{8.2.9}$$

where and below terms on the right-hand sides take their values at $\boldsymbol{x} = \boldsymbol{0}$. Noticing that ∂_y does not act on y_P. Conceive a circular arc locally tangent to the wall at x_s, one finds $\nabla^2 = \partial_x^2 - \kappa\partial_y + \partial_y^2$, where $\kappa = \partial_x \boldsymbol{e}_x \cdot \boldsymbol{e}_y$ is the wall curvature. Then the adherence condition, the on-wall versions of the Navier-Stokes equation $\nabla p = \mu\nabla^2 \boldsymbol{u}$, and continuity equation imply that at $y = 0$ there is

$$u = v = 0, \quad \partial_y v = \kappa u - \partial_x u = 0, \tag{8.2.10a}$$

$$\mu\partial_x^2 u = \partial_x p + \kappa\tau, \quad \mu\partial_y^2 v = -\partial_x \tau, \tag{8.2.10b}$$

where quantities on the right-hand side are evaluated at s with $\tau = 0$. Thus, in the linearized approximation (8.2.7) of (8.2.9), there is

$$a = \partial_x \tau, \quad b = \frac{1}{2}\partial_x p, \quad c = 0, \quad d = -\frac{1}{2}\partial_x \tau, \tag{8.2.11}$$

so that $ad - bc < 0$. Therefore, the separation point is a semi-saddle as we anticipated from the flow pattern.

Moreover, the initial slope $\tan\theta$ of the separation streamline can be inferred from (8.2.7):

$$\tan\theta = -3\frac{\partial_x \tau}{\partial_x p} = 3\frac{\nu \partial_x \omega}{\sigma}, \tag{8.2.12}$$

where $\omega = -\tau/\mu$ at the wall (Problem 8.3). Thus, flow separation is a boundary coupling effect of the shearing process (measured by $\partial_x \tau$) and compressing process (measured by $\partial_x p$). In the asymptotic state of $Re \to \infty$ or $\nu \to 0$, the separation streamline will leave the wall tangentially with $\theta \to 0^+$, in agreement with our result from vortex-sheet dynamics (Sect. 5.2.2).

The above Taylor expansion can be extended to higher orders, since by the Navier-Stokes and continuity equations normal derivatives of (u, v, p) of any order can always be expressed by on-wall quantities and their tangential derivatives (Wu et al. 2006).

2. Three-dimensional dynamic system and separation line. The wall curvature is a 2×2 symmetric tensor, say $\mathbf{K} = b_{\alpha\beta}\boldsymbol{e}_\alpha\boldsymbol{e}_\beta$ with $\alpha, \beta = 1, 2$. Denote tangent components of any on-wall vector field by subscript π, since $\nabla^2 = \nabla_\pi^2 - \kappa_B\partial_3 + \partial_3^2$ where $\kappa_B = \text{tr}\mathbf{K} = -\nabla_\pi \cdot \boldsymbol{e}_3 = b_{11} + b_{22}$ is twice of the *mean curvature* of the wall, similar to the Taylor expansion (8.2.9) in two dimensions, by the Navier-Stokes equation and continuity equation we now have

$$u_1 = u_2 = u_3 = 0, \quad \partial_3 u_3 = -\nabla_\pi \cdot \boldsymbol{u}_\pi = 0, \tag{8.2.13a}$$

$$\mu\partial_3^2 \boldsymbol{u}_\pi = \nabla_\pi p + \kappa_B \boldsymbol{\tau}, \quad \mu\partial_2^2 u_3 = \partial_3 p = -\nabla_\pi \cdot \boldsymbol{\tau}. \tag{8.2.13b}$$

Then by (8.2.5) the linear near-wall Taylor expansion of $\dot{\boldsymbol{x}} = \mu\boldsymbol{u}/x_3$ at a point $P(\boldsymbol{x})$ is given by

$$\begin{aligned}\dot{\boldsymbol{x}}_P &= \boldsymbol{\tau} + \boldsymbol{x} \cdot \nabla_\pi \boldsymbol{\tau} + \frac{\mu}{2}x_3(\partial_3^2 \boldsymbol{u}_\pi + \boldsymbol{e}_3\partial_3^2 u_3) \\ &= \boldsymbol{\tau} + \boldsymbol{x} \cdot \nabla_\pi \boldsymbol{\tau} + \frac{1}{2}x_3(\kappa_B \boldsymbol{\tau} + \nabla_\pi p - \boldsymbol{e}_3\nabla_\pi \cdot \boldsymbol{\tau}),\end{aligned} \tag{8.2.14}$$

by which the three-dimensional linearized dynamic system for $(\dot{x}_1, \dot{x}_2, \dot{x}_3)$ can be easily written down. Observe that boundary vorticity $\boldsymbol{\omega}$ and skin-friction $\boldsymbol{\tau} = \mu\boldsymbol{\omega} \times \boldsymbol{e}_3$ form a pair of two-dimensional vector fields over the body surface ∂B, which are orthogonal to each other everywhere except at isolated fixed points with vanishing $\boldsymbol{\tau}$ and $\boldsymbol{\omega}$. Figure 8.8 indicates that, in both separation zones $\boldsymbol{\tau} \neq \mathbf{0}$, and inside each zone the $\boldsymbol{\tau}$-lines are converging asymptotically to a *separation line*, which itself is a $\boldsymbol{\tau}$-line and from which initiates a unique *separation stream surface* as the skeleton of the separated free shear layer. Consequently, as a sharp contrast to two-dimensional

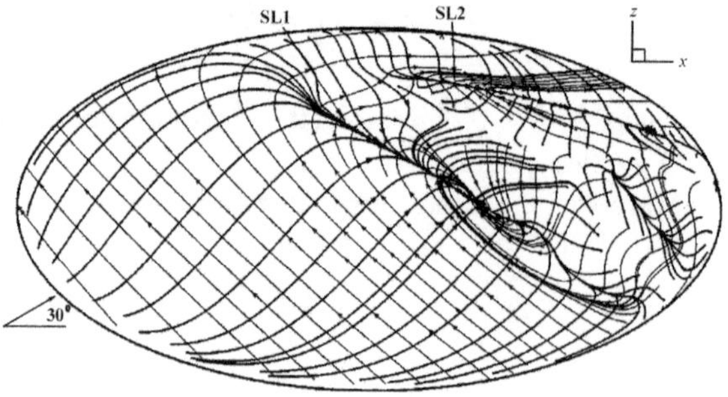

Fig. 8.8 Orthogonal pair $(\boldsymbol{\tau}, \boldsymbol{\omega})$ on prolate spheroid. Thick lines are $\boldsymbol{\tau}$-lines and thin lines are $\boldsymbol{\omega}$-lines. From Wu et al. (2000)

separation line which itself is a zero-$\boldsymbol{\omega}$ line, in three dimensions *at ordinary points with* $\boldsymbol{\tau} \neq \mathbf{0}$, *the separation line is orthogonal to* $\boldsymbol{\omega}$*-lines.* Note that the tangent pressure gradient and its caused boundary vorticity flux (BVF) $\boldsymbol{\sigma}_p = -\boldsymbol{e}_3 \times \nabla p$, see (3.4.24c), also form a pair of orthogonal on-wall vector fields.

Similar to intrinsic streamline coordinates inside the fluid, the orthogonality of $(\boldsymbol{\tau}, \boldsymbol{\omega})$ on body surface naturally suggests choosing $\boldsymbol{e}_1 = \boldsymbol{\tau}/\tau$ and $\boldsymbol{e}_2 = \boldsymbol{\omega}/\omega$, where $\tau = |\boldsymbol{\tau}|$ and $\omega = |\boldsymbol{\omega}|$. We call this intrinsic frame the $\boldsymbol{\tau}$*-frame*. The *on-wall* curvatures of the $\boldsymbol{\tau}$-line and $\boldsymbol{\omega}$-line are given by[3]

$$\kappa_1 = (\partial_1 \boldsymbol{e}_1) \cdot \boldsymbol{e}_2 = -(\partial_1 \boldsymbol{e}_2) \cdot \boldsymbol{e}_1, \tag{8.2.15a}$$

$$\kappa_2 = (\partial_2 \boldsymbol{e}_2) \cdot \boldsymbol{e}_1 = -(\partial_2 \boldsymbol{e}_1) \cdot \boldsymbol{e}_2. \tag{8.2.15b}$$

Then an inspection of (8.2.15a) reveals immediately that Lighthill's $\boldsymbol{\tau}$-line converging criterion is nothing but a simple geometric criterion

$$\kappa_2 > 0, \tag{8.2.16}$$

as can be observed from Fig. 8.8 where, as the $\boldsymbol{\tau}$-lines converge toward the separation line, the $\boldsymbol{\omega}$-lines make U-turns as they cross the separation lines. But, in Lighthill's argument the variation of $\boldsymbol{\tau}$ along the small fluid tube was neglected. An exact criterion is to ensure the fluid is upwelling away from the wall with $u_{3P} = x_3^2 \partial_3^2 u_3/2 > 0$. Because in $\boldsymbol{\tau}$-frame

$$\mu \partial_3^2 u_3 = -\nabla_\pi \cdot \boldsymbol{\tau} = \tau \kappa_2 - \partial_1 \tau = \mu(\omega \kappa_2 - \partial_1 \omega),$$

Lighthill's criterion (8.2.16) should be modified to (Wu et al. 2000)

$$\kappa_2 - \partial_1 (\ln \tau) = \kappa_2 - \partial_1 (\ln \omega) > 0. \tag{8.2.17}$$

[3]The true curvature of a $\boldsymbol{\tau}$-line in space involves not only the projection of $\partial_1 \boldsymbol{e}_1$ onto $\boldsymbol{e}_2$ but also that onto $\boldsymbol{e}_3$, the latter being an effect of wall curvature.

3. Separation pattern at fixed points. The $\boldsymbol{\tau}$-line converging criterion alone cannot pick up a single separation line from a bundle of mutually converging τ-lines. Lighthill (1963) argues that a unique feature of separation line is that it should initiate from a fixed point and terminate at another. We thus need to examine the initial separation behavior at a fixed point of the $(\boldsymbol{\tau}, \boldsymbol{\omega})$ field, say S, where the $\boldsymbol{\tau}$-frame does not exist and one may instead examine the separation behavior on a plane tangent to the wall and use local Cartesian coordinates (x, y, z) with $\boldsymbol{e}_z = \hat{\boldsymbol{n}}$. The x-axis may be chosen to be tangent to the separation line slightly downstream S. Then by (8.2.14) the linearized dynamic system (8.2.5) reads

$$\begin{bmatrix} \dot{x} \\ \dot{y} \\ \dot{z} \end{bmatrix} = \begin{bmatrix} \tau_{x,x} & \tau_{x,y} & \frac{1}{2}p_{,x} \\ \tau_{y,x} & \tau_{y,y} & \frac{1}{2}p_{,y} \\ 0 & 0 & -\frac{1}{2}\nabla_\pi \cdot \boldsymbol{\tau} \end{bmatrix} \begin{bmatrix} x \\ y \\ z \end{bmatrix}, \tag{8.2.18}$$

where the coefficients can be found to satisfy (Zhang 1985; Xia and Deng 1991)

$$\tau_{x,x} > 0 \tag{8.2.19a}$$

$$\tau_y = 0, \quad \tau_{y,x} = 0, \tag{8.2.19b}$$

$$\tau_{y,y} < 0, \tag{8.2.19c}$$

$$\nabla_\pi \cdot \boldsymbol{\tau} = \tau_{x,x} + \tau_{y,y} < 0. \tag{8.2.19d}$$

Here, (8.2.19a) holds as τ increases from zero at S, (8.2.19b) ensures the condition that separation line is a $\boldsymbol{\tau}$-line, (8.2.19c) ensures that neighboring $\boldsymbol{\tau}$-lines converge toward separation line so that their y-component changes sign across the separation line, and (8.2.19d) is the upwelling condition due to (8.2.13b). From these conditions it is easily seen that *the fixed point where a separation line initiates is a saddle*, as argued by Lighthill (1963), and the fixed-point behavior in three orthogonal planes must be that shown in Fig. 8.9, indicating that from the two separation lines grows a *separation stream surface*, on which all streamlines must be from S (Problem 8.4). This type of separation pattern starting from a saddle of the $\boldsymbol{\tau}$-field, is called *closed separation*. Opposite to closed separation, there may also appear *open separation*, which will be discussed in Sect. 8.4.5.

In contrast, if a separation line ends at another fixed point of the $(\boldsymbol{\tau}, \boldsymbol{\omega})$ field, there must be $\tau_{x,x} < 0$ and hence $q > 0$, indicating that *the fixed point where a separation line terminates must be a node (normal node or spiral)*.

8.3 Steady Boundary-Layer Separation

The generic separation criteria based on Taylor expansion discussed in Sect. 8.2 are insufficient for the study of boundary-layer separation, since they cannot reveal the dynamic processes in a *finite* (though small) separation zone. Neither can the classic

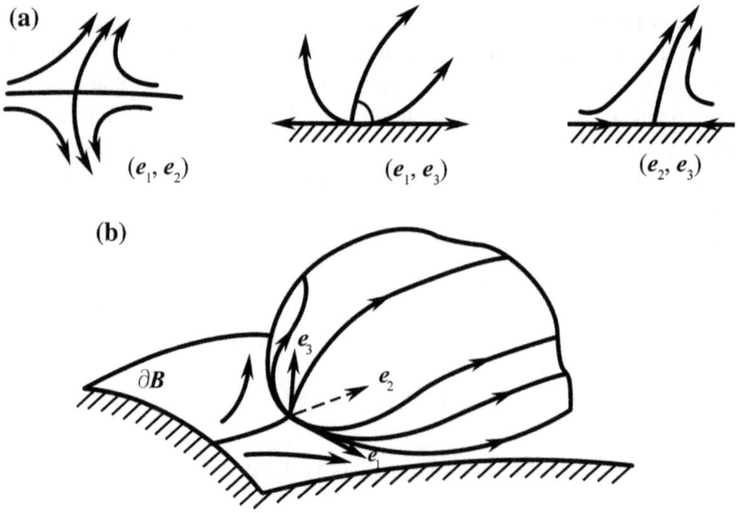

Fig. 8.9 Closed separation initiated from a saddle of the τ-field

boundary-layer theory do, which holds under the condition $v = O(Re^{-\frac{1}{2}})$. Once $v \gg Re^{-\frac{1}{2}}$ as fluid particles move up in the separated shear layer, these equations break down. Mathematically, since boundary-layer equation (4.2.10) is simplified to parabolic type, as an expense it cannot "feel" and response any changes of downstream flow condition such as the reversed stream at $x > x_s$. Consequently, the marching of boundary-layer solution along the wall will encounter a new singularity at $x = x_s$, proved by Goldstein (1948), and must blow up.

To fully understand the physical mechanisms of boundary-layer separation process, we have to proceed to the second level, the *boundary-layer separation*, which is the strong form of flow separation specifically at large Reynolds numbers. For this purpose a generalization of Prandtl's theory is necessary.

8.3.1 Deck Structure and Scale Analysis

Consider two-dimensional steady flow. Prandtl's method has taught us how to regularize the singularity of boundary-layer equation in separation zone: just like inserting a thin but finite layer to replace the singularity along the wall, we should now insert a thin but finite layer of thickness $l \ll 1$ *normal to the wall* in a neighborhood of $x = x_s$, which intersects the boundary layer to form a small but finite *deck-like structure*.

Similar to the scale analysis in Prandtl's theory that leads to the estimate (4.2.8) and then the whole set of boundary layer equations, we now analyze the scales in the deck region by examining the physical processes there. Let all variables have been made dimensionless by global scales U and L. First, the strong upwelling $v(x, y) \gg Re^{-\frac{1}{2}}$ as $x \to x_s$ in the separation zone interacts the external pressure gradient, which causes a viscous response in a sublayer adjacent to the wall. Let this response occur in a *lower deck* of normal thickness $y \sim Re^{-1/2}\hat{\delta}$, where $\hat{\delta}$ is to be determined. Since adjacent to the wall the streamwise velocity profile can be represented by a simple shear flow, we have $u(y) \sim \hat{\delta}$. Then by (4.2.10), in the boundary-layer scale the balance between the inertial force and an *interactive pressure increment* Δp as well as the viscous force requires that

$$\frac{\hat{\delta}^2}{l} \sim \frac{\Delta p}{l} \sim \frac{\hat{\delta}}{\hat{\delta}^2},$$

which gives $\Delta p = O(\hat{\delta}^2)$ and $l = O(\hat{\delta}^3)$. Finally, to determine $\hat{\delta}$, we note that the appearance of the lower deck raises the rest of the boundary layer (the "*main deck*") up by an additional displacement thickness $\hat{\delta}$. In global scale, the slope of this displacement is of $O(Re^{-1/2}\hat{\delta}/l)$. On the other hand, across the main deck the pressure remains unchanged, so that a $\Delta p = O(\hat{\delta}^2)$ propagates all the way to the outer edge of the boundary layer and alters the external potential flow in a zone called *upper deck*. The flow therein is inviscid and irrotational without preferred direction, so the upper-deck thickness should be of the same order as its streamwise length l. In the upper deck the displacement slope must be balanced by the interactive pressure, yielding $Re^{-1/2}\hat{\delta}/l \sim \hat{\delta}^2$. Therefore, we find $\hat{\delta} = O(Re^{-1/8})$, $l = O(Re^{-3/8})$, and $\Delta p = O(Re^{-1/4})$. In this way, a *triple-deck structure* as shown in Fig. 8.10 is established.

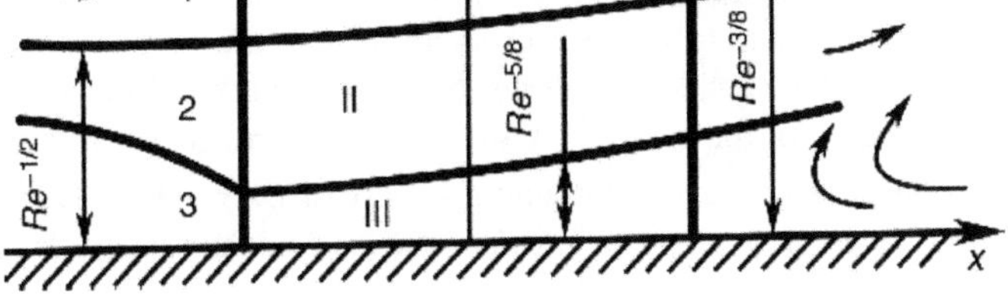

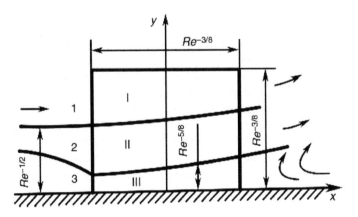

Fig. 8.10 Triple-deck structure and its scales. After Sychev et al. (1998)

8.3.2 *Triple-Deck Equations and Self-induced Pressure Gradient*

After expressed by respective local scales of the three decks, the Navier-Stokes equations in each deck can be accordingly simplified. The solution of one deck should match those at neighboring decks and/or boundary conditions at the wall and outer edge of upper deck by Prandtl's matched asymptotic expansion. We skip the lengthy derivation (e.g., Sychev et al. 1998; Wu et al. 2006) and just present the canonical form of the final governing equations, in terms of lower deck scales $X \sim Re^{3/8}x$, $Y \sim Re^{5/8}y$ and normalized by an $O(1)$ rescaling (subscripts denote partial derivatives):

$$UU_X + VU_Y = -P_X + U_{YY}, \tag{8.3.1a}$$

$$U_X + V_Y = 0, \tag{8.3.1b}$$

$$P = \frac{1}{\pi}\text{pv}\int_{-\infty}^{\infty}\frac{A'(\xi)d\xi}{X-\xi}, \tag{8.3.1c}$$

$$U \sim Y + A(X) \quad \text{as} \quad Y \to \infty, \tag{8.3.1d}$$

$$U = V = 0 \quad \text{at} \quad Y = 0. \tag{8.3.1e}$$

By the way, if the outer flow is supersonic, (8.3.1c) will be replaced by a simple algebraic relation $P = -A'(X)$.

The underlying physics of these equations is as follows. The flow in the lower deck governed by (8.3.1a) is most active, which is viscous, rotational, and nonlinear. It looks the same as Prandtl's equation (4.2.10) but with *self-induced pressure* (or *interactive pressure*) P to be determined by the coupling with upper-deck flow rather than a known exterior pressure P_e. The main-deck equation is analytically solved and leads simply to (8.3.1d). The flow there is effectively inviscid yet still rotational, which only plays a passive role, being displaced by the lower deck so that the slope of the streamlines is transported from the lower deck to the upper deck. The flow in the upper deck is both inviscid and irrotational, where the localized interactive pressure and displacement slope are related by linear integral equation (8.3.1c) (or $P = -A'(X)$ for supersonic external flow), to be coupled with (8.3.1a). This coupling eventually makes the triple-deck problem *locally elliptic* and solvable under conditions prescribed at all side boundaries of the whole triple deck.

The critical feature of the triple-deck theory is the appearance of the local self-induced pressure increment Δp. Although it is only of $O(Re^{-\frac{1}{4}})$, it leads to *a high narrow peak of streamwise pressure gradient and BVF* in the separation zone:

$$\Delta\sigma = \frac{\partial \Delta p}{\partial x} \sim \frac{\Delta p}{l} = O(Re^{1/8}). \tag{8.3.2}$$

Quantitatively, since $P_Y = 0$, the normal derivative of (8.3.1c) can be applied to $Y = 0$ to give the interactive BVF. In global scale, this sharp BVF peak reads

$$\Delta\sigma(x) = \Delta p'(x) = Re^{1/8}\mathrm{pv}\int_{-\infty}^{\infty} \frac{A''(\xi)d\xi}{X-\xi}, \tag{8.3.3}$$

which exists whenever $A(X)$ has curvature. This sharp peak can be easily identified numerically and hence can serve as a convenient marker of boundary-layer separation or other sudden changes of flow conditions.

The triple-deck structure has been found to exist universally in not only boundary-layer separation zone but also many other cases where the flow condition has sudden change: at a wing trailing edge, a discontinuity of wall slope or merely curvature, etc., from incompressible to hypersonic flows. Of these cases, our main concern is the process of an incompressible boundary layer breaking away to become a free separated shear layer. Recall that in Sect. 5.2.2 we mentioned the difficulty in using inviscid Kirchhoff's free-streamline theory to determine the separation point x_s: no choice of the sign of constant k in (5.2.8) was physically acceptable. This dilemma was resolved by Sychev in 1972 (see Sychev et al. 1998), who recognized that k should be taken a positive and Re-dependent function, such that

$$k(Re) \to 0^+ \quad \text{as} \quad Re \to \infty. \tag{8.3.4}$$

Then, in the limiting state one has a *smooth separation* but at finite $Re \gg 1$ there is a non-singular self-induced adverse pressure gradient. By inserting a thin vertical layer around x_s and performing singular perturbation, Sychev arrived at the same universal triple deck structure, with $k = O(Re^{-1/16})$. Moreover, because (8.3.1) is of elliptic type, a downstream boundary condition as $X \to \infty$ should be imposed, of which the asymptotic form as $Re \to \infty$ is exactly Kirckhoff's free streamline (free vortex sheet) leaving x_s tangentially. The formulation is thereby completed.

This triple-deck formulation of incompressible boundary-layer separation from a smooth surface was first confirmed numerically to be well-posed by Smith (1977). Figure 8.11 shows the Re-dependence of skin friction in the separation zone computed from the triple-deck theory.

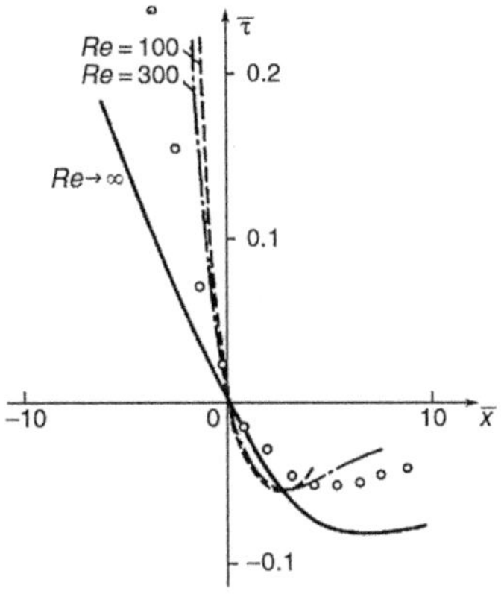

Fig. 8.11 Triple-deck solution of skin-friction (*solid line*), compared with Navier-Stokes solution (*dashed* and *dashed-dot lines*) and experiment (*circles*). From Sychev et al. (1998)

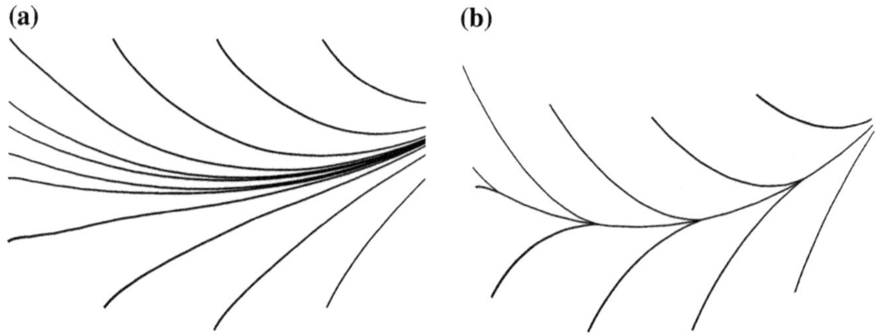

Fig. 8.12 Separation line as **a** asymptotic line (by Navier-Stokes equation) and **b** envelope (by boundary-layer equations) of neighboring vector lines

8.3.3 Three-Dimensional Triple Deck

In three dimensions, by Lighthill's criterion (8.2.3) the separation line is a $\boldsymbol{\tau}$-line toward which neighboring $\boldsymbol{\tau}$-lines converge asymptotically. But in boundary-layer approximation this asymptotic line degenerates to an envelope of neighboring lines (Wu et al. 2000), which is a singular line because at each point thereon with $\boldsymbol{\tau} \neq \mathbf{0}$ one may draw three lines (Fig. 8.12). We thus need to replace this singular line by a triple-deck "strip" of width of $O(Re^{-3/8})$ in the $\boldsymbol{\omega}$-line direction at each point of the separation line. This is sketched in Fig. 8.13.

Fig. 8.13 Three-dimensional triple-deck structure. From Wu et al. (2006)

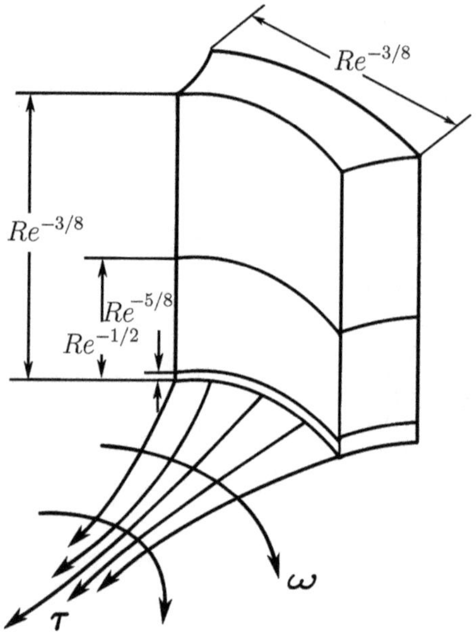

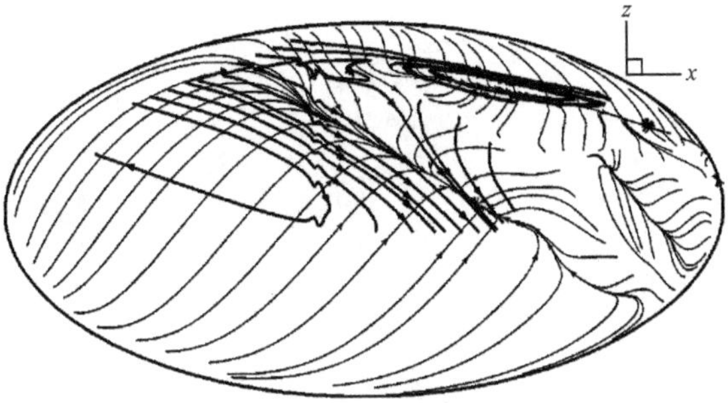

Fig. 8.14 Three-dimensional boundary-layer separation is characterized by the almost alignment of the BVF lines (*thick*) and $\boldsymbol{\tau}$-lines (*thin*). From Wu et al. (2000)

Since along a separation line the flow quantities have only mild $O(1)$ variation, now the self-induced pressure gradient must be along $\boldsymbol{\omega}$-lines, or the strong BVF peak of $O(Re^{1/8})$ must be along the separation $\boldsymbol{\tau}$-line. Since there may remain a BVF of $O(1)$ along the $\boldsymbol{\omega}$-lines as in attached boundary layer, *in the separation zone the resultant BVF vector lines will be roughly aligned to the separation line.* This alignment of BVF-lines and $\boldsymbol{\tau}$-lines can hardly happen elsewhere, and can serve as a warning of three-dimensional boundary-layer separation. Numerical example of the separation from the prolate spheroid confirms this assertion, see Fig. 8.14.

Moreover, unlike generic three-dimensional separation for which weak condition (8.2.17) is sufficient, boundary-layer separation happens only if this condition is significantly strengthened. In terms of the $\boldsymbol{\tau}$-frame, along the separation line (a $\boldsymbol{\tau}$-line) there is $\partial_1 = O(1)$, but across it (along an $\boldsymbol{\omega}$-line) the triple-deck scale implies $\partial_2 \sim l^{-1} = O(Re^{3/8})$. Thus, by (8.2.15b), (8.2.16) and (8.2.17) merge to a single criterion (Wu et al. 2000)

$$\kappa_2 = O(Re^{3/8}), \tag{8.3.5}$$

because $\partial_1 \ln \omega = O(1)$ is negligible. Condition (8.3.5) signifies a very large curvature or sharp turn of the $\boldsymbol{\omega}$-line in boundary-layer separation zone and is also a criterion for three-dimensional boundary layer to break away. This feature is seen in Fig. 8.8 as well, which can be clearly distinguished from some upstream converging $\boldsymbol{\tau}$-lines before boundary layer separates.

A practical example of the BVF peak in a three-dimensional triple-deck zone concerns a high-speed rotor blade, which was experimentally found to suffer from cracks at fixed locations. The flow still remained attached; numerically computed pressure distribution detected no trouble of the blade design. But the BVF does, see Fig. 8.15. The three narrow radial BVF peaks exposed the existence of triple-deck structure due to blade curvature discontinuities, which are precisely where cracks had appeared. Once the curvature is made continuous the crack problem is gone (Wu et al. 2009).

With the rapid development of computational fluid dynamics, today there is no longer a need for solving a triple-deck problem and embedding it into the separation zone; but the clear physical picture revealed by the triple-deck analysis can in no way be obtained by any pure numerical computations. Rather, since the triple-deck zone

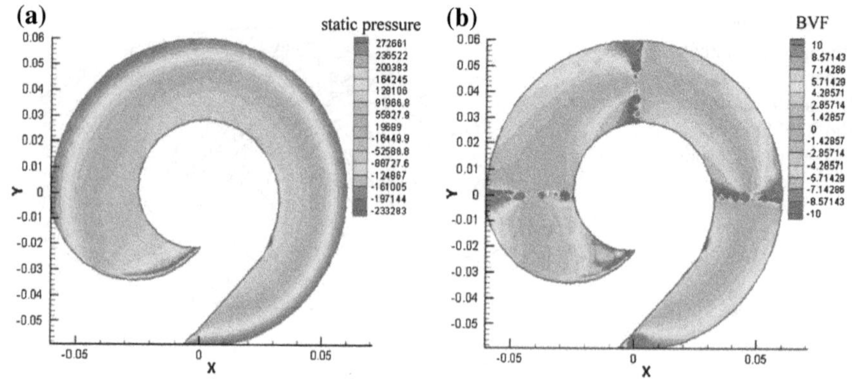

Fig. 8.15 Axial view of the distributions of pressure (**a**) and boundary vorticity flux (**b**) over the surface of a leading blade of a high-speed centrifugal pump. Only the latter can reveal the physical root of the cracks. From Wu et al. (2009)

is a physical reality, for computing Navier-Stokes solutions that can fully resolve the flow in such a zone one has to design the grid density according to the length scales in the deck zone.

8.4 Steady Separated Flows

Having discussed two- and three-dimensional generic flow separation and its strong form, boundary-layer separation, we proceed to consider global separated flows. Of the two types of separated flows mentioned in Sect. 8.1, the closed-bubble type for steady flows with certain symmetries will be discussed first. The type with free shear layers is more commonly encountered in practice but much more complicated; it will be treated in the rest of this section.

8.4.1 Steady Separated Bubble Flow

Closed-bubble separated flow (not to be confused with closed separation defined in Sect. 8.2.3) can exist if the flow is *two-dimensional* or *axisymmetric*, as well as *steady*. In the limit of infinite Reynolds number, this kind of flows can be described by an asymptotic theory. Although in a three-dimensional space such flows are unstable, a more realistic flow could appear as a perturbation of these highly symmetric flows.

Our main concern is the behavior of the vorticity inside the bubble at $Re \gg 1$. The central result is the *Prandtl-Batchelor theorem* initiated by Prandtl (1904) and fully rationalized by Batchelor (1956).

The smaller the viscosity ν is, the longer time is needed to reach a steady state after the motion starts. The correct Euler limit of a steady flow is established by *first taking* $t \to \infty$ such that the unsteadiness vanishes; and *then setting* $\nu \to 0$. Thus, in the final steady state the flow is governed by (assume $\rho = 1$)

$$-\nu\nabla \times \boldsymbol{\omega} = \boldsymbol{\omega} \times \boldsymbol{u} + \nabla H, \quad H = p + \frac{1}{2}q^2. \tag{8.4.1}$$

During the process as $t \to \infty$, the vorticity in separated shear layer is continuously diffused inward to the bubble and finally saturated as diffusion stops. Thus, the flow in the bubble must be rotational. The sheet vorticity is supplemented by the outer flow. Therefore, except the highly viscous regions near closed vortex sheet, the flow is *circulation-preserving*:

$$\boldsymbol{u} \times \boldsymbol{\omega} = \nabla H, \tag{8.4.2}$$

which is to be solved in the region away from closed vortex sheet. If the sheet is open to incoming flow, the boundary condition for solving (8.4.2) is obtainable from the vorticity or stagnation enthalpy on each streamline that extends to infinity. However, for a bubble flow one has to find different conditions for the solution. The unique nature of these conditions is: they come from the viscous equation (8.4.1) and exist only when ν is not identically zero, but ν does not appear explicitly. Consequently, the conditions are *invariant*: they are valid for arbitrary ν, so that in the Euler limit they still hold and provide additional information to the solution of (8.4.2) in the inner region of the bubble not very close to the surrounding vortex layer (*core region* for short). The solution of (8.4.2) obtained thereby is certainly the *relevant Euler solution* defined in Sect. 5.5.

The desired conditions have been found by Batchelor (1956) for two-dimensional and axisymmetric flows, who then obtained the relevant vorticity solution of (8.4.2) in the core region up to a constant factor to be determined by matching with the solution of cyclic vortex layer. Here we confined ourselves to two-dimensional flow. For extension to axisymmetric flow with improved proof see Wu et al. (2006). The result is stated as

Prandtl-Batchelor Theorem. *Under the steady Euler limit, for two-dimensional flow enclosed by vortex sheets, away from the sheets the vorticity is constant:*

$$\omega = \omega_0. \tag{8.4.3}$$

Proof For steady two-dimensional flow in a bubble, there exists stream function ψ, and in the Euler limit (8.4.2) implies that the flow is generalized Beltrami flow with (Sect. 3.1.2)

$$\boldsymbol{\omega} = \boldsymbol{e}_z\omega(\psi), \quad \boldsymbol{\omega} \times \boldsymbol{u} = \omega\nabla\psi, \quad \nabla H = H'(\psi)\nabla\psi, \quad \omega(\psi) = -H'(\psi).$$

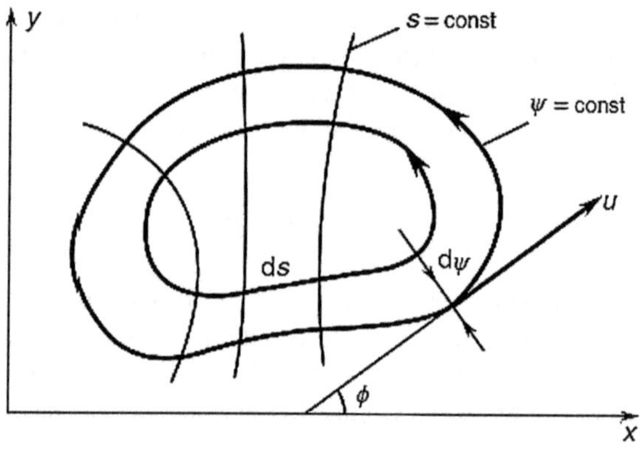

Fig. 8.16 Orthogonal curvilinear coordinates (s, ψ)

Thus, a single additional scalar condition is needed to solve (8.4.2). Because all streamlines inside the bubble are closed, provided $\nu \neq 0$, integrating (8.4.1) along any streamline C_s yields

$$\oint_{C_s} (\nabla \times \boldsymbol{\omega}) \cdot d\boldsymbol{x} = 0. \tag{8.4.4}$$

This condition is *invariant* as ν varies. In the asymptotic case with $\nu \to 0$, we can apply the Euler-limit behavior (8.4.2) to (8.4.4). For two-dimensional flow, we use streamline coordinates (s, ψ) along C_s, with unit vectors $(\boldsymbol{t}, \boldsymbol{n})$ and $d\psi = q dn$ due to (3.1.20), see Fig. 8.16. Then since

$$\nabla \times \boldsymbol{\omega} = \boldsymbol{t}\frac{\partial \omega}{\partial n} = \boldsymbol{t} q \frac{d\omega}{d\psi},$$

(8.4.4) leads to

$$\frac{d\omega}{d\psi}\oint_{C_s} q ds = \frac{d\omega}{d\psi}\Gamma_s = 0.$$

If $\Gamma_s = 0$, there must be $q = 0$ at all points of C_s; such a streamline, if exists, must be in the near-boundary viscous region. Therefore, in the core region there must be $d\omega/d\psi = 0$, which proves (8.4.3).

The physics behind the Prandtl-Batchelor theorem is simple. If ω varies across streamlines, there must be an inward or outward vorticity diffusive flux. But at the center of closed streamlines there is no vorticity source or sink; so in steady state the diffusion cannot exist.

Note that (8.4.3) is a direct extension of the Rankine vortex (Sect. 6.2.1) to arbitrary core shape. If the closed boundary is entirely stationary, the asymptotic steady state in the bubble must be $\boldsymbol{u} = \boldsymbol{0}$ and $\omega = \boldsymbol{0}$. In a nontrivial Prandtl-Batchelor flow, therefore, at least a part of boundary vortex sheet must be in tangent motion, which can have a variable velocity.

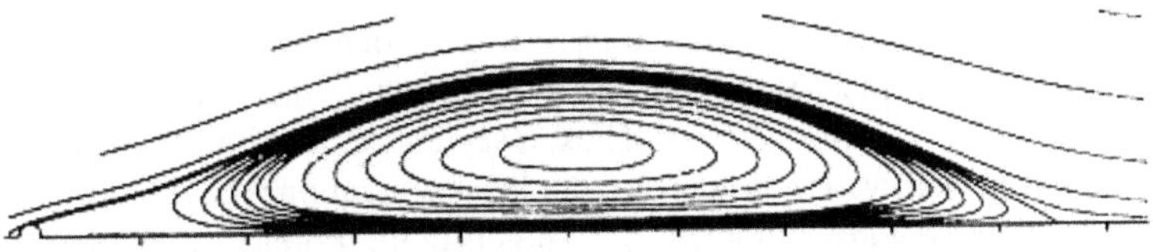

Fig. 8.17 Numerical solution of steady global wake behind a *circular cylinder* (the small semicircle at the *left* end of the plot) at $Re_D = 600$. From Fornberg (1985)

The Prandtl-Batchelor theorem provides the basis of a complete theory of two-dimensional flow with steady separated vortex bubble at large Re. This problem (and other similar problems) can be fully solved by first calculating the location and strength of the vortex sheet surrounding the bubble, and then perturb the sheet to a cyclic viscous vortex layer (the bottom row of Fig. 5.13). For details see the review of Chernyshenko (1998).

It will be seen in Fig. 8.31 below that as the Reynolds number $Re_D = UD/\nu$ (based on diameter D) increases to about 50 the wake behind a circular cylinder starts to be spontaneously unsteady and vortex shedding occurs. However, the *mathematical existence* of a steady but unstable wake cannot be excluded. Numerically, careful Navier-Stokes calculations (Fornberg 1985) which specifically eliminate the possibility of unsteadiness and asymmetry have shown that steady wake is indeed a Navier-Stokes solution, see Fig. 8.17. The vorticity in the core region of the bubble is indeed approximately constant as predicted by the Prandtl-Batchelor theorem.

8.4.2 Fixed-Point Index and Topology of Vector Field

Now we focus on free vortex-layer type of separated flows. Due to the big complexity and variety of such flows, relevant quantitative theories have to stay at highly simplified level. Even in the limit $Re \to \infty$ with known separation location, say a sharp edge, the calculation of the nonlinear evolution of a free vortex sheet has to rely on high-fidelity numerical simulation. Most of this type of complex flows, such as that shown in Fig. 8.5 and circular-cylinder flow at large Re, can only be solved by computational fluid dynamics, which is beyond this book. Nevertheless, there is a very powerful method to determine the *qualitative* structure of free vortex-layer separated flow, which is the *topological theory*. It can be applied to not only steady flows but also unsteady flows, provided that the fixed points of an unsteady velocity field can be identified.[4]

[4]But, for unsteady flow this identification itself may not be an easy job. Besides, although some examples will be presented in Sects. 8.4.4 and 8.5.3 on the topology analysis of unsteady streamline patterns, caution is necessary regarding the role of these streamline patterns in interpreting the evolution of unsteady vortical structures. See relevant observation of Sect. 8.5.2.

In Sect. 8.2.3 we have found that, in three-dimensional flow separation, if a separation line initiates from a fixed point and ends at another, then the fixed points must be a saddle and a node, respectively. Thus, a separation line implies a saddle-node connection. This observation brings us from local separation process to global separated flow property regarding how the fixed points of the $\boldsymbol{\tau}$-field are connected. Then, the total number of fixed points of different types and their connections are the topological property of the flow field, which for a given solid body may change as the flow condition varies, from simple to complex. This knowledge helps rationally interpret experimentally or numerically obtained visualization results of separated flows over various configurations (e.g., Lighthill 1963; Tobak and Peake 1982; Délery 2001).

Rewrite system (8.2.6) as

$$\dot{x} = \tilde{u}(x, y), \quad \dot{y} = \tilde{v}(x, y), \tag{8.4.5}$$

which defines a two-dimensional vector field $(\tilde{u}, \tilde{v})$. Let C be a sufficiently small single closed loop that does not meet any fixed point of (8.4.5). Let a point move along C a cycle counterclockwise, see Fig. 8.18, such that the vector $(\tilde{u}, \tilde{v})$ turns j rounds over an angle $2\pi j$. The integer j can be expressed by

$$j = \frac{1}{2\pi}\oint_C d\left(\tan^{-1}\frac{\tilde{v}}{\tilde{u}}\right) = \frac{1}{2\pi}\oint_C \frac{\tilde{u}d\tilde{v} - \tilde{v}d\tilde{u}}{\tilde{u}^2 + \tilde{v}^2}. \tag{8.4.6}$$

From Fig. 8.18 it is evident that, if C is a closed path of (8.4.5) enclosing $(\tilde{u}, \tilde{v}) = (0, 0)$ then $j = \pm 1$, while if C does not enclose any fixed point then $j = 0$. This assertion is independent of the shape of C since it can be continuously changed to another closed loop. We call j the *index* of the fixed point, which is a topological feature of the vector field.

To compute j for various isolated fixed points, let the origin be at a hyperbolic fixed point so that it suffices to consider the linearized system (8.2.7). Take a small circle of radius δ surrounding the origin, such that $(x, y) = \delta(\cos\theta, \sin\theta)$, and substitute into the linearized version of (8.4.6). We then obtain

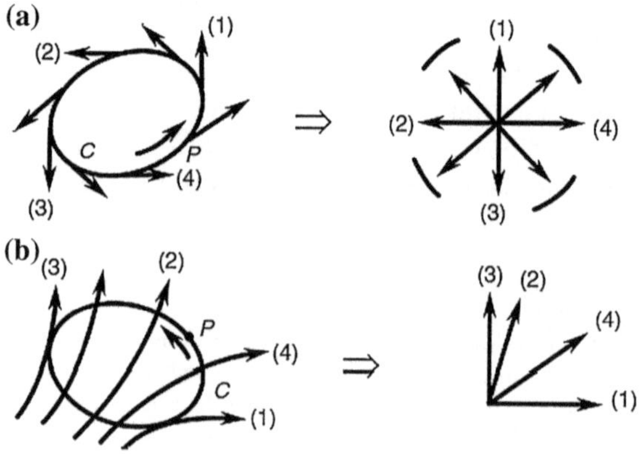

Fig. 8.18 Schematic interpretation of the index of a fixed point

$$j = \frac{ad - bc}{|ad - bc|}, \tag{8.4.7}$$

where a, b, c and d are the coefficients in (8.2.7). Thus, Fig. 8.6 shows that the index of any nodes (source, sink, or spiral) is $+1$, while that of saddles is -1.

On a two-dimensional surface in the three-dimensional space, the sum of the indices of fixed points of a vector field must obey certain rules. For example, on a sphere $\sum j = 2$, on a torus it is 1, and on an infinitely extended flat plate it is 0, etc. Then the topological rule is:

Any surface topologically equivalent to these shapes have the same $\sum j$.

This topological rule implies a simple relation between the total number of nodes and saddles, denoted by $\sum_N$ and $\sum_S$, respectively. For separated flows, we may thereby identify the possible number and nature of fixed points of the $\boldsymbol{\tau}_w$-field on the body surface and the $\boldsymbol{u}$-field in the space.

First, consider the $(\boldsymbol{\tau}, \boldsymbol{\omega})$-field on body surface. For the simplest attached flow over a singly-connected closed surface which is topologically equivalent to a sphere, there is $\sum_N = 2$ and $\sum_S = 0$. Indeed, for a fully attached flow over a three-dimensional body surface, there can only be a front stagnation point where the oncoming flow hits the body first, and a rear stagnation point where the flow leaves the body. Both stagnation points are nodes, satisfying $\sum_N = 2$. But if a closed separation from a saddle point happens and hence the flow topology is altered, to keep the same $\sum j$, as a general topological rule there must be

$$\sum_N - \sum_S = 2. \tag{8.4.8a}$$

Next, if a three-dimensional body B is connected to a plane P extending to up- and downstream infinity (e.g., a half-wing model mounted on a flat plate in a wind tunnel), then on the surface similar consideration implies

$$\left(\sum_N - \sum_S\right)_{B+P} = 0. \tag{8.4.8b}$$

Moreover, assume a plane cuts one or more solid bodies in the flow field so that the sectional flow on the plane has m isolated and finite holes and the connectivity of the sectional flow is $n = m + 1$. Let N' and S' be the semi-node and semi-saddle on the sectional-flow boundary (Fig. 8.19). Their j must be $\pm 1/2$. Again starting the reasoning from the simplest attached flow, we find on the plane there is

$$\left(\sum_N + \frac{1}{2}\sum_{N'}\right) - \left(\sum_S + \frac{1}{2}\sum_{S'}\right) = 1 - n. \tag{8.4.8c}$$

which for a single hole (say a two-dimensional flow over a body) is -1.

Fig. 8.19 Fixed points on a sectional plane of a three-dimensional flow

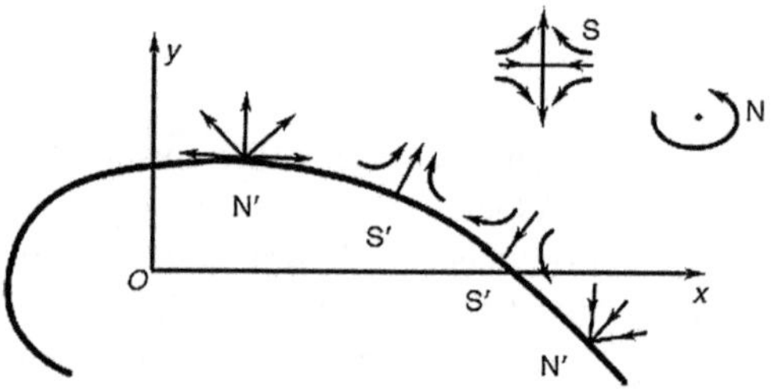

In practice, one may only be able to obtain an "incomplete" skin friction field in an arbitrary region $\mathcal{D}$ of the surface, enclosed by a *penetrable* boundary $\partial\mathcal{D}$ across which vector lines move freely inward and outward. To interpret the skin-friction topology in $\mathcal{D}$, a more general rule than (8.4.8a) and (8.4.8b) is necessary, which considers not only isolated fixed points in $\mathcal{D}$ but also the relevant inflow and outflow across $\partial\mathcal{D}$. A penetrable boundary has both inflow and outflow segments. If at a point Z on $\partial\mathcal{D}$ the neighboring inflow and outflow segments are divided, Z is called a *switch point* (also known as *tangent point* at which a skin friction line or a streamline is tangent to $\partial\mathcal{D}$), which has two types as sketched in Fig. 8.20. By following a skin friction line (or a streamline), if the vector in a sufficiently small neighborhood of Z moves inward first and then outward across $\partial\mathcal{D}$, Z is said to be negative and denoted by Z^-; if the vector moves inward first and then outward, Z is said to be positive and denoted by Z^+.

After the role of switch points is taken into account, the generalized topological rule for a two-dimensional vector field in a region enclosed by a singly-connected penetrable boundary is the *Poincare-Bendixson index formula* (P-B formula in short), which is essentially a conservation law for the numbers of isolated fixed points and boundary switch points, and reads (Hartman 1964; Izydorek et al. 1996; Liu et al. 2011)

$$\sum_N - \sum_S = 1 + \frac{1}{2}\left(\sum_{Z^+} - \sum_{Z^-}\right). \tag{8.4.9}$$

Fig. 8.20 Positive and negative switch points at boundary $\partial\mathcal{D}$

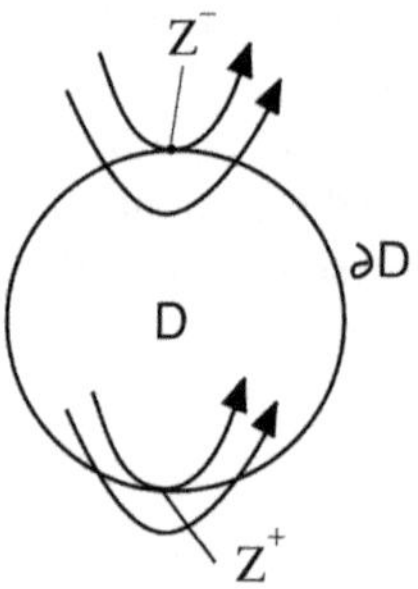

When the inflow and outflow across a closed $\partial \mathcal{D}$ remain qualitatively invariant, (8.4.9) indicates that the nodes and saddles enclosed by a boundary must occur or disappear in pairs. As special cases, (8.4.8a) and (8.4.8b) can be derived from the P-B formula.

8.4.3 Topological Diagnosis of Separated Flows

In experimental studies of separated flows, to observe skin-friction topology, a conventional method is brushing oil to a model surface in wind tunnel tests. However, it is very difficult to identify some subtle topological features in complicated separated flows from oil-streak images only. With the help of (8.4.8), a rational determination can be made on the types of isolated fixed points. The oil-visualization method can be combined with other visualization techniques inside the flow field to gain a three-dimensional picture. For example, Fig. 8.21 shows schematically a typical flow pattern over a slender delta wing at a large angle of attack, of which the visualized pattern is qualitatively clarified by the topological interpretation based on (8.4.8a) and (8.4.8c). At the smoothed apex the separation line still initiates from a saddle, indicating a closed separation.

Recently, high-resolution diagnostics of skin-friction topology is developed based on surface flow visualizations. Based on *global luminescent oil-film* (GLOF) visualization, the $\boldsymbol{\tau}$-field can be quantitatively determined from a time sequence of GLOF images as an inverse problem via a variational method (Liu (2013) and references therein). Figure 8.22 shows a typical GLOF image and extracted $\boldsymbol{\tau}$-lines on a delta wing of sweeping angle $\chi = 65°$ at $\alpha = 10°$. The reattachment lines associated with the primary leading edge vortices and the secondary separation lines between the primary and secondary vortices can be clearly identified in the plot of the $\boldsymbol{\tau}$-lines.

Figure 8.23 shows a typical GLOF image and $\boldsymbol{\tau}$-lines of a junction flow around a square cylinder mounted on a flat plate (Liu 2013). The separation and attachment lines associated with the horseshoe vortex around the cylinder are clearly identified. The separation line in the front of the cylinder is originated from the saddle S_1. The primary spiraling nodes N_1 and N_2 connected through the saddle S_2 occur behind the cylinder. Several subtle fixed points can be identified by zooming the high-resolution $\boldsymbol{\tau}$-field in the selected local regions.

A very interesting example of topological analysis is related to the flow structures of the Jupiter's *Great Red Spot* (GRS). From the cloud images of GRS obtained by Galileo spacecraft in 1996, Liu et al. (2012) extracted the high-resolution velocity field of GRS by using the physics-based *optical flow method* they developed. The flow field is approximately decomposed into a high-speed anti-cyclonic near-elliptical collar and an inner region where the flow is considerably slow and complex. Figure 8.24 shows the mosaics of GRS and the coarse-grained flow fields in these two regions. It is found that the cyclonic source node N_1 is directly associated with the cyclonic rotation near the center of GRS. Since the complex velocity field in the inner region is enclosed by near-elliptical streamlines in the collar, by (8.4.9) the

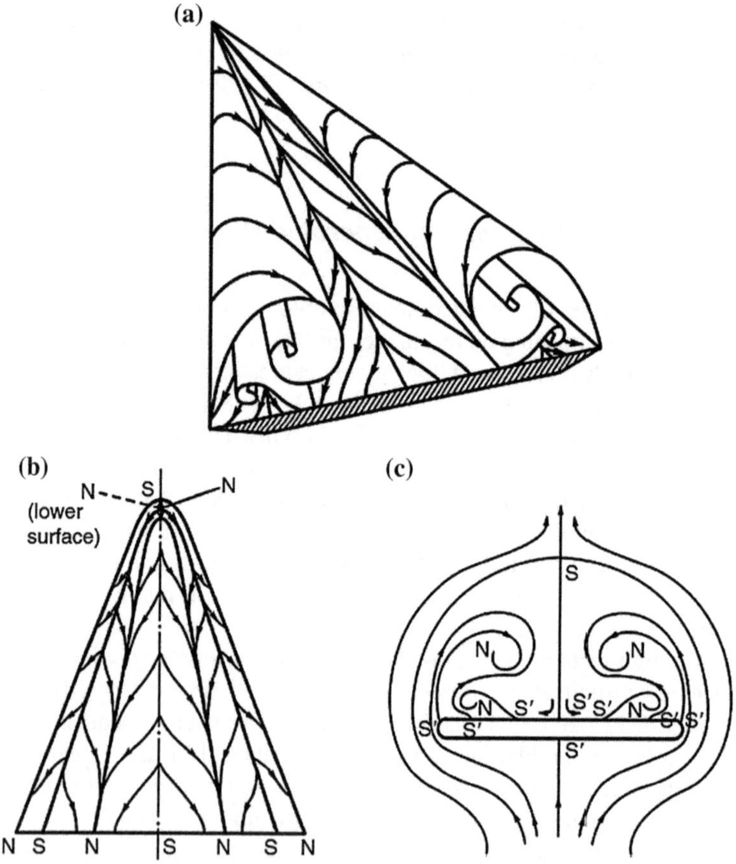

Fig. 8.21 Separated flow on a slender delta wing. **a** Schematic stream surfaces, from Délery (2001). **b** Skin-friction line topology, where the wing apex is rounded off to show the saddle point there (cf. Lighthill 1963). **c** Conceptual cross-flow topology

number of nodes minus the number of saddles in the inner region of GRS must equal one. This topological constraint implies that at least one node in the inner region is necessary for the persistence of GRS. The node N_1 is considered as such a long-lived seed node which could be related to the convection instability.

8.4.4 Structural Stability

We now proceed to ask how a topological structure varies as flow parameters. This question leads to the concept of *structural stability* and *bifurcation*. Let $\lambda = (\lambda_1, \lambda_2, \ldots, \lambda_n)$ be a set of flow conditions (e.g., Mach number, Reynolds

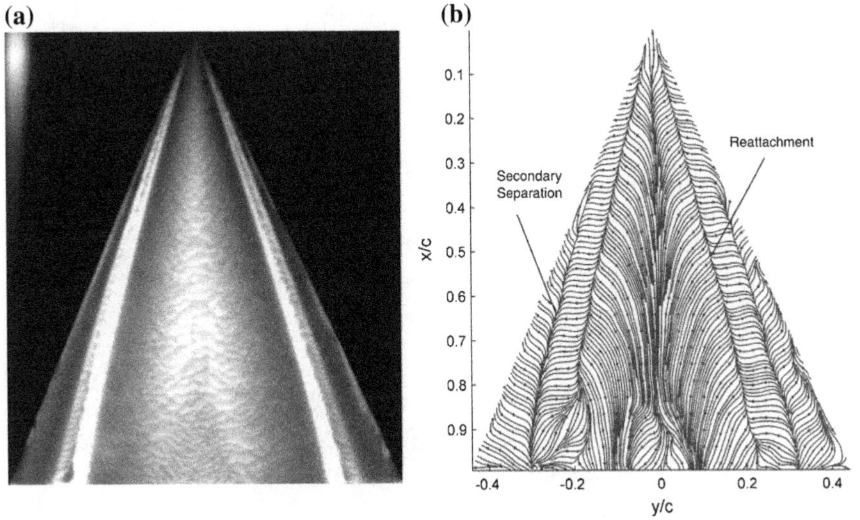

Fig. 8.22 GLOF image (**a**) and extracted $\boldsymbol{\tau}$-field (**b**) of separated delta-wing flow with $\chi = 65°$ and $\alpha = 10°$. From Woodiga and Liu (2009)

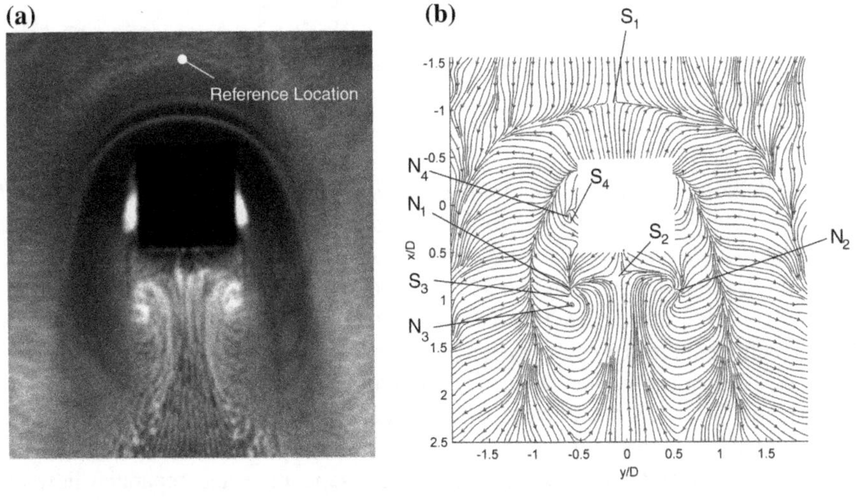

Fig. 8.23 Juncture flow around square cylinder mounted on a flat plate. **a** GLOF image, and **b** extracted $\boldsymbol{\tau}$-field. From Liu (2013)

number, geometric parameters, angle of attack, etc.). If for a fixed λ_0, an infinitesimal change $\delta\lambda$ does not change the topological structure of the *phase portrait*, then this portrait is *structurally stable*, otherwise it is structurally unstable. For two-dimensional vector fields, the structural stability is answered by the following theorem, showing again the importance and robustness of hyperbolic fixed points:

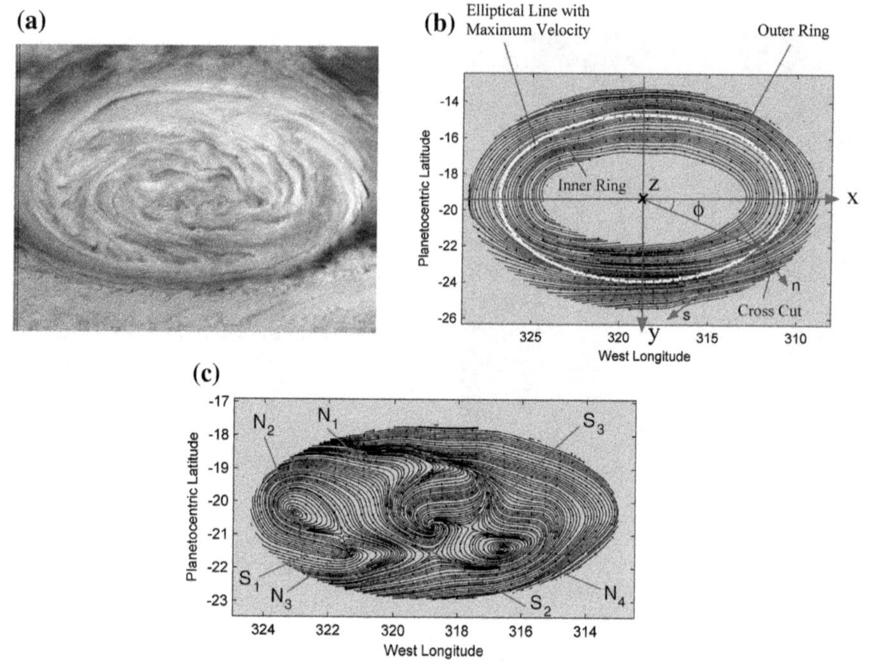

Fig. 8.24 Great Red Spot of Jupiter. **a** Mosaics, **b** high-speed collar, and **c** inner-region topology. From Liu et al. (2012)

The Peixoto Theorem. *A two-dimensional vector field has stable topological structure if and only if (1) the vector field has only finite number of hyperbolic fixed points and single limit cycles; and (2) there is no saddle-to-saddle trajectory.*

To understand the theorem, we add that if the vector field has periodic solution $\boldsymbol{x}(t+T) = \boldsymbol{x}(t)$, then its pathline is a closed loop; if in a small neighborhood of a closed pathline there is no other closed pathline, then that pathline is called a *limiting cycle*, of which an example was seen in the Sullivan vortex (Fig. 6.2).

The conditions in the Peixoto theorem help identify possible real separated flows which are constantly exposed to some disturbances. In a thorough investigation of steady and three-dimensional generic separation and separated flow by using dynamic-system theory, and in accordance with the Peixoto theorem, Surana et al. (2006) have shown that, except saddle-node and saddle-spiral connections as we have seen, a closed separation line may also terminate at an attracting *limit cycle* or just be such a limit cycle. They proved that any robust closed separation patterns, which deform smoothly but survive under small disturbance to the flow field, can only be these four kinds, see the sketches in Fig. 8.25.

Unfortunately, there is no analogue of the Peixoto theorem for dimensions greater than 2. Neither can one decompose a generic three-dimensional flow to three plane flows and then use the Peixoto theorem to find its structural stability except certain

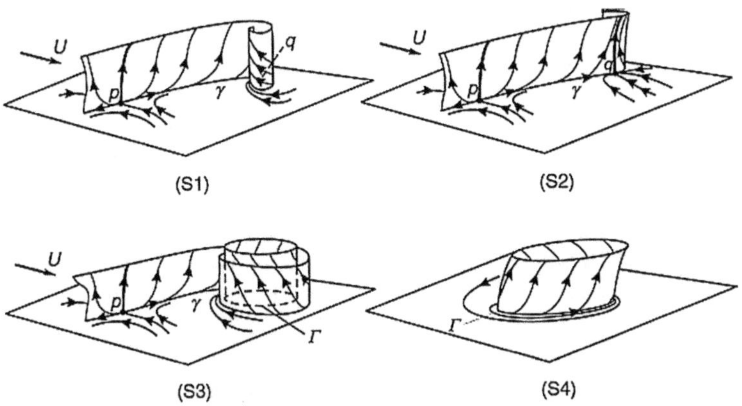

Fig. 8.25 The four robust separation patterns *S1* to *S4*. The *separation line* is denoted by γ. In terms of the fixed points of the $\boldsymbol{\tau}$-field on the wall, for *S1* and *S2* γ initiates at a saddle $\boldsymbol{p}$ and terminates at a spiral ($\boldsymbol{q}$ in *S1*), a node ($\boldsymbol{q}$ in *S2*), or a limit cycle Γ in *S3*. In *S4* γ is simply a stable limit cycle Γ. From Surana et al. (2006)

special cases where a flow has a symmetry or a conservative quantity. Actually, the fact that even a steady three-dimensional flow can have chaotic streamlines, say the ABC flow (3.1.16), already excludes the general possibility of constructing a three-dimensional flow solely from observations of lower-dimensional subspaces.

Once a flow field is structurally unstable, a small disturbance may cause a *topological bifurcation*. If the bifurcation is induced by a change of parameter λ, we say a *parametric bifurcation*. A bifurcation can be *local* if its consequence is confined to a local region. This happens when λ (say the Reynolds number) is near a critical value corresponding to a nonhyperbolic fixed point, such as a change from Figs. 8.4a to 8.7. Or, it can be *global* if its consequence is not local, for example opening a bubble to a free vortex layer (see Sect. 8.4.5). For more discussion on the concept of bifurcation see Wu et al. (2006). Notice that the two-dimensional closed-bubble flow as seen in Fig. 8.17 is structurally unstable due to the saddle-saddle connection. It can exist only after all disturbances and asymmetries are carefully excluded.

8.4.5 Open Separation with Boundary-Layer Breaking Away

In generic three-dimensional flow separation, although one may identify a specific $\boldsymbol{\tau}$-line as separation line that connects a saddle and a node, say, one can hardly judge whether the appearance of that separation line has strong effect on the global flow performance. Actually, even the quite weak $\boldsymbol{\tau}$-line convergence criterion (8.2.16) or upwelling criterion (8.2.17) may not hold all the way along this line. From Fig. 8.8 one may indeed observe opposite situation.

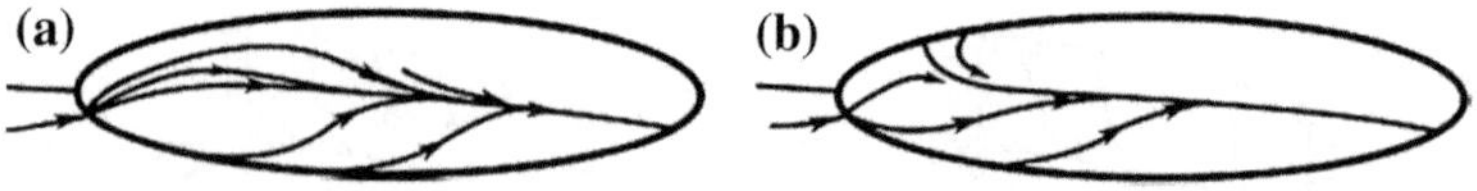

Fig. 8.26 Open and closed separation from prolate spheroid. From Wang (1982)

What matters in practical applications is separated flow caused by *boundary-layer breaking away*. Figures 8.5 and 8.14 show that this may well happen half way from an ordinary point of a generic separation line, where a narrow trip of boundary-layer separation zone is formed, characterized by (8.3.5). This situation is beyond the coverage of the four types of generic separated-flow patterns given in Fig. 8.25, but rather indicates that in addition to closed separation, there exists another separation pattern called *open separation*. It is not associated with saddles, and the flow topology can be the same as that of fully attached flow. Figures 8.5 and 8.14 also show that a primary open separation may induce a secondary open separation with an *open reattachment* in between. It has been found (Wang et al. 1990) that for the flow over such a prolate spheroid, when the incidence gradually increases from zero, the separation pattern changes from closed type to open type and back to closed type again. Two situations are shown in Fig. 8.26.

A sufficiently strong open separation must lead to free vortex-layer separated flow and hence is always associated with strong separated vortices. In contrast, a closed separation may lead to a separated flow of either bubble type or free vortex-layer type. The latter equally leads to strong separated vortices; of which examples include the prolate-spheroid separated flow in Fig. 8.26b, the leading-edge vortices shown in Figs. 8.21 and 8.22, etc.

8.5 Unsteady Separation and Separated Flow

Compared with steady separation, either generic separation or boundary-layer separation, unsteady separation involves more complicated time-dependent phenomena. Of course so do global unsteady separated flows. The importance of this field has now been increasingly recognized and one's interest in it has been rapidly growing in recent decades. In this introductory section, we outline the basic physics and some relevant theoretical results of unsteady separation process, and then illustrate unsteady separated flows by some examples of practical interest. These case studies fall into two typical classes of unsteady flows: transient flow and oscillatory flow, with Stokes' first and second problems (Sect. 4.1) as their respective prototypes.

8.5.1 A Highlight of Unsteady Separation

A fundamental kinematic fact of unsteady flow is that the patterns of streamlines, pathlines, and streaklines of the same flow can be very different (Sect. 1.1.1). This simple fact makes local separation criteria and the triple-deck theory developed for steady separation, all in terms of the behavior of streamlines or stream-surfaces in separation zone, need to be modified. Early efforts toward understanding unsteady boundary-layer separation were made in terms of Eulerian description. For two-dimensional unsteady flow, Moore, Rott, and Sears in 1950s proposed that instead of Prandtl's steady separation criterion in terms of on-wall zero-$\boldsymbol{\tau}$ point, unsteady separation should be marked by a zero-shear point inside the fluid, which moves at local flow velocity. Namely, in boundary-layer approximation, if $\boldsymbol{x}_s(t) = (x_s(t), y_s(t))$ is the separation point, then

$$\frac{\partial x_s}{\partial t} = u_s(t), \quad \left.\frac{\partial u_s}{\partial y}\right|_{y=y_s(t)>0} = 0, \tag{8.5.1}$$

which is called the MRS criterion. The point $\boldsymbol{x}_s(t)$ is known as the MRS point, which is actually a moving saddle with neighboring flow being irrotational. But the MRS criterion is difficult to apply. Other efforts involving attempt to extend the triple-deck theory for steady boundary-layer separation to unsteady, in Eulerian description, have also turned out to be unsuccessful.

Actually, the first issue in the study of unsteady separation is to determine what quantities should be chosen to describe relevant unsteady physical process. To gain some feeling, consider a model problem of unsteady closed-bubble separated flow, derived from a truncated Taylor expansion of the Navier-Stokes and continuity equations near a flat plate:

$$\left.\begin{aligned} u(x,y,t) &= -y + 3y^2 + x^2 y - \frac{2}{3}y^3 + \beta xy \sin nt, \\ v(x,y,t) &= -xy^2 - \frac{1}{2}\beta y^2 \sin nt, \end{aligned}\right\} \tag{8.5.2}$$

where β is a parameter. Figure 8.27 shows its streamlines and pathlines for $n = 2\pi$ and $\beta = 3$ (Haller 2004).[5] A set of pathlines initially aligned to the wall evolves to form an upwelling, then a singular-looking tip, and then a sharp *spike*. It is this pathline spike that signifies a *generic unsteady separation*. Evidently, the patterns of instantaneous streamlines have no direct relevance to the identification of unsteady separation.

The example of Fig. 8.27 indicates that, since separation involves the behavior of material fluid particles or material fluid layer, unsteady separation has to be analyzed in terms of *Lagrangian description* that views the flow as a dynamic system of infi-

[5] The times for the 6 plots in the original figure of Haller have been modified to be consistence with (8.5.2). The plots display the instant positions of the released particles at each individual time.

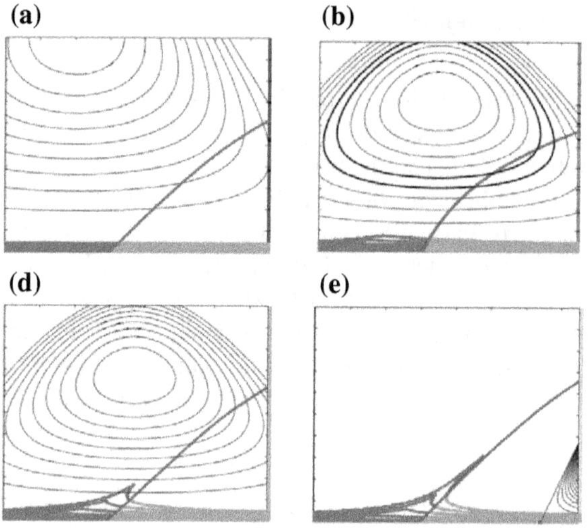

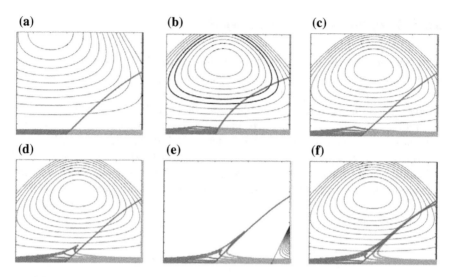

Fig. 8.27 Time evolution of pathlines and streamlines for the periodic separation-bubble model (8.5.2). Particles with different colors were released from different side of the predicted separation profile at $t = 0$. From **a** to **f**: $t = 0.15, 8.40, 10.10, 15.10, 18.75$, and 25.10, respectively. The time-dependent curve serving as the asymptotic line of the pathlines is the separation profile predicted by the theory. The curve serving as the asymptotic line of pathlines during separation will be discussed below. From Haller (2004)

nite degrees of freedom. Along this line, a theory for unsteady generic separation based on non-autonomous dynamic system has been developed by Haller (2004) and coworkers. Based on the Taylor expansion of near-wall Navier-Stokes flow (a common approach in steady and unsteady generic separation), Haller (2004) was able to determine the envelope (a material curve) of the pathline spikes that attracts fluid particles released from its both sides and ejects them into the main stream.

On the other hand, in terms of the Lagrangian form of boundary-layer equations (also a dynamic system), Van Dommenlen and Shen (1982) have found that unsteady boundary-layer separation is marked by the appearance of a material singularity, which forms a vertical barrier so that the forward motion of all fluid elements are blocked there. By the continuity, the elements must extend unboundedly along the normal, causing the boundary layer to blow up. But so far the theory has not reached a stage that can be conveniently applied in complex-flow diagnosis. In the rest of this section we discuss a few unsteady separated flows of practical significance.

8.5.2 Formation of Airfoil Circulation in Starting Flow

The viscous transient process by which the airfoil gains its circulation in a starting flow is a typical example of unsteady separation and separated flow, which is critical in the determination of wing lift (Sect. 9.2). We now study this process based on a numerical simulation of Zhu et al. (2015) for this laminar transient flow, who consider an accelerating uniform incoming flow of incompressible fluid,

$$U_\infty(t) = \begin{cases} \sin(\pi t/2) & \text{if } t < 1.0, \\ 1.0 & \text{if } t \geq 1.0 \end{cases}$$

over a NACA-0012 airfoil at angle of attack $\alpha = 6°$. Flow quantities are made dimensionless by ρ, chord length, and final incoming velocity. For $0 < t \leq 1.0$, the Reynolds number varies as $Re(t) = \sin(\pi t/2) \times 10^5$. The evolution of computed flow patterns is shown in Fig. 8.28 (only parts of flow field near trailing edge are displayed), which shows instantaneous vorticity contours, selected streamlines, pathlines, and streaklines at different times.

To obtain deeper physical insight into the flow patterns shown in Fig. 8.28, we use vorticity formulation. Figure 8.29 displays the on-wall distributions of dynamic quantities, including on-wall normal pressure gradient (NPG) $\partial_n p$ given by (2.3.7b), boundary vorticity ω_B, boundary vorticity flux (BVF) $\sigma = \partial_s p$ which by (2.3.7a) equals tangent pressure gradient (TPG), and boundary enstrophy flux (BEF) $\eta = \omega_B \sigma$ defined by (3.4.23). These on-wall dynamic mechanisms exist for both steady and unsteady flows. Note that since the orientation of instantaneous streamlines adjacent to the wall is almost time-independent, we may use (2.3.6b) to observe the near-wall normal pressure gradient. In dimensionless form it reads

$$\frac{1}{\rho}\frac{\partial p}{\partial n} = -\kappa q^2 - \frac{1}{Re(t)}\frac{\partial \omega}{\partial s}, \tag{8.5.3}$$

where $\boldsymbol{n}$ points into the airfoil and κ is streamline curvature. Setting $q = 0$ yields (2.3.7b).

The time-history of airfoil circulation is shown in Fig. 8.30 for $t \leq 1.0$. It will approach asymptotically to its saturated steady value $-\pi\alpha$ as $t \to \infty$. After $t = 3$, the laminar wake at $Re = 10^5$ becomes unstable and Γ-curve starts to oscillate. Its mean value approaches -0.32 at $t = 30$.

The rich information contained in the above figures enables a thorough understanding of this transient flow and deserves a careful analysis. Below we just highlight the key physical processes and events. Detailed discussions are left to Problem 8.8.

1. Right after the airfoil gains a nonzero incoming velocity and before the viscous diffusion is fully balanced by advection, the flow pattern is very similar to the corresponding inviscid potential flow (Fig. 8.28a). Some fluid particles near lower surface turn rapidly around the trailing edge (TE) and move upstream. Thus, in addition to the front semi-saddle at leading edge, there is a rear semi-saddle point S' on its upper surface. This initial flow pattern is associated with an inviscid effect and a viscous effect: a very strong suction peak at trailing edge (TE) due to (8.5.3), which by (8.2.2) implies a pair of strong BVF peaks near TE; and a thin attached boundary layer with no circulation, having $\omega < 0$ (clockwise) upstream S' and $\omega > 0$ downstream S' on upper surface and over the lower surface.

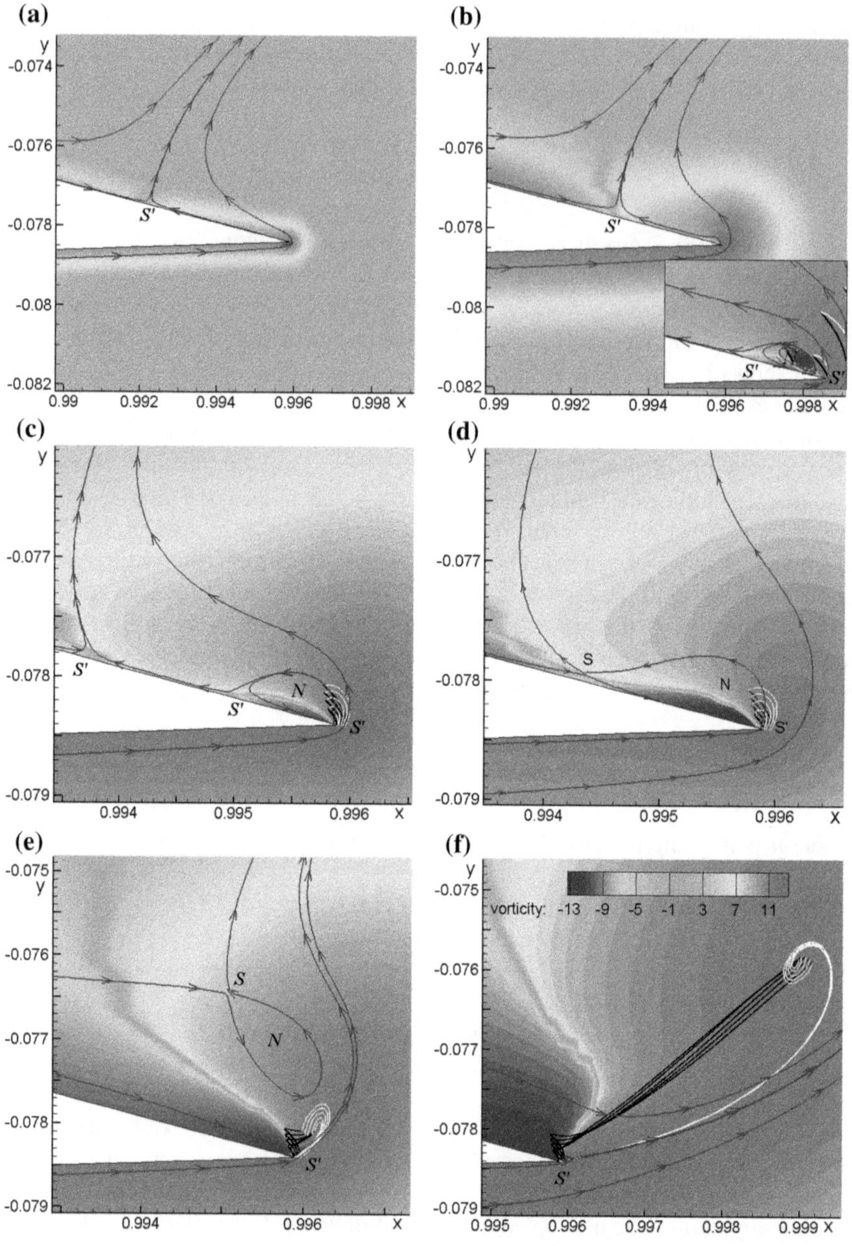

Fig. 8.28 Time variation of accelerated starting flow over NACA-0012 airfoil at $Re = 10^5$ when the flow becomes steady and $\alpha = 6°$. Vorticity contours along with pathlines (*black*), streaklines (*white*), and streamlines (*gray*). Only the rear half of flow field and curves near trailing edge are shown. From **a** to **f**: $t = 0.002, 0.020, 0.026, 0.029, 0.050, 0.100$. The corresponding vorticity ranges are $(-72, 667)$, $(-206, 2272)$, $(-206, 2272)$, $(-248, 2512)$, $(-689, 3068)$, $(-1041, 2446)$, respectively. From Zhu et al. (2015)

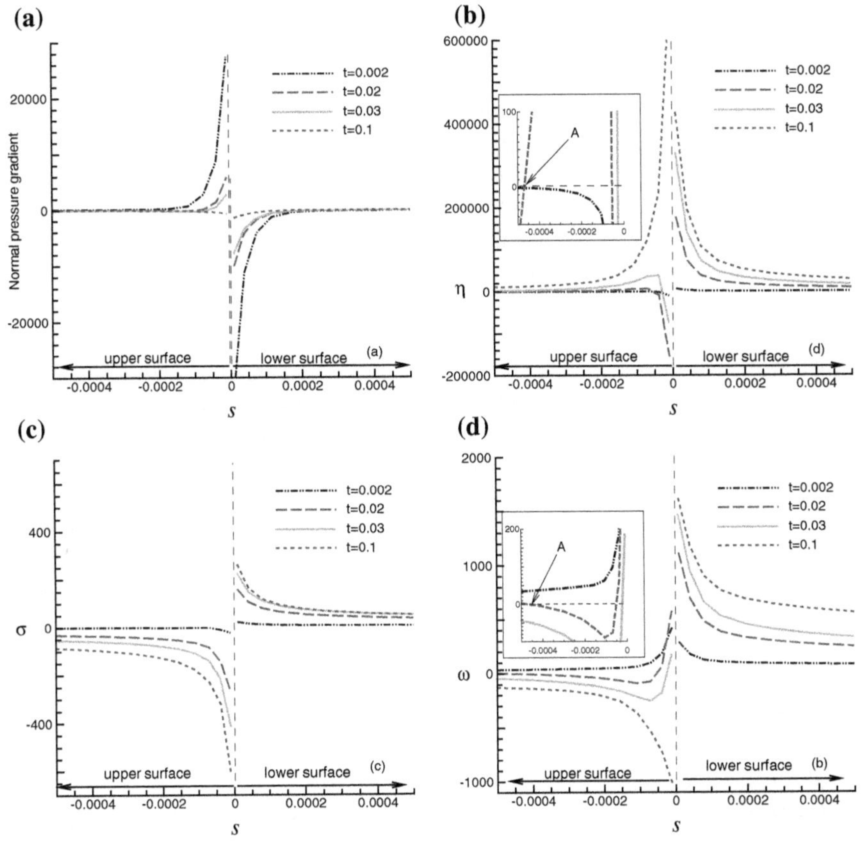

Fig. 8.29 Distributions of boundary normal pressure gradient (**a**), on-wall vorticity ω_B (**b**), boundary vorticity flux σ (**c**), and boundary enstrophy flux η (**d**) at selected times. From Zhu et al. (2015)

Fig. 8.30 Time history of airfoil circulation (including the circulation in attached bubble seen in Fig. 8.28a–d) up to $t = 1.0$. From Zhu et al. (2015)

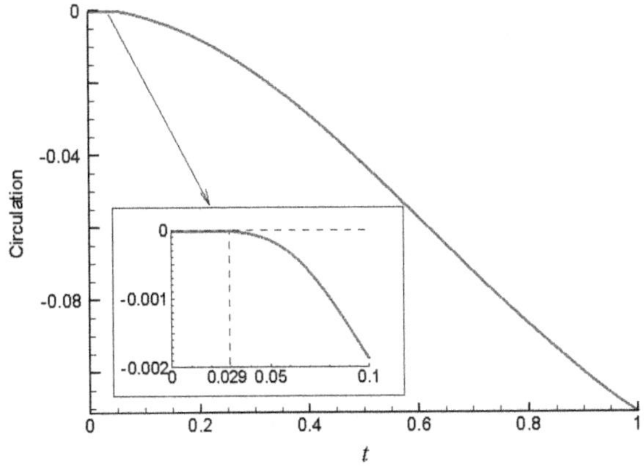

Then, the dominant dynamic mechanism causing the subsequent flow evolution can summarized as what we call the *dynamic* (σ, ω) *interaction*, which is a special form of boundary $(p, \boldsymbol{\omega})$-coupling and consists of a few processes: BVF (=TPG) creates new vorticity, which is diffused into the fluid and advected downstream, which in turn alters existing ω_B depending on the sign and strength of BEF η.

2. At $t = 0.02$ ($Re(t) \approx 3140$), typical high-Re viscous flow behavior starts to show up, with a strongly attached boundary layer of $\omega > 0$ on lower surface and an attached separated bubble of mainly $\omega > 0$ on upper surface near TE. The bubble is associated with two new semi-saddles and a node, indicating a generic unsteady separation.

3. A critical event occurs at $t = 0.029$ ($Re(t) \approx 4554$), in which several qualitative changes take place. The first S' runs to meet the new semi-saddle at upstream boundary of the bubble to form a new saddle S. The near-wall flow becomes fully attached with only an S' at TE. Pathlines coming from TE stop moving upstream but turn sharply up, indicating unsteady boundary-layer separation. The bubble is detached. The strong lower boundary layer of $\omega > 0$ is about to leave TE and shed into wake to become a free shear layer. The airfoil starts to gain a circulation.

4. After $t = 0.03$, the separated shear layer rolls into a starting vortex which is well shaped at $t = 0.10$. At this time the streaklines and streamlines tend to leave the lower surface tangentially, so both the suction peak at TE and strong BVF peaks nearby start to reduce.

5. As the incoming flow ceases to accelerate at $t = 1.0$, the flow becomes fully attached and quasi-steady (Figure not shown), where the peaks of $\partial_n p$, BVF, and boundary enstrophy flux near TE disappear. After that the flow evolves smoothly toward final steady state.

The formation process of circulation for an impulsively started airfoil is qualitatively the same as the above accelerating airfoil, but takes much shorter time because both vortex-sheet strength and BVF at trailing edge are singular at $t = 0^+$.

8.5.3 Separated Flow Over Circular Cylinder

The incompressible flow past a stationary and nominally two-dimensional circular cylinder is a classic prototype of unsteady separated complex flows. It possesses almost the entire complexity of shearing process, such as spontaneous unsteady separation, free shear layer and its rolling up, vortex interactions, various shear instabilities, transition to three-dimensional flow and to turbulence, and unsteady turbulent separated flow. After over a century of effort since Strouhal observed in 1878 that the frequency f of vortex shedding is proportional to U/D with the proportionality constant now being known as the *Strouhal number* $St = fD/U$, and Kármán in 1911 constructed the famous vortex street model and estimated the drag, *the problem of bluff body flow still remains almost entirely in the empirical, descriptive realm of knowledge* (Roshko 1993). Our concern here is confined to an outline of the basic phenomena and the formation process of vortex shedding.

The major difficulty in understanding cylinder flow lies in its strong and complicated dependence on the Reynolds number in the regime of $Re \gg 1$, due to a series critical transition events: from attached flow to steady and then unsteady separated flow, from two-dimensional to spontaneous three-dimensional flow, and from laminar flow to turbulence. The transition events first occur locally somewhere in the flow field and then develop to global behavior. This very complicated Re-dependence of the flow patterns are summarized schematically in Fig. 8.31 for side view and top view. It can also be reflected by the curves versus $Re \in (10, 10^7)$ of St, time-averaged C_D and *base suction coefficient* $-Cp_b = (p_b - p_\infty)/(0.5\rho U^2)$,[6] for which a detailed review see Wu et al. (2006) and references therein.

In the variation of cylinder-flow patterns as Re increases, the near-wake *vortex formation region* is crucial for the entire wake and deserves a careful study, since it is the region where vortex shedding is initiated at *all* Reynolds numbers.

At $Re \gg 1$, the interaction of the upper and lower separated shear layers plays an important role in this vortex-formation region. Consider the upper shear layer which rolls into a vortex with $\omega < 0$ of increasing strength due to the continuous feeding of vorticity from the upstream boundary layer. Suppose the vortex is stronger than that from the lower shear layer due to an asymmetric disturbance. It then entrains the lower-side fluid with positive and zero vorticity across the wake to enter the upper side. This causes a vorticity cancellation, which eventually cuts off the feeding sheet so that the vortex with $\omega < 0$ has to shed downstream. The lower shear layer will then roll into a stronger vortex and the process is repeated to form a cycle.

The entrained fluid from opposite side may enter different regions of the other side in a delicate balanced manner, see the sketch of Fig. 8.32. Fluid a enters the vortex and weakens it; fluid b enters the feeding shear layer and may cuts it; and fluid c moves back toward the cylinder where it is cancelled in the next half of the cycle. Therefore, the vorticity in the formation region and shed vortices (denoted by V in the figure) is considerably weaker than that in the shear layers. The end-point location of the vortex formation region coincides roughly with the overall location in the wake where the vortex strength is a maximum. As Re increases, the vortex formation region shrinks, while the base suction coefficient grows (a larger drag), which is progressively dominated by the *Reynolds stress* in the shear layers.[7]

The vortex-street formation process via shear-layer interaction can be further clarified by topological analysis of the type of Sect. 8.4.4. In two-dimensional steady separated flow there can only be saddles and centers, and the saddle-saddle connection is structurally unstable. As the energy level increases and is saturated, the closed bubble opens and instantaneous "alleyways" of fluid can penetrate the separated flow region. A series of instantaneous streamline patterns in different phases of a period, from an impulsively started cylinder motion and viewed in the frame fixed to

[6]Cp_b is the pressure coefficient at the downstream end b of the body, which reflects the sensitivity of the flow pattern to Re more adequately than that of C_D.

[7]The concept of Reynolds stress is mainly introduced in the study of averaged flow behavior, see Sect. 11.1.2. This concept can equally applied to unsteady laminar flows, where the time average of products of fluctuating velocities can also be viewed as a Reynolds stress.

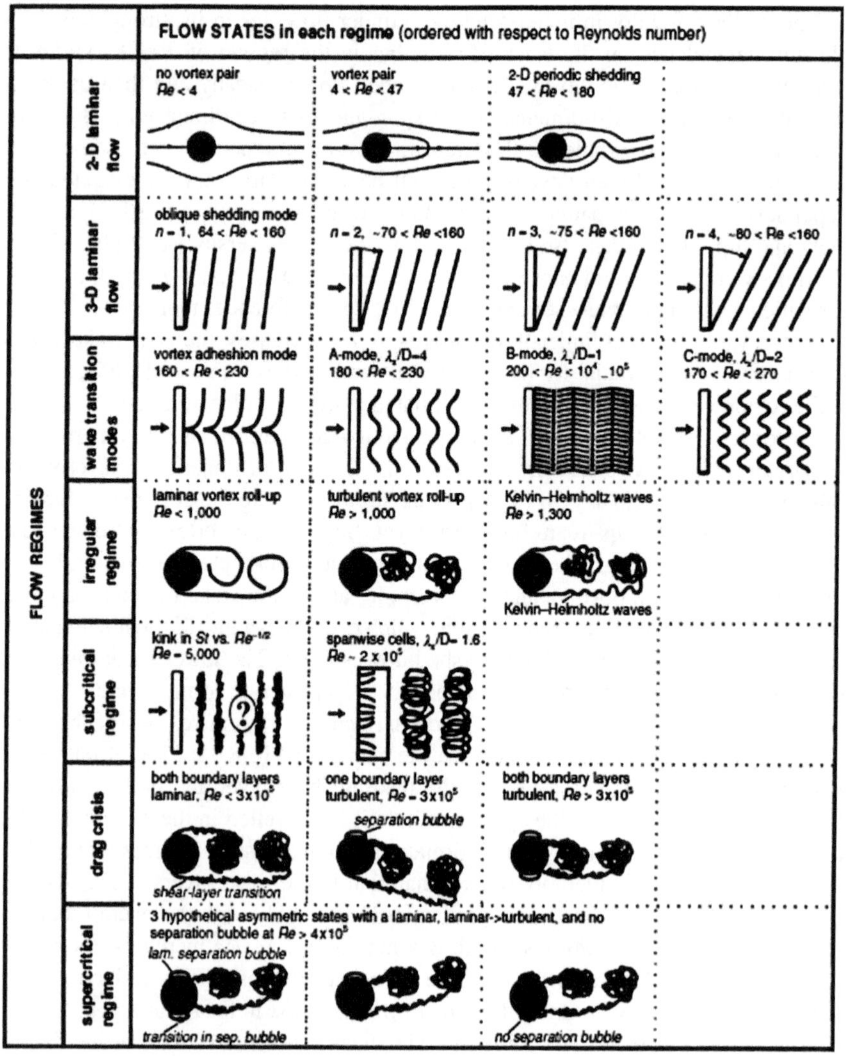

Fig. 8.31 Schematic patterns of unsteady bluff-body flow in different regimes. Modified from Noack (1999)

the cylinder, is shown in Fig. 8.33. The strength of vortex A with $\omega < 0$ is growing in (a)–(d); but in (e) a new saddle forms at the lower side, which cuts off the vorticity feeding to vortex A, making it shed away, and meanwhile forms a new vortex with $\omega > 0$. Actually steps (e)–(h) are just half-cycle difference from (a)–(d), respectively.

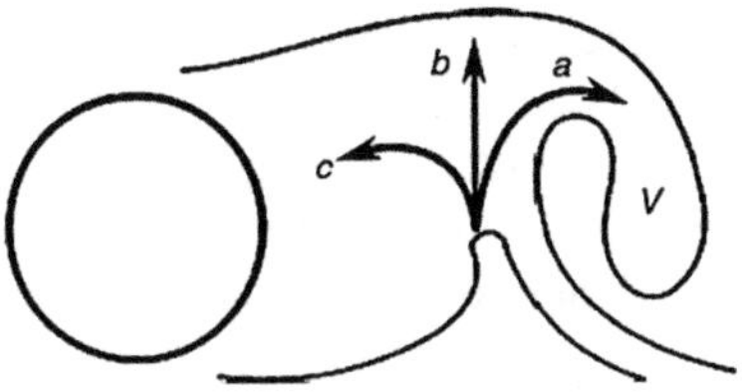

Fig. 8.32 Sketch of vortex formation region behind bluff body. *Arrows* show reverse flow (c) and entrainment (a) and (b). Reproduced from Gerrard (1966)

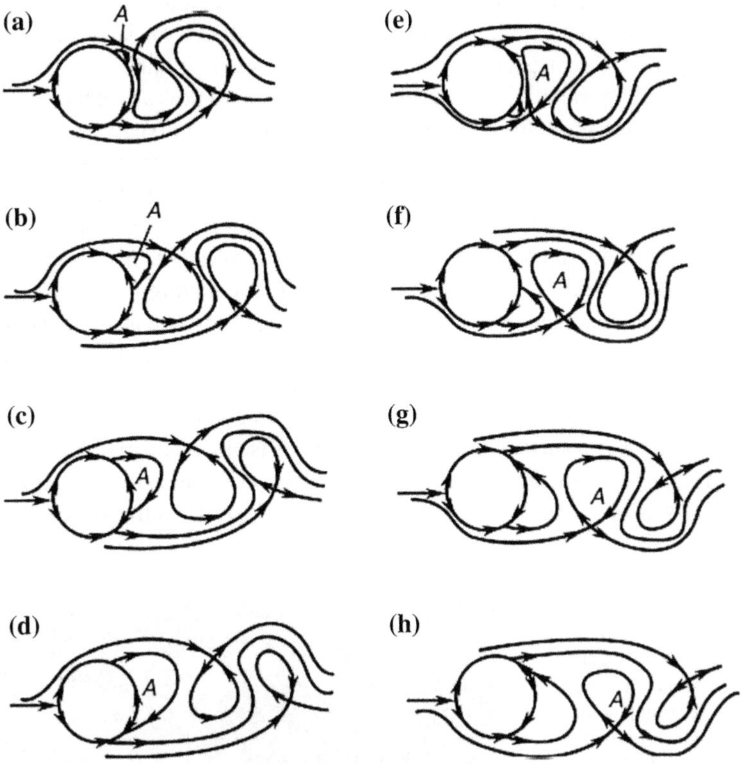

Fig. 8.33 Sketch of vortex shedding from circular cylinder by the topology of instantaneous streamlines. Plotted are only those streamlines which leave or terminate at saddle points. Notice the generation and evolution of vortex A. From Perry et al. (1982)

8.5.4 Falling Disk in Still Water

We now consider freely falling thin circular disks in still water. It adds further complication to the above two examples of complex unsteady separated flows over fixed bodies, because the vortical structures produced by the disk motion have strong reaction to the motion. Our discussion is based on an accurate experimental study of Zhong et al. (2011, 2013) and Lee et al. (2013).

Let the diameter and thickness of the disk be d and h, respectively. Falling disk-flows are jointly governed by three dimensionless parameters: aspect ratio λ, dimensionless moment of inertia I^*, and Reynolds number Re, defined by

$$\lambda = \frac{h}{d}, \quad I^* = \frac{\pi \rho_d}{64 \rho_f}\lambda, \quad Re = \frac{Ud}{\nu},$$

where ρ_d and ρ_f are densities of disk and fluid, respectively. For a thin disk with $\lambda < 0.1$, only I^* and Re are important. At $Re < 100$, a disk of small I^* may fall steadily with a recirculation zone (a ring-like vortical bubble) followed by a streamwise trailing wake. The azimuthal symmetry of the flow about the central axis is lost at $Re \approx 100$, when the one-threaded wake bifurcates into two branches. As Re increases to a critical value between 150 and 160, periodic hairpin-like vortices are shed from the recirculation zone and plane symmetry is recovered. The periodically shed vortices cause the disk to oscillate as Re further increases, forming a plane zigzag motion. Since then the disk motion and the evolution of its wake vortices interact each other and form a complicated *fluid-solid coupling*.

In the zigzag mode, the disk undergoes a gliding motion like a simple pendulum. Its wake vortical structures can be identified from a sequence of flow visualization pictures, of which one is shown in Fig. 8.34. As the disk swings toward its extreme location, it decelerates and pitches up, forming a recirculation zone near the edge that attaches to the bottom surface. As the disk further tilts up, another recirculation zone is formed at the upper surface. When the disk swings back, the two recirculation zones of opposite vorticities detach and form a *hairpin vortex*. Thus the wake consists of

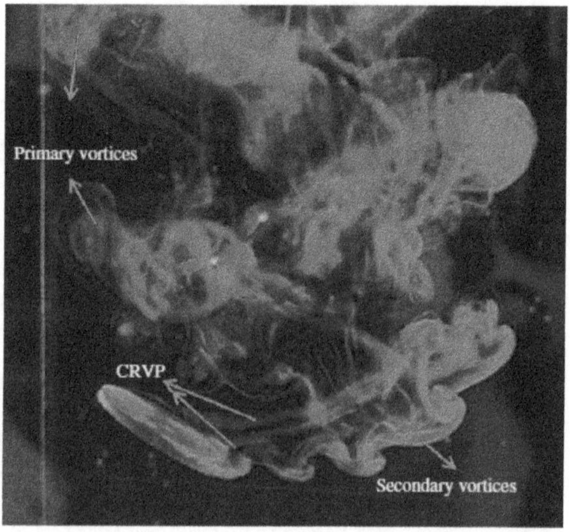

Fig. 8.34 Visualization of wake structures behind a zigzag disk at $Re \approx 2200$. The disk is at its left extreme position. CRVP stands for counter-rotating vortex pair. From Zhong et al. (2013)

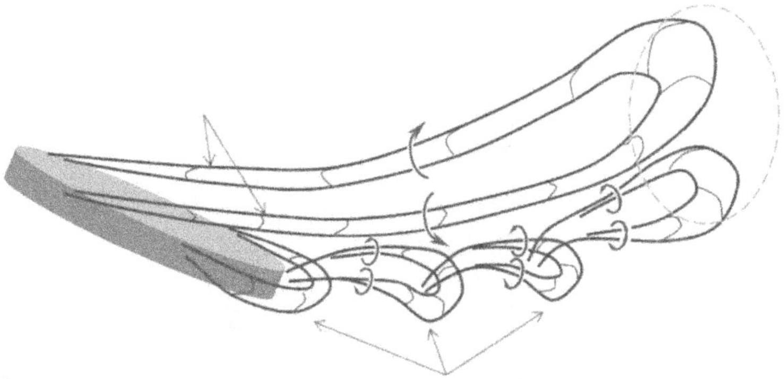

Fig. 8.35 A sketch of counter-rotating vortex pairs (CRVP), secondary hairpins and primary vortices in zigzag disk motion. From Zhong et al. (2013)

two rows of vortices of opposite signs spaced evenly in the vertical direction, which are bridged by a counter-rotating vortex pair (CRVP). These structures are sketched in Fig. 8.35.

As the moment of inertia I^* of the disk is reduced from $O(10^{-2})$ to $O(10^{-3})$, the zigzag mode may transit to spiral mode, and become fully spiral mode. In experiments another transition process, zigzag-spiral-zigzag, has also been observed. On a spiralling disk moving around a vertical axis, then, the unsteady separation process is more complicated as seen from Fig. 8.36. The horizontal velocity near the outside edge of the disk is higher than that near the inside edge, and hence more vorticity is produced there, fed into the wake via outside-edge separation, and scrolled into a helical roll. The flow inside this roll has large downward velocity that causes separation from the smooth disk surface, which is connected to the inside-edge separated vortex. These together generate an upright vortex inside the helicoidal vortex, which carries streamwise vorticity with opposite sign of the latter.

The fluid-solid coupling nature of the falling-disk flow implies that, for a complete understanding of the observed phenomena, one should not only explain the wake structures on the fluid side, but also explain the motion of the solid side. In the experiments of Zhong et al., all six degrees of freedom of the disk motion were quantitatively measured. This enables the authors to analyze the physical constituents of the force and torque exerted to the disk by observed vortices, by the known momentum and angular momentum equations for rigid-body motion, with external forces and torques from the surface stresses of viscous fluid and gravity. The former includes the virtual-mass effect (Sect. 2.6.3), normal and shear stresses (pressure and skin friction); for details see Lee et al. (2013). It has been found that vortex-induced stresses sometimes dominate the disk motion.

We remark that the falling-disk problem is just a relatively simple case of complex unsteady separated flow with fluid-solid coupling. This kind of flows exists widely in nature and technology, in particular in bird/insect flight and fish swimming, where solid bodies such as animal wings and tails are highly deformable and interact actively the separated flow.

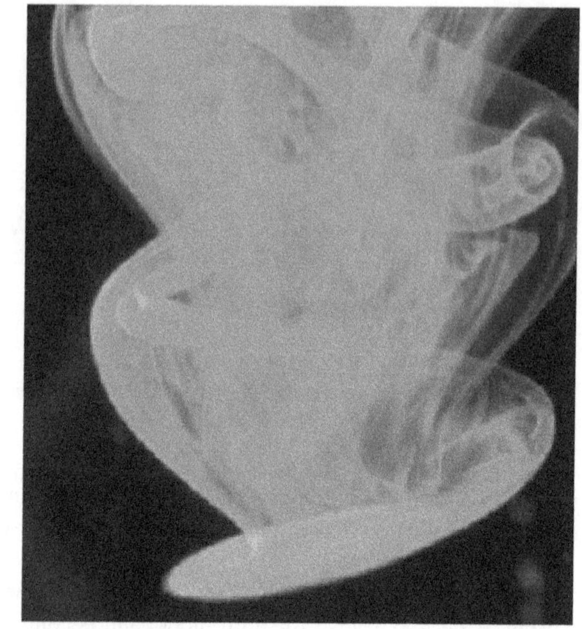

Fig. 8.36 Separation of vortical structure from near a spiraling disk, indicating two separation locations at the disk edge and the upper disk surface. From Lee et al. (2013)

8.6 Problems for Chapter 8

8.1. Consider a rotationally symmetric viscous flow over a sphere of radius a at small Reynolds number $Re = Ua/\nu = \epsilon \ll 1$ or of $O(1)$. It is known that an approximate solution of the flow, expressed by the Stokes stream function and in terms of spherical coordinates (R, θ, ϕ), is

$$\psi = \frac{1}{4}(R-1)^2 \sin^2\theta \left[\left(1+\frac{3\epsilon}{8}\right)\left(2+\frac{1}{R}\right) - \frac{3\epsilon}{8}\left(2+\frac{1}{R}+\frac{1}{R^2}\right)\cos\theta\right]. \tag{8.6.1}$$

Prove: The solution predicts a ring-like separated vortex bubble when $Re \geq 8$ (cf. Fig. 8.3).

Hint: Examine the behavior of the streamline with $\psi = 0$.

8.2. In Problem 8.1, the vector potential $\boldsymbol{\psi}$ of velocity is $\boldsymbol{\psi} = (0, 0, \psi_\phi)$ with $\psi_\phi = \psi/R$. Derive the boundary vorticity $\boldsymbol{\omega}_B$, boundary vorticity flux $\boldsymbol{\sigma} = \boldsymbol{\sigma}_p + \boldsymbol{\sigma}_{\text{vis}}$, and $\boldsymbol{\sigma}_p$. Then find values of θ at which these quantities change sign. Compare these values and explain the physical implication of the result.

8.3. Prove (8.2.12).

8.4. Use (8.2.19) to prove the assertion that if in three dimensions a separation line is initiated from a fixed point of the $\boldsymbol{\tau}$-field, then the fixed point must be a saddle. Then confirm Fig. 8.9.

8.5. In the neighborhood of a τ-field saddle where a closed three-dimensional separation starts (Fig. 8.9), find the inclination angle of the separation stream surface on the plane perpendicular to the separation line. Compare the result with (8.2.12) in two-dimensional separation.

8.6. Compare the time needed for the formation mechanism of a bubble flow (Sect. 8.4.1) and that of concentrated vortices by the rolling up of free vortex layers (Sect. 5.4), and explain the underlying physics.

8.7. Give a physical interpretation of the formation, vortical structures, and separation/reattachment processes in the flow pattern shown in Fig. 8.21.

8.8. Carefully inspect Figs. 8.28, 8.29, and 8.30 jointly, and use the concept of (σ, ω) interaction to explain the detailed formation process of airfoil circulation in starting flow. Specifically, pay attention to (but not limited by) the following issues:

(a) What are the respective roles of streamlines, pathlines, and streaklines in the understanding of this transient process? Why the end points of the pathline of a fluid particle released at $\boldsymbol{X} = \boldsymbol{x}(t_0)$ in a period $t_0 \leq t \leq t_1$, say, coincides the end points of the streakline formed by a sequence of particles released at the same $\boldsymbol{x}(t_0)$ in the same period, but in the middle the two lines may diverse? Is there any instantaneous MRS point defined by (8.5.1) during the evolution?

(b) Compare the time scales by which the wall pressure is established and the BVF-generated vorticity is diffused to the upper edge of boundary layer. Why initially fluid particles can move around the sharp trailing edge?

(c) Explain the physics underlying the curves in Fig. 8.29. Describe the respective roles of the positive and negative BVF peaks in flow evolution. Try to explain why the BVF curves remain almost invariant during $t = 0.02$ to $t = 0.10$, while other curves vary as time significantly.

(d) Explain the flow structure in the attached bubble and the effects of its appearance and disappearance on the flow near trailing edge. Why at the bottom of bubble there appears a sublayer of $\omega < 0$?

(e) Give a detailed description of various critical changes of the flow pattern from $t = 0.03$ to 0.05. Can you find some correspondence between the pathline behavior during this period and the existing theories of unsteady separation outlined in Sect. 8.5.1?

(f) The Kutta condition for inviscid and steady airfoil flow has different statements, see a footnote in Sect. 5.2.2. Although those statements merge to a single one in steady state, can you find their corresponding events in the present viscous transient flow and identify when they take place?

Chapter 9
Vortical Fluid-Dynamic Force and Moment

A body moving through a viscous fluid will experience a reaction force (and moment) by its generated flow. This is the central problem of aero- and hydro-dynamics, as encountered in the study of various air or water vehicles, turbo or rocket engines, as well as fish swimming or bird/insect fly.

Standard formulas for fluid-dynamic forces have been given in Sect. 1.3.2, which however are not yet optimal as they tell too little about the physical causes of the forces. Indeed, since both air and water have very small viscosities, the forces are dominated by the normal stress due to pressure; but pressure is a global variable dependent on the entire flow. No *localized* dynamic mechanism can be identified from a distributed pressure field itself, nor from a velocity field.

Actually, as said in the context of (1.3.4a), standard formulas only provide a room for further exploration on how various local processes and flow structures determine the forces. To this purpose, an effective strategy is to cast standard formulas to "non-standard" ones that can explicitly exhibit these contributions. Historically, it was precisely this strategy that had led to the most brilliant achievements of classic aerodynamics. The classic theories have now been further extended and unified to cover various complex vortical flows. Given a set of numerically solved complex flow data, modern vortical force theory enables one to quantitatively pinpoint the contribution of every local process and structure to total forces, and thereby to rationally optimize configuration design and flow control.

Since this book is focused on shearing process, we shall be confined to incompressible flow with $M \to 0$. We first discuss in details the physics of lift generation in steady flow in Sect. 9.1, which is followed by an outline of classic steady aerodynamics in Sect. 9.2. We then proceed to classic unsteady aerodynamics in Sect. 9.3. Section 9.4 presents the foundation of modern vortical aerodynamic force theory, of which various applications will only be briefly mentioned; for details the reader is referred to Wu et al. (2006).

J.-Z. Wu et al., *Vortical Flows*, DOI 10.1007/978-3-662-47061-9_9

9.1 Origin of Lift

The very most fundamental issue in the science of flight is the origin of lift on a wing. Of various lift-generation mechanisms which are practically available today, the most basic one is the lift on a fixed wing making forward flight with constant velocity, of which a thorough understanding should be the first task of aerodynamics. This is the focus of the present section. Unless otherwise stated, in this chapter we use wind-coordinate system fixed to a wing, with (x, y, z) along the incoming-flow, spanwise, and vertical-up directions, respectively. Two-dimensional flow occurs on an (x, z)-plane.

We consider a steady incompressible flow with uniform free-stream velocity $\boldsymbol{U} = U\boldsymbol{e}_x$ over a stationary airfoil of chord length c at angle of attack α. Figure 9.1 shows the experimentally measured mean lift coefficient $C_l = L/(\frac{1}{2}\rho U^2 c)$ and drag coefficient $C_d = D/(\frac{1}{2}\rho U^2 c)$ for NACA-0012 airfoil as α varies. It is seen that in a range $\alpha < 12°$ there is an enormous lift but a very little drag due to skin friction. It is this huge *lift-drag ratio* L/D that ensures our flight. For $\alpha > 12°$ the lift drops and drag increases rapidly, which is due to boundary-layer separation and massively separated flow shown in Fig. 8.1. Our task is to explain this result.

The lift has two unique and essential features: *(1) it is a lateral force perpendicular to the wing's motion direction, and hence does not do work and consume power, and (2) its existence requires that the wing has a velocity.* How these features are realized involves complicated mechanisms and had been sought for centuries, along with various controversies, misconceptions, and even wrong "theories", of which some are still widely spread today, see the comprehensive review and comments of McLean (2012).

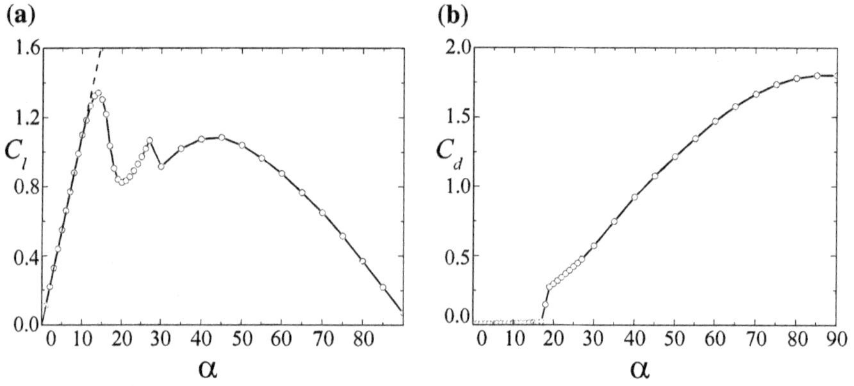

Fig. 9.1 Experimental result of time-averaged lift (**a**) and drag (**b**) coefficients of NACA-0012 airfoil at $Re = 5 \times 10^6$ (*solid line with circles*) versus angle of attack α (Sheldahl and Klimas 1981), compared with two-dimensional airfoil theory (*dashed line* of slope 2π)

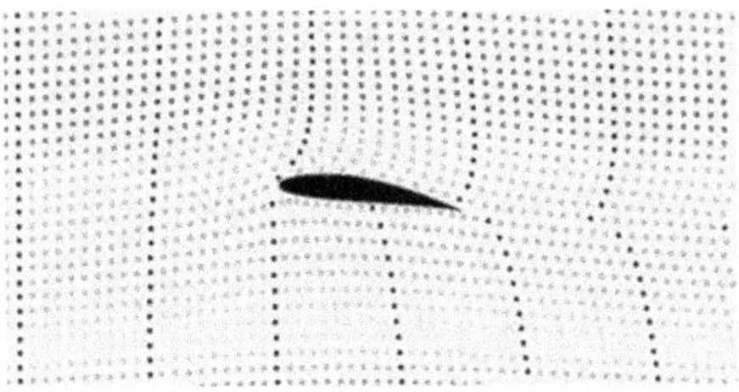

Fig. 9.2 The time-lines (*black dots*) and streamlines (*colored dots*) over a Kármán-Trefftz airfoil at $\alpha = 8°$, computed by potential flow theory with Kutta condition at trailing edge. From *Lift*, Wikipedia. Animation online

For example, a popular naive interpretation is like this: two fluid particles that separate at front stagnation point of an airfoil should meet again at the trailing edge. Since the particle over the upper surface has to run longer distance than that over the lower surface, it has to move faster, and hence with lower pressure by the Bernoulli equation. Then the pressure difference provides a lift. However, this interpretation is just a fallacy. The numerical calculation shown in Fig. 9.2 clearly indicates that the upper particles run far more faster and lead far ahead of the lower ones.

Actually, the conceptual confusions on the lift-generation mechanisms reflect the inherent complication of the problem. For a thorough understanding one has to be fully armed with modern fluid dynamics.

9.1.1 Inviscid Circulation Theory and Criticisms

The right track started at the beginning of the 20th century, and below we make a critical review of the major milestones in the search for the origin of lift.

1. Lift and circulation of two-dimensional airfoil. With remarkable physical insight, F.W. Lanchester was the first to recognize the importance of vortices in the lift generation starting from 1894. According to Giacomelli and Pistolesi (1934), "*there are two great ideas conceived by him: the idea of circulation as the cause of lift, and the idea of tip vortices as the cause of that part of the drag, known today as the induced drag.*" But Lanchester's contribution had long been ignored since he did not put his ideas in rigorous mathematic form. Then, Kutta in 1902 derived the lift for a circular-arc airfoil with requirement that the flow has to leave the trailing edge smoothly, and Joukowski in 1906 derived the famous lift-circulation formula

$$L = \rho U \Gamma, \tag{9.1.1}$$

with drag $D = 0$ (the d'Alembert paradox). Their results are now known as **Kutta-Joukowski lift theorem**.

It is of interest to recall how the aforementioned "non-standardized strategy" was employed by Joukowski in deriving this formula mentioned in Sect. 2.6.4. A key step was to use (1.3.6) instead of (1.3.4a) as the starting point for steady total force, and also to use Bernoulli's equation to remove pressure:

$$\begin{aligned} \boldsymbol{F} &= -\int_{\Sigma} (p\boldsymbol{n} + \rho \boldsymbol{u}\boldsymbol{u} \cdot \boldsymbol{n}) dS \\ &= \rho \int_{\Sigma} \left(\frac{1}{2} q^2 \boldsymbol{n} - \boldsymbol{u}\boldsymbol{u} \cdot \boldsymbol{n} \right) dS, \quad q = |\boldsymbol{u}|, \end{aligned} \tag{9.1.2}$$

where Σ is a fixed control contour enclosing a control volume V including the airfoil. Compared with the body-surface integral (1.3.4a), the control-surface integral now provides the desired room to include flow-field behavior in the flow, which Joukowski's found is the vorticity and circulation. This was done by setting $\boldsymbol{u} = \boldsymbol{U} + \boldsymbol{v}$ and letting Σ be sufficiently large, so that the quadratic terms of disturbance velocity $\boldsymbol{v}$ can be neglected. This step leads to (2.6.26):

$$\begin{aligned} \boldsymbol{F} &= \rho \boldsymbol{U} \cdot \int_{\Sigma} (\boldsymbol{v}\boldsymbol{n} - \boldsymbol{n}\boldsymbol{v}) dS \\ &= \rho \boldsymbol{U} \times \int_{\Sigma} \boldsymbol{n} \times \boldsymbol{v} dS = \rho \boldsymbol{U} \times \int_{V} \omega \boldsymbol{e}_z dV \end{aligned} \tag{9.1.3}$$

by the generalized Gauss theorem, from which (9.1.1) follows at once. As a check of this approach, we note that in two dimensions with $\Gamma \neq 0$, (2.4.19) shows that $|\boldsymbol{v}| = O(r^{-1})$ ($r = |\boldsymbol{x}|$) as $r \to$ "∞", confirming that its quadratic terms can indeed be dropped.

The magnitude of Γ in (9.1.1) is determined by the *Kutta condition* or *Joukowski hypothesis* at sharp trailing edge stated in Sect. 5.2.2. It is the circulation so determined that ensures a faster fluid speed on upper surface than that on lower surface, and hence a pressure jump across the airfoil. This results in the classic two-dimensional steady airfoil theory to be presented in Sect. 9.2, which predicts $dC_l/d\alpha = 2\pi$ as shown in Fig. 9.1, indicating a very excellent agreement with experiments.

2. Vortex force of three-dimensional wing. Then, in developing finite-span wing aerodynamic theory, Prandtl (1918) found an elegant three-dimensional generalization of (9.1.1). Actually, recall identity (3.3.5a), i.e.,

$$\int_{\mathcal{D}} \boldsymbol{u} \times \boldsymbol{\omega} dV = \int_{\partial \mathcal{D}} \left(\frac{1}{2} q^2 \boldsymbol{n} - \boldsymbol{u}\boldsymbol{u} \cdot \boldsymbol{n} \right) dS \tag{9.1.4}$$

for any fluid domain $\mathcal{D}$, it follows from (9.1.2) at once that the force is precisely the Lamb-vector integral, which Prandtl (1918; see also Saffman 1992) calls *vortex force*:

$$\boldsymbol{F} = \rho \int_{V} \boldsymbol{u} \times \boldsymbol{\omega} dV, \tag{9.1.5}$$

where we have continued $\boldsymbol{u}$ into the body by adherence condition, which does not affect the result. The conceptual beauty of this formula lies in the fact that the Lamb vector is the *only* force constituent (per unit mass) for low Mach-number flow that has the aforementioned two essential features of lift: it acts only if $\boldsymbol{u} \neq \mathbf{0}$ and its direction is perpendicular to $\boldsymbol{u}$. Similarly, for the total moment there is

$$\boldsymbol{M} = \rho \int_V \boldsymbol{x} \times (\boldsymbol{u} \times \boldsymbol{\omega}) dV. \tag{9.1.6}$$

Therefore, the finding of (9.1.1) and (9.1.5) brought vorticity and vortex dynamics to the center of modern aerodynamics. We remark that here V is an arbitrary subspace of V_{st} in which the flow is steady (Sect. 2.4.3) and which in turn is a subspace of V_∞. If V could be extended to V_∞, then there would be $\boldsymbol{F} = \mathbf{0}$ and $\boldsymbol{M} = \mathbf{0}$ due to (3.3.6a) and (3.3.6b), respectively. But we have said in Sect. 2.4.3 that in V_∞ any flow must be unsteady at any finite $t < \infty$. The total force and moment for such unsteady flow in V_∞ will be discussed in Sect. 9.3.

To recover (9.1.1) from (9.1.5) for two-dimensional flow, we follow the elegant kinematic argument of Prandtl (1918) and von Kármán and Burgers (1935). Write $\boldsymbol{u} = \boldsymbol{U} + \boldsymbol{v}$ with $\boldsymbol{v}$ being solely induced by vorticity field, such that

$$\boldsymbol{F} = \rho \boldsymbol{U} \times \int_V \boldsymbol{\omega} dV + \rho \int_V \boldsymbol{v} \times \boldsymbol{\omega} dV. \tag{9.1.7}$$

This is a linear-nonlinear splitting of the Lamb-vector integral. The second term represents a nonlinear $(\boldsymbol{\omega}, \boldsymbol{v})$-coupling inherent in the Lamb vector. Because in inviscid theory the flow outside the body is assumed irrotational, in both (9.1.1) and (9.1.5) the vorticity in V has to be idealized as *bound vortex* located inside the body. Thus, one may extend V to the entire free space V_∞ and use the free-space Biot-Savart formula (3.1.35), yielding

$$\int_{V_\infty} \boldsymbol{v} \times \boldsymbol{\omega} dV = \int_{V_\infty} \int_{V'_\infty} \frac{(\boldsymbol{\omega}' \times \boldsymbol{r}) \times \boldsymbol{\omega}}{2(n-1)\pi r^n} dV' dV, \quad \boldsymbol{r} = \boldsymbol{x} - \boldsymbol{x}'$$

where V_∞ and V'_∞ are the same space and

$$(\boldsymbol{\omega}' \times \boldsymbol{r}) \times \boldsymbol{\omega} = \boldsymbol{r}(\boldsymbol{\omega} \cdot \boldsymbol{\omega}') - \boldsymbol{\omega}'(\boldsymbol{\omega} \cdot \boldsymbol{r}).$$

Now, if we replace $\boldsymbol{r}$ by $-\boldsymbol{r}$, the above first term will change sign, but its double integral is symmetric with respect to $\boldsymbol{x}$ and $\boldsymbol{x}'$, making it unchanged and hence can only be zero. Hence, the nonlinear term amounts to

$$\int_{V_\infty} \boldsymbol{v} \times \boldsymbol{\omega} dV = -\int_{V_\infty} \int_{V'_\infty} \frac{\boldsymbol{\omega}'(\boldsymbol{\omega} \cdot \boldsymbol{r})}{2(n-1)\pi r^n} dV' dV, \tag{9.1.8}$$

which is generically nonzero for $n = 3$. However, for $n = 2$ there is $\boldsymbol{r} \cdot \boldsymbol{\omega} \equiv 0$ and hence the nonlinear effect disappears completely, indicating that *in two-dimensional steady flow the nonlinear* $(\boldsymbol{\omega}, \boldsymbol{v})$*-coupling has no effect on the vortex force.* Thus (9.1.1) is recovered as an exact special corollary of (9.1.5). Note that although this purely kinematic derivation of (9.1.1) makes Joukowski's far-field linear expansion redundant, the two approaches are closely related through identity (9.1.4), of which the boundary integral is to be linearized. Thus, the far-field approach also implies the $(\boldsymbol{\omega}, \boldsymbol{v})$-decoupling but implicitly.

For convenience, we call the above vortical theory on lift developed by Kutta, Joukowski and Prandtl the *circulation theory* as a whole.

3. Criticisms on inviscid circulation theory. A remarkable historical fact is that the circulation theory did not immediately receive full acceptance. Its starting equations such as (9.1.2) rely crucially on the use of Bernoulli integral at Σ that holds only for inviscid potential flow. But in potential flow Kelvin's circulation conservation theorem $D\Gamma/Dt = 0$ should apply, evidently contradicting the appearance of Γ and $\boldsymbol{\omega}$ around airfoils and wings in the inviscid circulation theory. As a result, the circulation theory was strongly rejected by some leading English mathematical physicists (Darrigol 2005; Bloor 2011). Here, we may add another limitation of the inviscid theory, which has the same root: As long as the Bernoulli equation is applied to a contour surrounding the airfoil, either at ∂B or Σ as in all the above derivations, it requires the flow to be wake-free, and hence classic circulation theory can only coexist with d'Alembert paradox $D = 0$ in steady flow, which however contradicts the reality that (9.1.1) can coexist with a nonzero viscous drag, see (9.1.14) below.

The debate on whether Kelvin's theorem was violated continued for about two decades, till one eventually realized that *it is only in terms of viscous-flow theory can the circulation theory and Kutta condition, along with a thorough physical understanding, find a profound basis*. In particular, Glauert (1926) clearly states that the circulation theory holds for viscous flow with $\mu \neq 0$ but $\mu \to 0$. In this case, (9.1.1) can also coexist with viscous drag.

The ending of the debate around (9.1.1) and (9.1.5), however, is just the first necessary but insufficient step toward eliminating today's widespread chaotic status in the interpretation of lift. The content of the origin of lift should include not only lift formulas re-derived from viscous-flow theory, but also the whole physical scenario of events responsible for lift generation, especially the key dynamic causality therein. Unfortunately, one's search for the viscous origin of lift has so far still not been accomplished. In this regard, as stressed by McLean (2012), in the search of origin of lift what one should focus on are relevant *dynamic causal mechanisms* rather than any kinematic relations. McLean has used this causality criterion to easily disqualify many misconceptions, and pointed out that neither can the inviscid circulation theory satisfy this criterion despite of its successful predictions.

9.1.2 Viscous Circulation Theory

The drawbacks of inviscid circulation theory imply that it is necessary to reconstruct it within viscous-flow theory. This is easily understood, because the theory is essentially about the shearing process which is inherently of viscous nature. The viscous origin of the Kutta condition has been clarified in Sect. 8.5.2, and below we address the rest key issues.

1. Viscous vortex-force theory in three dimensions. We first re-derive (9.1.5) by viscous-flow theory, for which Kelvin's circulation-conservation theorem does not apply.

Return to the viscous force formula (1.3.4b), where for steady flow there is

$$\boldsymbol{a} = \nabla \cdot (\boldsymbol{uu}) = \boldsymbol{\omega} \times \boldsymbol{u} + \nabla \left(\frac{1}{2} q^2 \right),$$

from which we recall that (9.1.4) follows at once. Now, in (1.3.4b) the volume integral of $\nabla(q^2/2)$ over V_f can be cast to surface integral of $\boldsymbol{n}q^2/2$ over $\partial V_f = \partial B + \Sigma$, so by $\boldsymbol{u} = \boldsymbol{0}$ on ∂B, for incompressible and steady flow there is

$$\boldsymbol{F} = \rho \int_{V_f} \boldsymbol{u} \times \boldsymbol{\omega} dV + \int_{\Sigma} [(P_\infty - P)\boldsymbol{n} + \boldsymbol{\tau}] dS, \tag{9.1.9}$$

where $P = p + \rho q^2/2$ is the total pressure with P_∞ being its constant value at upstream infinity. When the Reynolds number is large, the $\boldsymbol{\tau}$-integral can well be dropped as long as Σ is away from wing's boundary layers. Thus (9.1.9) is reduced to

$$\boldsymbol{F} = \rho \int_{V_f} \boldsymbol{u} \times \boldsymbol{\omega} dV + \int_{\Sigma} (P_\infty - P)\boldsymbol{n} dS, \quad Re \gg 1. \tag{9.1.10}$$

Here, we may well replace V_f by $V = V_f + B$. $P_\infty - P \geq 0$ measures the total-pressure loss in the vortical wake. It appears only at downstream face of Σ that can be selected as a *wake plane* W with $\boldsymbol{n} = \boldsymbol{e}_x$. Then the second term of (9.1.10) has no contribution to the lift but represents a drag of viscous origin, known as *form drag* or *profile drag* D_{form}:

$$D_{\text{form}} = \int_W (P_\infty - P) dS. \tag{9.1.11}$$

On the other hand, the vortex force

$$\boldsymbol{F}_{\text{vf}} \equiv \rho \int_V \boldsymbol{u} \times \boldsymbol{\omega} dV \tag{9.1.12}$$

gives the entire lift and, in three dimensions, another drag component known as the *induced drag* D_{in}:

$$L = \rho \int_V \boldsymbol{e}_z \cdot (\boldsymbol{u} \times \boldsymbol{\omega}) dV, \quad D_{\text{in}} = \rho \int_V \boldsymbol{e}_x \cdot (\boldsymbol{u} \times \boldsymbol{\omega}) dV. \tag{9.1.13}$$

Finally, let the Reynolds number further increase toward infinity. Then there is $P = P_\infty$ and the form drag vanishes, confirming that (9.1.5) indeed holds for $\mu \neq 0$ but $\mu \to 0$, yielding lift and induced drag. Note that in this asymptotic limit (9.1.2) and Bernoulli equation hold everywhere as in inviscid potential flow. This asymptotic limit was the basis of the entire classic circulation theory.

2. Viscous circulation theory in two dimensions. The problem of lift and drag in two-dimensional flow with $Re \gg 1$ was first analyzed by Taylor (1926). We state his main result as

Theorem *For viscous, incompressible and steady (or time-averaged steady) two-dimensional attached flow over a streamlined stationary body at $Re \gg 1$, the lift and drag are given by*

$$L = \rho U \Gamma_\Sigma, \quad \Gamma_\Sigma = \oint_\Sigma \boldsymbol{u} \cdot d\boldsymbol{x} = \int_V \omega dV, \tag{9.1.14a}$$

$$D = \int_W (P_\infty - P) dz, \tag{9.1.14b}$$

where W is a "wake line" of normal $\mathbf{n} = \mathbf{e}_x$.

Denote the two terms of (9.1.7) by $\boldsymbol{F}_0$ and $\boldsymbol{F}_1$. Evidently, (9.1.14a) follows the contribution of $\boldsymbol{F}_0$ in (9.1.7), and (9.1.14b) follows directly from (9.1.11) since there is no induced drag. Thus, To prove the theorem we just need to show $\boldsymbol{F}_1 = \boldsymbol{0}$ again, which however requires great care. Due to the existence of vortical wake, the flow at Σ is no longer irrotational. Thus neither Prandtl's derivation of (9.1.1) for inviscid flow given above, nor Joukowski's original derivation based on (9.1.2), is still applicable.[1] Different approaches by near-field or far-field analysis are necessary. Here we introduce a near-field analysis as follows (Liu et al. 2015, Zhu et al. 2015).

The domain V in (9.1.7) is a subspace of V_{st} in which the flow is steady. Because in the entire free space V_∞ (which contains the far-away starting vortex system) (3.3.6a) holds, we can consider the problem in a complement space $V_W = V_\infty - V$ instead, which extends to downstream infinity but has wake line W at an arbitrary $x_W \in V_{\text{st}}$ as its upstream boundary (cf. Fig. 2.12). Thus, with $\boldsymbol{u}' = (u', w)$, the component form of $\boldsymbol{F}_1$ in (9.1.7) can be re-expressed as

$$L_1 = -\rho \int_{V_W} \omega u' dV, \quad D_1 = \rho \int_{V_W} \omega w dV. \tag{9.1.15}$$

Let us cut V_W into infinitely many slices of wake-line integrals with x_W increasing from trailing edge to arbitrarily far downstream. Then, if we can prove

[1] Taylor's (1926) proof of (9.1.14a) is invalid since he relied on (9.1.2).

$$\int_W \omega u' dz = 0, \quad \int_W \omega w dz = 0 \quad \text{for arbitrary } W, \tag{9.1.16}$$

we may well assert $\boldsymbol{F}_1 = \mathbf{0}$. Note that x_W may even go out of $V_{\rm st}$ and reach the very remote starting vortex located around $x_{\rm start}(t) = O(Ut)$, where $t = 0$ is the time the flow starts. The starting vortex can be represented by a point vortex of circulation $\Gamma_S = -\Gamma_\Sigma$, of which the induced velocity at any point in $V_{\rm st}$ is of the order of $\Gamma_S/(2\pi U t)$ that can be smaller than any $\delta > 0$ as $t \to \infty$, and hence is negligible. Therefore, it suffices to set the upper limit of x_W within the steady-flow space $V_{\rm st}$.[2]

For clarity we first assume that the centerline of wake vortex layer is the x-axis. Denote by δ^+ and δ^- the thicknesses of the upper and lower halves of the wake layer at $z = \pm 0$, respectively. Then the wake-line integrals in (9.1.16) are over $z \in [-\delta^-, \delta^+]$. To study these integrals, we use the vortical form of the steady momentum equation,

$$\rho \boldsymbol{\omega} \times \boldsymbol{u} = -\nabla P - \mu \nabla \times \boldsymbol{\omega}, \tag{9.1.17}$$

where P is the total pressure. Also assume that the flow has $Re \gg 1$ and is attached to streamlined body, so that we can treat the wake vortex layer in boundary-layer approximation (§ 4.2). Thus, in the (x, z)-plane there is $\nabla \times \boldsymbol{\omega} = -\boldsymbol{e}_x \partial_z \omega$, and the component form of (9.1.17) reads

$$\rho w \omega = -\partial_x P + \mu \partial_z \omega, \tag{9.1.18a}$$

$$\rho u \omega = -\partial_z P, \quad u = U + u'. \tag{9.1.18b}$$

Outside the layer we have $\omega = 0$ and $P = P_0$. Integrating these equations along z yields

$$\rho \int_{-\delta^-}^{\delta^+} \omega w dz = \partial_x \int_{-\delta^-}^{\delta^+} (P_\infty - P) dz + \mu [\![\omega]\!] = 0, \tag{9.1.19a}$$

$$-\rho \int_{-\delta^-}^{\delta^+} \omega u dz = [\![P]\!] = 0 = \frac{1}{2}[\![u^2]\!] = \overline{u}[\![u]\!] = \overline{u} \int_{-\delta^-}^{\delta^+} \omega dz. \tag{9.1.19b}$$

In deriving (9.1.19a) we used the fact $\partial D/\partial x = 0$ with D given by (9.1.14b), and in deriving (9.1.19b) we used the fact $\omega = \partial_z u$, $u^2 \gg w^2$, and $[\![p]\!] = 0$, with $\overline{u} = (u^+ + u^-)/2 \neq 0$ being the mean velocity. We thus have proven

$$\int_W u \omega dz = 0 \quad \text{and} \quad \int_W w \omega dz = 0, \tag{9.1.20a}$$

$$\int_W \omega dz = 0. \tag{9.1.20b}$$

[2] In practice the wake vortex layer is unstable and may become spontaneously unsteady, say by rolling up into row of vortices. Then the wake flow can be steady only in the time-averaged sense. Below we assume the average is already taken.

Since the first equation of (9.1.20a) along with (9.1.20b) times U ensures the $u'\omega$-integral also vanishes, (9.1.16) is proven.

Finally, the above results can be extended to a mildly curved wake vortex layer, where (x, z) are replaced by (s, n), the curvilinear coordinates along and normal to the wake centerline $z_c = f(x)$. Let $\tan\theta = f'(x)$ with $|\theta| < \pi/2$, (u_s, u_n) and (u', w) are related by a one-to-one mapping

$$\begin{bmatrix} u_s \\ u_n \end{bmatrix} = \begin{bmatrix} \cos\theta & \sin\theta \\ -\sin\theta & \cos\theta \end{bmatrix} \begin{bmatrix} U + u' \\ w \end{bmatrix}, \tag{9.1.21}$$

from which the desired extension follows easily.

Now, the physics behind (9.1.20) deserves noticing. The first equation of (9.1.20a) reconfirms an observation of Taylor (1926) that, since

$$\frac{d\Gamma_\Sigma}{dt} = \frac{\partial}{\partial t}\int_V \omega dV + \oint_\Sigma \omega \boldsymbol{u} \cdot \boldsymbol{n} dl, \tag{9.1.22}$$

so for steady flow the total vorticity flux must be zero. By this result Taylor asserted that *Γ_Σ is independent of the size of Σ* (as long as its downstream face W has normal $\boldsymbol{n} = \boldsymbol{e}_x$). We refer this assertion as the *Taylor criterion*, which has been observed experimentally by Bryant et al. (1926).

Taylor's criterion is enhanced by (9.1.20b), which further asserts that the total vorticity of the entire wake in $V_{\rm st}$ must also vanish. Physically, vorticity in boundary layers are continuously created at the body surface and keeps shedding into wake. Since Γ_Σ and hence Γ_S are time-independent, the rates of vorticity creation and shedding must be the same, for otherwise a nonzero total vorticity in the wake would alter Γ_S and violate the balance. Actually, (9.1.20b) has been rigorously confirmed by an analytical far-field theory (Liu et al. 2015).

It should be stressed that (9.1.20) holds only in subsonic steady flow. In high-transonic and supersonic flows shock-boundary layer interaction or shock-shock interaction may result in a vortical wake with nonzero total vorticity and its flux.

Like in three dimensions, when $\mu \to 0$ the boundary layers and wake vortex layer degenerate to bound vortex sheets and free vortex sheets. Condition (9.1.20b) implies no vortex sheet can shed from trailing edge, in agreement with the analysis of vortex-sheet dynamics (§ 5.2.2). It also indicates that classic Kutta-Joukowski lift formula can well be viewed as the limiting case of viscous flow with $\mu \to 0$.

The radical difference of the circulation theory for viscous flow with $\mu \neq 0$ but $\mu \to 0$ and its classic version for inviscid flow with $\mu \equiv 0$ is now evident. The viscous circulation theory can be completely free from all reasonable criticisms particularly on a lack of causality mechanisms since neither the Bernoulli integral nor the Kelvin circulation-conservation theorem is used (none of them is applicable as long as $\mu \neq 0$). Therefore, it can serve as a perfect basis for developing modern vortical aerodynamics to calculate the force and moment in unsteady viscous complex flows over arbitrarily moving and deforming bodies (Wu et al. 2006, Liu et al. 2014).

9.1.3 Further Issues

1. Existence and physical carrier of wing circulation. Recognizing the viscous-flow basis of circulation theory also reveals what is the physical carrier of the nonzero Γ and ω around an airfoil or a wing. By the total-vorticity conservation (Sect. 3.3.1) in the infinite space V_∞, once the airfoil generates a nonzero Γ after the onset of its motion, there must be a vortex of $-\Gamma$ produced at the trailing edge and left to the fluid, which is the *starting vortex* whose formation process has been discussed in Sect. 8.5.2. As asserted in Sect. 2.4.3, steady-flow formulas (9.1.1) and (9.1.5) hold when the starting vortex moves to sufficiently far downstream so that its influence on the near field of airfoil becomes negligible. While the airfoil circulation was difficult to visualize in early 20th century, the starting vortex right after its formation was indeed observed by Lanchester and Prandtl, among others, which confirmed the *existence* of airfoil circulation.

In potential-flow theory, however, the airfoil vorticity had nowhere to identify but could only be idealized as a *point vortex* inside the airfoil, which is not a real physical entity. Consequently, within the framework of potential-flow theory one could only infer the existence of airfoil circulation from that of observable starting vortex, like one infers the existence of an object only by identifying its shadow.

It was Prandtl's 1904 seminal work on boundary-layer theory that revealed the true carrier of circulation and vorticity must be airfoil's boundary layers, of which the limiting case at $Re \to \infty$ is attached vortex sheets. Surprisingly, it is the *asymmetry* of the very thin upper and lower boundary layers that leads to a nonzero Γ with sufficiently strong downwash, and so to lift a huge aircraft like Airbus 380 into the sky!

2. Origin of vorticity. The identification of the physical carrier of wing vorticity field naturally leads one to further trace the physical origin of the vorticity in boundary layers. This is in turn due to tangent pressure gradient (TPG) as seen from (2.3.7a), which was recognized by Lighthill (1963) as the key mechanism of vorticity creation, and of which extensive discussion has been made in Sect. 3.4.4.

It is of interest that Lighthill's theory came almost 60 years after Prandtl's 1904 boundary layer theory. This late appearance is related to the fact discussed in Sect. 2.6.1 that (2.3.7a) does not surface in conventional momentum formulation. Its crucial role becomes explicit only in vorticity formulation, of which one of the roles is shedding more light on flow separation, especially on the causal mechanisms in the formation of airfoil circulation (Chap. 8).

The above long-time missing story plays a critical role in the whole scenario of the events in the lift generation, because:

1. It is a typical local dynamic process with clear cause-and-effect mechanism. It takes time for the newly generated vorticity to enter the fluid (by diffusion) and alter the near-wall vorticity distribution. Without this process, the causality mechanism of the entire circulation theory would still be lacking.

2. It brings one back to pressure distribution over airfoil surface and stresses that the vorticity or circulation alone cannot fully explain the origin of lift. At large Reynolds number, it is TPG (*rather than pressure distribution itself*) at the airfoil that is the primary dynamic driver of lift generation, but its precise role is creating vorticity via viscosity and no-slip condition.

We may accurately identify the causality mechanisms in (2.3.7a). When a body moves in a fluid, it produces normal stress p as the first causal mechanism since pressure is a result of molecules collisions in equilibrium state. This is an inviscid process with time scale $t_p \sim c/a$, where c is the chord length of the airfoil and a is the speed of sound. In incompressible flow $t_p \to 0$. We have seen that pressure alone can by no means fully explain lift generation. The appearance of viscous shear stress is a necessary mechanism, which comes after a short relaxation time (Sect. 1.2.1). Strictly, TPG causes BVF; but they occur almost simultaneously since the molecule relaxation time can be neglected in the local equilibrium assumption used throughout the continuous fluid dynamics.

Then, the vorticity generated at the wall by BVF is diffused into the fluid and advected downstream by TPG. The advection time scale is $t_a \sim c/U$ so that $t_p/t_a = M$ with M being the Mach number. The diffusion time scale is $t_\omega = \delta^2/\nu$ where δ is the effective diffusion distance. If $\delta \sim cRe^{-n}$ is the boundary-layer thickness with $Re = Uc/\nu$ and n an exponent, there is $t_p/t_\omega \sim MRe^{2n-1}$, so for incompressible flow with $M \to 0$ and $n \leq 1/2$, there must be $t_p/t_\omega \ll 1$. Thus, the vorticity in a boundary layer (including boundary vorticity ω_B) is an accumulated effect of BVF in both space and time.

Note that (2.3.7a) represents a causal mechanism by which existing ω_B reacts to the pressure field through the Neumann condition of the Poisson equation for p at boundary. But since $\partial/\partial s = O(1)$ except at sharp edge, $\partial_n p$ in most part of the airfoil is of $O(Re^{-n})$ and negligible. In this case the causality relation between p and ω is a one-way mechanism, corresponding to the prescribed pressure distribution in boundary-layer approximation.

3. Lift and downwash. By Newton's third law, an upward lift must be associated with a downward motion of a massive fluid body behind the wing, known as *downwash*, which has been used by some investigators to explain the origin of lift. It is therefore of interest to examine the quantitative *lift-downwash relation* which, as suggested by (9.1.13), can be clarified within inviscid-flow theory.

A force (including the force exerted by the wing and pressure force at the outer boundary of fluid) causes the rate of change of total momentum, which for steady flow is measured by the change of uniform incoming fluid momentum $\rho \boldsymbol{U}$ to a distributed $\rho \boldsymbol{u}$-field at the rate $\nabla \cdot (\rho \boldsymbol{u}\boldsymbol{u})$. Therefore, in the lift-downwash relation what appears can only be the rate of change of downwash rather than fluid's downward velocity. The mathematic formula expressing this physical fact is nothing else but (9.1.2), and hence its vertical component is precisely the desired lift-downwash relation:

$$L = \rho \int_{\Sigma} \left(\frac{1}{2} q^2 n_z - w\boldsymbol{u} \cdot \boldsymbol{n} \right) dS, \quad n_z = \boldsymbol{n} \cdot \boldsymbol{e}_z, \tag{9.1.23}$$

which is all what the lift-downwash argument can tell (see also Problem 9.1).

As a unique property of shearing process, however, lift force does not do work to fluid; so (9.1.23) looks much less clear-cut than the action-reaction in rocket propulsion (a compressing process), which is simply jet-and-thrust. In particular, (9.1.23) cannot reveal the physical mechanism by which downwash is formed, nor can it alone lead to any concise, exact, and predictive lift theory.

It is now clear that both lift-downwash argument and circulation theory are based on exactly the same (9.1.2). Just like velocity and vorticity that are purely related kinematically through the Biot-Savart law with no dynamic causal mechanism, lift and downwash are also in a reciprocal or circular causal relation. But the latter has made one more step that turns out to be of great significance, because unlike widely distributed velocity field, at $Re \gg 1$ the Lamb vector is a highly condensed vector field inside and even narrower than wing boundary layers (if the flow is attached) as seen clearly in Sect. 4.2.5. The localized boundary layers and Lamb vector therein should be one's focus in the exploration of causality mechanisms, since any dynamic cause can always be nailed down to local processes expressible by local differential equations.

9.2 Classic Steady Vortical Aerodynamics

In this section we review classic vortical aerodynamic theories for a thin wing, based on (9.1.5) for steady flow. Unlike modern force theories that can fully utilize CFD-provided numerical Navier-Stokes solutions, these classic theories have to derive linearized flow equations and solve them analytically. We focus on underlying physics and theoretical modelings, and skip technical details.

Classic wing aerodynamics considers only *attached flow* at $Re \to \infty$, for which we saw in Chap. 5 that all vorticity is confined to vortex sheets attached to the upper and lower surfaces of the wing and a free wake vortex sheet γ_w. For a thin wing, the attached sheet can be modeled by a *bound vortex sheet* γ_b. Thus, all these early theories amount to constructing simplified (linearized) model problems by in-depth physical insight, and solving unknown vortex sheet strength γ.

9.2.1 Steady Lift on Airfoil

In two dimensions at $Re \to \infty$, since for steady flow no free vortex sheet sheds from the trailing edge, the Kutta-Joukowski formula (9.1.1) simply becomes

$$L = \rho U \Gamma_b = \rho U \int_0^c \gamma_b(s) ds, \tag{9.2.1}$$

where c is the chord length. What remains now is the determination of the Γ_b, of which the analytical approach is outlined below.

1. Conformal mapping. Set $Z = x + iz$ on the flow plane (x, z), we have analytic complex velocity potential

$$W(Z) = \phi + i\psi, \quad \frac{dW}{dZ} = u - iw,$$

and transform the airfoil onto a unit circle $\zeta = e^{i\theta}$ on the ζ plane with an incoming flow $Ue^{i\alpha}$ at incidence α and *arbitrary circulation* Γ:

$$\begin{aligned}\frac{dW}{d\zeta} &= U\left(e^{-i\alpha} - \frac{e^{i\alpha}}{\zeta^2}\right) + \frac{i\Gamma}{2\pi\zeta} \\ &= -ie^{-i\theta}\left[2U\sin(\alpha - \theta) - \frac{\Gamma}{2\pi}\right] \quad \text{on circle.}\end{aligned} \tag{9.2.2}$$

We can always map the trailing edge of the airfoil to $\zeta = 1$, thus the *Kutta condition* (Sect. 5.2.2) requires $dW/d\zeta = 0$ at $\theta = 0$. This gives

$$\Gamma = 4\pi U \sin\alpha, \tag{9.2.3a}$$

$$\frac{dW}{d\zeta} = 4iUe^{-i\theta}\cos\left(\alpha - \frac{\theta}{2}\right)\sin\frac{\theta}{2}. \tag{9.2.3b}$$

Therefore, by (9.2.1), since Γ is invariant during the mapping, the total transverse force on the airfoil is $L = 4\pi\rho U^2 \sin\alpha$, independent of the airfoil shape. However, the lift coefficient

$$C_l = \frac{L}{\frac{1}{2}\rho U^2 c} = \frac{8\pi}{c}\sin\alpha \tag{9.2.4}$$

depends on the airfoil chord length c which in turn depends on the specific mapping function $\zeta(Z)$. For example, the *Joukowski mapping*

$$Z = \zeta + \frac{1}{\zeta} \tag{9.2.5}$$

maps a unit circle $\zeta = e^{i\theta}$ to a flat plate with $Z = 2\cos\theta$, which has chord length $c = 4$. Thus, for the flat plate at angle of attack α, there is

$$C_l = 2\pi\alpha \quad \text{for } \alpha \ll 1. \tag{9.2.6}$$

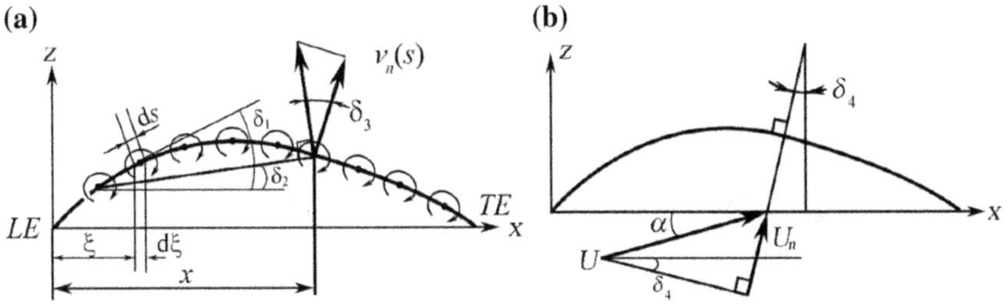

Fig. 9.3 Geometric relations for the determination of (**a**) v_n induced by a vortex-sheet element, and (**b**) U_n

This is approximately the lift coefficient of all thin airfoils (see below). A very thick airfoil may have much larger lift coefficient but suffers from early flow separation.

2. Thin airfoil theory. For an arbitrary airfoil, to avoid seeking for mapping function, consider a uniform flow over a thin airfoil with small thickness and camber at small angles of attack, such that the problem can be linearized. Then, an airfoil-flow problem can be split to the superposition of a *thickness problem* and a *lifting problem*. The former deals with the flow over a thin wing with symmetric thickness distribution along the *camber line* at zero angle of attack α by distributing sources and sinks and solving their strength. The latter deals with a curved airfoil without thickness at $\alpha \neq 0$, which is our present concern. Replace the airfoil by its camber line $z = f(x)$ (the maximum distance between the camber line and the chord line is known as the "camber" of the airfoil). Then the incoming velocity $\boldsymbol{U}$ has a normal component $U_n(x)$ on the camber. To ensure the no-through condition, the camber line is replaced by a *bound vortex sheet* with strength $\gamma(s) = [\![v_s]\!]$, where suffix s denotes tangent component. The no-through condition implies that the normal velocity $v_n(x)$ induced by distributed $\gamma(s)$ at every point of the camber line just cancels $U_n(x)$, i.e., $v_n + U_n = 0$ on the camber. The relevant geometry of the problem is sketched in Fig. 9.3. Note that the γ-distribution also represents the distribution of pressure difference on upper and lower surface of the airfoil. In fact, as shown in (5.2.6) we have $[\![p]\!] = -\rho\bar{u}\gamma$,[3] where the mean local velocity can be simplified to

$$\bar{u} = U_s + \frac{1}{2}(v_s^+ + v_s^-) \simeq U$$

because γ itself is already a small quantity. Therefore, we simply have

$$[\![C_p]\!] = \frac{[\![p - p_\infty]\!]}{\frac{1}{2}\rho U^2} = -\frac{2\gamma}{U}. \tag{9.2.7}$$

The induced velocity $\boldsymbol{v}$ is obtained by applying the Biot-Savart formula (3.1.37) for vortex sheet which in two dimensions reads

[3]The sign difference from (5.2.6) is due to the change of the flow plane from (x, y) to (x, z).

$$\boldsymbol{v}(\boldsymbol{x}) = \int_S \frac{\gamma(\boldsymbol{\xi})\boldsymbol{e}_y \times \boldsymbol{r}}{2\pi r^2} ds(\boldsymbol{\xi}), \quad \boldsymbol{r} = \boldsymbol{x} - \boldsymbol{\xi}. \tag{9.2.8}$$

Referring to Fig. 9.3 and remembering that α and camber are both small. Then, in linear approximation, the no-through condition $U_n + v_n = 0$ can be imposed on the x-axis, yielding the basic *integral equation for thin airfoil theory*:

$$\frac{1}{2\pi} \int_0^c \frac{\gamma(\xi)d\xi}{x - \xi} = U\left(\alpha - \frac{dz}{dx}\right), \tag{9.2.9}$$

where $\gamma(c) = 0$ is required by the Kutta condition that no vortex sheet leaves the trailing edge. The bound vortex sheet has been approximately located at $z = 0$. The singular integral in (9.2.9) requires taking the Cauchy principal value.

An elegant way of solving (9.2.9) for unknown $\gamma(\xi)$ is to use a transformation

$$\xi = \frac{c}{2}(1 - \cos\theta), \quad d\xi = \frac{c}{2}\sin\theta d\theta, \quad 0 \le \theta \le \pi. \tag{9.2.10a}$$

Then by setting $x = (c/2)(1 - \cos\theta_0)$, (9.2.9) is cast to

$$\frac{1}{2\pi} \int_0^\pi \frac{\gamma(\theta)\sin\theta}{\cos\theta - \cos\theta_0} d\theta = U\left(\alpha - \frac{dz}{dx}\right). \tag{9.2.10b}$$

Since the effects on $\gamma(\theta)$ due to α and camber dz/dx can be separated, we write $\gamma(\theta) = \gamma_1(\theta) + \gamma_2(\theta)$ such that γ_1 and γ_2 correspond to a flat plate at $\alpha \neq 0$ and a camber with $\alpha = 0$, respectively. The flow in the former must have singularity at the leading edge as shown by the flow pattern in Fig. 9.4a, where no second Kutta condition can be imposed.[4] It can be proved that this singularity implies an infinite suction peak there (a thrust) that will cancel the x-component of the pressure force on the plate to ensure no net drag. Due to the singularity, only γ_2 can be expressed by a Fourier series; while γ_1 has to ensure $\gamma(\pi) = 0$ (at $x = c$). These two aspects can be taken care of by setting $\gamma_1 \sim \cot(\theta/2)$. Consequently, we write

$$\gamma(\theta) = 2U\left(A_0 \cot\frac{\theta}{2} + \sum_{n=1}^{\infty} A_n \sin n\theta\right), \tag{9.2.11a}$$

$$\gamma d\xi = cU\left[A_0(1 + \cos\theta) + \sum_1^{\infty} A_n \sin n\theta \sin\theta\right] d\theta. \tag{9.2.11b}$$

Substituting this into (9.2.10b) then yields the solution (see Problem 9.3)

$$A_0 = \alpha - \frac{1}{\pi}\int_0^\pi \frac{dz}{dx} d\theta_0, \quad A_n = \frac{2}{\pi}\int_0^\pi \frac{dz}{dx} \cos n\theta_0 d\theta_0. \tag{9.2.12}$$

[4]In practice there will be a small separated bubble attached behind the leading edge to make the velocity finite there.

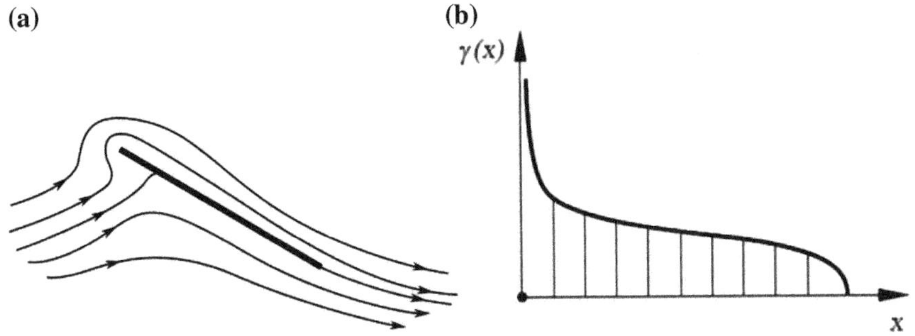

Fig. 9.4 Flow over a flat plate at $\alpha \ll 1$. **a** Flow pattern. **b** γ-distribution

For the flat plate we simply have

$$\gamma = 2U\alpha \cot\frac{\theta}{2} = 2U\alpha\sqrt{\frac{c-x}{x}},$$

of which a typical distribution is shown in Fig. 9.4b.

In calculating the total force and moment only A_0, A_1, and A_2 are involved and we do not need to compute all A_n. For the lift we have

$$L = \rho U \int_0^c \gamma(x)dx = \frac{\pi}{2}\rho U^2 c(2A_0 + A_1),$$

so that by (9.2.12) the lift coefficient reads

$$C_l \equiv \frac{L}{\frac{1}{2}\rho U^2 c} = 2\pi(\alpha - \alpha_0), \tag{9.2.13a}$$

where

$$\alpha_0 = \frac{1}{\pi}\int_0^\pi \frac{dz}{dx}(1-\cos\theta)d\theta \tag{9.2.13b}$$

is the angle of attack with zero lift. For a flat plate we recover $C_l = 2\pi\alpha$, and in any case $dC_l/d\alpha = 2\pi$. Figure 9.1 has shown that this prediction is in excellent agreement with experimental data for $\alpha \ll 1$ when the flow remains attached. The aerodynamic moment property can be similarly analyzed and is omitted here.

9.2.2 Steady Lifting-Line Theory

In three dimensions, a free wake vortex sheet does exist but has zero Lamb vector $\gamma \times \overline{\boldsymbol{u}}$. Hence, denote $\boldsymbol{u} = \boldsymbol{U} + \boldsymbol{v}$ again, (9.1.5) is reduced to

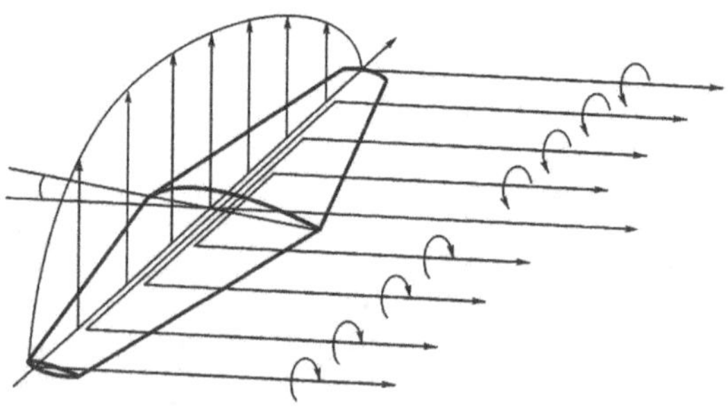

Fig. 9.5 Prandtl's model of vortex system generated by a thin wing

$$\boldsymbol{F} = \rho \int_{S_b} \boldsymbol{u} \times \gamma_b dS = \rho \boldsymbol{U} \times \int_{S_b} \gamma_b dS + \rho \int_{S_b} \boldsymbol{v} \times \gamma_b dS, \tag{9.2.14}$$

where S_b is the area of the wing. Note that the free wake vortex sheet induces a velocity over the wing that joins the determination of γ_b distribution and has non-negligible effect on $\boldsymbol{F}$, which makes three-dimensional wing theory more complex than airfoil theory.

Consider a straight wing of span $b = 2s$ and assume the flow is symmetric with respect to y, so only lift and drag is to be studied. For a large aspect-ratio wing, one may assume that each sectional flow on an (x, z)-plane is still two-dimensional, with sectional lift L_{sec} given by (9.2.1). But since the sectional chord length $c(y)$ depends on spanwise location y, so must be $\Gamma(y)$ and $L_{\text{sec}}(y)$. Then Helmholtz's first vorticity-tube theorem implies that the bound vortex along the span can no longer be represented by a single vorticity tube. Rather, it has to be a bundle of vorticity tubes that turn to the flow direction at different y-locations of the trailing edge, and shed into wake to form a free *trailing vortex sheet*. As said before, this free trailing sheet will affect $\Gamma(y)$. Generically, the trailing vortex system is a part of the nonlinear problem to be solved, and no exact result can be found for the second term of (9.2.14). To proceed analytically, further simplified assumptions are necessary (Prandtl 1918):

(1) the attached vortex sheets can be represented by a single *bound line vortex* of variable circulation $\Gamma(y)$ along the span (the *lifting line*), such that S_b shrinks to a line segment at $x = 0$ with $y \in [-s, s]$ and $\gamma dS = \Gamma(y)dy$. This lifting vortex has no induced velocity to itself, and $\boldsymbol{v}$ comes solely from the induction of the trailing vortex;

(2) the self-induced rolling up and deformation of the free trailing vortex sheet are negligible, so that it lies in the (x, y)-plane with $\gamma = -d\Gamma/dy\boldsymbol{e}_x$. It induces only a vertical disturbance velocity $\boldsymbol{v} = w\boldsymbol{e}_z$ with $w < 0$, known as *downwash*.

The simplified model vortex system is shown in Fig. 9.5, and (9.2.14) yields the formulas of Prandtl's lifting-line theory:

$$L = \rho U \int_{-s}^{s} \Gamma(y)dy = \int_{-s}^{s} L_{\text{sec}}(y)dy, \tag{9.2.15a}$$

$$D_{\text{in}} = -\rho \int_{-s}^{s} w(y)\Gamma(y)dy, \tag{9.2.15b}$$

where $L_{\text{sec}} = \rho U \Gamma(y)$ is the sectional lift. Namely, the first term of (9.2.14) alone gives a pure lift $L\boldsymbol{e}_z$, while the second term alone is an *induced drag* $D_{\text{in}}\boldsymbol{e}_x$, which exists only in three-dimensional flow. The vertical velocity w comes from the Biot-Savart formula (3.1.37) for vortex sheet, in which we now set $\boldsymbol{r} = -x'\boldsymbol{e}_x + (y - y')\boldsymbol{e}_y$ and obtain (see Problem 3.4)

$$w(y) = -\frac{1}{4\pi} \int_{-s}^{s} \frac{d\Gamma}{dy'} \frac{dy'}{y - y'}. \tag{9.2.16}$$

There remains again the determination of $\Gamma(y)$. The downwash $w(y)$ causes an induced angle of attack $\alpha_i(y) = w(y)/U$, so that the local effective angle of attack at y is

$$\alpha_e(y) = \alpha + \alpha_i = \alpha + \frac{w(y)}{U} \leq \alpha. \tag{9.2.17}$$

Therefore, let

$$C_l(y) = \frac{L_{\text{sec}}(y)}{\frac{1}{2}\rho c(y) U^2}, \quad a(y) \equiv \frac{\partial C_l}{\partial \alpha} \tag{9.2.18}$$

be the lift coefficient of the wing strip and the slope of the C_l-α curve with chord length $c(y)$, there is

$$\Gamma(y) = \frac{1}{2} acU \sin\alpha_e \simeq \frac{1}{2} acU\alpha_e.$$

Hence, a substitution of (9.2.17) and (9.2.16) into this expression yields the *basic equation of lifting-line theory*:

$$\Gamma(y) = \frac{1}{2} acU \left[\alpha(y) - \frac{1}{4\pi U} \int_{-s}^{s} \frac{d\Gamma(y')}{dy'} \frac{dy'}{y - y'} \right], \tag{9.2.19}$$

where the dependence of α on y may be due to wing's bending and twist. In general this linear integral-differential equation still has to be solved approximately, say by matrix method.

To estimate the effect of wing planform on α_i and the induced drag, we use *Glauert's transformation*

$$y = s\cos\theta, \quad 0 \leq \theta \leq \pi$$

along the span. Since $\Gamma = 0$ at $\theta = 0, \pi$, we can write

$$\Gamma = U \sum_{n=1}^{\infty} A_n \sin n\theta, \quad d\Gamma = U \sum_{n=1}^{\infty} n A_n \cos n\theta d\theta.$$

Then (9.2.15a) gives $L = \pi\rho U^2 s A_1/2$, depending on A_1 only. Thus, *the lift is always the same as if the load distribution is elliptical*:

$$\Gamma(y) = U A_1 \sin\theta = \Gamma_0 \sqrt{1 - \frac{y^2}{4b^2}}, \tag{9.2.21}$$

where $\Gamma_0 = A_1 U$ is the circulation at $y = 0$. Define

$$\text{Æ} \equiv \frac{b^2}{S_b} \tag{9.2.22}$$

as the *aspect ratio*, it follows that

$$C_L \equiv \frac{L}{\frac{1}{2}\rho U^2 S_b} = \pi \frac{\text{Æ}\Gamma_0}{2bU}, \tag{9.2.23}$$

which holds for all planforms of the wing.

On the other hand, the induced drag is found to be (Problem 9.4)

$$C_{D_i} = \frac{C_L^2}{\pi \text{Æ}}(1+\delta), \quad \text{where } \delta = \sum_{n=2}^{\infty} n \left(\frac{A_n}{A_1}\right)^2 \geq 0. \tag{9.2.24}$$

Clearly, all A_n's for $n \geq 2$ only cause the increase of induced drag. Therefore, *the elliptical load distribution has the minimum induced drag* with constant w and α_i. This is the most famous result of the lifting-line theory. When the wing planform is not far from elliptic, δ is quite small.

Prandtl's conclusion that an elliptic wing is optimal holds only under the assumption of flat trailing vortex sheet, which as seen in Fig. 5.10 is not the real case. The trailing vortex sheet must roll up into a pair of concentrated trailing vortices due to its self-induction. During this process the vortex sheet moves downward due to self induction, which in turn causes their mutual induction in the y-direction to shorten their distance, known as the *side wash*.

9.3 Classic Unsteady Vortical Aerodynamics

9.3.1 Vortical Impulse Theory

Steady flow is merely a simplified model of generic flows. Most flows in aerodynamics are more or less unsteady, and (9.1.5) is no longer applicable. Instead, we have to turn to another force formula (3.3.23) based on *vortical impulse* (or *vorticity moment*) for unsteady flow over arbitrarily moving body at any Re in a strictly unbounded space, developed independently by Burgers (1921), Wu (1981), and Lighthill (1986), see also Wu (2005):

$$\begin{aligned} \boldsymbol{F} &= -\rho\frac{d\boldsymbol{I}}{dt} + \rho\frac{d}{dt}\int_B \boldsymbol{u}_B dV && (9.3.1a)\\ &= -\rho\frac{d\boldsymbol{I}_f}{dt} + \frac{\rho}{n-1}\frac{d}{dt}\int_{\partial B} \boldsymbol{x}\times(\hat{\boldsymbol{n}}\times\boldsymbol{u}_B)dS, && (9.3.1b)\end{aligned}$$

which holds at any Reynolds number. The above second expression comes from a transformation of velocity integral in B by identity (A.2.23), with $\hat{\boldsymbol{n}}$ being the unit normal outward from B and $\boldsymbol{I}_f$ the impulse of fluid only. It is more convenient when the body is flexible, since it avoids assuming artificially introduced $\boldsymbol{u}_B$ inside B.

Similarly, due to (3.3.6b), for the moment one has

$$\begin{aligned} \boldsymbol{M} &= -\rho\frac{d\boldsymbol{L}}{dt} + \rho\frac{d}{dt}\int_B \boldsymbol{x}\times\boldsymbol{u}dV && (9.3.2a)\\ &= -\rho\frac{d\boldsymbol{L}_f}{dt} + \frac{\rho}{2}\frac{d}{dt}\int_{\partial B} x^2\hat{\boldsymbol{n}}\times\boldsymbol{u}_B dS, && (9.3.2b)\end{aligned}$$

where $\boldsymbol{L}$ is the integral of $-x^2\boldsymbol{\omega}/2$ given in Sect. 3.3.3.

Below we discuss some special properties of the impulse theory.

1. Linear superposition. Unlike (9.1.5), the impulse theory is a formally kinematic result and enjoys the full application range for any viscous incompressible flow. In particular, the impulses $\boldsymbol{I}$ and $\boldsymbol{L}$ *depend linearly* on $\boldsymbol{\omega}$, so that each can be treated as a superposition of impulses of every individual vortical structures. Actually, this formal linearity even permits a direct generalization of the above formulas to the total force and moment acting on a multi-body system.

In the impulse theory, nonlinearity and kinetics appear only if d/dt is shifted into the integral as we saw in the derivation of (3.3.19), where evidently the Lamb vector does not enjoy this decomposition property. It can be said that the impulse theory is characterized by retaining operator d/dt outside integrals.

2. Steady vortical force in impulse theory. The inherent flow unsteadiness in V_∞ and its reflection in the impulse theory by no means implies that the theory cannot be applied to study steady aerodynamics. Rather, such a study can well be done by using unsteady-flow reasoning. For example, for flow over an airfoil in the (x, z)-plane, at

$Re \to \infty$ the entire vortex system generated by a moving airfoil at constant velocity $-\boldsymbol{U} = -U\boldsymbol{e}_x$ is simply a vortex couple of circulation $\pm\Gamma\boldsymbol{e}_y$ consisting the bound vortex moving at speed U and starting vortex which is at rest, since no vortex sheet can shed from the airfoil to connect the couple (Sect. 5.1.2). Then the vortical impulse of the couple is $\boldsymbol{I}_\infty = -\boldsymbol{e}_z\Gamma r$, where r is the distance of the vortices (Sect. 3.3.3) that is elongated at constant rate $dr/dt = U$ due to the continuous moving of the airfoil. Thus, the steady-lift formula (9.1.1) follows from (9.3.1) at once (Wu 1981):

$$\boldsymbol{F} = \boldsymbol{e}_z\rho\Gamma\frac{dr}{dt} = \rho U\Gamma\boldsymbol{e}_z. \tag{9.3.3}$$

3. Relation with virtual-mass force and moment. The virtual-mass force and moment are generated by an accelerating body and can be calculated by solving an *acyclic* (*non-circulatory*) velocity potential ϕ_{ac} (Sect. 2.6.3). In viscous flow this effect comes from shearing process rather than outside it. To see this, let Δt be a characteristic time scale over which a continuously accelerating body changes from one motion state to another, somewhat like a computational time step. The no-slip condition will then create an acyclic shear layer adjacent to the body surface, of which the thickness is of $O(\sqrt{\nu\Delta t})$ (cf. Sect. 4.1.2), much thinner than a steady boundary layer. Such a super-thin layer can well be treated as an unsteady vortex sheet. It is this acyclic vortex sheet caused by body acceleration that is responsible for the virtual mass effect. As long as the fluid has viscosity (no matter how small), an accelerating body is always dressed in an attached and time-dependent acyclic vortex sheet as a part of its attached shear layers.

The above physical reasoning can be proven by impulse theory. Due to its linear decomposition feature, we can divide the vorticity field into an "*additional vorticity*" $\boldsymbol{\omega}_a$ termed by Lighthill (1986) that will be diffused and advected to the interior of the fluid, being either carried along with the body or shed into the wake, and an *acyclic* vortex sheet always attaching to the accelerating body surface. For the latter alone the external flow is irrotational, so it has strength

$$\boldsymbol{\gamma} = \hat{\boldsymbol{n}} \times (\nabla\phi_{\mathrm{ac}} - \boldsymbol{u}_B). \tag{9.3.4}$$

Thus in (9.3.1b) there is

$$\begin{aligned}\boldsymbol{I}_f &= \boldsymbol{I}_a + \frac{1}{n-1}\int_{\partial B} \boldsymbol{x} \times [\hat{\boldsymbol{n}} \times (\nabla\phi_{\mathrm{ac}} - \boldsymbol{u}_B)]dS \\ &= \boldsymbol{I}_a - \int_{\partial B} \phi_{\mathrm{ac}}\hat{\boldsymbol{n}}dS - \frac{1}{n-1}\int_{\partial B} \boldsymbol{x} \times (\hat{\boldsymbol{n}} \times \boldsymbol{u}_B)dS,\end{aligned}$$

where $\boldsymbol{I}_a$ is the vortical impulse contributed by $\boldsymbol{\omega}_a$ in V_f, and the $\phi_{\mathrm{ac}}\hat{\boldsymbol{n}}$-integral is precisely the non-circulatory potential impulse $\boldsymbol{I}_{\partial B}$ defined by (3.3.10a). By (3.3.16a), its rates of change is just the ϕ_{ac}-caused pressure force exerted by ∂B, namely the virtual-mass force. Accordingly, (9.3.1) and (9.3.2) are cast to a neater form (Lighthill 1986):

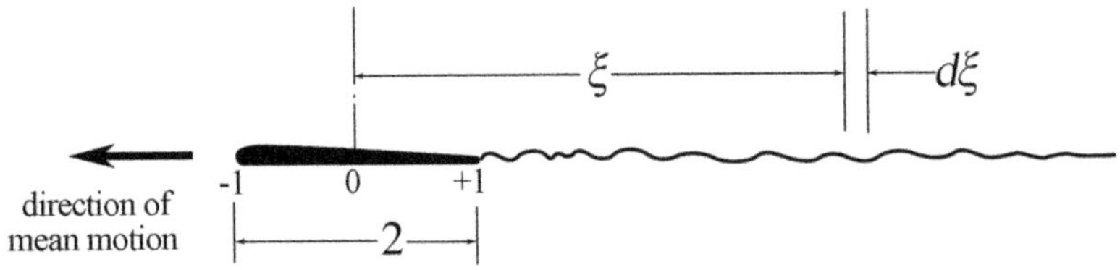

Fig. 9.6 Diagram for notation in the study of unsteady airfoil theory. Based on von Kármán and Sears (1938), Fig. 1

$$\boldsymbol{F} = -\rho \frac{d}{dt}(\boldsymbol{I}_a + \boldsymbol{I}_{\partial B}), \tag{9.3.5a}$$

$$\boldsymbol{M} = -\rho \frac{d}{dt}(\boldsymbol{L}_a + \boldsymbol{L}_{\partial B}). \tag{9.3.5b}$$

Later we shall see an ingenious specification of (9.3.5).

9.3.2 Force and Moment on Unsteady Thin Airfoil

Consider now a thin airfoil moves unsteadily with mean velocity $\boldsymbol{U} = -U\boldsymbol{e}_x$ in an otherwise still two-dimensional fluid at $Re \to \infty$. Recall that steady airfoil flow has no wake vortex sheet, and steady three-dimensional wing flow has a wake vortex sheet with $\gamma \times \overline{\boldsymbol{u}} = \boldsymbol{0}$, now the unsteady vortex sheet does exist and has nonzero Lamb vector. Hence, it affects airfoil force and moment not only by inducing a change of γ_b but also by a direct contribution. Yet the location and strength distribution of γ_w are to be solved in general. These issues add complexity to the problem, let alone all flow variables are time dependent.

The most widely known work on unsteady airfoil flow is due to Theodorson (1935), who constructed a linear theory for flow over an oscillatory airfoil by decomposing the force and moment into circulatory part and acyclic virtual-mass part. The theory has been the standard basis of classic wing flutter analysis. Shortly after, in a very elegant work von Kármán and Sears (1938, K-S for short) applied the impulse theory to achieve a general linear theory of thin airfoil in arbitrary non-uniform motion and thoroughly elucidated its underlying physics. Their approach is highlighted below.

As in steady wing theory, the airfoil is treated as a flat plate. As in lifting-line theory, the bound and free vortex sheets are assumed on the x-axis. Let the chord length be 2 and the origin of coordinates be at midpoint of the airfoil, we use $x \in [-1, 1]$ and $\xi \in (1, \infty)$ to denote the locations of bound sheet $\gamma(x)$ and free wake sheet $\gamma(\xi)$, respectively, see the sketch of Fig. 9.6.

By (9.3.1) and (9.3.2), the total lift and moment (about $x = 0$) exerted to the airfoil are given by (by convention, the moment is positive nose-down, i.e., counterclockwise)[5]

$$L = -\rho \frac{d}{dt} \int_{-1}^{1} \gamma(x) x dx - \rho \frac{d}{dt} \int_{1}^{\xi_m} \gamma(\xi) \xi d\xi, \tag{9.3.6a}$$

$$M = -\frac{\rho}{2} \frac{d}{dt} \int_{-1}^{1} \gamma(x) x^2 dx - \frac{\rho}{2} \frac{d}{dt} \int_{1}^{\xi_m} \gamma(\xi) \xi^2 d\xi, \tag{9.3.6b}$$

where $\xi_m = \xi(t)$ is the location of the starting vortex (the downstream end of the wake sheet) at time t. We take $\xi_m = \infty$ for simplicity.

Now, we write $\gamma(x) = \gamma_0(x) + \gamma_1(x)$, where $\gamma_0(x)$ is a bound vortex sheet that would be produced by the motion of the airfoil as if the wake had no effect, and $\gamma_1(x)$ is solely induced by the wake. Since in steady flow $\gamma(\xi)$ is absent, $\gamma_0(x)$ is actually a "*quasi-steady*" sheet, namely the flow at any instance is conceived to be frozen and keep steady. Thus, γ_0 is obtainable from steady airfoil theory. Its integral,

$$\int_{-1}^{1} \gamma_0(x) dx = \Gamma_0, \tag{9.3.7}$$

gives a lift $\rho U \Gamma_0$. Then, the sheet $\gamma_1(x)$ is solely induced by the unsteady wake sheet $\gamma(\xi)$. The total-circulation conservation theorem requires that

$$\Gamma_0 + \int_{-1}^{1} \gamma_1(x) dx + \int_{1}^{\infty} \gamma(\xi) d\xi = 0. \tag{9.3.8}$$

1. Bound vortex sheet induced by a wake point vortex. The integrals in (9.3.6) can be written as the sums of the moments of distributed discrete point vortices of strength $\Gamma_k = \gamma_k dx$. Consider first how a single vortex $\Gamma' = \gamma(\xi) d\xi$ at a wake point ξ induces a vortex sheet $\gamma_1(x)$ around the airfoil. This is done by mapping the flat plate on the complex z-plane to a unit circle on the complex z'-plane, with Γ' at η, see Fig. 9.7:

$$2z = z' + \frac{1}{z'}.$$

The no-through condition on the circle requires introducing an image vortex of circulation $-\Gamma'$ at $1/\eta$. This vortex couple does not change total circulation but induces a tangent velocity $v_{\theta 1}$ at the circle (positive in clockwise direction), which tends to infinity at the trailing edge. The Kutta condition requires adding a pure rotating flow (induced by a point vortex at the center of the circle) to cancel the infinity. This gives a tangent velocity v_θ that vanishes at $x = 1$. Then, in the vortex-induced circulation element $d\Gamma_1 = \gamma_1(x) dx$, γ_1 is found to be

[5]All variables here are functions of t. We save this independent variable below for neatness.

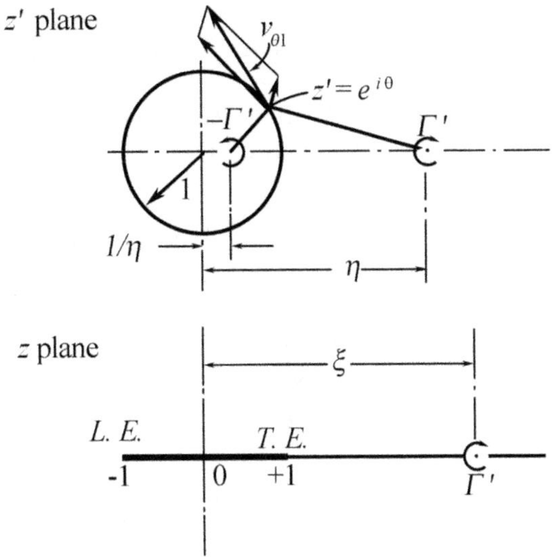

Fig. 9.7 Conformal mapping of the airfoil and a wake vortex. Based on von Kármán and Sears (1938), Fig. 2

$$\gamma_1(x) = \frac{1}{\pi}\frac{\Gamma'}{\xi - x}\sqrt{\frac{1-x}{1+x}}\sqrt{\frac{\xi+1}{\xi-1}}. \tag{9.3.9}$$

The integrals of the zeroth, first, and second moments of $\gamma_1(x)$ will yield the circulation Γ_1, y-momentum L_1, and angular momentum M_1, respectively. Note that Γ_1 equals the circulation of the point vortex for satisfying the Kutta condition; it is found to be

$$\Gamma_1 = \int_{-1}^{1} \gamma_1(x)dx = \int_1^\infty \left\{\sqrt{\frac{\xi+1}{\xi-1}} - 1\right\}\gamma(\xi)d\xi. \tag{9.3.10}$$

Then the total-circulation preservation (9.3.8) is cast to an integral equation for $\gamma(\xi)$, known as *the Wagner integral equation* (Wagner 1925):

$$\Gamma_0 + \int_1^\infty \sqrt{\frac{\xi+1}{\xi-1}}\gamma(\xi)\mathrm{d}\xi = 0. \tag{9.3.11}$$

2. Lift and moment. In (9.3.6a) the total impulse generated by the vortex sheets is

$$\begin{aligned} I &= \int_{-1}^{1} \gamma_0(x)x\mathrm{d}x + \int_{-1}^{1} \gamma_1(x)x\mathrm{d}x + \int_1^\infty \gamma(\xi)\xi\mathrm{d}\xi \\ &= \int_{-1}^{1} \gamma_0(x)x\mathrm{d}x + \int_1^\infty \gamma(\xi)\sqrt{\xi^2-1}\mathrm{d}\xi, \end{aligned} \tag{9.3.12}$$

where the second line follows by substituting (9.3.9) with $\Gamma' = \gamma(\xi)d\xi$ into the second term of the first line. Then the only unknown sheet strength is $\gamma(\xi)$ in the wake.

In calculating $-dI/dt$, to deal with possible discontinuity of the wake sheet (e.g., in gust-wing interaction), K-S turn to consider the derivative of a general integral

$$A = \int_1^\infty \gamma(\xi) f(\xi) d\xi,$$

where γ in the wake is stationary relative to the fluid. If X is the instantaneous distance of γ from the origin, it remains to be function of X. But after Δt the airfoil moves through a distance $U\Delta t$, there is $\xi = X + U\Delta t$. When $\gamma(X)$ is finite and if $f(1) = 0$, this trick leads to a simple formula

$$\frac{dA}{dt} = U \int_1^\infty \gamma(\xi) f'(\xi) d\xi. \tag{9.3.13}$$

Therefore, by the Wagner equation (9.3.11) we obtain the lift formula:

$$L = \rho U \Gamma_0 - \rho \frac{\mathrm{d}}{\mathrm{d}t} \int_{-1}^{1} \gamma_0(x) x \mathrm{d}x + \rho U \int_1^\infty \frac{\gamma(\xi)}{\sqrt{\xi^2 - 1}} \mathrm{d}\xi. \tag{9.3.14}$$

For steady flow (9.3.14) is reduced to (9.2.1). The moment about $x = 0$ can be similarly obtained. In summary, the lift and moment are the sums of three parts:

(a) Pseudo-steady force and moment

$$L_0 = \rho U \Gamma_0, \quad M_0 = \rho U \int_{-1}^{1} \gamma_0(x) x dx;$$

(b) Wake-induced bound-vortex force and moment

$$L_1 = -\rho \frac{d}{dt} \int_{-1}^{1} \gamma_0(x) x dx, \quad M_1 = -\frac{\rho}{2} \frac{d}{dt} \int_{-1}^{1} \gamma_0(x) \left(x^2 - \frac{1}{2} \right) dx;$$

(c) Free-wake caused force and moment

$$L_2 = \rho U \int_1^\infty \frac{\gamma(\xi) d\xi}{\sqrt{\xi^2 - 1}}, \quad M_2 = -\frac{1}{2} \rho U \int_1^\infty \frac{\gamma(\xi) d\xi}{\sqrt{\xi^2 - 1}}.$$

Observe that if the airfoil motion produces no circulation Γ_0 and hence $\gamma(\xi) = 0$ as well, then the lift will be solely from L_1, indicating that the wake-induced bound vortex sheet is precisely the virtual-mass effect due to the body acceleration. This assertion has been explicitly proved by K-S. Thus, in unsteady aerodynamics there remains only solving the Wagner integral equation (9.3.11) for $\gamma(\xi)$, which has to be done along with initial-boundary conditions for specific problems.

The most significant feature of the K-S theory is that of various formulations of unsteady aerodynamics at $Re \to \infty$ proposed in 1920s–1930s, the *Kárḿan-Sears triple-decomposition* is the only one that remains effective for fully nonlinear unsteady aerodynamic problems. This valuable feature has been confirmed by McCune and Tavares (1993), Wu (2007), and Hou et al. (2007), who applied the K-S formulation to hyper maneuvering delta wing and two-dimensional flexible and flapping wing, respectively. Both using vortex-sheet model at $Re \to \infty$ and solving the nonlinear evolution of free wake sheet numerically.

3. Oscillatory thin airfoil. Kármán and Sears (1938) have applied the above general linearized unsteady airfoil theory to several cases of practical importance. Here we outline one, where the wing performs periodic oscillatory motion. For clarity we recover dimensional variables with the airfoil chord length set as c.

Due to elasticity, a wing often has *bending* and *torsional* fluctuation modes. As a sectional profile, an airfoil often experiences two oscillation modes: *heaving* normal to the plate and *pitching* about a point thereon, making its normal velocity w take the forms

$$w_b(t) = A_0 U e^{i\nu t}, \quad w_p(x,t) = \Omega_0 x e^{i\nu t}, \tag{9.3.15}$$

respectively. Here, A_0 and Ω_0 are constants and ν is the circular frequency. As said before, the main calculation is only L_2 and M_2, in which there appears an important complex *Theordorson function* first introduced by Theodorson (1935) in his pioneering work on *flutter*:

$$C(\omega) \equiv \frac{K_1(i\omega)}{K_0(i\omega) + K_1(i\omega)}, \quad \omega \equiv \frac{\nu c}{2U}, \tag{9.3.16}$$

where K_0 and K_1 are the modified Bessel functions of the second kind with imaginary argument $i\omega$, and ω is the aerodynamic *reduced frequency* (not to be confused with vorticity) characterizing the air reaction to the wing oscillation. The Theodorson function $C(k)$ is essentially a transfer function, of which the real and imaginary parts account for attenuation of lift amplitude and phase lag in lift response, respectively. These two parts are plotted as functions of ω in Fig. 9.8, in the form of "vector diagram".

For the heaving mode, by the K-S triple decomposition, the three parts of force and moment are (see also Sears 1941):

$$L_0 = \pi\rho c U^2 A_0 e^{i\nu t}, \quad L_1 = \frac{i\omega}{2} L_0, \quad L_2 = -L_0[1 - C(\omega)], \tag{9.3.17a}$$

$$M_0 = \frac{1}{4} L_0 c, \quad M_1 = 0, \quad M_2 = \frac{1}{4} L_2 c, \tag{9.3.17b}$$

and hence

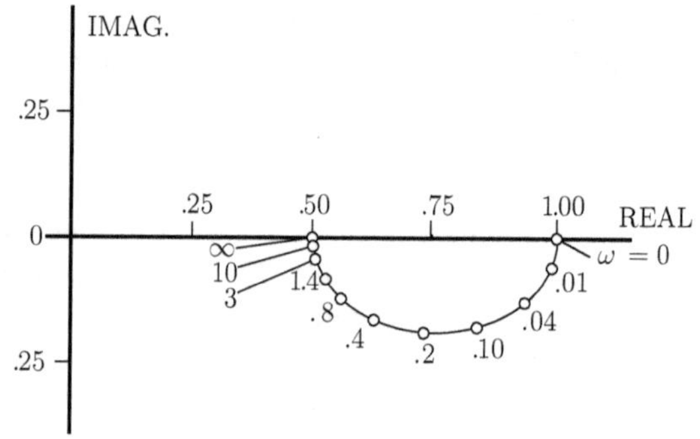

Fig. 9.8 Vector diagram of the real and imaginary parts of $C(\omega)$ as functions of ω. From Sears (1941)

$$L = L_0\left[C(\omega) + \frac{i\omega}{2}\right], \quad M = \frac{c}{4}L_0C(\omega). \tag{9.3.17c}$$

Similarly, for the pitching mode there is

$$L_0 = \frac{\pi}{4}\rho c^2 U\Omega_0 e^{i\nu t}, \quad L_1 = 0, \quad L_2 = -L_0[1 - C(\omega)], \tag{9.3.18a}$$

$$M_0 = 0, \quad M_1 = -\frac{i\omega}{16}L_0c, \quad M_2 = \frac{1}{4}L_2c, \tag{9.3.18b}$$

and hence[6]

$$L = L_0C(\omega), \quad M = \frac{L_0c}{4}\left[C(\omega) - \left(1 + \frac{i\omega}{4}\right)\right]. \tag{9.3.18c}$$

The physics of L_0, M_0 and L_1, M_1 in (9.3.17) and (9.3.18) are straightforward. For example, for a normal translational motion of the flat plate, L_1 is the product of the *acceleration* $\dot{w}$ (dot denotes time derivative) and the virtual mass $\rho\pi c^2/4$. Similarly, for pitching motion L_2 is the product of the angular acceleration $\dot{\Omega}$ and the virtual moment of inertia. In harmonic oscillation the acceleration appears as $i\nu$ times velocity, and hence L_2 is characterized by leading L_0 and M_0 by a phase angle 90°.

Figure 9.9 is a typical diagram for the lift of a heaving oscillating airfoil. It shows how L is composed of the three parts for a fixed ω. L_0 is in phase with velocity w and appears as a horizontal vector. L_1 is always directed vertically, leads w by 90°. In contrast, L_2 tends to reduce L_0 and cause it to lag behind w. Consequently, the total force has a phase angle φ.

Figure 9.10 shows the vector diagrams for the magnitudes and phase angles of total L and M at various ω, scaled by L_0 and M_0, where M_0 is replaced by $\pi\rho U^2 c\Omega_0 e^{i\nu t}$. Take the bending mode as example, as ω increases from zero (steady motion), the amplitude of L at first steadily decreases and its phase lags slightly behind w due to the effect of L_2. But with further increase of ω, L_1 becomes very large and dominates

[6]Here M is defined as positive if it makes the nose up (the moment vector along positive y-direction). In aerodynamics the practical convention is to define $M > 0$ if it makes the nose down. Then the sign of M in present formulas should be changed.

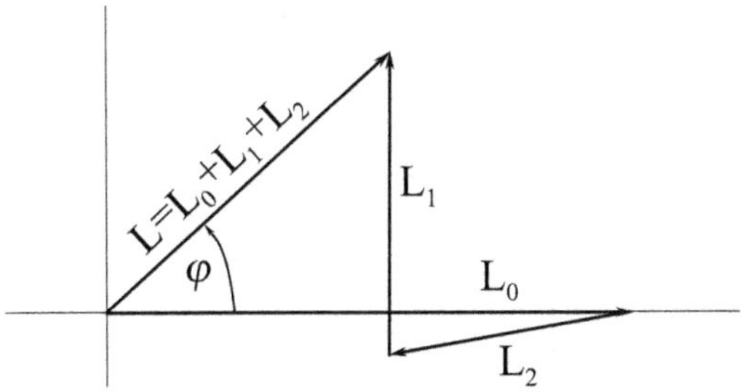

Fig. 9.9 Typical vector diagram for the lift of an oscillating airfoil. From Kármán and Sears (1938)

the total lift. In the limit $\omega \to \infty$ we have $L_0 + L_2 = L_0/2$ but $L_1 \to \infty$ that leads w by 90°. Similar discussion can be made for M/M_0 as well as L and M of pitching mode.

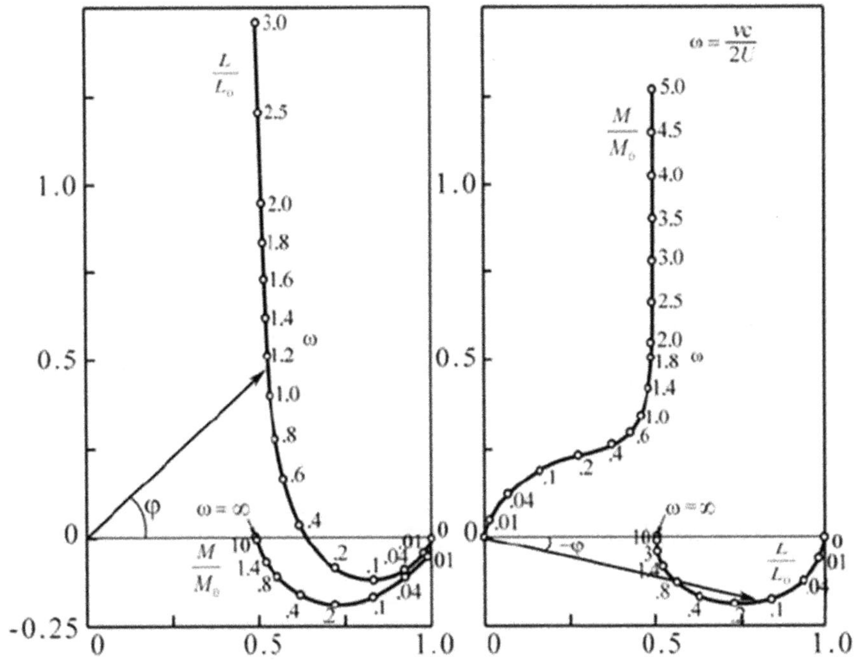

Fig. 9.10 Vector diagrams for L and M of bending mode (*left*) and pitching mode (*right*) of oscillating airfoils. From Kármán and Sears (1938)

9.4 A General Formulation of Vortical-Force Theory

Having reviewed the classic theories of vortical aerodynamics, we now make a survey of general nonlinear theory for vortical aerodynamic force and moment, in which the vortex-force theory and impulse theory are unified into a single formulation. We focus on a typical external-flow problem: a material body of volume B moves arbitrarily in a viscous fluid. The body may have arbitrarily deformable material boundary ∂B which has specified velocity distribution $\boldsymbol{u}_B(\boldsymbol{x}, t)$, as in cases of fish swimming and insect flight in external biofluiddynamics, nonlinear fluid-solid coupling, and flow control by flexible walls, etc. The fluid volume V_f is bounded internally by ∂B and externally by a control surface Σ. The latter may have arbitrary velocity $\boldsymbol{v}(\boldsymbol{x}, t)$ or extend to infinity where the fluid is at rest or in uniform translation. The flow domain is sketched in Fig. 1.12.

9.4.1 Pressure Removal

As exemplified in Sect. 9.1, the first step toward vortical-force theory is to eliminate pressure from standard force formula (1.3.4b),

$$\boldsymbol{F} = -\int_{V_f} \rho \boldsymbol{a} dV + \int_{\Sigma} (-p\boldsymbol{n} + \boldsymbol{\tau}) dS, \tag{9.4.1}$$

and transform its acceleration integral. This can be done by different ways. Similar to the derivation of vortical impulse $\boldsymbol{I}$ and potential impulse $\boldsymbol{I}_\phi$ in Sect. 3.3.3, the way we follow now is to properly use a set of vectorial integral identities listed in Appendix A.2.3, which we call *derivative moment transformation* (DMT for short).[7] The DMT identities most frequently used below for any vector field $\boldsymbol{f}$ and scalar field ϕ which are differentiable in any n-dimensional domain $\mathcal{D}$, $n = 2, 3$, are (A.2.23) and (A.2.25).

Let $n = 2, 3$ be the spatial dimension and $k = n - 1$, by (A.2.23) there is

$$\begin{aligned} \boldsymbol{F} = &-\frac{\rho}{k} \int_{V_f} \boldsymbol{x} \times (\nabla \times \boldsymbol{a}) dV + \frac{\rho}{k} \int_{\partial V_f} \boldsymbol{x} \times (\boldsymbol{n} \times \boldsymbol{a}) dS \\ &+ \int_{\Sigma} (-p\boldsymbol{n} + \boldsymbol{\tau}) dS, \quad \boldsymbol{\tau} = \mu \boldsymbol{\omega} \times \boldsymbol{n}. \end{aligned}$$

Here, the integral over ∂V_f can be split to integrals over ∂B and Σ, then the latter can cancel the Σ-integral of $-p\boldsymbol{n}$ by the incompressible Navier-Stokes equation (2.6.2b) and (A.2.25), yielding

---

[7] For a different approach to similar goal see Sect. 11.2 of Wu et al. (2006) where it is called *projection method*. In literature it is also called *auxiliary-function method* or *force element method*.

$$\boldsymbol{F} = -\frac{\rho}{k}\int_{V_f} \boldsymbol{x} \times (\nabla \times \boldsymbol{a}) dV + \boldsymbol{F}_\Sigma + \boldsymbol{F}_B, \tag{9.4.2}$$

where

$$\boldsymbol{F}_\Sigma = \int_\Sigma \boldsymbol{\tau} dS - \frac{\mu}{k}\int_\Sigma \boldsymbol{x} \times [\boldsymbol{n} \times (\nabla \times \boldsymbol{\omega})] dS, \tag{9.4.3a}$$

$$\boldsymbol{F}_B = \frac{\rho}{k}\int_{\partial B} \boldsymbol{x} \times (\boldsymbol{n} \times \boldsymbol{a}) dS \tag{9.4.3b}$$

are an explicitly viscous contribution of finite Σ to the force and the contribution of body acceleration, respectively. The key volume integrand in (9.4.2) is, of course,

$$\nabla \times \boldsymbol{a} = \boldsymbol{\omega}_{,t} + \nabla \times (\boldsymbol{\omega} \times \boldsymbol{u}) = \nu \nabla^2 \boldsymbol{\omega}. \tag{9.4.4}$$

With (9.4.4), it is evident that all integrals of (9.4.2) contain vorticity, and hence converge very fast as Σ recedes to infinity. The convergence problem of momentum integral of unbounded flow as discussed in Sect. 3.3.3 is therefore completely avoided.

Equation (9.4.4) gives two alternative dynamic expressions of $\nabla \times \boldsymbol{a}$ in terms of vorticity advection and diffusion, respectively. This has led to different forms of vortical-force formulas, which we outline below.

9.4.2 Advection Form of Vortical Force

1. General formulas. Substituting the first expression of (9.4.4) into (9.4.2) and using (A.2.23), we obtain one of the general formulas of the *advection form* of viscous vorticity aerodynamics:

$$\begin{aligned}\boldsymbol{F} = \rho\int_{V_f} \boldsymbol{u} \times \boldsymbol{\omega} dV + \frac{\rho}{k}\int_{\partial V_f} \boldsymbol{x} \times [\boldsymbol{n} \times (\boldsymbol{u} \times \boldsymbol{\omega})] dS + \boldsymbol{F}_\Sigma \\ - \frac{\rho}{k}\int_{V_f} \boldsymbol{x} \times \boldsymbol{\omega}_{,t} dV + \boldsymbol{F}_B.\end{aligned} \tag{9.4.5a}$$

Note that inner-boundary integrals over ∂B can be removed by continuing the velocity field into the body B. Denote $\boldsymbol{l} = \boldsymbol{\omega} \times \boldsymbol{u}$ in (9.4.5a) and split the boundary integral of $\boldsymbol{x} \times (\boldsymbol{n} \times \boldsymbol{l})$ into those over ∂B and Σ. The sum of the former and $\boldsymbol{F}_B$ can be cast kinematically to volume integrals over B by using (A.2.23):

$$-\frac{\rho}{k}\int_{\partial B} \boldsymbol{x} \times [\hat{\boldsymbol{n}} \times (\boldsymbol{a} - \boldsymbol{l})] dS = \rho\int_B (\boldsymbol{a} - \boldsymbol{l}) dV - \frac{\rho}{k}\int_B \boldsymbol{x} \times \boldsymbol{\omega}_{,t} dV,$$

where $\hat{\boldsymbol{n}} = -\boldsymbol{n}$ is the outward normal vector from body surface. Putting this back to (9.4.5a) leads to an alternative form of it:

$$\begin{aligned} \boldsymbol{F} = \rho \int_V \boldsymbol{u} \times \boldsymbol{\omega} dV + \frac{\rho}{k} \int_\Sigma \boldsymbol{x} \times [\boldsymbol{n} \times (\boldsymbol{u} \times \boldsymbol{\omega})] dS + \boldsymbol{F}_\Sigma \\ - \frac{\rho}{k} \int_V \boldsymbol{x} \times \boldsymbol{\omega}_{,t} dV + \int_B \boldsymbol{a} dV, \end{aligned} \tag{9.4.5b}$$

where $V = V_f + B$. Recall that generically $\boldsymbol{\omega}$ and $\boldsymbol{l}$ are discontinuous across the body surface.

The moment formula (1.3.7b) can be similarly transformed (Problem 9.7). Note that in (9.4.5) the vorticity appears at its lowest (zeroth) order of spatial derivative, and hence in the volume integrals $\boldsymbol{u} \times \boldsymbol{\omega}$ and $\boldsymbol{\omega}_{,t}$ have widest possible distribution in the flow field. Thus, the advection form is a powerful tool for identifying the contribution of every individual vortical structures in V_f to the total force.

Since most of practical flows are turbulent at $Re \gg 1$, one often needs to take the Reynolds average of (9.4.5). Interestingly, Marongiu and Tognaccini (2010) have shown that the formula keeps exactly the same as (9.4.5) if all quantities therein are simply replaced by their mean values (denoted by overline):

$$\begin{aligned} \overline{\boldsymbol{F}} = \rho \int_{V_f} \overline{\boldsymbol{u}} \times \overline{\boldsymbol{\omega}} dV + \frac{\rho}{k} \int_{\partial V_f} \boldsymbol{x} \times [\boldsymbol{n} \times (\overline{\boldsymbol{u}} \times \overline{\boldsymbol{\omega}})] dS + \overline{\boldsymbol{F}}_\Sigma \\ - \frac{\rho}{k} \int_{V_f} \boldsymbol{x} \times \overline{\boldsymbol{\omega}}_{,t} dV + \boldsymbol{F}_B, \end{aligned} \tag{9.4.6}$$

where $\overline{\boldsymbol{F}}_\Sigma$ can often be neglected. The authors have used RANS formula (9.4.6) to analyze two-dimensional unsteady mean turbulent flow over a NACA-0012 airfoil performing pitching oscillation at $Re = 10^6 \sim 10^7$, with small and stall angles of attack that belong to unsteady attached flow and separated flow (dynamic stall), respectively. The results are in good agreement with that by standard surface-stress integral. But (9.4.6) permits examining the behaviors and contributions of the volume integral and boundary integral (wake integral) of $\overline{\boldsymbol{u}} \times \overline{\boldsymbol{\omega}}$ as well as the moment of $\overline{\boldsymbol{\omega}}_{,t}$ separately, and thereby to reach more detailed diagnosis.

2. Constituents of steady aerodynamic force. Equation (9.4.5a) or (9.4.5b) is the central result for the general vortical forces. The key quantity therein is the Lamb-vector integral, a *transverse* force as it locally does not do work. To understand its properties, we assume a uniform incoming flow $\boldsymbol{U} = U\boldsymbol{e}_x$ such that the flow in V is steady. Then the second line of (9.4.5b) vanishes. If $Re \gg 1$ and Σ is sufficiently away from the body, $\boldsymbol{F}_\Sigma$ is negligible. Thus (9.4.5) is reduced to

$$\boldsymbol{F} = \boldsymbol{F}_{\text{vf}} + \frac{\rho}{n-1} \int_\Sigma \boldsymbol{x} \times [\boldsymbol{n} \times (\boldsymbol{u} \times \boldsymbol{\omega})] dS \tag{9.4.7a}$$

$$= \boldsymbol{F}_{\text{vf}} + \int_\Sigma (P_\infty - P) \boldsymbol{n} dS \quad \text{for } Re \gg 1, \tag{9.4.7b}$$

where $\boldsymbol{F}_{\text{vf}}$ is the vortex force defined by (9.1.12), and the second expression comes from comparing with (9.1.10), which implies

$$\int_{\Sigma} (P_\infty - P) \boldsymbol{n} dS = \frac{\rho}{n-1} \int_{\Sigma} \boldsymbol{x} \times [\boldsymbol{n} \times (\boldsymbol{u} \times \boldsymbol{\omega})] dS, \tag{9.4.8}$$

as can be directly proved for the case of $Re \gg 1$ by using DMT (Problem 9.9). Equation (9.4.7a) indicates that *for steady flow at large Reynolds numbers the aerodynamic force is solely from the Lamb vector.*

As explained in Sect. 9.1.2, the Σ-integrals of both (9.4.7a) and (9.4.7b) are nonzero only on a wake plane W which can be chosen to have $\boldsymbol{n} = \boldsymbol{e}_x$. We see from (9.4.8) that, for steady flow, the total-pressure deficit in the wake is due to the Lamb vector that sheds into wake from wing's boundary layers and separated shear layer. This total-pressure deficit is a viscous effect and disappears if $\mu \to 0$. On the other hand, recall that, provided the wake plane W is perpendicular to the incoming flow, for $n = 3$ the vortex force $\boldsymbol{F}_{\text{vf}}$ provides a lift and an induced drag, see (9.1.13); while for $n = 2$ it provides a pure lift, see (9.1.14a). Therefore, we can now make a general classification of the force constituents for a steady and viscous incompressible aerodynamic flow at large Reynolds numbers:

(1) The vertical component of the vortex force is responsible for the entire lift, which exists even in the limit $\mu \to 0$;

(2) The streamwise component of the vortex force is responsible for the entire induced drag, which exists in three dimensions only, even in the limit $\mu \to 0$;

(3) The wake integral of Lamb-vector moment, the second term of (9.4.7a) or (9.4.7b), which is responsible for the form drag that exists only if $\mu > 0$.

Note that every component of $\boldsymbol{F}$ must be independent of the choice of V and Σ, hence so must be L and $D_{\text{in}} + D_{\text{form}}$; but the partition of D_{in} and D_{form} may depend on the location x_W of the wake plane W (Marongiu et al. 2013).

The above classification has been numerically tested by Marongiu et al. (2013) for an elliptic wing, who found it in excellent agreement with the corresponding on-wall stress integrals and theoretical result of $D_{\text{in}} = C_L^2/\pi Æ$ (see (9.2.24) with $\delta = 0$) for induced drag.

We have seen that a typical steady free vortex-layer separated flow occurs on slender delta wings at large angles of attack, as sketched in Figs. 5.11 and 8.21. It has been a common concept that the leading-edge vortices (LEVs) enhance the lift at large angles of attack, but so far this picture can only be supported by approximate theory (see (9.5.7) in Problem 9.10). In contrast, the accurate formula (9.4.7a) asserts that high peaks of Lamb vector dominate the force. How the two understandings are related has been studied numerically by Yang et al. (2007). The authors found that, although Lamb-vector peaks do appear in the core of LEVs, they are in pair with opposite signs and thus cancel each other after integration. Instead, the dominating Lamb-vector peaks are in the strong attached boundary layers and initial segments of free shear layers separated from leading edges. The two physical pictures are unified, though, since the strong attached upper-surface boundary layer, before separation, are the result of induction of LEVs.

3. Vortical impulse theory for infinite and finite domains. In contrast to the above-discussed steady flow in $V \subset V_{\text{st}}$, assume now V extends to the entire space V_∞ and contains all vorticity including starting vortex system. Then the flow is inherently unsteady (Sect. 2.4.3) and the Lamb-vector integral over V_∞ vanishes (Sect. 3.3.2). Thus the first line of (9.4.5b) disappears, and the force formula (9.3.1) by time rate of impulse is recovered. Therefore, classic vortex-force theory and infinite-domain impulse theory are two mutually exclusive special cases of (9.4.5).

Now, we may transform the local unsteady term in (9.4.5b) by the *Reynolds transport theorem* (1.1.39a). Then since

$$\frac{d}{dt}\int_V \boldsymbol{x}\times\boldsymbol{\omega}dV = \int_V \boldsymbol{x}\times\boldsymbol{\omega}_{,t}dV + \int_\Sigma \boldsymbol{x}\times\boldsymbol{\omega}u_n dS,$$
$$\boldsymbol{x}\times(\boldsymbol{n}\times\boldsymbol{l}) = \boldsymbol{x}\times(\boldsymbol{\omega}u_n - \boldsymbol{u}\omega_n),$$

we obtain a finite-domain extension of the impulse theory:

$$\boldsymbol{F} = -\frac{d\boldsymbol{I}_V}{dt} + \int_V \boldsymbol{u}\times\boldsymbol{\omega}dV + \frac{1}{k}\int_\Sigma \boldsymbol{x}\times\boldsymbol{u}\omega_n dS + \int_B \boldsymbol{a}dV + \boldsymbol{F}_\Sigma, \tag{9.4.9}$$

where $\boldsymbol{F}_\Sigma$ is given by (9.4.3a). This formula is convenient since practical computed or measured flow data are always confined in a finite domain.

Caution is required, however, to understand the simultaneous appearance of time rate of impulse and vortex force in (9.4.9). If we simply substitute (3.3.19) directly into (9.4.9), all volume integrals would be canceled, and the remaining boundary integrals can be regrouped to return to a *boundary form* of force formulas to be discussed in Sect. 9.4.4, thus we could go to nowhere but merely a check of algebra. But we have stressed that in impulse theory d/dt has to be kept outside the integral. More importantly, the Lamb-vector integral over V is exactly a negative Lamb-vector integral over $V_{\text{out}} = V_\infty - V$ occupied by the fluid domain out of volume V. Thus, the vortex force in (9.4.9) represents the contribution of all vortex systems outside V. In particular, if $\mathcal{V}_f$ encloses a *compact vortex system* with $\boldsymbol{\omega} = \boldsymbol{0}$ on ∂V, then as seen in Sect. 3.3.3 we may write $\boldsymbol{u} = \boldsymbol{v} + \nabla\phi_e$ with $\boldsymbol{v}$ and $\nabla\phi_e$ being the velocities induced by $\boldsymbol{\omega}$ in V and V_{out}, respectively, and (9.4.9) yields

$$\boldsymbol{F} = -\frac{d\boldsymbol{I}_V}{dt} + \int_V \nabla\phi_e\times\boldsymbol{\omega}dV + \int_B \boldsymbol{a}dV \tag{9.4.10a}$$
$$= -\frac{d\boldsymbol{I}_V}{dt} - \int_{V_{\text{out}}} \boldsymbol{u}\times\boldsymbol{\omega}dV + \int_B \boldsymbol{a}dV. \tag{9.4.10b}$$

Impulse theory has been found to be a powerful tool in studies of biological locomotion. Sun and Wu (2004) have used Navier-Stokes computation to demonstrate that a closed vortex loop is formed in each stroke of flapping wing, so that they used (3.3.12) to infer the rate of change of the vector area spanned by the vortex loop and then used (9.3.1) to calculate lift and drag during the first stroke. The result is in very good agreement with standard surface-stress integral (1.3.4a).

In a theoretical-numerical study of viscous and unsteady wake generated by flapping plates, Li and Lu (2012) have used the finite-domain impulse theory (9.4.9), especially (9.4.10), to find that, in relatively slow forward motion, the vortical wake of flapping wings consists of two rows of almost isolated vortex rings, and only the pair of rings nearest to the trailing edge of the wing has nonzero $d\boldsymbol{I}/dt$. This finding motivated the authors to propose a very simple way to estimate the thrust and propulsion efficiency with good accuracy.

9.4.3 Diffusion Form of Vortical Force

While the above advection form of vortical-force theory is in the mainstream of aerodynamic analysis, the vortical-force theory has still two more forms, the *diffusion form* and *boundary form*. These are unique to vorticity dynamics and have special interesting characteristics and applications.

Substituting the second expression of (9.4.4) into (9.4.2), we obtain the *diffusion form* of viscous vortical aerodynamics (Wu and Wu 1993, 1996; Wu et al. 2006, 2007):

$$\boldsymbol{F} = -\frac{\mu}{n-1}\int_{V_f} \boldsymbol{x} \times \nabla^2 \boldsymbol{\omega} dV + \boldsymbol{F}_B + \boldsymbol{F}_\Sigma, \tag{9.4.11}$$

where $\boldsymbol{F}_B$ and $\boldsymbol{F}_\Sigma$ are given by (9.4.3b) and (9.4.3a). This form is unique to vorticity dynamics and shows that for any viscous steady or unsteady flow $\boldsymbol{F}$ is solely expressible by the moment of vorticity diffusion plus vorticity-related boundary integrals. It also holds for compressible flow if μ is constant. Since all terms in (9.4.11) but the known $\boldsymbol{F}_B$ depend only linearly on $\boldsymbol{\omega}$, its RANS form also remains unchanged.

Since the vorticity appears at the second-order of spatial derivative, the volume integrand must be highly concentrated to those vortical structures in narrow near-field regions. Indeed, in a numerical simulation of Wu et al. (2007) for two-dimensional unsteady and separated circular-cylinder flow, the volume integral of (9.4.11), denoted by $\boldsymbol{F}_f$, was found to provide about 90 % of the total force, and its integrand is dominated by near-wall vortical structures, far before the Kármán vortex street is formed. Its peak value is of $O(Re^{1/2})$ and concentrates in a thin near-wall layer of thickness $O(Re^{-1/2})$ with very strong shear. Away from this narrow peak region the same integrand has an $O(1)$ distribution over a near-wake (back-flow) region of area of $O(1)$. This observation motivated Fiabane et al. (2011) to study the respective Reynolds-number dependence of the integrand of $\boldsymbol{F}_f$. They found that, despite a constant difference, the time-averaged drag due to thin shear layer of $O(Re^{1/2})$ and wall-friction have almost the same Re-dependence, while that due to back-flow structures and wall-pressure have nearly the same dependence of wake vortex bubble length. This important discovery suggests for the first time that the two kinds of near-field vortical structures are directly responsible for the on-wall friction and pressure, respectively.

9.4.4 Boundary Form of Vortical Force

1. General formulas. The aforementioned strong correlation between on-wall stresses and near-field vortical structures can be further traced by the boundary form of vortical aerodynamics, which expresses the force and moment by the boundary vorticity flux (BVF), of which the general formulas are given by (3.4.24). First, $\boldsymbol{F}_B$ and $\boldsymbol{F}_\Sigma$ in (9.4.11) can be written as

$$\boldsymbol{F}_B = \frac{1}{n-1}\int_{\partial B} \rho \boldsymbol{x} \times \boldsymbol{\sigma}_a dS \tag{9.4.12a}$$

$$\boldsymbol{F}_\Sigma = \frac{1}{n-1}\int_{\Sigma} (\rho \boldsymbol{x} \times \boldsymbol{\sigma} + \boldsymbol{\tau}) dS. \tag{9.4.12b}$$

Next, owing to identity

$$-\boldsymbol{x} \times \nabla^2 \boldsymbol{\omega} = \nabla \cdot (\nabla \boldsymbol{\omega} \times \boldsymbol{x}) + \nabla \times \boldsymbol{\omega},$$

$\boldsymbol{F}_f$ can be integrated to a boundary integral:

$$\boldsymbol{F}_f = -\frac{1}{n-1}\int_{\partial V_f} (\rho \boldsymbol{x} \times \boldsymbol{\sigma} + \boldsymbol{\tau}) dS. \tag{9.4.13}$$

But since $\partial V_f = \partial B + \Sigma$, by (9.4.12b) there is

$$\boldsymbol{F}_f + \boldsymbol{F}_\Sigma = -\frac{1}{n-1}\int_{\partial B} (\rho \boldsymbol{x} \times \boldsymbol{\sigma} + \boldsymbol{\tau}) dS. \tag{9.4.14}$$

Thirdly, for $n = 3$ only, by (A.2.26) there is

$$-\int_{\partial B} \boldsymbol{\tau} dS = -\int_{\partial B} \rho \boldsymbol{x} \times \boldsymbol{\sigma}_{\text{vis}} dS,$$

in terms of the BVF caused by viscous effect. Thus, by (9.4.14) and (9.4.12a), we can use (3.4.24) to recognize that $\boldsymbol{F}$ can be solely expressed by two *stress-related* constituents of BVF (Wu 1987; Wu et al. 2006):

$$\boldsymbol{F} = -\rho \int_{\partial B} \boldsymbol{x} \times \left(\frac{1}{2}\boldsymbol{\sigma}_p + \boldsymbol{\sigma}_{\text{vis}} \right) dS \quad \text{for } n = 3. \tag{9.4.15a}$$

In two-dimensional flow on the (x, z)-plane with $\boldsymbol{\omega} = \omega \boldsymbol{e}_y$ and $\boldsymbol{\sigma}_{\text{vis}} = \boldsymbol{0}$, the integral of $\boldsymbol{\tau}$ cannot be expressed by the moment of any constituents of BVF. Rather, it can be expressed by a moment of *tangent* vorticity diffusion, such that

$$\boldsymbol{F} = \rho \oint_{\partial B} \left(-\boldsymbol{x} \times \boldsymbol{e}_y \sigma_p + \boldsymbol{x} \nu \frac{\partial \omega}{\partial s} \right) ds \quad \text{for } n = 2, \tag{9.4.15b}$$

where s is the arclength of the boundary loop that increases in counter-clockwise direction.[8] Equation (9.4.15) is the main formula of *boundary form*, which is also unique to vorticity dynamics and can be directly extended to compressible flow by replacing $\rho\boldsymbol{\sigma}_p$ by $\rho\boldsymbol{\sigma}_\Pi = \boldsymbol{n} \times \nabla\Pi$.

As stressed before, the wall stresses $p\boldsymbol{n}$ and $\boldsymbol{\tau}$ appearing in (1.3.4a) are not directly related to localized structures. They are spatial-temporal accumulated effects of the BVF. But in turn the BVF is mainly caused by on-wall pressure gradient via no-slip condition. In this way, the local vortical structures behind the inherent physical chain between standard formulas (1.3.4a) and (1.3.4b) are clarified, and the correlation between on-wall stresses and near-field vortical structures found by Fiabane et al. (2011) is rationally explained.

So far we have seen three equivalent forms of vortical-force formulas, the advection form, diffusion form, and boundary form. They are of course equivalent in that for the same flow they all give the same forces. But each of them captures a specific stage of vorticity evolution. As one goes from advection form to diffusion form, one traces from vortical structures in the entire flow field to the upstream near-field vorticity; and as one goes from diffusion form to boundary form, one further traces the near-field vorticity to the vorticity creation at the body surface.

2. BVF-based optimal airfoil design. Since the boundary form involves only on-wall variables, it is especially useful in on-wall flow diagnosis and configuration optimization of various external and internal complex flow problems of practical interest. For many successful examples, the reader is referred to reviews of Wu et al. (2006, 2009, 2010). Below we just give an elementary example of optimal design problem (Wu et al. 2006) to show some unique feature in applying the boundary form.

Consider a uniform incoming flow $\boldsymbol{U} = U\boldsymbol{e}_x$ over an airfoil contour C on an (x, z)-plane, and assume the flow is attached and steady. For airfoil design at $Re \gg 1$ we may ignore the skin friction, so (9.4.15b) gives

$$L = -\rho \oint_C x\sigma_p ds, \quad D = \rho \oint_C z\sigma_p ds. \tag{9.4.16}$$

The single-valueness of p over C requires

$$\oint_C \sigma_p ds = -\oint_C \frac{\partial p}{\partial s} ds = 0. \tag{9.4.17}$$

We know that on the upper surface and right downstream of front stagnation point there is a strong favorable pressure gradient, with $\sigma_p > 0$ and $|\sigma_p| \gg 1$ there. As x increases, σ_p becomes milder and then changes sign. The sign distribution of σ_p at

[8]Equation (9.4.15a) and (9.4.15b) can also be derived by directly transforming the pressure force in (1.3.4a) by DMT identity (A.2.25), or by shrinking the fluid volume V_f in (9.4.5) to zero, leaving only boundary integrals with Σ coinciding ∂B but with opposite normal-vector direction.

lower surface can also be estimated by the same physical consideration along with constraint (9.4.17). This leads to Fig. 9.11a.

Now, set for convenience the origin of (x, z)-coordinates at mid-chord point. Then (9.4.16) provides an *extra criterion* to flow diagnosis: both $\sigma_p > 0$ for $x < 0$ and $\sigma_p < 0$ for $x > 0$ are favorable; and to gain larger lift these σ_p's should peak at as larger $|x|$ as possible, and vise versa for unfavorable σ_p. This implies a better airfoil contour shown in Fig. 9.11b.

In an effort to delay stall angle of attack and enhance maximum lift coefficient of a helicopter rotor foil VR-12 (Zhu 2000), the above unique extra criterion plays a key role. After using (9.4.16) to optimize the σ_p distribution, a new airfoil marked by F3-VR-12 is obtained; the lift, drag, and moment coefficients of the two airfoils computed by a N-S solver are compared in Fig. 9.12.

9.5 Problems for Chapter 9

9.1. Consider a rectangular control surface Σ enclosing a two-dimensional airfoil, of which the front face S_0 is at $x = 0$ with $\boldsymbol{n} = -\boldsymbol{e}_x$ where $\boldsymbol{u} = U\boldsymbol{e}_x$, rear face (a wake plane) W at $x = L \geq c$ with $\boldsymbol{n} = \boldsymbol{e}_x$, upper side face S_u at $z = z_u$ with $\boldsymbol{n} = \boldsymbol{e}_z$, and lower side face at $z = z_l$ with $\boldsymbol{n} = -\boldsymbol{e}_z$. Denote the jump of the flow quantities at upper and lower faces by $[\![\cdot]\!]$, and let c be the chord length of the airfoil so that $\boldsymbol{u}^* = \boldsymbol{u}/U$ and $\boldsymbol{x}^* = \boldsymbol{x}/c$ are non-dimensional velocity and coordinates.

(1) Prove that, as an *identity* in two dimensions, the air downwash associated with lifting the airfoil is related to Γ by

$$\Gamma^* = \frac{\Gamma}{Uc} = \frac{1}{2}\int_0^{L/c} [\![u^{*2} - w^{*2}]\!] dx^* - \int_W u^* w^* dz^*. \tag{9.5.1}$$

Discuss the physical implication of each integral on the right of (9.5.1). Point out the dominant mechanism thereof.

(2) Observe that this identity is actually the z-component of a vector identity, which also has an x-component identity. What is that?

(3) Describe how the downwash is provided by the vortex system generated by a finite-span wing. Can you obtain an identity similar to (9.5.1)?

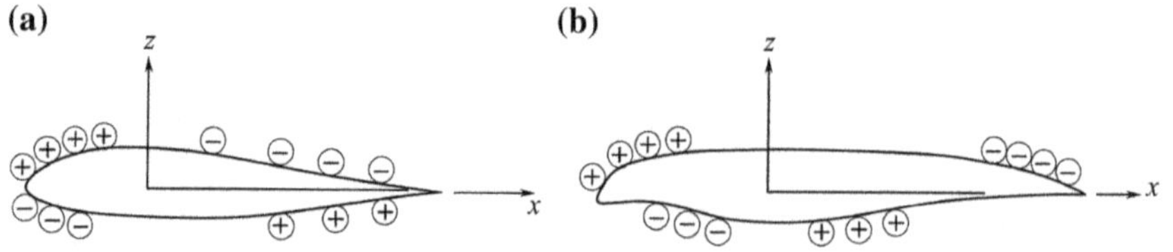

Fig. 9.11 **a** A qualitative estimate of the sign distribution of σ_p on an airfoil contour. **b** A better airfoil design with larger lift based on (9.4.16)

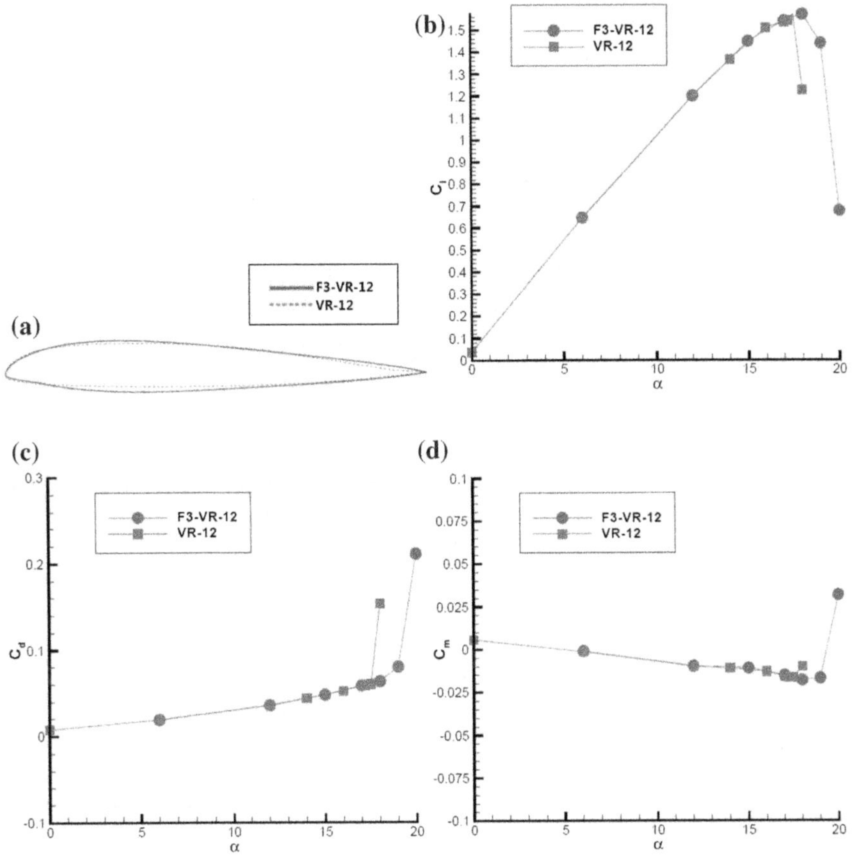

Fig. 9.12 A BVF-based optimal design of helicopter rotor blade airfoil for improving the performance. **a** Airfoil shapes. **b–d** C_l, C_d, and C_m. *Squares* original VR-12 airfoil; *circles* optimal-designed airfoil F3-VR-12. From Zhu (2000)

9.2. Use Fig. 9.3 to show:

(1) the velocity $dv_n(x)$ normal to the camber induced by a vortex-sheet element $\gamma(\xi)ds$ at ξ is

$$dv_n = -\frac{\gamma(\xi)ds}{2\pi r}, \quad v_n \simeq -\frac{1}{2\pi}\int_0^c \frac{\gamma(\xi)d\xi}{x-\xi},$$

where $r \simeq x - \xi$ and $ds \simeq d\xi$;

(2) on the chamber line of a thin airfoil there is

$$U_n \simeq U\left(\alpha - \frac{dz}{dx}\right). \tag{9.5.2}$$

Then derive (9.2.9).

9.3. Give a detailed derivation of (9.2.13). In so doing (and in the next problem) you may need trigonometry identities

$$\int_0^\pi \frac{\cos n\theta d\theta}{\cos\theta - \cos\phi} = \pi \frac{\sin n\phi}{\sin\phi}, \quad n = 0, 1, 2, ..., \tag{9.5.3a}$$

$$\int_0^\pi \frac{\sin n\theta \sin\theta d\theta}{\cos\theta - \cos\phi} = -\pi\cos n\phi, \quad n = 1, 2, ... \tag{9.5.3b}$$

9.4. Show that in terms of Glauert's transformation the induced angle of attack is given by

$$\alpha_i(\theta) = -\frac{1}{2b}\sum_{n=1}^{\infty} nA_n \frac{\sin n\theta}{\sin\theta}. \tag{9.5.4}$$

Then derive (9.2.24).

9.5. According to Kármán-Sears theory (Sect. 9.3.2) and its later nonlinear extension, for a general unsteady aerodynamic force we have a triple decomposition (not necessarily confined to the case with $Re \to \infty$)

$$\boldsymbol{F} = \boldsymbol{F}_0 + \boldsymbol{F}_1 + \boldsymbol{F}_2.$$

Discuss how many constituents appear in (1) two-dimensional steady airfoil flow, (2) three-dimensional steady wing flow, and (3) three-dimensional unsteady wing flow. Explain why and how.

9.6. Complete the derivation of (9.4.5).

9.7. Prove that, as the counterpart of the general force formula (9.4.5), for the total moment there is (neglect viscous terms on Σ)

$$\boldsymbol{M} = \rho\int_V \boldsymbol{x}\times\boldsymbol{l}dV + \frac{\rho}{2}\int_\Sigma x^2\boldsymbol{n}\times\boldsymbol{l}dS$$
$$+\rho\int_V \frac{1}{2}x^2\boldsymbol{\omega}_{,t}dV + \rho\int_B \boldsymbol{x}\times\boldsymbol{a}_B dV. \tag{9.5.5}$$

9.8. The Kutta-Joukowski lift formula $L = \rho U\Gamma$ can be proved by various different approaches. Except those earlier controversial proofs based on inviscid-flow theory, For viscous flow with $\mu \neq 0$ but $\mu \to 0$, we have provided three proofs in Sects. 9.1.2 and 9.3.1, and Problem 5.1. Here are some more.

(a) Derive it from finite-domain impulse theory (9.4.10b) for two-dimensional steady flow at $Re \to \infty$. Compare your derivation with that leading to (9.3.3).

(b) Consider a uniform incoming flow past a three-dimensional wing of span b. Use (9.3.1) to show $L \approx \rho U\Gamma b$. Explain why this impulse-based formula, applied to unsteady flow, yields the lift for steady flow.

(c) Compare the respective advantages and disadvantages of all these proofs. Can you construct yet another one based on viscous-flow theory with $\mu \neq 0$ but $\mu \to 0$?

9.9. Prove (9.4.8) directly by using a DMT identity. Then:

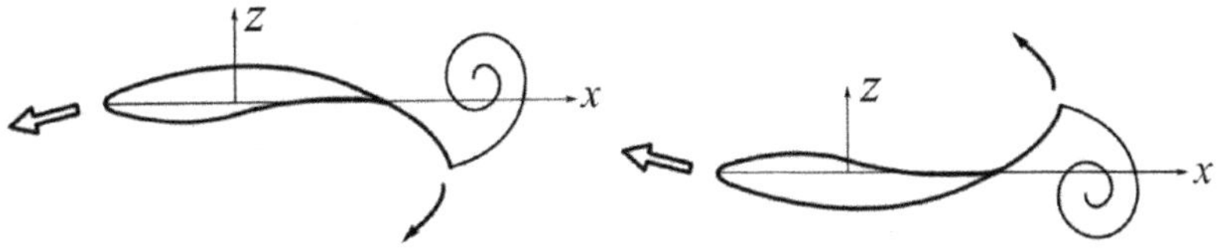

Fig. 9.13 A qualitative assessment of the effect of unsteady vorticity moments on the total force and moment

(1) Explain the physical implication of this result, focusing on the physical origin of $P_\infty - P$.

(2) Select the downstream face of Σ as a wake plane W at a fixed $x = x_W$ with $\boldsymbol{n} = \boldsymbol{e}_x$. Let $\boldsymbol{l} = \boldsymbol{\omega} \times \boldsymbol{u}$ and denote its tangent components on W by $\boldsymbol{l}_\pi$. Show that (9.4.8) yields a neater *form-drag* formula in terms of wake-plane integral:

$$D_{\text{form}} = \int_W (P_\infty - P)dS = \frac{\rho}{n-1}\int_W \boldsymbol{x}\cdot\boldsymbol{l}_\pi dS. \tag{9.5.6}$$

9.10. Consider three-dimensional steady flow (including separated flow) with constant incoming velocity $\boldsymbol{U} = U\boldsymbol{e}_x$ over a wing without yawing (i.e., the flow is symmetric with respect to (x, z)-plane). Let the control surface be defined as in Problem 9.9, write $\boldsymbol{u} = \boldsymbol{U} + \boldsymbol{v}$, and neglect the effect of $\boldsymbol{v} \times \boldsymbol{\omega}$.

(a) Show that the wing lift and form drag are approximately given by

$$L \approx \rho U \int_W y\omega_x dS, \tag{9.5.7a}$$

$$D_{\text{form}} \approx \frac{\rho}{2}\int_W (z\omega_y - y\omega_z)dS. \tag{9.5.7b}$$

(b) Can the linear approximation of the Lamb vector give an approximate induced drag? Why?

(c) Use (9.5.7) to make a qualitative diagnosis of the slender delta-wing flow at large angles of attack α shown in Fig. 8.21: which vortical structures lead higher lift and smaller form drag, and which do the opposite? In your view, for given wing sweeping angle and α, what would be the optimal delta-wing vortex system for larger L/D_{form}, and to this end how to reshape the wing?

9.11. Consider the starting stage of a two-dimensional model-fish locomotion by swimming its caudal fin as sketched in Fig. 9.13. Assume the caudal fin motion is very rapid so that only the unsteady terms in (9.4.5) and (9.5.5) are significant. Use these formulas to identify the qualitative property of the thrust, side force, and moment. Explain the physics associated with these forces and moments.

9.12. Compare the structures of vortex systems in steady wing flow and flapping-wing flow, focusing on how they produce different downwash.

Chapter 10
Vortex Instability, Breakdown, and Transition to Turbulence

In the last two chapters of this book we will consider the late-stage evolution of vortical flows: vortex instability and transition to turbulence. This involves more complicated processes than those we have seen in previous chapters, and relevant studies require more advanced theoretical, numerical and experimental methods. This book only makes an introductory discussion of these topics, with emphasis on the underlying physics and vortical structures during the evolution. In this chapter we focus on instability and breakdown (or *burst*) of axial vortices, including transition to turbulence in vortex-ring flows. Shear-layer instability and vortical structures during transition and in fully developed turbulence will be addressed in Chap. 11.

To be self-contained, and as a preparation of these discussions, this chapter starts from a review of the basic concepts and analysis methods of vortical-flow stability (or instability) to be used in this and next chapters. It is confined to a linearized analysis. A similar preparation on the concept of turbulence will be given in the beginning of Chap. 11.

10.1 Basic Concepts of Vortical-Flow Stability

Flow stability is an important area of fluid mechanics that explores how a flow loses its stability and becomes turbulent eventually. Many complicated phenomena appearing in vortical flows can only be explained in terms of vortex instability. Miscellaneous vortices of various scales may occur in the complicated process as a flow loses its stability and gradually becomes turbulent. Typical examples include the Taylor vortices between two concentric circular cylinders, streamwise and hairpin vortices in boundary layers, the Görtler vortices in concave-wall boundary layers, etc.

A laminar flow that one is interested in its stability is called a *basic flow*, which is subjected to disturbances under natural or laboratory conditions. If the disturbances decay as they evolve in space and time so that the flow recovers its original basic-flow state, the flow is said to be stable. If the disturbances continuously grow so

J.-Z. Wu et al., *Vortical Flows*, DOI 10.1007/978-3-662-47061-9_10

that the basic flow develops to another state or becomes turbulent, the flow is said unstable. Abundant literatures are available related to hydrodynamic stabilities, for example the books of Schmid and Henningson (2001) and Drazin (2002). Detailed materials of vortical-flow stability can be found in the books of Yin and Sun (2003) and Alekseenko et al. (2007).

The most mature and popular method in the study of flow stability is linear analysis for sufficiently small disturbances. Since vortical flows are dominated by the Reynolds number, in linear theory one may find a critical Reynolds number $Re_{\rm cr}$ such that for $Re > Re_{\rm cr}$ the flow is unstable to certain infinitesimal disturbances, while if $Re = Re_{\rm cr}$ the flow is at least unstable to infinitesimal disturbance of one frequency. Then, one writes the velocity field of a disturbed flow as $\boldsymbol{U}(\boldsymbol{x}) + \boldsymbol{u}(\boldsymbol{x}, t)$, where $\boldsymbol{U}$ is the basic flow, which is usually steady and preferably an exact solution, and $\boldsymbol{u}$ is the disturbance velocity governed by a homogeneous linear partial differential equations. Then the stability problem amounts to an *eigenvalue problem* of the disturbance velocity. A linear theory can give sufficient condition for instability, but a linearly stable flow may not be nonlinearly stable.

10.1.1 Normal-Mode Analysis

Consider a basic flow in Cartesian coordinates (x, y, z). The following expression of the disturbance velocity $\boldsymbol{u}$ can be obtained:

$$\boldsymbol{u}(\boldsymbol{x}, t) = \Re\hat{\boldsymbol{u}}(y, \alpha, \beta)\mathrm{e}^{i(\alpha x + \beta z - \omega t)}, \tag{10.1.1}$$

where $\Re$ represents the real part, α and β are the wave numbers along x and z directions, respectively. Substituting (10.1.1) into the linearized disturbance equations, one obtains a *dispersion relation* for the eigenvalues of α, β and ω:

$$\mathcal{F}(\alpha, \beta, \omega; Re, \ldots) = 0. \tag{10.1.2}$$

Here, as long as just one mode grows then the flow must be unstable. Thus, it suffices to consider a single most amplified mode. This approach is known as the *normal-mode analysis*. The governing disturbance equations and analyses for the stability of various columnar vortices will be discussed in the rest of this chapter.

1. Orr-Sommerfeld equation. We now illustrate the linear stability theory by the most important and classic example, where the basic flow is a *parallel shear flow* $\boldsymbol{U} = U(y)\boldsymbol{e}_x$ in Cartesian coordinates. A complete set of linearized disturbance equations can be conveniently expressed for normal velocity v and normal vorticity η, governed by the linearized version of the y-components of the vorticity equation and its curl, respectively (cf. Wu et al. 2006, Sect. 4.5):

$$\left(\frac{\partial}{\partial t} + U\frac{\partial}{\partial x} - \nu\nabla^2\right)\nabla^2 v = \frac{d^2U}{dy^2}\frac{\partial v}{\partial x}, \tag{10.1.3a}$$

$$\left(\frac{\partial}{\partial t} + U\frac{\partial}{\partial x} - \nu\nabla^2\right)\eta = -\frac{dU}{dy}\frac{\partial v}{\partial z}. \tag{10.1.3b}$$

When these equations are nondimensionalized, the viscosity becomes $1/Re$. The boundary conditions are

$$v = \frac{\partial v}{\partial y} = 0, \quad \eta = 0, \quad \text{at } y = y_1, y_2, \tag{10.1.3c}$$

where, for channel flow between parallel plates, free shear layer, and boundary layer, one sets $(y_1, y_2) = (-1, 1), (-\infty, \infty)$, and $(0, \infty)$, respectively. In normal-mode analysis we substitute a single mode

$$(v, \eta)(\boldsymbol{x}, t) = (\hat{v}, \hat{\eta})(y, t)e^{i\alpha x + i\beta z}, \tag{10.1.4}$$

into (10.1.3). Denoting $D = \partial/\partial y$ and $k^2 = \alpha^2 + \beta^2$, in matrix operator form we obtain

$$\frac{\partial}{\partial t}\begin{bmatrix}\hat{v}\\ \hat{\eta}\end{bmatrix} = -i\begin{bmatrix}L_{os} & 0\\ L_c & L_{sq}\end{bmatrix}\begin{bmatrix}\hat{v}\\ \hat{\eta}\end{bmatrix}, \tag{10.1.5}$$

and

$$L_{os} = -\frac{1}{(D^2 - k^2)}\left[\frac{(D^2-k^2)^2}{iRe} - \alpha U(D^2 - k^2) + \alpha U''\right], \tag{10.1.6a}$$

$$L_c = \beta DU, \tag{10.1.6b}$$

$$L_{sq} = \alpha U - \frac{D^2 - k^2}{iRe} \tag{10.1.6c}$$

where $U' = dU/dy$, are called the *Orr-Sommerfeld operator*, the *coupling operator*, and the *Squire operator*, respectively, since the differential equations for $\hat{v}$ and $\hat{\eta}$ are known as the *Orr-Sommerfeld equation* and *Squire equation*, respectively, which are coupled by L_c for three-dimensional disturbances.

Within the normal-mode framework, there is a **Squire theorem**: *parallel shear flows first become unstable to a two-dimensional disturbance wave at a value of the Reynolds number smaller than any value for which three-dimensional disturbances can grow*. Thus, to find a sufficient condition for flow instability, one has only to consider two-dimensional disturbances. In this case $L_c = 0$ and $\eta = 0$, and one needs work with the Orr-Sommerfeld operator. Then, let the stream function $\psi(x, y, t)$ take the form

$$\psi(x, y, t) = \varphi(y)e^{i\alpha(x-ct)} \tag{10.1.7}$$

and substitute it into (10.1.6a). We obtain the conventional form of the *Orr-Sommerfeld equation* (O-S equation for short):

$$(i\alpha Re)^{-1}(D^2 - \alpha^2)^2\varphi = (U - c)(D^2 - \alpha^2)\varphi - U''\varphi, \tag{10.1.8a}$$

under boundary conditions

$$\varphi = D\varphi = 0 \quad \text{at } y = y_1, y_2. \tag{10.1.8b}$$

This problem has long been the very basis of linear instability analysis of free shear layers and boundary layer, as will be discussed in the next chapter. Recall that in two dimensions there is $\nabla^2\psi = -\omega_z$, (10.1.8a) is essentially an equation for disturbance vorticity. The central task is to find the four independent solutions of (10.1.8) and thereby obtain the neutral stability boundary in parametric space. This used to be a very challenging task in pre-computer era. In other vortical-flow stability analysis one faces the same task.

Many vortical flows can be approximately treated inviscid (that is $Re \to \infty$); it is true in particular if no solid boundary appears. Then the O-S equation (10.1.8a) degenerates to

$$(U - c)(\varphi'' - \alpha^2\varphi) - U''\varphi = 0, \quad \varphi = 0 \text{ at } y = y_1, y_2, \tag{10.1.9}$$

called the *Rayleigh equation*, of which studies have led to important information about the criteria of stability/instability of inviscid vortical flows. For example, for inviscid parallel shear flow, from (10.1.9) one may prove the **Rayleigh-Fjørtoft theorem**: *The necessary condition for the flow to be unstable is that the mean velocity profile has an inflexion point or vorticity has an extremum inside the flow.*

2. Temporal/spatial modes and absolute/convective instability. Within normal-mode analysis one may consider either *temporal mode* or *spatial mode*. In temporal-mode analysis, the growth of disturbances with time is the concern. Then α and β are given real values, and the eigenvalue ω is complex, $\omega = \omega_r + i\omega_i$. Thus, for a given mode, the velocity disturbance can be expressed as

$$\boldsymbol{u}(x, t) = \widehat{\boldsymbol{u}}(y, \alpha, \beta)e^{\omega_i t} \cdot e^{i(\alpha x + \beta z - \omega_r t)}, \tag{10.1.10}$$

where $\widehat{\boldsymbol{u}}$ is the amplitude and ω_r is the frequency. As time goes on, $\widehat{\boldsymbol{u}}$ will vary while the disturbance is propagating as a travelling wave. ω_i is the growth rate of the disturbance. The flow is stable or asymptotically stable if $\omega_i < 0$, and *neutrally stable* if $\omega_i = 0$. When $\omega_i > 0$, the disturbance grows exponentially with time and the flow is unstable.

In the spatial-mode analysis, the spatial development of disturbances is the concern. Then frequency ω and wave number β are chosen real, but eigenvalue $\alpha = \alpha_r + i\alpha_i$ is complex. Then (10.1.10) is replaced by

$$\boldsymbol{u}(\boldsymbol{x}, t) = \widehat{\boldsymbol{u}}(y, \beta, \omega)e^{-\alpha_i x} \cdot e^{i(\alpha_r x + \beta z - \omega t)}. \tag{10.1.11}$$

The flow is linearly stable or neutrally stable in its spatial development if $\alpha_i \geq 0$. When $\alpha_i < 0$, the disturbance grows exponentially in space and the flow is unstable.

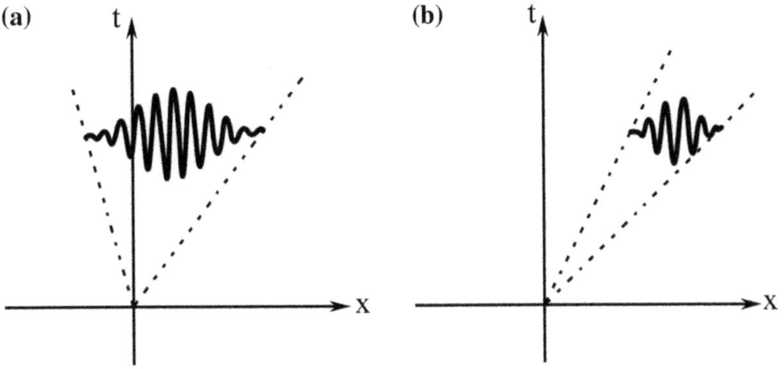

Fig. 10.1 Sketches of the impulse response. **a** Absolutely unstable. **b** Convectively unstable

In the stability analysis of a flow, one may be more interested in the flow response to an initial and localized disturbance, i.e. the spatio-temporal development of the response to an impulsive disturbance (*impulsive response*), which is more inherently related to the physical stability concept. This has led to the theory of absolute and convective instability, which has revealed more comprehensive and profound flow instability behavior both qualitatively and quantitatively.

Absolute instability (AI) is referred to the situation where local disturbances can propagate both upstream and downstream (Fig. 10.1a) so that an unstable flow is gradually contaminated everywhere by a point-source input. In contrast, *convective instability* (CI) is referred to the situation where local disturbances propagate and develop only in downstream direction (Fig. 10.1b) so that the flow can eventually recover its original undisturbed state at the location where the disturbance is initially introduced.

Strictly, this spatial-mode analysis is applicable only to a flow that is stable or convectively unstable. For an absolutely unstable flow, the region influenced by an impulsive disturbance will be large enough after certain time, leading to an onset of synchronized self-sustained oscillation, or self-excited resonance, which can bury the initial disturbances. In this case, observing spatial propagation of an initial disturbance is obviously meaningless. Thus, the AI/CI theory is a powerful tool to clearly distinguish the temporal and spatial development of disturbances.

One of the purposes of stability analysis is for flow control, where the AI/CI analysis is of crucial importance. Many experimental and numerical studies have shown that an unsteady forcing with a very small power input may cause essential changes of the flow state, which is therefore an efficient method for flow control. Here, to answer the question such as how and where to impose the forcing and what kind of disturbance modes to pose for the maximum absolute growth rate, a clear understanding of the specific AI/CI character of the flow is required.

10.1.2 Nonmodal Analysis and Transient Growth

In above normal-mode analyses a single representative mode is sufficient for finding linear instability. Despite the simplicity and wide applications of this approach, however, difficulties have been encountered. While the critical Reynolds numbers predicted by normal-mode analysis agree very well with experiments in some special flows, the agreement was found poor for most vortical flows, such as the plane Couette flow, plane Poiseuille flow, boundary layers, free shear layers, and concentrated vortices.

Studies in recent decades have revealed that the property of the linear operator in a disturbance equations, i.e., whether it is symmetric, normal, or self-adjoint plays a crucial role and is of close relevance to the aforementioned discrepancies. These concepts are generalization of the corresponding concepts of matrices in linear algebra to functional space. Consider a linear operator L. Its *adjoint* operator $L^\dagger$ is its complex-conjugate transverse, defined by the inner-product integral relation

$$\int_{\mathcal{D}} \boldsymbol{v} \cdot (L\boldsymbol{u}) dV = \int_{\mathcal{D}} \boldsymbol{u} \cdot (L^\dagger \boldsymbol{v}) dV$$

over the domain $\mathcal{D}$ where L is defined for any vectors $\boldsymbol{u}$ and $\boldsymbol{v}$. L is called *normal* if $LL^\dagger = L^\dagger L$, which includes *symmetric* and *self-adjoint* operators (Boundary conditions should be understood implicitly built in the operator L). For a normal and compact L, its eigenfunctions form a complete orthogonal set, so that the solution of the linear disturbance equation can be expanded in terms of the eigenfunctions of L as used in (10.1.1). This is the case where normal-mode analysis works well, for example the Bénard flow and narrow-gap Taylor-Couette flow.[1]

If L is non-normal, however, although its eigenfunctions may still form a complete set, they can be non-orthogonal. In this case, even if all individual disturbance modes decay, a linear combination of some of them may lead to *transient algebraic growth*. For example, let $\boldsymbol{u}_1$ and $\boldsymbol{u}_2$ be two normal modes such that $\langle \boldsymbol{u}_1, \boldsymbol{u}_2 \rangle$ vanishes due to orthogonality ($\langle \cdot, \cdot \rangle$ denotes inner-product integral), then their contributions to the kinetic energy can be superposed:

$$\langle \boldsymbol{u}_1 + \boldsymbol{u}_2, \boldsymbol{u}_1 + \boldsymbol{u}_2 \rangle = \langle q_1^2 + q_2^2 \rangle, \quad q_\alpha = |\boldsymbol{u}_\alpha|, \quad \alpha = 1, 2.$$

Thus, if the kinetic energy of all normal modes decays, so must be that of any disturbance. But if these modes are non-orthogonal, then

$$q^2 = |\boldsymbol{u}_1 + \boldsymbol{u}_2|^2 = q_1^2 + q_2^2 + 2q_1 q_2 \cos\theta,$$

where $\theta \neq \pi/2$ is the angle between $\boldsymbol{u}_1$ and $\boldsymbol{u}_2$ in functional space. Thus, assuming θ is time-independent, there is

[1] For Bénard flow and Taylor-Couette flow and their stability analysis see, e.g., Drazin (2002).

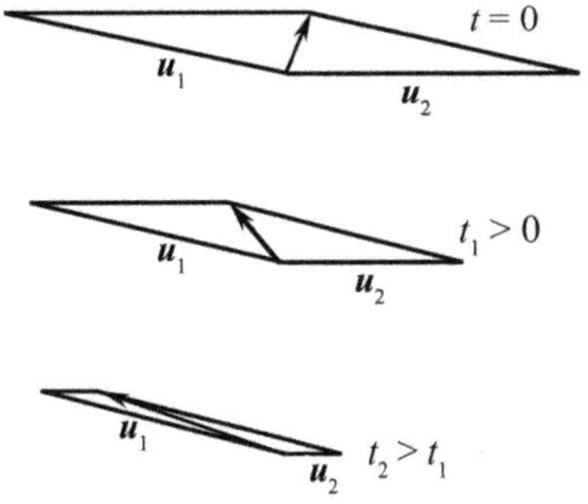

Fig. 10.2 Sketch illustrating transient growth due to nonorthogonal superposition of two eigenvectors that decay at different rates as time evolves

$$\dot{q^2} = \dot{q_1^2} + \dot{q_2^2} + 2(\dot{q_1}q_2 + q_1\dot{q_2})\cos\theta.$$

Therefore, if $\theta > \pi/2$, q^2 may have algebraic growth even if both $\dot{q_1^2}$ and $\dot{q_2^2}$ are negative. But this growth can happen only in a finite time, because it will eventually tend to zero as q_1^2 and q_2^2 continue to decay as $t \to \infty$. This situation is sketched in Fig. 10.2. Thus, if the operator in a linearized disturbance-flow equation is non-normal, one cannot judge the stability simply through a single-mode analysis.

To have a further intuitive feeling of how the linearized N-S equation (10.1.5) permits transient growth, we follow Reshotko (2001) (with minor change) to consider a simple "model solution". We just approximate the coupling operator $L_c = \beta DU$ by a constant γ, and replace operators L_{os} and L_{sq} by their eigenvalues such that

$$L_{os}\hat{v} = -\lambda\hat{v}, \quad L_{sq}\hat{\eta} = -\mu\hat{\eta},$$

with initial conditions at $t = 0$ denoted by suffix 0, where $\lambda = \lambda_r + i\lambda_i$ and $\mu = \mu_r + i\mu_i$. Then the "solution" of (10.1.5) reads

$$\hat{v} = \hat{v}_0 e^{-\lambda t}, \quad \hat{\eta} = -i\gamma\hat{\eta}_0\left(\frac{e^{-i\lambda t} - e^{-i\mu t}}{\lambda - \mu}\right) + \hat{\eta}_0 e^{-i\mu t}. \tag{10.1.12}$$

We choose $\lambda_i, \mu_i < 0$ such that both normal modes represent damping disturbances. But the coupling term via γ is different. For $t \ll 1$, it is linearly increase in time; but for $t \gg 1$ it decays exponentially. This is seen most clearly if $\lambda = \mu$, for which the coupling term takes the form of $te^{-i\mu t} = te^{-i\mu_r t}e^{\mu_i t}$, that is, a harmonic wave with linear growth followed by exponential decay.

The importance of transient growth lies in the fact that the operators in linearized Navier-Stokes equation (10.1.5) and the O-S equation (10.1.8) are intrinsically non-normal for vortical flows. Thus the Squire theorem is not universally true. Only in certain rare exceptional cases can those operators degenerate to normal ones. The most famous landmark examples of such cases are the *Taylor-Couette flow* (Problem 10.8) between two coaxial rotating cylinders with narrow gap, and the *Bénard convection* in a plane horizontal fluid layer heated from below. For details of both see, e.g., Drazin (2002). This was why for these flows the normal-mode predictions of initial instability growth were quickly confirmed by experiments. But in most other situations, for a complete linear stability analysis one has to return to (10.1.5) to include three-dimensional disturbance. The transient growth has been found to have very significant effect on the transition of wall-bounded shear layers.

10.1.3 Receptivity

In addition to its eventual decay, the transient disturbance growth discussed above differs in another aspect from the normal-mode instability. In the latter, as long as a mode is unstable, the disturbance must increase exponentially (within linearized theory) no matter how small the initial disturbance could be. But for the former the amplitude of initial disturbance has to be finite.

Actually, either normal-mode or transient growth is in action, realistic vortical flows in nature and technology are never in idealized quiet circumstance. Rather, they are always exposed in certain external disturbances, which may lead to a *forced excitation* of some instability mechanism of the vortices. Similarly, in one's study of flow instability by experiment or computation, one always imposes a finite-amplitude initial disturbance to trigger the instability as soon as possible. In all these cases there appear the *receptivity* problem, which is different from instability problem.

We have seen that instability problem deals with the spontaneous evolution of disturbance waves. It is described mathematically by an eigenvalue problem with *homogeneous* equation and boundary condition. Thus, instability waves are eigen waves with *indeterminate* amplitude. In contrast, in receptivity problem, one studies the means by which a specific forced disturbance enters the flow and its signal evolution in the disturbed flow (Reshotko 1984). Thus, receptivity is described mathematically by an initial-boundary value problem with *inhomogeneous* equations and/or initial-boundary conditions. The amplitude of excited instability wave depends on that of incoming waves.

In this book we shall not go into receptivity problem; but it should be born in mind that *realistic instability problems are always preceded by a receptivity process.* For a review on the role of receptivity in vortical-flow control by unsteady excitation see Wu et al. (1991).

10.2 Instability of Axisymmetric Columnar Vortices

Consider now the instability of an isolated and axisymmetric vortex without axial stretching. Its velocity profiles of the basic flow take the form $\boldsymbol{U} = (0, V(r), W(r))$ in cylindrical coordinates (r, θ, z). We study its linear normal-mode instability including temporal, spatial, and temporal-spatial modes.

Denote the disturbance velocity by $\boldsymbol{u} = (u, v, w)$. By (6.1.1), the linearized disturbance equations read

$$\frac{\partial u}{\partial r} + \frac{u}{r} + \frac{1}{r}\frac{\partial v}{\partial \theta} + \frac{\partial w}{\partial z} = 0, \tag{10.2.1a}$$

$$\frac{\partial u}{\partial t} + \frac{V}{r}\frac{\partial u}{\partial \theta} + W\frac{\partial u}{\partial z} - 2\frac{Vv}{r} = -\frac{\partial p}{\partial r} + \nu\left(\nabla^2 u - \frac{u}{r^2} - \frac{2}{r^2}\frac{\partial v}{\partial \theta}\right), \tag{10.2.1b}$$

$$\frac{\partial v}{\partial t} + u\frac{dV}{dr} + \frac{V}{r}\frac{\partial v}{\partial \theta} + W\frac{\partial v}{\partial z} + \frac{uV}{r} = -\frac{1}{r}\frac{\partial p}{\partial \theta} + \nu\left(\nabla^2 v - \frac{v}{r^2} + \frac{2}{r^2}\frac{\partial u}{\partial \theta}\right), \tag{10.2.1c}$$

$$\frac{\partial w}{\partial t} + u\frac{dW}{dr} + \frac{V}{r}\frac{\partial w}{\partial \theta} + W\frac{\partial w}{\partial z} = -\frac{\partial p}{\partial z} + \nu\nabla^2 w, \tag{10.2.1d}$$

where

$$\nabla^2 = \frac{\partial^2}{\partial r^2} + \frac{1}{r}\frac{\partial}{\partial r} + \frac{1}{r^2}\frac{\partial^2}{\partial \theta^2} + \frac{\partial^2}{\partial z^2}.$$

Thus a normal-mode disturbance takes the form

$$(u, v, w, p) = \left(\hat{u}(r), \hat{v}(r), \hat{w}(r), \hat{p}(r)\right) e^{i(kz + n\theta - \omega t)}, \tag{10.2.2}$$

where k and $n = 0, \pm 1, \pm 2, \dots$ are axial and circumferential wave numbers, respectively. The case with $n = 0$ implies axisymmetric disturbance wave known as "sausage mode"; all cases with $n \neq 0$ are *spiral modes*, for example $n = \pm 1$ implies *bending wave* and $|n| \geq 2$ implies "fluted mode". These instability waves of columnar vortices are known as *Kelvin waves*, as Kelvin was the first to study the instability of columnar vortices by normal-mode analysis, and have been extensively studied (e.g., Saffman 1992).

10.2.1 Stability of Pure Vortices

Stretch-free pure vortices were discussed in Sect. 6.2, where we saw that steady pure vortices can only be inviscid with basic flow velocity $\boldsymbol{U} = (0, V(r), 0)$. Based on the angular-momentum conservation, Rayleigh in 1916 found that the criterion

$$\frac{1}{r^3}\frac{dC^2}{dr} > 0, \quad C \equiv rV, \tag{10.2.3}$$

is a sufficient condition for the vortex to be stable to an axisymmetric disturbance. By using Sturm-Liouville theory, Synge in 1933 proved that (10.2.3) is actually a sufficient and necessary condition. To understand this criterion, recall that for pure vortices the radial pressure gradient is solely balanced by the centrifugal force as seen from (6.1.10a):

$$\rho \frac{C^2}{r^3} = \frac{\partial p}{\partial r}. \tag{10.2.4}$$

The Rayleigh criterion (10.2.3) can be explained by a simple observation. Conceive that a fluid particle at r_1 with C_1 is moved to r but keeps its C_1. If the vorticity distribution is such that

$$\rho \frac{C_1^2}{r^3} < \frac{\partial p}{\partial r} = \rho \frac{C^2}{r^3},$$

then by (10.2.4) the restoring force will drive the particle back to r_1 (Alekseenko et al. 2007). The stability mechanism of a pure vortex to axisymmetric disturbances is solely due to the centrifugal force.

Similar to the *Rayleigh-Fjøtoft theorem* for parallel shear flow, we can also find that the necessary condition for a pure vortex to be unstable to *non-axisymmetric* disturbances is the existence of mean vorticity extremum, i.e. the mean vorticity gradient becomes zero at certain location inside the flow field.

10.2.2 Temporal Instability of Swirling Vortices

Compared with pure vortices, swirling vortices are much more common in vortical flows as mentioned in Sect. 6.3. Except highly simplified theoretical model like the Burgers vortex (Sect. 6.3.1), most of swirling vortices have variable axial velocity $W(r)$, implying axial shearing between swirling fluid layers of different r. Thus, unlike pure vortices, the instability mechanism of swirling vortices involves complicated coupling between centrifugal instability and shear instability, known as the *Kelvin-Helmholtz instability* (to be addressed in Chap. 11).

For a swirling vortex with variable axial velocity $W(r)$, Howard and Gupta (1962) derived an equation for inviscid instability analysis:

$$\gamma^2 D(SD_* u) - u\left\{\gamma^2 + \gamma r D\left[S\left(\frac{D\gamma}{r} + 2n\frac{V}{r^3}\right)\right] - 2kS\frac{V}{r^2}(krD_*V - nDW)\right\} = 0, \tag{10.2.5}$$

where $D = d/dr$, $D_* = D + 1/r$, $\gamma(r) = nV/r + kW - \omega$ is the Doppler frequency, and $S = r^2/(n^2 + k^2r^2)$.

Unfortunately, no general solution of the *Howard-Gupta equation* (10.2.5) can be found for three-dimensional disturbances. Hence, so far the results obtained from this equation are either general criteria but only for special disturbances or at asymptotic limit, or the detailed instability features have to be worked out case by case for specific swirling flows.

For the general criteria, as an extension of the Rayleigh criterion (10.2.3) for pure vortices, Howard and Gupta (1962) found a sufficient but not necessary condition for a swirling flow to be stable to axisymmetrical disturbances:

$$\frac{1}{r^3}\frac{dC^2}{dr} \geq \frac{1}{4}\left(\frac{dW}{dr}\right)^2. \tag{10.2.6}$$

Then, based on (10.2.5) and using an energy consideration, Leibovich and Stewartson (1983) have carried out an asymptotic analysis for $|n| \gg 1$ and achieved a finite upper limit of the growth rate ω_i for inviscid disturbances as $n \to \infty$. Within the normal-mode framework, they obtained a sufficient condition for a columnar vortex to be unstable to small disturbances:

$$V\frac{d\Omega}{dr}\left[\frac{d\Omega}{dr}\frac{dC}{dr} + \left(\frac{dW}{dr}\right)^2\right] < 0, \quad n \to \infty, \tag{10.2.7}$$

where $\Omega = V/r$. Emanuel (1984) has proven that this result is a direct extension of the Rayleigh criterion (10.2.3) to swirling vortices with disturbances not necessarily axisymmetric.

As an example of applying (10.2.5) to special flows, we now consider the temporal instability of a widely used model for swirling vortices, the q-vortex defined by (6.4.15) through (6.4.17) (Sect. 6.4.3), where the constant W^* and indefinite sign in $w^*(r^*)$ permit modeling both swirling jets and wakes. In temporal normal-mode these merely alter the frequency ω_r but not affect whether the flow is stable; so a Galilean transformation can remove W^* so that the basic flow (6.4.15) can be simply written as

$$U = 0, \quad V = \frac{q}{r}\left(1 - e^{-r^2}\right), \quad W = e^{-r^2}. \tag{10.2.8}$$

The main results on the temporal-mode stability of this q-vortex are as follows. Firstly, so far it has never been found that to an axial disturbance ($n = 0$) the flow can become unstable. For $n > 0$, i.e., an increase of rotation by the disturbance, even a weak rotation will stabilize all the modes as long as $q > 0.08$. In other words, the rotation increase helps the flow to resist further disturbances. In contrast, a swirling flow is unstable to negative azimuthal wave numbers $n < 0$, and the temporal growth rate ω_i will increase as $|n|$. For a given n, if q is gradually increased, the flow will become stable again to all the disturbance modes after $q > 1.5$. A typical result is shown in Fig. 10.3. Secondly, the maximum unstable region in the parameter (k, q)-plane corresponds to the flow with $n = -1$, for which extremely complicated flow pattern may occur.

Thirdly, for most vortical flows of interest in engineering or geophysical fluid dynamics, centrifugal force is indeed the source to cause instability and it is essentially an inviscid mechanism. Therefore, it had long been confident that viscosity acts only on dissipation as an stabilizing effect. However, Khorrami's (1991) calculation revealed two purely viscous instability modes in the q-vortex which, due to

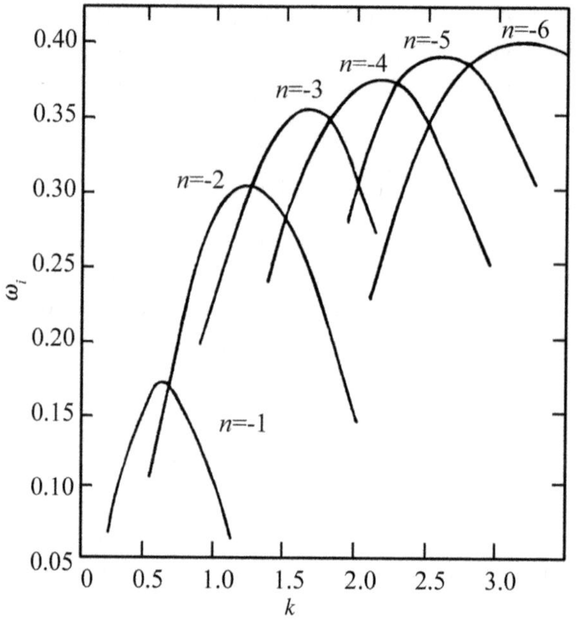

Fig. 10.3 Variation of the growth rate ω_i as the axial wave number k with different azimuthal wave numbers n for a q-vortex with $q = 0.8$. From Lessen et al. (1974)

the existence of centrifugal force, are different from the viscous instability of plane shear layers.

Very recently, Feys and Maslowe (2014) have studied the linear stability for the Moore-Saffman vortex (Sect. 6.5.2, M-S vortex for short), which is an improvement of Batchelor vortex or q-vortex for modeling the trailing vortex profiles and contains a wing loading parameter n. Recall that for $1 \geq n > 0.44$[2] all axial velocity profiles $W_n(r)$ of the M-S vortex are wake-like, among which $n = 1$ returns to the q-vortex and $n = 0.5$ corresponds to elliptical wing loading; but $W_n(r)$ becomes jet-like for $n < 0.44$ (Fig. 6.5). The temporal instability calculation of Feys and Maslowe shows that the M-S vortex is unstable for $1 \geq n \geq 0.44$ with wake-like $W_n(r)$. The maximum growth rates increase with azimuthal wave number and are progressively larger as n decreases, which are approximately 22 % greater than those for q-vortex. The largest growth rate occurs at $n = 0.44$. When $n < 0.44$ the amplification factor of the M-S vortex decreases rapidly and the vortex is stable for $n \leq 0.25$.

For comparison with the stability character of q-vortex, Feys and Maslowe introduce a swirling number q to the M-S vortex. The disturbance growth rate also depends on this parameter as well as the azimuthal wave number. The critical value $q = q_c$, above which the vortex is stable, is found more than 60 % larger than that for the q-vortex.

---

[2] Here n is the wing-load parameter rather than azimuthal wave number.

10.2.3 Absolute and Convective Instability of Swirling Vortices

The temporal-mode analysis of swirling vortices instability has later been developed to the absolute instability (AI) and convective instability (CI) analysis. Recall that for AI/CI analysis Galilean transformation is not permitted. Thus, to consider the AI/CI characters of a q-vortex, we use (6.4.17) and write

$$U(r) = 0, \quad V(r) = \frac{q}{r}(1 - e^{-r^2}), \quad W(r) = a + e^{-r^2}, \tag{10.2.9a}$$

which has two parameters and one can define a Reynolds number:

$$a = \frac{W_\infty}{\Delta W}, \quad q = \frac{\Omega(0) r_0}{\Delta W}, \quad Re = \frac{\Delta W r_0}{\nu}, \quad \Delta W = W(0) - W_\infty, \tag{10.2.9b}$$

where r_0 is the radius of the vortex core, and $\Omega(0)$ and $W(0)$ are the angular velocity and axial velocity at the axis, respectively. Thus, a and q characterize the axial velocity distribution and the level of swirl for a q-vortex. We stress that, while for temporal-mode analysis a can be eliminated by a Galilean transformation, it does have important influence on the AI/CI behavior and the spatial propagation of disturbances. As discussed in Huerre and Monkewitz (1990), it is precisely in the situation where Galilean invariance is broken that absolute-convective instability acquires physical significance. Associated with this is the fact that the sign of azimuthal wave number n of the disturbance also matters. Consequently, the value and sign of parameter a of a typical q-vortex distinguish several different flow patterns as shown in Fig. 10.4:

$$\begin{aligned}
a < -1, &\quad \text{co-flow wake;} \\
a = -1, &\quad \text{wake with } W(0) = 0; \\
-1 < a < 0, &\quad \text{counter-flow wake or jet;} \\
a = 0, &\quad \text{jet with } W_\infty = 0; \\
a > 0, &\quad \text{co-flow jet.}
\end{aligned}$$

In the (a, q)-space, the half planes $a > -0.5$ and $a < -0.5$ represent the jet side and wake side, respectively. Figure 10.5 plots the AI/CI boundaries for $n = \pm 1, -2, \ldots, -7$ at $Re = 667$, which divide the (a, q)-plane into three distinct regions: a stable (S) region, an AI region, and two CI outer regions at both the jet side and wake side. Recall that if $\Omega(0) > 0$ there is $q > 0$ for jet flow with $\Delta W < 0$ and $q < 0$ for wake flow with $\Delta W < 0$. For the latter the temporal-mode unstable azimuthal wavenumber n is positive. But it can be shown that the instability equation is invariant under transformation $(q, -n) \to (-q, n)$; so the AI/CI boundaries with $n = 1$ for a wake-type q-vortex can still be read off from Fig. 10.5 by taking $q > 0$ and $n < 0$.

A close look of Fig. 10.5 shows that the transitional helical mode from CI to AI is very sensitive to the wake-like or jet-like nature of the flow, as well as to the level of swirl. For wake-like flow, the critical transitional mode is always $n = -1$. But on

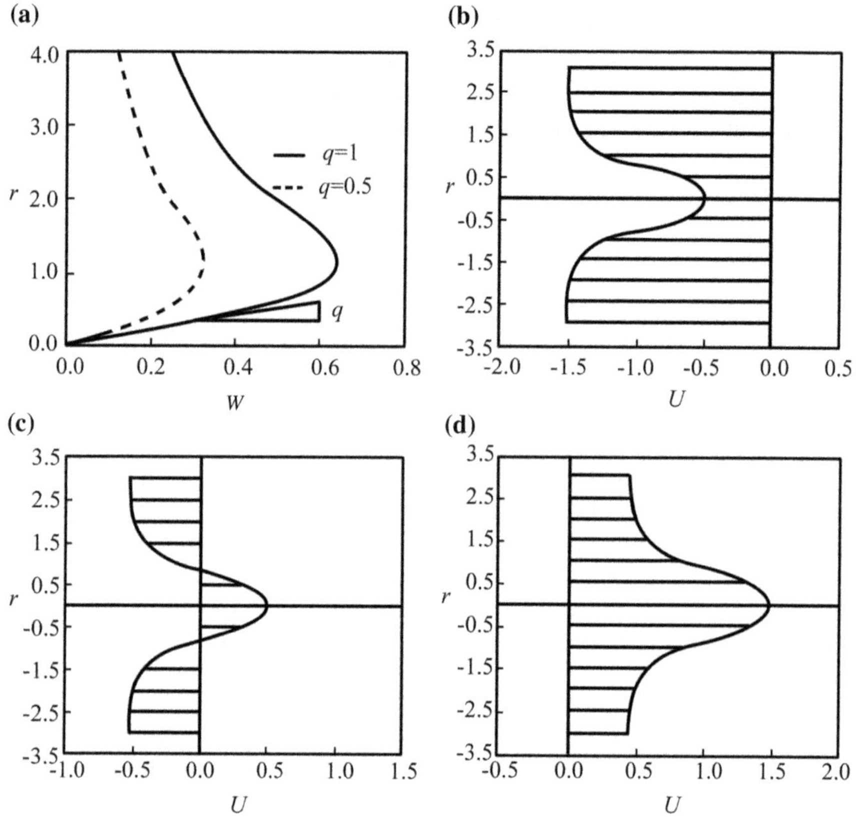

Fig. 10.4 Velocity profiles of a q-vortex. **a** Azimuthal velocity profile $V(r)$; **b** axial velocity profile $W(r)$ for co-flowing wakes ($a < -1$); **c** counter-flowing wakes or jets ($-1 < a < 0$); **d** co-flowing jets ($a > 0$). From Olendraru et al. (1999)

the jet side ($a > -0.5$), the critical transitional mode may have different azimuthal wave numbers $n = -1, -2, \ldots$. Moreover, when $q = 0$, there is no swirl and the flow is a pure wake or jet, and there is a small AI region in a narrow range of a. An increase of q will cause significant enlargement of the AI region.

Note that due to the limitation of the numerical method applied, one can hardly determine the neutral boundary accurately. Once the viscous effect is also involved the situation is changed. As shown in Fig. 10.6 for the helical mode $n = -1$ and different Re, the increase of Re also enlarges the AI region significantly. But, since both AI and CI regions are subregions of the temporal instability, and the q-vortex will be linearly stable if $q > 1.5$, the AI regions for any Re have a common upper bound at $q \simeq 1.5$.

The absolute growth rate ω_i^0 of the q-vortex is illustrated in Fig. 10.7 for wake-like flow with different (a, q) and $n = 1, 2$. Initially, an increase of q leads to a corresponding increase of ω_i^0, which then starts to decrease for larger q. Thus, for each pair of (a, n), there is a maximum ω_i^0, denoted by $\omega_{i,max}^0$. It is evident that the

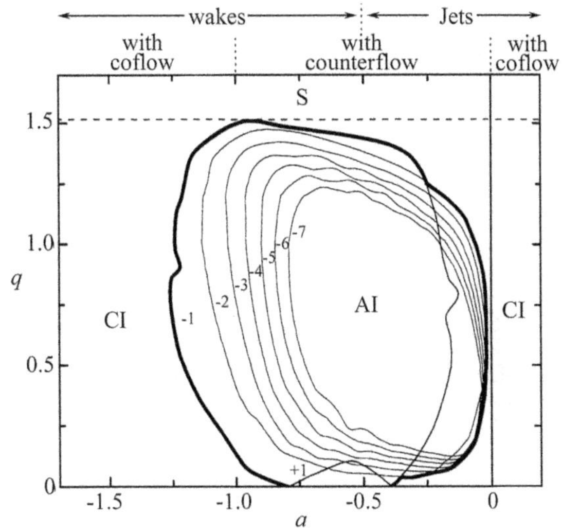

Fig. 10.5 AI/CI boundaries of the helical mode $n = \pm 1, -2, \ldots, -7$ in the (a, q)-plane for $Re = 667$. From Delbende et al. (1998)

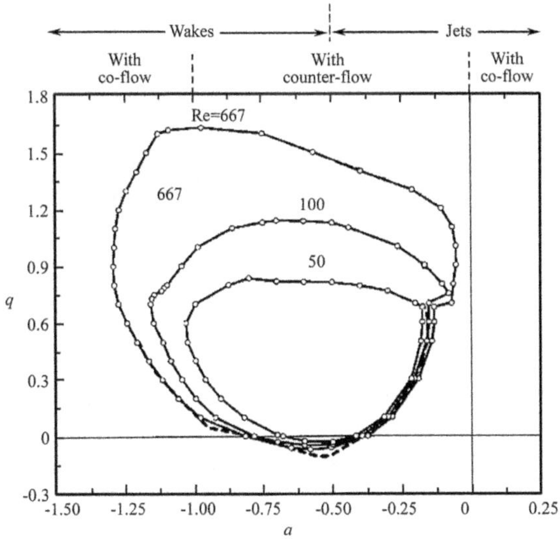

Fig. 10.6 AI/CI boundaries of the helical mode $n = -1$ in the (a, q)-plane at different Reynolds numbers. From Olendraru and Sellier (2002)

axial flow has a strong effect on the absolute growth rate, which reaches the largest value when there is a counter-flow ($a = -0.8$), and this rate reduces as the axial velocity deficit a does.

So far our discussions of instability mechanisms of swirling vortices have been solely based on normal-mode analysis. But due to the coexistence of shear and swirl in all swirling vortices, the operators in the linear disturbance equations of these vortices are always non-normal. Thus, nonmodal instability with transient algebraic growth must occur, which has been studied since 1990s as reviewed in Wu et al. (2006, Sect. 9.3.4).

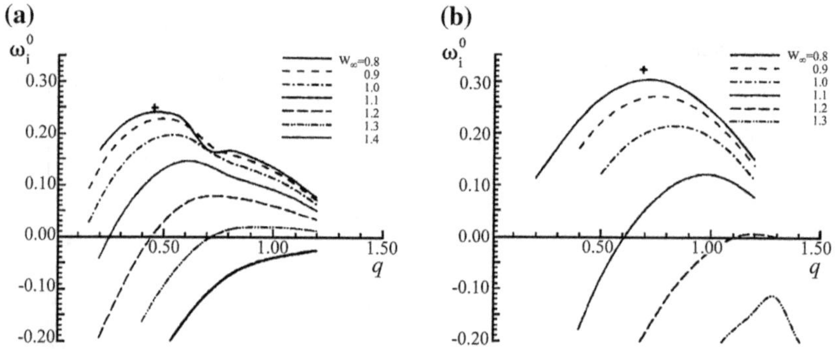

Fig. 10.7 The variation of absolute growth rate ω_i^0 for wake-type q-vortex. **a** $n = 1$; **b** $n = 2$. The sign + marks the corresponding maximum temporal growth rate $\omega_{i,max}$, which is the upper bound of all ω_i^0. From Yin et al. (2000)

10.2.4 Instability of Trailing Vortex Pair

In Sect. 6.5 we mentioned the importance of understanding the behavior and stability of aircraft trailing vortices. For the study of their stability, the basic flow is simplified as a pair of parallel and counter-rotating vortices. In this model, denote the strength, core radius, and distance of vortex pair by $\pm\Gamma$, a, and b, respectively. In the aircraft wind-coordinate system used in Chap. 9, in addition to a convective velocity in the vortex axis U_0, the self-induction of the vortex pair causes a downwash velocity W_0. According to Prandtl's lifting-line theory, W_0 is related to U_0 by (e.g., Prandtl and Tietjens 1934)

$$\frac{W_0}{U_0} = \frac{C_L}{4\pi \mathit{AR}} \left(\frac{b_0}{b}\right)^2, \tag{10.2.10}$$

where C_L is the lift coefficient of the wing, AR is its aspect ratio, and b_0 its wing span. The basic flow pattern is sketched in Fig. 10.8.

There are three distinct instability length scales as the trailing vortex pair travels downstream. The first (smallest-scale) instability is found immediately behind a delta wing, which scales on the thickness of the two shear layers separating from the upper and lower surfaces of the wing trailing edge. The second (short-wave) instability, at an intermediate distance downstream, scales on the primary vortex core radius a. The third (long-wave) instability far downstream represents the classical Crow instability, scaling on the distance between the two primary vortices b. We discuss the last two.

1. Long-wave instability. Figure 10.9 shows the observed *long-wave instability* pattern of trailing vortices. A growing symmetric waviness at a wavelength several times of b is developed, with the plane of the wavy disturbance inclined by about 45° with respect to the vortex-pair plane. From the figure it is evident that this long-wave instability is triggered by *bending waves*, i.e., $n = \pm 1$ in (10.2.2), which deform each

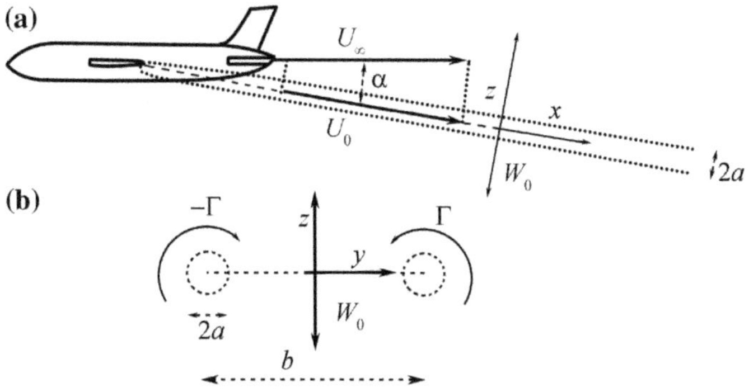

Fig. 10.8 Vortex-pair basic flow: **a** side view; **b** view in a plane orthogonal to the axes of the vortices. From Fabre et al. (2000b)

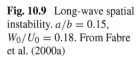

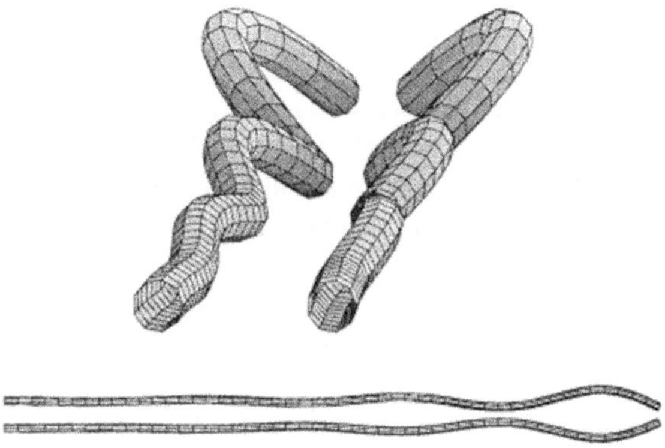

Fig. 10.9 Long-wave spatial instability. $a/b = 0.15$, $W_0/U_0 = 0.18$. From Fabre et al. (2000a)

rectilinear vortex into a sinusoidal filament. The mechanism of disturbance growth was explained by Crow (1970) and is known as the *Crow instability*.

Crow modelled the vortex pair by two infinitely long filaments subjected to a harmonic disturbance of wavelength $2\pi/k \gg a$, and used the Biot-Savart formula with two-term *localized induction approximation* as given by (3.1.40) to calculate the vortex mutual- and self-induced velocities. He found that three mechanisms are involved in the disturbance development: (i) the self-induction of each individual vortex, (ii) the mutual induction between the two vortices, and (iii) the action of the strain field. Crow showed that (i) is a stabilizing factor but (iii) is destabilizing. The resultant instability is due to the *resonance* of a long bending wave on the vortex filament with the external strain field imposed by the other vortex filament. In fact, if the filament is steady under its own self-induced velocity field (i.e. crests do not rotate or translate), then the uniform strain field induced by the other vortex will advect the displaced portions of the filament away from their undisturbed positions, with velocities proportional to their displacement and hence the wave grows. The measured linear growth rate of the long-wave instability is in good agreement with Crow's predictions (Fig. 10.10).

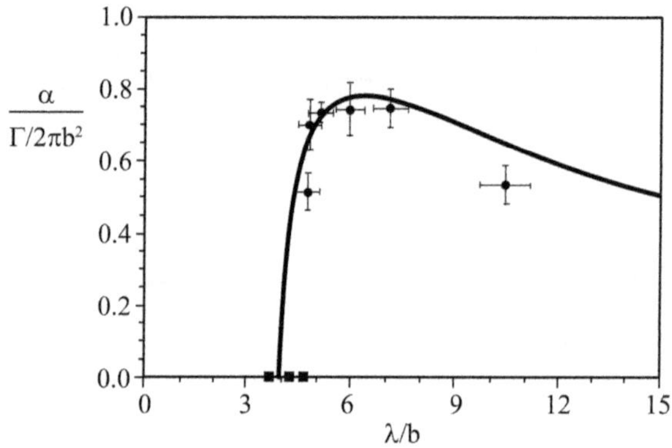

Fig. 10.10 Growth rate of the Crow instability as a function of axial wavelength, where α is the growth rate, Γ is the circulation, b is the vortex spacing, and λ/b is the normalized wavelength. From Hunt and Vassilicos (2000)

2. Short-wave instability and coupling with long-wave instability. The fact that each vortex of the pair is in a strain field induced by the other vortex not only causes long-wave Crow instability, but also makes the cross shape of vortex cores and surrounding streamlines deform from circular to slightly elliptical. Such a strained vortex has been found to have invariably three-dimensional instability, and has linear growth rates scaled with the strain rate. This short-wave instability with $(a/b)^2 \ll 1$ and $|ka| = O(1)$ is also known as *elliptical instability*, discovered by several authors in 1970s, see reviews of Kerswell (2002), Yin and Sun (2003), and Wu et al. (2006).

Elliptical instability is found to arise in the neighborhood of a critical wavenumber k_c, where a resonance condition occurs between two Kelvin waves of one isolated vortex and the straining field induced by the other vortex. Figure 10.11 shows a flow visualization of the vortex pair at various evolution stages. In the beginning of vortex-pair formation, the vortices are straight and uniform along their axes. Later, two different instability behaviors occur. The large-scale symmetric deformation with axial wavenumber of about $6b$ corresponds to the Crow instability, while the growing disturbance with $\lambda \ll b$ is due to the short-wave instability which, unlike the Crow instability, will change the vortex-core structure. The amplitude is larger in the regions where the Crow instability brings the two vortices closer with stronger mutually induced strain rate. A close look at Fig. 10.11c indicates that the initial reflectional symmetry of the flow with respect to the vortex-pair plane is lost, and the displacements of the two vortices are out of phase. Thus, the two instabilities are closely coupled in their evolution.

If only the Crow instability acts, the vortex pair will experience a vortex-reconnection process to form a periodic array of vortex rings, which persist for a long time. Most of the initial circulation is retained in the long-life and large-scale structures. But, when long- and short-wave instabilities coexist, their mutual interaction significantly complicates the late-stage evolution pattern. The mutually-induced shift velocity of the vortex pair due to long-wave instability is a constant; but with short-wave instability this velocity will shortly become much smaller and finally tend to a constant again, about only 1/3 of its initial value. This reduction of the

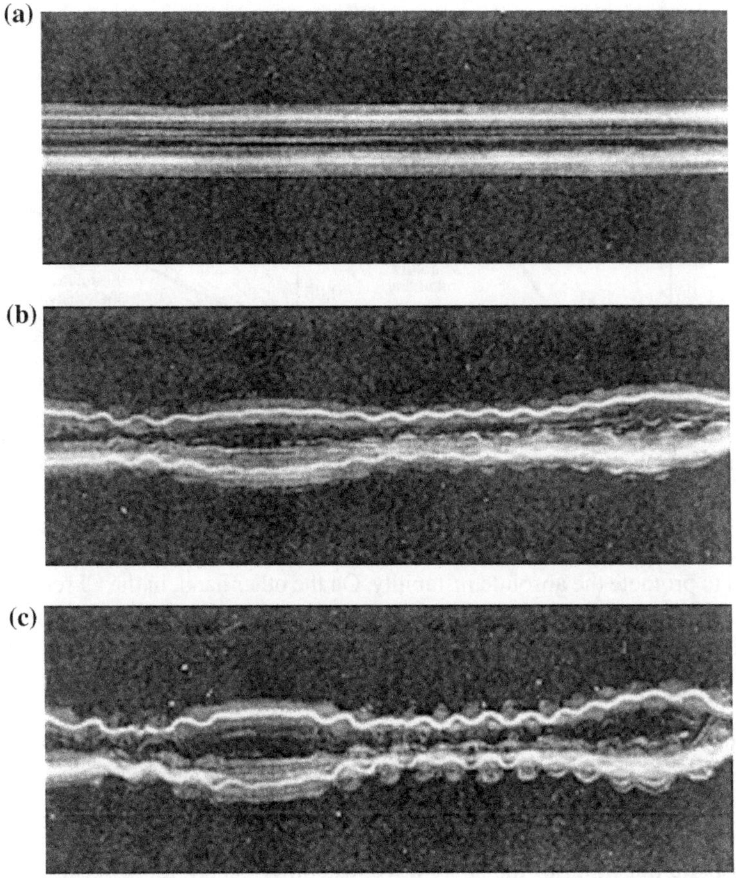

Fig. 10.11 Visualization of vortex pair evolution under the combined action of long-wavelength and short-wavelength instabilities. Illustration of the Rankine vortex deformation induced by the principal modes. **a** $t^* = 1.7$; **b** $t^* = 5.6$; **c** $t^* = 6.8$. From Leweke and Williammson (1998)

shift velocity is an indication of the breakdown of the complicated and organized structures formed by both instabilities.

3. AI/CI character of long- and short-wave instabilities. The spatial-temporal development of both long- and short-wave instabilities of trailing vortices has been examined by Fabre et al. (2000a, b). The dispersion relation for spatio-temporal mode analysis has the form

$$D\left(k, \omega; \frac{a}{b}, \frac{W_0}{U_0}\right) = 0.$$

At the long-wave asymptotic end with $(a/b)^2 \ll 1$, the AI/CI regions are shown in Fig. 10.12a in the parametric $(W_0/U_0, a/b)$-plane. As W_0/U_0 increases the flow changes from CI to AI, and in the limit of $a/b \to 0$ a finite ratio $W_0/U_0 > 0.14$ is

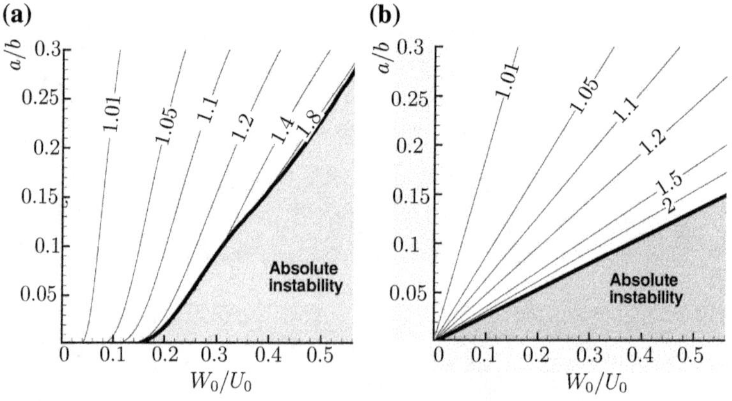

Fig. 10.12 The AI/CI regions of a vortex pair in parameter space $(W_0/U_0, a/b)$. **a** Long-wave instability; **b** short-wave instability. The left part of each plot is convective region, where the ratio of spatial to temporal growth rates is displayed. From Fabre et al. (2000b)

needed to promote the absolute instability. On the other hand, in the CI region on the left part of the figure one may perform a spatial-mode analysis. Then the ratio of the spatial and temporal growth rates can be compared, also shown in the CI region of Fig. 10.12. The spatial growth is seen increasingly larger than the temporal growth as the AI region is closer.

At the opposite short-wave asymptotic end with $|ka| = O(1)$, the AI/CI boundary is shown in Fig. 10.12b. Qualitatively similar to the long-wave case, the flow is convectively unstable when $W_0/U_0 \to 0$, and its spatial growth rate is larger than the temporal one as W_0/U_0 increases. Except for very small ratio a/b, the convective/absolute transition of the short-wave instability occurs for W_0/U_0 larger than that for the long-wave instability.

The AI/CI character of trailing vortices is of great value in aeronautical application. Referring to (10.2.10), assume $\mathcal{R} \gg 1$ and the wing load has elliptic distribution, then $b/b_0 = \pi/4$ and $W_0/U_0 \simeq 0.13C_L/\mathcal{R}$. For a common transport aircraft in the landing state there is $C_L \approx 2$, $\mathcal{R} \approx 10$, and $W_0/U_0 \approx 0.026$. Thus, by Fig. 10.12 the trailing vortices are convectively unstable, and spatial-mode analysis can be applied to study their evolution. The obtained maximum spatial growth rate of the disturbance is close to the maximum temporal growth rate. But, for wings with small $\mathcal{R}$ or high-lift devices, the trailing vortices may exhibit absolute instability.

10.3 Vortex Breakdown

Vortex breakdown is a sudden and abrupt structural change in the evolution of an axial vortex, typically associated with the appearance of nonlinear instability. Different patterns of vortex breakdown may occur, such as nearly-axisymmetric bubble type, spiral type, and double-helix type. Figure 10.13 shows a typical photo of the bubble-type vortex breakdown, and Fig. 10.14 is a famous breakdown photo because both bubble and spiral types appear simultaneously.

Vortex breakdown plays an important favorable or unfavorable role in nature and technology. It has been observed in tornadoes, dust devils, and water spouts. In a combustion chamber the recirculation zone after the breakdown of the swirling flow may support and stabilize the flame, and enhance the mixing of fuel and air. The trailing vortices behind large aircrafts could be destructed by breakdown, and thereby their threat to a following aircraft would be reduced during take-off and landing. However, the breakdown of the leading-edge vortices of a slender-wing aircraft at large angles of attack may significantly decrease nonlinear vortex lift, increase the drag, and deteriorate the controllability. Figure 10.15 shows another example in hydraulic engineering. Strong vortex can be formed in inlet cone of the draft tube of a Francis turbine at certain off-design conditions, which may subject to breakdown in either spiral or bubble type, associated with very harmful strong pressure fluctuation.

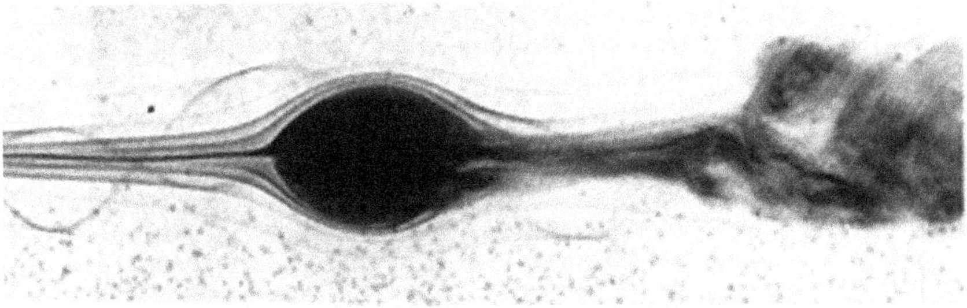

Fig. 10.13 A bubble-type vortex breakdown. Reproduced from van Dyke (1982)

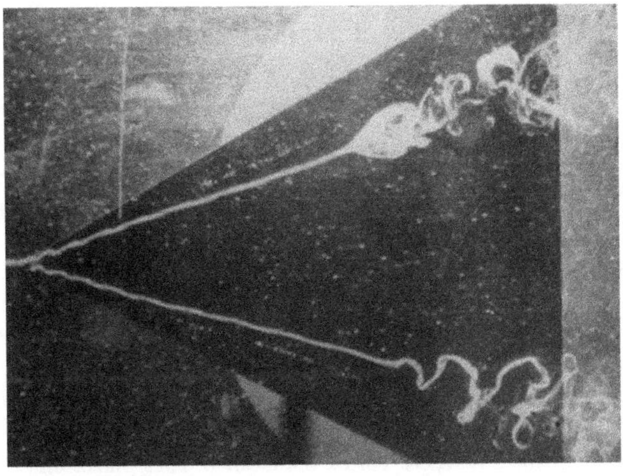

Fig. 10.14 Two types of vortex breakdown on the leeside of a delta wing

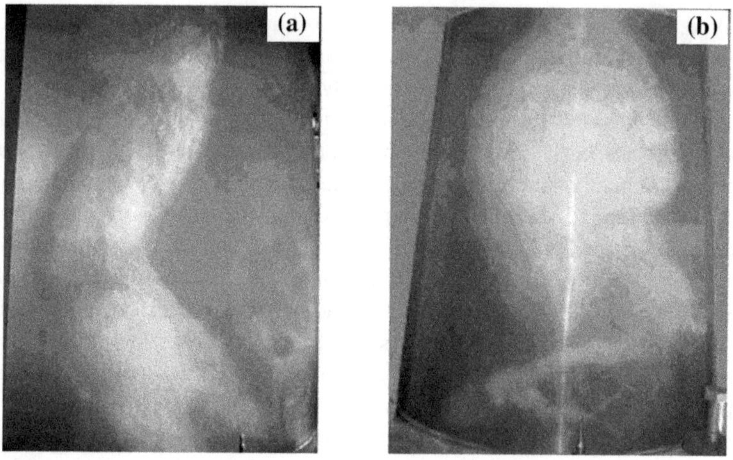

Fig. 10.15 Vortex breakdown in the draft-tube cone of a Francis turbine model (*photos* taken at harbin electric machinery Co. by Q. D. Cai). **a** Spiral type and **b** bubble type

As a typical and extremely complicated nonlinear-dynamics phenomenon, a clear understanding of the vortex breakdown at the fundamental level is of significant value. But owing to its inherent complexity, so far a unified and commonly accepted theory on the physics and topological structure of vortex breakdown has not yet appeared. This situation is also responsible for the lacking of technical means in achieving effective and efficient control of vortex breakdown. Nevertheless, our current knowledge has shed some light on its basic mechanism and control principle, as discussed below.

10.3.1 Breakdown in Terms of Vorticity Dynamics

There have been various models proposed toward a rational understanding of vortex breakdown, among which the one in terms of vorticity dynamics (Brown and Lopez 1990) is physically most intuitive. According to this interpretation, the origin of vortex breakdown comes from the interaction of pressure gradient, axial vorticity ω_z, and azimuthal vorticity ω_θ.

In the discussion of vortex-core dynamics of a swirling flow (Sect. 6.4.2), we have shown that axial pressure gradient (and axial acceleration wdw/dz at the vortex axis) has two constituents, imposed at the outer edge of the core by the exterior boundary condition and caused by the rotational effect of the swirl, respectively. Even under an externally imposed favorable pressure gradient there can still be an adverse pressure gradient (and axial deceleration) at the vortex axis. Actually, in practical circumstances where vortex breakdown takes place, either in a vortex generator with a straight circular pipe followed by an expanding section[3] or on the lee side of a slender wing at large angles of attack, the axial deceleration must cause an outward radial flow due to the continuity. Consequently, the balances of momentum and angular momentum will force a redistribution of ω_z and ω_θ, see (6.1.4), of which the detailed process is as follows.

First, from the observed velocity distribution upstream the breakdown point, it is known that in the early stage of the flow evolution the axial vorticity ω_z is dominant, while ω_θ takes a weak and positive value, that induces a positive axial velocity w as seen from the Biot-Savart formula for axisymmetric flow and applied to the axial velocity at $r = 0$,

$$w(0, z) = \frac{1}{2}\int_0^\infty \int_{-\infty}^\infty \frac{r^2\omega_\theta}{[r^2 + (z - z')^2]^{3/2}} dr dz'. \tag{10.3.1}$$

However, to have an internal stagnation point at the vortex axis that signifies the onset of vortex breakdown and is followed by a recirculating zone, the axial velocity w must change sign, associated with a sign change of ω_θ. Therefore, as pointed out by Brown and Lopez (1990) based on their numerical simulation, the generation of negative ω_θ (hence the swirling flow is of wake-type if $u = 0$) is a necessary condition for the appearance of the internal stagnation point and recirculating zone.

Next, to explain how ω_θ becomes negative, notice that by (6.1.8b) and (6.1.3) there is

$$\frac{D}{Dt}\left(\frac{\omega_\theta}{r}\right) = \frac{1}{r^4}\frac{\partial C^2}{\partial z} = -\frac{u}{r^3}\frac{\partial C^2}{\partial \psi}. \tag{10.3.2}$$

where $C = rv$ and ψ is the Stokes stream function. But, as sketched in Fig. 10.16, an expansion of the flow tube will cause an increase of C; and, in most practical cases with vortex breakdown, one observes $dC^2/d\psi \geq 0$. This is a tilting effect of the vorticity. Therefore, a positive radial velocity $u > 0$ as asserted above will cause a decrease of ω_θ. If this trend continues, ω_θ will eventually become negative.

Then, the distributed azimuthal vorticity in an axisymmetric flow can be viewed as a bundle of thin vortex rings, each having a length $l = 2\pi r$ that is enlarged by the

Fig. 10.16 The generation the axial gradient of the circulation. $\Gamma = 2\pi C$ is circulation of circles centered at axis

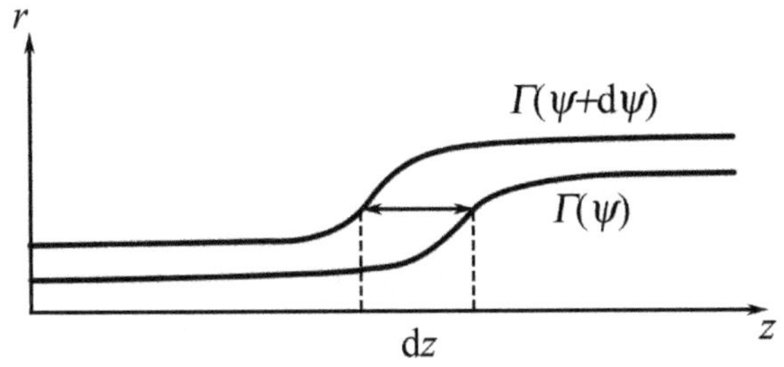

[3]Most of experimental studies of vortex breakdown were conducted in this kind of apparatus.

outward flow $u > 0$. Therefore, once appears, the negative ω_θ will be enhanced due to stretching, which will in turn induce a stronger negative axial velocity at the axis.

We thus see a *positive-feedback mechanism*: an adverse pressure gradient $\partial_z p > 0$ reduces the axial velocity w and causes an outward radial velocity u, associated with a reduction of the azimuthal vorticity ω_θ due to the vorticity tilting; once ω_θ becomes negative, it is enhanced by the stretching effect due to $u > 0$, and hence causes a further reduction of w to make it become zero and even negative. This nonlinear positive-feedback mechanism finally leads to the formation of an internal stagnation point and vortex breakdown.

Only the above positive-feedback mechanism, however, cannot stabilize the breakdown bubble at an equilibrium position. This mechanism has to be counteracted by a *negative-feedback mechanism*. In fact, the latter does exist, again due to the vorticity stretching and tilting. Once is formed, the breakdown bubble will force the external fluid to move outward like passing a bluff body, making the vorticity tubes tilted outward too and stretched as well. These two effects enhance the outward radial velocity from the vortex axis, which supplements the fluid momentum far from the axis to suppress the external axial adverse pressure gradient. It is the coexistence and dynamic balance of the positive and negative feedback mechanisms that forms a complete vorticity-dynamics interpretation of the vortex breakdown.

10.3.2 Breakdown in Term of AI/CI

We have seen the AI/CI analysis of vortices enables the hydrodynamic stability theory to proceed from local temporal-mode to spatial-temporal mode. The theory may also help understand the mechanism of the breakdown development after its sudden onset, especially the mode selection in the breakdown zone, i.e., under what conditions there appear axisymmetric bubble breakdown, and helical or double-helical breakdown.

Loiseleux et al. (1998) have examined the relation between vortex breakdown and the AI/CI characters of a pre-breakdown vortex, for the latter they used the Rankine vortex as the basic flow, with axial velocity profile being a plug flow. The shape of AI domain is qualitatively similar to its counterpart for q-vortex (Figs. 10.5 and 10.6). The authors found that for $n = -2$, when the external axial velocity becomes zero and the swirl number $q = 1.61$ (or the Rossby number $Ro = q^{-1} = 0.62$), the flow becomes absolutely unstable. A comparison with experimental data indicates that the AI/CI criterion does provide a reasonable estimate of vortex breakdown onset.

Yin et al. (2000) have used available experimental data of vortex breakdown to conduct an AI/CI numerical analysis, where the basic flow is assumed as a q-vortex and a modified q-vortex, respectively, which fit the experimental profiles quite well. The result shows $\omega_i^o > 0$ for all tested cases of both experiments, that is, the flow in the near wake of breakdown bubble is indeed absolutely unstable. They then applied the AI/CI theory to predict the dominant frequency f^0 in a breakdown bubble. They found that the Strouhal number (non-dimensionalization of f^0 by the core radius and characteristic axial velocity of the basic flow) predicted by AI/CI analysis takes

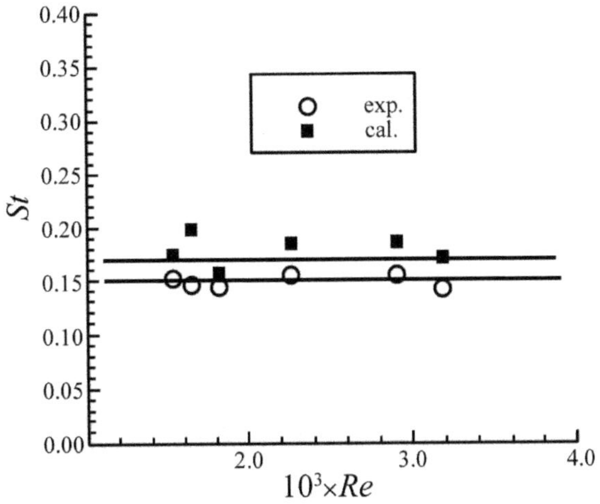

Fig. 10.17 Dominant Strouhal number in a breakdown bubble. $St = 0.171$: theoretical predication. $St = 0.15$: average of measured values. From Yin et al. (2000)

about the same value around 0.17, in reasonable agreement with the experimentally observed value around 0.15, see Fig. 10.17. All these observations strongly suggest that the absolute instability should be a common character in vortex breakdown.

Let us return to the example of vortex breakdown in the draft-tube of a Francis turbine (Fig. 10.15). Based on Reynolds-averaged Navier-Stokes simulation of the draft-tube flow, Zhang et al. (2009) have identified a strong swirling vortex in the inlet cone under certain off-design conditions, which suffers strong and robust spiral vortex breakdown. They found that before breakdown this vortex can also be modelled by a q-vortex defined by (10.2.9) with variable (a, q) along its axis, and these parameters fall deeply into the AI zone (cf. Figs. 10.5 and 10.6), causing a *global instability* of the vortex. Therefore, the basic strategy of draft-tube vortex control should be enhancing the forward mass flux near vortex axis to shift the values of (a, q) out of the AI zone. This can be realized by a water injection, and more radically by careful optimal design of the runner blades right upstream the draft tube. Model experiments and numerical simulations have proved this strategy is successful (Wu et al. 2010). This control strategy is also a confirmation of the intimate relation between vortex breakdown and AI/CI instability.

10.4 Vortex Ring Instability and Transition

After the linear instability properties of a single rectilinear vortex are identified, it is quite difficult to further study the nonlinear evolution and transition to turbulence of a single rectilinear vortex. This is due first to the infinite axial extension of columnar vortices, and also to the requirement of total-vorticity conservation by which columnar vortices appear naturally in pair or, as a theoretical model, in an array

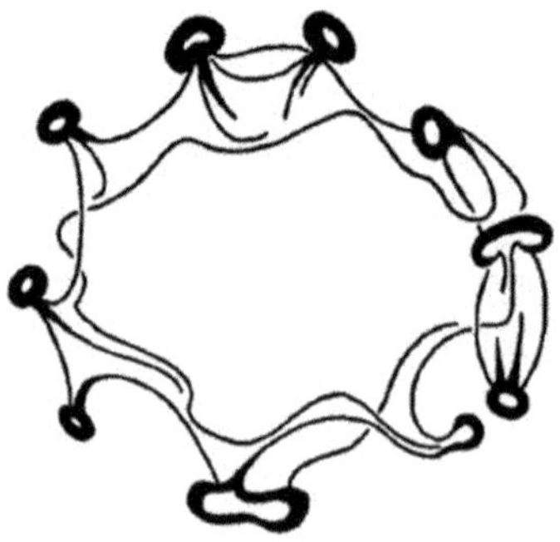

Fig. 10.18 Sketch of the nonlinear breakdown of a drop of ink in water. From Tennekes and Lumley (1972)

of infinitely many pairs.[4] However, these delicate issues do not exist for compact vortex rings which are physically more natural objects. Their analytical solutions with various integral invariants, as given in Chap. 7, may serve as basic-flow models. Ultimately, vortex-ring instability and transition to turbulence are frequently encountered important phenomena in nature and many engineering applications. Figure 10.18 shows how a vortex ring formed by a drop of ink in still water loses its stability and evolves nonlinearly to form small vortex rings which in turn could experience the same process. This is a vivid example of the *cascade process* typical in transition to turbulence, to be discussed in the next chapter.

In this section we outline some relevant results on the vortex-ring instability and transition.

10.4.1 Linear Instability: Single Vortex Ring

Early studies of vortex-ring stability may be traced back to late 1880s using Kelvin waves, which suggested that inviscid vortex rings is neutrally stable. It is not until half century later that the rings were experimentally observed to have azimuthal instability, which was then predicted by instability theories and confirmed by flow visualizations after another 40 years. For the review of these progresses see Shariff and Leonard (1992) and Dazin et al. (2006a).

A vortex ring has certain aspects in common with a vortex pair as in trailing vortices. Not only they move by self-induction, but also they are both in a straining field. Thus, for a vortex ring one should find counterparts of Crow instability and elliptical instability as well as their interaction seen in trailing vortices. This comparison has led Widnall et al. (1974) to give the first interpretation of vortex-ring instability mechanism. By applying the localized induction approximation (3.1.40) to a thin vortex ring, they showed that the primary instability wave is azimuthal one (the bending mode) and the direction of instability growth is at 45° with the vortex

[4]For a numerical simulation of the nonlinear development of instability modes on such a model see Faddy and Pullin (2005).

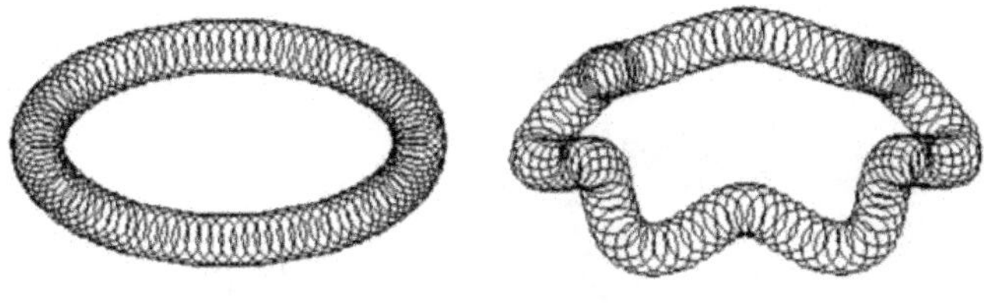

Fig. 10.19 Geometry of original and disturbed vortex rings. From Dazin et al. (2006a)

motion direction, see Fig. 10.19, which is evidently a counterpart of Fig. 10.9. However, scaled by ring radius R, the wavelength of the ring Crow instability is much smaller than the case of trailing vortices.

Assuming the basic flow is thin-core inviscid vortex rings, Widnall and Tsai (1977) show that it is the strain field induced by the ring that causes the onset of instability. This mechanism is the same as in the Crow instability. They then make prediction of the most linearly amplified azimuthal wavenumber n and growth rate. But, vortex ring differs from columnar vortex pair in that now there is $n = kR$, where k is the wavenumber of steady infinitesimal disturbance wave. Saffman (1978) points out that since n is an integer, k will not be exactly the single wavenumber of that infinitesimal wave but must be in the band of wavenumbers around it. He thus develops a theory to predict the bandwidth, which also takes into account of the viscosity translational velocity (7.3.2) (but remains at inviscid stability analysis) and, like the Moore-Saffman model for columnar vortex formed by roll-up of a vortex sheet (Sect. 6.5.2), assuming the ring forms from vortex-sheet roll up. The short-wave elliptical instability leads to a narrow band grow independently, which determine the number of standing waves around the core azimuth.

These theoretical predictions based on linear instability analysis have been quantitatively tested by detailed experimental studies of Dazin et al. (2006a) using both laser induced fluorescence and particle image velocimetry. They confirmed the primary instability mechanism as due to the vortex-ring's self-induced strain field, and found that the viscous vortex-ring instability is initially non-viscous: the instability geometry and most unstable modes, as well as band of growing modes, are well predicted by inviscid theories. Among different basic-flow models, Saffman's (1978) model appears better. But the growth rate was overestimated by inviscid theories since it is strongly influenced by viscosity.

10.4.2 Nonlinear Instability and Transition: Single Vortex Ring

After the start of linear instability phase, nonlinear phenomena appear rapidly, but relevant studies have to mainly rely on experiments and numerical simulations. Dazin et al. (2006b) and Gan et al. (2011) have studied experimentally the nonlinear development of vortex-ring instability and transition to turbulence. Their findings are in agreement with and further detailed by several research groups using direct numer-

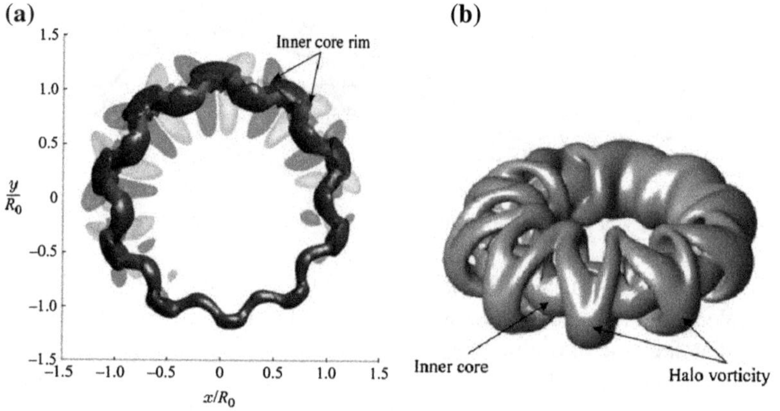

Fig. 10.20 Isosurface visualization of a vortex ring of core radius $a_0 = 0.2$ in nonlinear instability phase by direct numerical simulation with $Re = \Gamma_0/\nu = 7500$ at $t = 65$. **a** *Top view* of vorticity isosurfaces. *Dark surface* corresponds to the *inner* core region $|\omega| = 3.8$, *mid-grey* isosurface corresponds to $|\omega| = 0.8$, and *light grey* to $\omega_z = -0.8$ that visualizes the secondary structure. **b** Isosurface of the second invariant II of velocity gradient tensor, $II = -0.005$. Quantities here are made dimensionless by the initial ring radius R_0 and circulation Γ_0. From Archer et al. (2008)

ical simulation (see Archer et al. (2008) and references therein). Below we outline the evolution processes of vortical structures in nonlinear instability and transition phases of vortex rings.

At certain amplitude, the independently growing modes begin to interact each other nonlinearly, producing higher harmonics. But their growth is rapidly stopped by the development of low-order modes. Then there appears an $n = 0$ mode which, due to non-uniform vortex stretching and azimuthal pressure gradient, is associated with oscillatory azimuthal streams (along the core axis) around the vortex, namely oscillatory swirl. In this phase the inner vortex core with high vorticity peak and outer core with low vorticity ("halo" vorticity) evolve in different manners, the former is displaced by elliptical instability into a stationary wave pattern, while the latter is displaced in opposite direction. Meanwhile, secondary vortical structures develop all around the vortex in its peripheral zone due to the tilting and reorganization of the outer core, forming a series of loops of alternating signed vorticity that encompass the inner core, see Fig. 10.20. This pattern can be compared with Fig. 10.11c.

Then, as the loops in secondary structures trail outside the ring, they detach and reattach to their neighbors to form hairpin vortices that fill the wake, indicating the termination of secondary structures, see Fig. 10.21. This is followed immediately by transition to turbulence.

10.4.3 Instability and Transition: Multiple Vortex Rings

Like trailing vortices, vortex rings may appear in pair or more, with either the same-signed circulations as in the case of leap frogging shown in Fig. 7.5, or opposite-

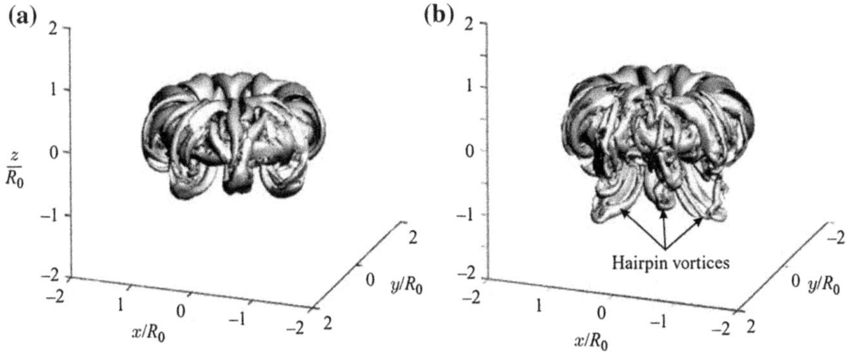

Fig. 10.21 Isosurface of $II = -0.25$ showing the termination of the secondary structures, forming hairpin vortices in the wake for the same vortex ring as in the preceding *figure*. **a** $t = 75.8$, **b** $t = 79.4$. From Archer et al. (2008)

signed circulations as what happens when a vortex ring impinges on a wall. This latter case is of significant interest since it provides the simplest scientific model for understanding and control of ring-wall interactions encountered in many engineering problems, for example the nearly ring-like spiral vortices shed from rotor-blade tips may impinge on the helicopter body, and deteriorate its flight performance. The instability of such multiple vortex rings is of course one's concern. Thus, many studies have been made on vortex-ring interaction with solid surfaces of various geometry at different inclinations. We close this chapter with a review of the instability and transition of a vortex ring in normal collision with a flat wall.

As a vortex ring of circulation Γ approaches a flat plane, its initial interaction with the wall can be modelled by that with an image ring of $-\Gamma$ at the other side of the wall. As the two rings approach each other, they induce an outward radial velocity to enlarge the rings and slow down their axial motion. Then, viscosity enters into play when the vortex ring is near the plane. The ring-induced tangent pressure gradient on the wall implies a boundary vorticity flux to generate new vorticity that forms secondary vortex ring of opposite circulation, which interacts with the primary ring. At sufficiently high Reynolds numbers, tertiary vortex ring may also be generated near the plane. This sequence is sketched in Fig. 10.22. The trajectories

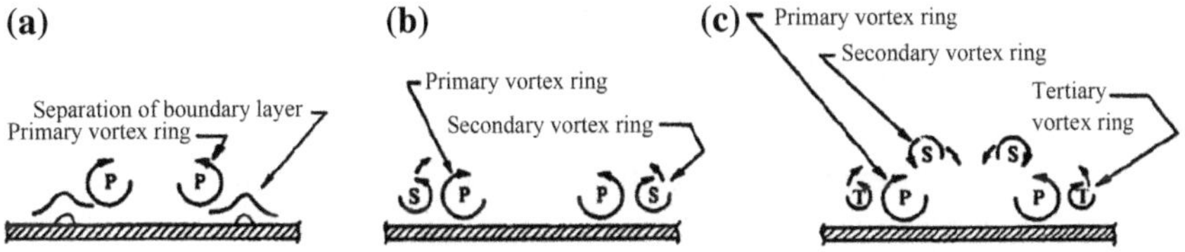

Fig. 10.22 The generation and motion of the secondary and tertiary vortices during the interaction of a vortex ring with a flat plane. From Walker et al. (1987)

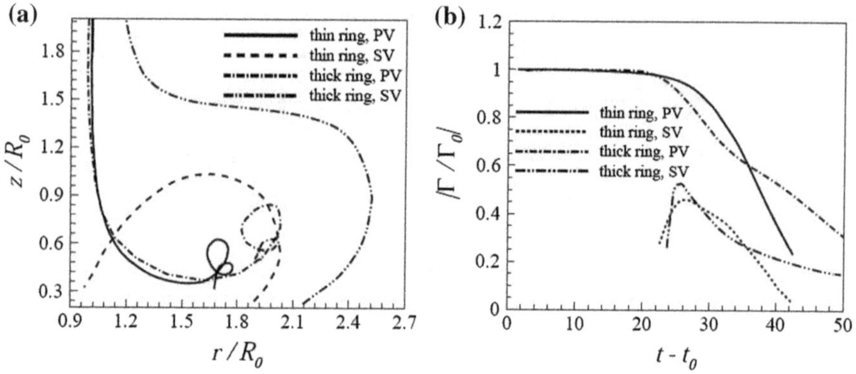

Fig. 10.23 The time evolution of the trajectories (**a**) and circulations (**b**) of two primary vortices (PV) of different thicknesses and corresponding secondary vortices (SV) near a normal wall. From Ren and Lu (2013)

and circulations of the primary and secondary vortex rings as time goes on are shown in Fig. 10.23 for two initial ring thicknesses, calculated by Ren and Lu (2013) using large-eddy simulation.

During the generation process of multiple vortex rings with different signs and magnitudes of circulations, each vortex ring experiences similar linear and nonlinear long- and short-wave instabilities as the single vortex ring, with also two types of secondary vortical structures: loops surrounding the vortex cores and hairpins shedding into wake. But the existence of multiple ring-like structures adds complexity to the instability mechanisms, causing the rings stretched and enhancing the growth of those structures.

Figures 10.24 and 10.25 display the evolution of the three-dimensional vortex structures, also calculated by Ren and Lu (2013). In addition to loops and hairpins, we see that the secondary ring moves around the primary ring after it is generated from the wall. The generation of large-scale hairpin vortices is associated with the stretching and deformation of the tertiary ring caused by the azimuthal instability. In the late stage of collision, the strong vortex-vortex and vortex-wall interactions lead to breakdown of the vortical structures into small-scale vortices. The turbulent kinetic energy grows rapidly during the transition process.

10.5 Problems for Chapter 10

10.1.[5] Derive the Rayleigh equation for the inviscid stability of the shear flow from the vorticity dynamics equation.

[5]Problems of this chapter are selected from Drazin (2002).

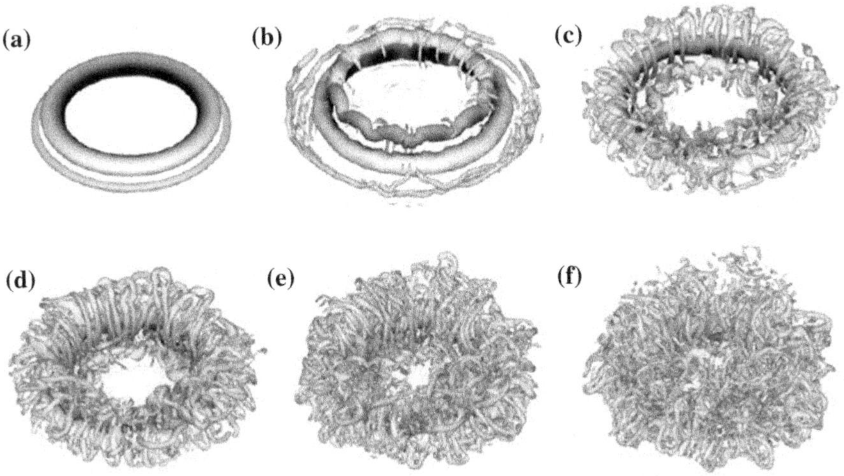

Fig. 10.24 The visualization of vortex structures evolution for a thin vortex ring by the Q-criterion (Jeong and Hussain 1995), where $Q \equiv (\|\mathbf{\Omega}\|^2 - \|\mathbf{D}\|^2)/2$ is chosen as 1.5, with $\mathbf{D}$ and $\mathbf{\Omega}$ being the symmetric and skew-symmetric parts of the strain rate tensor, respectively. **a** t = 22.5, **b** 27.5, **c** 32.5, **d** 35.0, **e** 37.5, **f** 40.0. From Ren and Lu (2013)

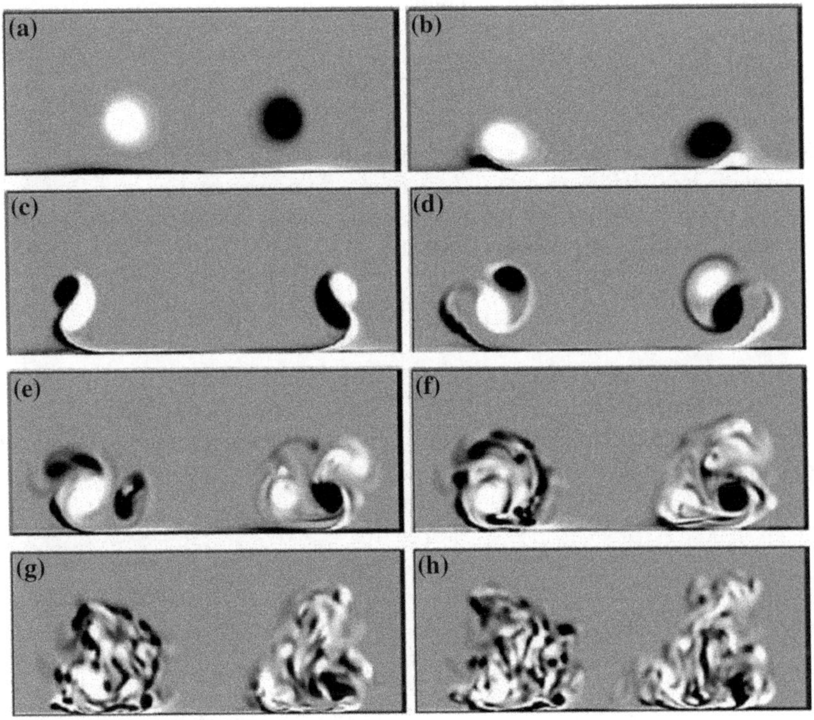

Fig. 10.25 The vorticity field for a thin vortex ring. **a** t = 20.0, **b** 22.5, **c** 25.0, **d** 27.5, **e** 32.5, **f** 35.0, **g** 37.5, **h** 40.0. From Ren and Lu (2013)

10.2. In Sect. 10.1.1 only the simplest basic flow is considered, that is, two layers with different velocities. Now consider the basic flow as

$$(\boldsymbol{U}, \rho) = \begin{cases} U_2 \boldsymbol{e}_x, \rho_2 & \text{for } y \geq 0, \\ U_1 \boldsymbol{e}_x, \rho_1 & \text{for } y < 0, \end{cases}$$

Prove that for this case, the eigenvalue relation is

$$s = -ik\frac{\rho_1 U_1 + \rho_2 U_2}{\rho_1 + \rho_2} \pm \left\{ \frac{k^2 \rho_1 \rho_2 (U_1 - U_2)^2}{(\rho_1 + \rho_2)^2} - \frac{kg(\rho_1 - \rho_2)}{\rho_1 + \rho_2} \right\}^{1/2}. \tag{10.5.1}$$

10.3. The moderation of Kelvin-Helmholtz instability by surface tension. Show that if there is surface tension γ between the two fluids, then the eigenvalue relation is

$$s = -ik\frac{\rho_1 U_1 + \rho_2 U_2}{\rho_1 + \rho_2} \pm \left\{ \frac{k^2 \rho_1 \rho_2 (U_1 - U_2)^2}{(\rho_1 + \rho_2)^2} - \frac{k^2}{\rho_1 + \rho_2} \left[\frac{g(\rho_1 - \rho_2)}{k} + k\gamma \right] \right\}^{1/2}. \tag{10.5.2}$$

Deduce that the flow is stable if and only if

$$(U_1 - U_2)^2 \leq 2\frac{\rho_1 + \rho_2}{\rho_1 \rho_2} [g\gamma(\rho_1 - \rho_2)]^{1/2},$$

and the wavelength of the least stable wave on the margin of stability is $\lambda = 2\pi/k = 2\pi[\gamma/g(\rho_1 - \rho_2)]^{1/2}$.

Using this model show that the wind generates waves on the sea if the difference of the basic air and water speeds is such that

$$|U_1 - U_2| > 6.6\, m/s,$$

and the least stable wave has length $\lambda = 0.017$ m speed 0.008 m/s.

($\rho_1 = 1020\,\text{kg/m}^3$, $\rho_2 = 1.25\,\text{kg/m}^3$, $g = 9.8\,\text{m/s}^2$, $\gamma = 0.074\,\text{N/m}$).

10.4. Study the inviscid instability of the unbound vortex sheet where the velocity profile of the basic flow is

$$U(z) = \begin{cases} 1 & \text{for } z \geq 0, \\ -1 & \text{for } z < 0, \end{cases}$$

Prove that the eigenvalue relation is

$$c_r = 0, \quad c_i = 1$$

and the normalized eigenfunction is

$$\phi = \begin{cases} \exp(-\alpha z) & \text{for } z \geq 0, \\ -i \exp(\alpha z) & \text{for } z < 0, \end{cases}$$

10.5. Consider the shear layer for which

$$U(z) = \begin{cases} z & \text{for } |z| < 1, \\ z/|z| & \text{for } |z| > 1, \end{cases}$$

Prove that the eigenvalue relation is

$$c^2 = (4\alpha^2)^{-1}[(1 - 2\alpha)^2 - e^{-4\alpha}],$$

and the normalized eigenfunction is

$$\phi = \begin{cases} \exp(-\alpha_s(z - 1)) & \text{for } (z \geq 1), \\ (\cosh \alpha_s)^{-1} \cosh \alpha_s z & \text{for } (|z| < 1), \\ \exp(\alpha_s(z + 1)) & \text{for } (z < -1) \end{cases}$$

where α_s is the root of $(1 - 2\alpha + e^{-2\alpha} = 0)$, and $\alpha_s \approx 0.64$.

10.6. Study the stability characteristics of an inviscid triangular jet, when $U(z) = 1 - |z|$ for $|z| < 1$ and $U(z) = 0$ for $|z| > 1$. Show that the eigenvalue relation for the sinuous mode (ϕ even) is

$$2\alpha^2 c^2 + \alpha(1 - 2\alpha - e^{-2\alpha})c - [1 - \alpha - (1 + \alpha)e^{-2\alpha}] = 0.$$

Hence show that this mode is unstable for $0 < \alpha < \alpha_s$ where $\alpha_s \simeq 1.833$ and $c_s = 0.367$.

For the varicose mode (ϕ odd) show that

$$c = (2\alpha)^{-1}(1 - e^{-2\alpha}).$$

10.7. Consider the piecewise linear channel flow given by

$$U(z) = \begin{cases} 1 - z & \text{for } 0 \leq z \leq 1, \\ 1 + z & \text{for } - \leq z \leq 0, \end{cases}$$

and determine the eigenvalue relation for waves in an inviscid fluid. What is the group velocity of the waves? Use the group velocity to show that short waves are found in the front of the dispersive disturbance and that long waves are in the back.

10.8. Consider two-dimensional disturbances bounded by rigid cylinders at $r = R_1$ and $r = R_2$, that is the disturbances are only the function of (r, θ, t). The linearized motion equations in the cylindrical coordinates are

$$\frac{\partial u_r}{\partial t} + \Omega \frac{\partial u_r}{\partial \theta} - 2\Omega u_\theta = -\frac{1}{\rho}\frac{\partial p}{\partial r}, \tag{10.5.3}$$

$$\frac{\partial u_\theta}{\partial t} + \Omega \frac{\partial u_\theta}{\partial \theta} + \left(\frac{dV}{dr} + \frac{V}{r}\right) u_r = -\frac{1}{\rho r}\frac{\partial p}{\partial \theta}, \tag{10.5.4}$$

$$\frac{\partial u_z}{\partial t} + \Omega \frac{\partial u_z}{\partial z} = -\frac{1}{\rho}\frac{\partial p}{\partial z}, \tag{10.5.5}$$

$$\frac{\partial u_r}{\partial r} + \frac{u_r}{r} + \frac{1}{r}\frac{\partial u_\theta}{\partial \theta} + \frac{\partial u_z}{\partial z} = 0, \tag{10.5.6}$$

where u_r, u_θ, u_z, p are the perturbation velocities and pressure, Ω is the angular velocity and $V(r) = r\Omega(r)$.

For the two-dimensional disturbances, stream function ψ may be introduced:

$$u_r = \frac{1}{r}\frac{\partial \psi}{\partial \theta}, \quad u_\theta = -\frac{\partial \psi}{\partial r}.$$

Consider the normal mode analysis, that is, let

$$\psi(r, \theta, t) = \phi(r) exp(st + im\theta).$$

1. Show that the stability equation is derived as

$$(s + im\Omega)(D_* D - \frac{m^2}{r^2})\phi - \frac{im}{r}(DZ)\phi = 0, \tag{10.5.7}$$

where $Z(r) = rD\Omega + 2\Omega$ is the vorticity of basic flow, $D = \frac{d}{dr}$, $D_* = \frac{d}{dr} + \frac{1}{r}()$.

2. Show that there is a similar Rayleigh's inflexion point theorem for two dimensional disturbances of rotating flow.

First multiply Eq. (10.5.7) with $r\phi^*/(s + im\Omega)$, integrate it from $r = R_1$ to $r = R_2$, in which ϕ^* is the complex conjugate of ϕ, we deduce that

$$\int_{R_1}^{R_2} r[|D\phi|^2 + \frac{m^2}{r^2}|\phi|^2 + \frac{im(DZ)|\phi|^2}{r(s + im\Omega)}]dr = (r\phi^* D\phi)\Big|_{R_1}^{R_2}. \tag{10.5.8}$$

Hence, show that if there is stability (i.e. $\Re s > 0$) then

$$m \int_{R_1}^{R_2} \frac{(DZ)|\phi|^2}{|s + im\Omega|^2} dr = 0. \tag{10.5.9}$$

What conclusion can you deduce from this equation?

10.9. As the simplest model of the pure vortices, study the inviscid instability of the rigidly rotating fluid column, where

$$V(r) = \Omega r.$$

Try to derive the eigenvalue relation for the disturbances.

Chapter 11
Vortical Structures in Transitional and Turbulent Shear Flows

While in Chap. 10 we focused on the instability and breakdown of axial vortices, in this chapter we turn to the instability of shear layers, both free and wall-bounded. Unlike axial vortices, however, shear-flow transition from laminar to turbulent and fully developed turbulent shear flows have been the main field of turbulence and investigated much more comprehensively and intensively, including the vortical structures therein, which will be our main concern.

11.1 Overview and Background

11.1.1 What Is Turbulence

It is well known that fluid turbulence has been the most difficult problem of macroscopic physics in more than a century. No consensus has been fully reached on what is turbulence and how it is formed. Nevertheless, modern understanding of the essential nature of turbulence has been focused on its three basic properties: *multiple scales*, spatio-temporally *chaotic* behavior, and various so-called *coherent structures*. One thing for certain is that, if one regards vortices as the sinews and muscles of fluid motions (Küchemann 1978), there is no doubt that vortices should also be the sinews and muscles of turbulence; no vortices, no turbulence. It may also be said that turbulence is the most complicated but most common vortical flows in the world.

Before proceed, we first make a brief physical survey of the above three basic properties and associated notions. This is a general background for the subsequent sections. For comprehensive physics, theory, and modelling of turbulence, see e.g., Pope (2000). Abundant literatures are also available related to the vortical structures in turbulence, for example Jiménez (2004).

Leonardo da Vinci was the first to make scientific observation of turbulent flow, and gave its modern name (*turbulenza*) 500 years ago. On his famous sketches of turbulence (one of which is shown in Fig. 11.1), da Vinci wrote: ... *The small eddies*

J.-Z. Wu et al., *Vortical Flows*, DOI 10.1007/978-3-662-47061-9_11

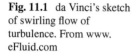

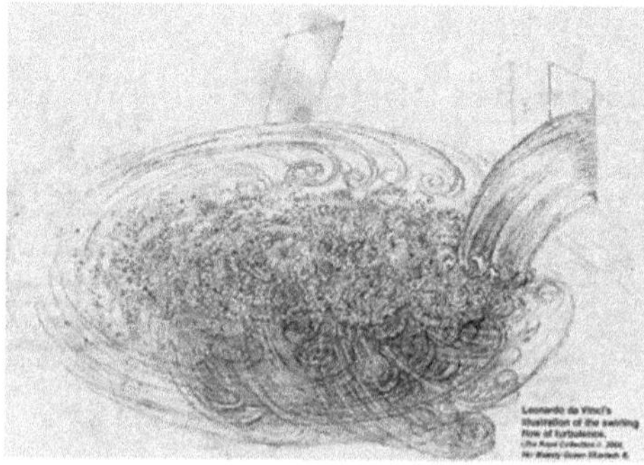

Fig. 11.1 da Vinci's sketch of swirling flow of turbulence. From www.eFluid.com

are almost numberless, and large things are rotated only by large eddies and not by small ones, and small things are turned by both small eddies and large (Gad-el-Hak 2000). This description already captures the sinews of turbulence as multi-scaled and interacting coherent structures!

1. Multiple scales. In preceding chapters we have discussed various aspects of laminar vortical flows with localized vortical structures. In turbulence they appear as the largest coherent structures that draw energy from the main stream. We have seen in Chap. 10 that at large Reynolds numbers $Re_L = UL/\nu$ based on global scales U and L, concentrated vortices are unstable, and we shall soon show that the same is true for attached and free shear layers. This leads to disturbance amplification and formation of smaller-scale vortical structures of scale $l \ll L$. This instability is mainly inviscid in nature.

In three-dimensional vortical flow, a typical inviscid mechanism of producing small-scale structures via instability is *vorticity-line stretching and tilting* as addressed in Sect. 3.2.3 and demonstrated in Fig. 3.10. Then, if based on l and typical velocity fluctuation u' (say the mean square root $\sqrt{|\boldsymbol{u}'|^2}$) there is $Re_l = u'l/\nu \gg 1$ as well, these smaller-scale vortical structures will still be inviscidly unstable, so that the destabilizing process continues onto even smaller structures, and hence turbulence appears. In this evolution from larger to smaller scales, as long as $Re_l \gg 1$, the total kinetic energy of fluid is almost preserved (Sect. 3.3.5). Thus, the energy gained at large scale L is almost conservatively passed down to smaller scales. This range of energy-preserving scales is called the *inertial range*. Eventually, this process reaches the smallest possible scale, on which vortical structures must be dissipated into heat by viscosity.

In turbulence theory, three scales are often used to separate these multiple length scales. One is the macro-scale, called the integral scale, which is the length scale of the large vortices of $O(10^{-1}) \sim O(1)$, but less than unity, times geometric scale of a flow field. In terms of the energy spectrum, the integral scale is often identified with the wavenumber corresponding to maximum energy.

At the opposite end is the Kolmogorov micro-scale or dissipation scale. Most of the viscous dissipation of energy occurs near the Kolmogorov micro-scale. Quantitatively, denote the energy dissipation rate at smallest scale by ϵ, what is relevant in the dissipation stage can only be ϵ and ν of dimensions L^2/T^3 and L^2/T, respectively. Thus a simple dimensional analysis yields immediately the dissipative scales of length η, time τ, and velocity v given by Kolmogorov in 1941 (known as the *K41 theory*):

$$\eta = \left(\frac{\nu^3}{\epsilon}\right)^{1/4}, \quad \tau = \left(\frac{\nu}{\epsilon}\right)^{1/2}, \quad v = (\nu\epsilon)^{1/4}. \tag{11.1.1}$$

Kolmogorov scale η represents statistically the smallest possible eddy structures before dissipated by viscosity (yet still much larger than molecule's free path) with $Re_\eta = v\eta/\nu = 1$, where v is the "dissipative velocity".

Between the integral and dissipative scales is Taylor's micro-scale λ. It is the statistical length of the small eddies at which turbulence may be supposed to become isotropic.

2. Cascade. The evolution process from large to small scales is a crucially important characteristic feature of turbulence and named *cascade* by Richardson (1922). Energy cascade is referred to the transfer of kinetic energy from large, macroscopic scales of motion, where it is presumed to be input to the flow, through successively smaller scales, ending with viscous dissipation and conversion to heat (thermal energy). The main physical mechanism proposed for this process is interaction of vortices: large vortices break into smaller ones that, in turn, break into yet smaller ones.[1] Richardson's (1922) rhyme vividly expresses this cascade process, as a development of da Vinci's observation and a pictorial explanation of the multi-scale nature of turbulence:

Big whorls have little whorls,
which feed on their velocity.
And little whorls have lesser whorls,
And so on to viscosity.

Figure 11.2 displays the physical mechanism of cascade and the aforementioned three length scales.

Since in a fully developed equilibrium turbulence the rate of kinetic energy gained at large scale is approximately the same as ϵ, we may take ϵ as a constant energy flux passed from large to small scales. Before the Kolmogorov dissipative scale is reached, viscosity does not enter into play. In this inertial range of length scale or of wave number $k \sim l^{-1}$, relevant quantities are only ϵ and k, say. In the wave-number *spectrum space*, the Fourier image $K(k)$ of turbulent fluctuation energy $|\boldsymbol{u}'|^2/2$ can only depend on k and ϵ. Thus another dimensional analysis yields the famous *five-third scaling law*

$$K = A\epsilon^{2/3}k^{-5/3}, \tag{11.1.2}$$

[1] The word "break" here does not mean the specific breakdown process of an axial vortex discussed in Sect. 10.3.

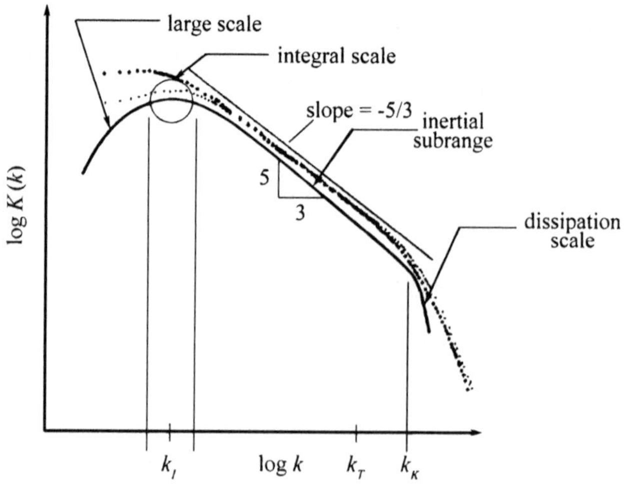

Fig. 11.2 Energy spectrum $K(k)$ versus wave number k in double logarithmic coordinates

also found in K41 theory. It has been extensively confirmed by enormous experiments and *direct numerical simulations* (DNS) at large Re over the past 70 years, see also Fig. 11.2.

The K41 theory holds in statistic sense and reflects the reality only approximately. As observed by da Vinci and by modern experiments and numerical simulations, large vortices and tiny ones can have direct interactions, for example the former induces neighboring small eddies to align along its circumferential direction like a bobbin. Further investigations have indicated that the Kolmogorov five-third law (11.1.2) cannot be universally true: as k increases vortex stretching implies increasingly thinner filament structures, among which the distances must become larger. Actually, both DNS and experiments have revealed the existence of isolated vortex filaments even in dissipation scale, but that happens very rarely. This phenomenon is called *turbulence intermittency*, and has led to an accurate revision of (11.1.2) discovered by She and Leveque (1994).

Turbulence also appears in large-scale motion of ocean and atmosphere, of which the vertical scale is much smaller compared to its horizontal extension, and can be approximately studied by two-dimensional turbulence model. Two-dimensional turbulence also have multi-scale coherent structures and associated cascade process. However, as k increases these structures are not enhanced by vortex-line stretching, but instead by the sharpening of vorticity gradient $\nabla\omega$. This leads to the merging of smaller-scale eddies to larger ones, with associated *inverse cascade* of kinetic energy from larger to smaller k's (e.g., Wu et al. 2006, Sect. 12.3.1).

Another view of this difference in two- and three-dimensional turbulence is as follows. In Sect. 3.3 we have met various integral invariants for inviscid (circulation-preserving) flows. A real flow has to obey simultaneously the constraint of several integral invariants. Thus, let n be the spatial dimension of a turbulence, when $n = 3$,

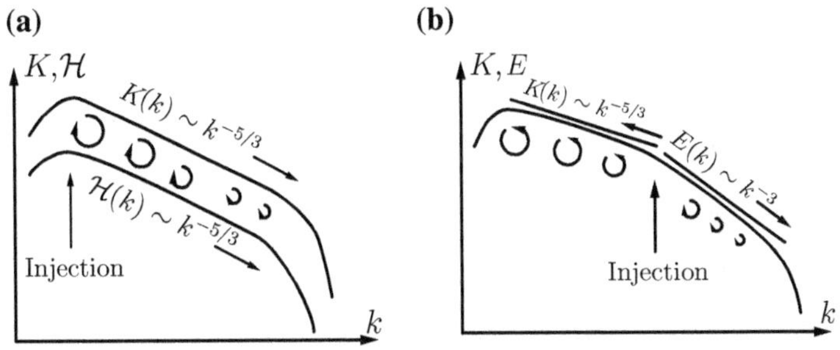

Fig. 11.3 Joint cascades of **a** $(K, \mathcal{H})$ pair in three dimensions and **b** (K, E) pair in two dimensions, as wave number k of turbulent eddies increases. Larger k implies smaller eddies. "Injection" points to the wave numbers where energy is injected into the turbulence. In the inertial range, while for $n = 3$ both K and $\mathcal{H}$ are cascading down to smaller scales at slope $-5/3$ (in log-log coordinates), for $n = 2$ K has *inverse cascade* to larger scales at slope $-5/3$ and E is cascading down at slope -3. Courtesy of Zuo-Li Xiao and Yan-Tao Yang

both total kinetic energy K and helicity $\mathcal{H}$ are invariants in the inertial range. By contrast, for $n = 2$ one does not have helicity but instead the total enstrophy E is invariant. While both K and E are non-negative, $\mathcal{H}$ does not have definite sign. This sign difference of $(K, \mathcal{H})$ pair and (K, E) pair has been found to be responsible for the observed distinct cascade behaviors of three-dimensional and two-dimensional turbulence, of which a schematic sketch is shown in Fig. 11.3. For detailed studies see, e.g., Chen et al. (2003) and Xiao et al. (2009) as well as references therein.

3. Chaos versus randomness. All turbulent flows are solutions of the Navier-Stokes equations which themselves are *deterministic*. But at large Re the N-S solutions are very sensitive to the initial data, so that except very short initial period any realization of turbulence is not repeatable. Two realizations at almost identical initial-boundary conditions will eventually differ each other enormously (the "butterfly effect"). For example, the motion and deformation pattern of an elliptic vortex ring by stretching and tilting due to self induction and instability shown in Fig. 3.10, or the exact shape of the knotted vortex filament shown in Fig. 3.14, is not completely repeatable.

As a consequence of the sensitivity to initial-boundary conditions, many realizations of the N-S solutions of turbulent flows must exhibit *chaotic* behavior. If an instability mechanism changes a steady laminar flow to a laminar periodic flow, a subsequent evolution will make it quasi-periodic (still laminar), and finally chaotic.[2] The last status looks like *random* events. While a purely random process can only be described by statistical approach, which was indeed the character of early-stage turbulence theory and has still been very useful today, it cannot cover all aspects of turbulence. In particular, coherent structures do obey quasi-deterministic equations as we shall see later, and each DNS of turbulence gives exact deterministic evolution of those apparently randomly appeared vortical structures.

[2]While turbulent flows all have chaotic behavior, the reverse is not true.

4. Coherent structures. Owing to the multi-scale and chaotic characters of turbulence, it is impossible to identify and define turbulent structures as clean as we do for those in laminar flow. Perhaps inevitably, terms like turbulent "eddy", "whorl", and "coherent structure" are all not well defined concepts. A *coherent structure* may not simply correspond to a single vortex; this term is used to indicate a fluid motion that has coherence over a spatial region and lasts for a reasonable period of time (Panton 2001). Coherent structures are identified as being easy to "see", may or may not be of fairly large scale, and are somewhat persistent. They are mainly groups of evolving vortical structures in turbulence which are spatially well-organized and contain essential portion of turbulence energy. They can be identified from the turbulent flow field as quasi-deterministic structures in the sense that they are more unpredictable in spatial or temporal phases than in size or in shape.

As a complicated, nonlinear dynamic system, a turbulence field consists of many time-dependent coherent structures of various characteristics, interacting with each other. Their generation, evolution, interaction and dissipation dominate the nature and statistics of turbulence fields. Thus, they are crucial entities for understanding and modelling of turbulence, as well as for controlling turbulent phenomena relevant to mixing, heat transfer, combustion, turbulent drag and acoustic noise, etc. To understand, prescribe, and control turbulent flows, we have to discuss these vortical structures, which is just what we will do in this chapter.

11.1.2 Mean Turbulent Flow

Despite the extreme complexity of turbulent flows, they are still governed by the Navier-Stokes equation, based on which deterministic numerical methods have been developed, such as DNS. But the cost is high and so far DNS is still limited to relatively low Reynolds-number flows.

In science and engineering, it is not always necessary to know everything about a particular realization of a turbulent flow, but some averaged properties of turbulent flows, e.g., the mean velocity profile, or the lift and drag of a flight vehicle, are important. This approach started from Reynolds' pioneering paper in 1894.

Experimental studies have always been the primary motivation in advancing one's understanding of turbulence and associated theories. Early stage of quantitative measurement mainly used hotwires to obtain temporal fluctuating signals of velocity at selected spatial points. Naturally, the corresponding theory was of statistic feature. As modern experimental techniques permitted quantitatively visualizing and scanning a distributed turbulent field, deterministic theories able to identify coherent structures have accordingly been developed. In what follows we briefly review the relevant governing equations to exemplify how deterministic and statistic approaches are combined.

1. Reynolds equation and Reynolds stress. Consider an incompressible turbulent velocity field. We separate at each point the velocity as a mean value and a fluctuation:

$$\boldsymbol{u}(\boldsymbol{x}, t) = \boldsymbol{U}(\boldsymbol{x}, t) + \boldsymbol{u}'(\boldsymbol{x}, t). \tag{11.1.3}$$

We substitute (11.1.3) into (2.6.2b), and take an ensemble averaging denoted by an overline with $\overline{\boldsymbol{u}'} = \boldsymbol{0}$. This leads to the governing equation of the mean velocity field $\boldsymbol{U}$, known as the *Reynolds averaging Navier-Stokes equation* (RANS equation):

$$\frac{\partial U_i}{\partial t} + U_j \frac{\partial U_i}{\partial x_j} = -\frac{1}{\rho}\frac{\partial P}{\partial x_i} + \frac{1}{\rho}\frac{\partial}{\partial x_j}\left(\mu \frac{\partial U_i}{\partial x_j} - \rho \overline{u'_i u'_j}\right). \tag{11.1.4}$$

Unlike (2.6.2b), (11.1.4) is not closed as it contains an extra unknown term

$$\tau'_{ij} = -\rho \overline{u'_i u'_j}, \tag{11.1.5}$$

known as the *Reynolds stress*. The governing equation for fluctuation $\boldsymbol{u}'$ follows from subtracting (11.1.4) from (2.6.2b), which also contains the Reynolds stress but with opposite sign. One may write down the equation for τ'_{ij}, but then there appears an unknown triple correlation of fluctuating velocity that cannot be expressed by τ_{ij}. Continuing this process will lead to an infinite hierarchy of equations of higher moments, none of which can be closed. This is the so-called *turbulence closure problem* not caused by nature but by the averaging approach.

The main concern in practical engineering problems of turbulent flows is the estimate of the Reynolds stress or its divergence, a *turbulent force* per unit mass

$$\boldsymbol{f} \equiv -\nabla \cdot (\overline{\boldsymbol{u}'\boldsymbol{u}'}) = -\overline{\boldsymbol{\omega}' \times \boldsymbol{u}'} - \frac{1}{2}\nabla \overline{|\boldsymbol{u}'|^2}. \tag{11.1.6}$$

To close the Reynolds equation, one has to appeal to *RANS models* that mimic τ'_{ij} in terms of mean quantities, to which early physical clues came from traditional statistic theory developed by Reynolds, Prandtl, Taylor and others. The theory emphasized the perceived randomness of turbulent flows and dominated the early years of development of RANS models.

2. Mean behavior of wall turbulence. To demonstrate the application of Reynolds' average, we follow Adrian (2007) to make a survey of the mean behavior of wall turbulence. The result is also a necessary preparation for the discussion of Sect. 11.3.

Steady, incompressible, two-dimensional in average, and fully developed smooth-wall turbulence constitutes the canonical forms of wall turbulence, including pipe flow, channel flow, and flat-plate boundary layer at zero pressure gradient (turbulent counterpart of Blasius boundary layer). Understanding the common feature of these flows is the basis of understanding more complex wall flows. Figure 11.4a sketches the Reynolds-averaged profiles of velocity $U(y)$, Reynolds shear stress $-\overline{u'v'}(y)$, and total shear stress

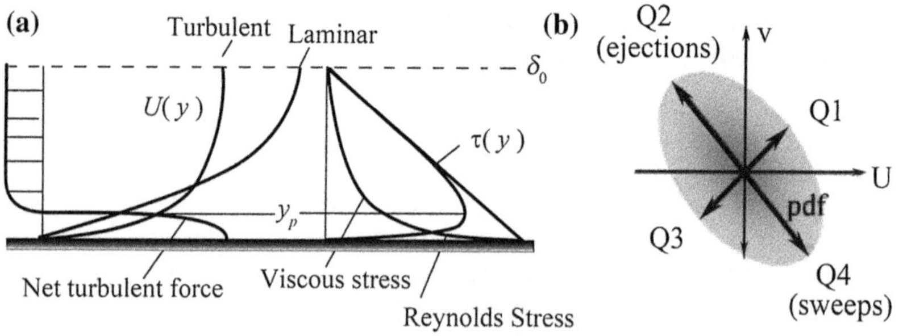

Fig. 11.4 **a** Sketch of profiles of total shear stress $\tau(y)$, Reynolds shear stress $-\overline{u'v'}$,and turbulence force $d(-\rho\overline{u'v'})/dy$ in fully developed turbulent pipe and channel flow. **b** Distribution of $-\overline{u'v'}$ on the (u', v')-plane. From Adrian (2007)

$$\tau(y) = -\rho\overline{u'v'} + \mu\frac{dU}{dy} \tag{11.1.7}$$

in such a channel flow. The profiles are qualitatively similar for turbulent boundary layer. The figure shows that in turbulence the mean velocity profile is flattened compared to laminar flow. Since $\partial U/\partial x = 0$, $\tau(y)$ decreases linearly from the wall friction τ_w to zero at the centerline, where also $dU/dy = 0$ and $\overline{u'v'} = 0$ due to symmetry. Thus, from $\tau(y)$ and flattened $U(y)$ one may infer the profile of $-\overline{u'v'}$ as seen in the figure: it is zero at $y = 0$, reaches a maximum at a location y_p, and then asymptotes to the linear curve as $dU/dy \to 0$. Moreover, by (11.1.6) the net turbulent force is $f_y = d(-\rho\overline{u'v'})/dy$, which must be negative and roughly constant above y_p, and positive below y_p.[3]

The behavior of Reynolds shear stress and turbulent force in Fig. 11.4 indicates that the mean transport of turbulent momentum accelerates $U(y)$ near the wall but retards it in the core flow. The former leads to higher skin friction τ_w than laminar flow. Associated with this feature, the probability density function of u', v' is biased toward events in the second and forth quadrants on the (u', v')-plane (simply called "Q2 events" and "Q4 events"), namely the averaged $u'v'$ is negative or u', v' are anti-correlated, see Fig. 11.4b. Namely, negative streamwise fluctuations are lifted up by positive normal fluctuations, referred to as *ejections*; while positive streamwise fluctuations are moved down toward the wall, referred to as *sweeps*. It should be stressed that, although $\overline{u'v'}$ is only function of y in the present canonical case, u' and v' are functions of both x, y, z and t, and each individual $Q2$ and $Q4$ event may actually be highly intermittent. It would be misleading to think the mean-flow behavior is caused by persistent fluctuations.

Now, let δ be the length scale of the wall shear layer (radius in pipe flow, half-width in channel flow, and a thickness of a boundary layer) and U_0 be the characteristic

---

[3]Reynolds shear stress $-\overline{u'v'}$ is one's main concern in the study of wall turbulence, because it affects the mean velocity profile. There are of course other components such as $-\overline{u'u'}$, $-\overline{u'w'}$, etc., which we do not consider here.

flow velocity, and assume $Re_\delta = \delta U_0/\nu \gg 1$. Prandtl postulated in 1925 that, close to the wall ($y/\delta \ll 1$) there is an inner layer in which $U(y)$ is solely determined by viscous scales of turbulent fluctuations, independent of U_0 and δ. Since in channel flow $\tau(y)$ is a linear function, τ_w and $-\overline{u'v'}$ are related and should be expressible in terms of these scales. This suggests one to introduce a *friction velocity* u_τ and a viscous length scale δ_v by

$$u_\tau = \sqrt{\frac{\tau_w}{\rho}}, \quad \delta_v = \frac{\nu}{u_\tau}, \tag{11.1.8}$$

by which a *friction Reynolds number* can be defined based on u_τ and δ:

$$Re_\tau \equiv \frac{u_\tau \delta}{\nu} = \frac{\delta}{\delta_\nu}. \tag{11.1.9}$$

The scales u_τ and δ_v form a *wall unit*, make x, y coordinates dimensionless to $x^+ = x/\delta_v$ and $y^+ = y/\delta_v$. Then a dimensional analysis and some physical assumptions lead to a partition of $u^+(y^+) \equiv U/u_\tau$ into four regions as confirmed by experiments and DNS (e.g., Pope 2000), as sketched in Fig. 11.5:

1. Viscous sublayer:

$$u^+ = y^+ \quad \text{for } y^+ < 5, \tag{11.1.10}$$

which is the leading term of Taylor expansion of $U^+(y^+)$.

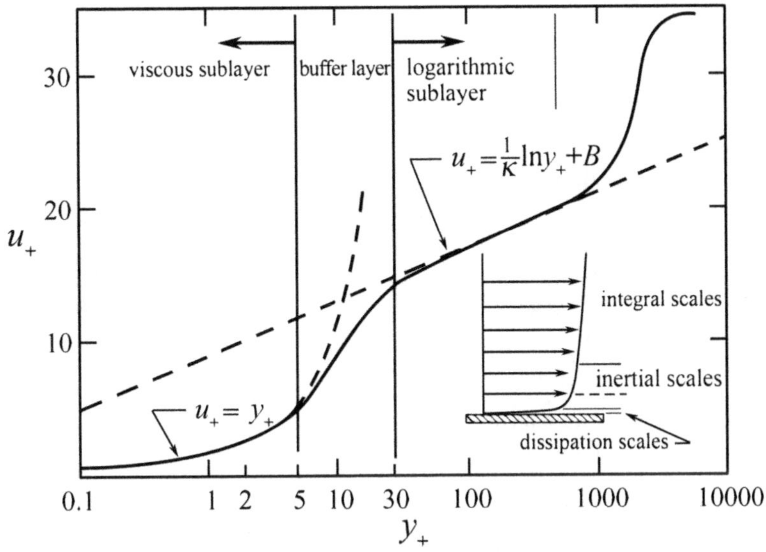

Fig. 11.5 Mean velocity profiles of the turbulent boundary layer

2. *Logarithmic layer*:

$$u^+ = \frac{1}{\kappa} \ln y^+ + B \quad \text{for } y^+ > 30, \tag{11.1.11}$$

where the viscosity has little effect but dominated by turbulent fluctuations. This log law was discovered by von Kármán in 1930, with κ being the *von Kármán constant*. κ and B have been experimentally determined to be about 0.41 and 5.2, respectively.

3. *Buffer layer*: This is a transition layer between the viscous sublayer and log layer at $5 < y^+ < 30$. Some researchers use a value larger than 30 for the beginning position.

4. *Outer layer* (or *defect layer*): $y/\delta > 0.15$, where turbulent fluctuations exhibit *intermittency* and profiles are independent of viscosity. The independent variable is scaled as y/δ.

We remark that the universality of (11.1.10) and (11.1.11) indicates that the wall units are correct scales in the inner layer. They exemplify that a good scaling can capture general physical behavior of the flow and is able to collapse various correlations taken from different data bases to single curves, and can thereby provide an envelope for which detailed studies about the underlying mechanisms must comply.

Having gained the basic concept of mean behavior of wall turbulence, it is natural to further ask (Adrian 2007): first, what special coherent structures are responsible for creating the Reynolds stress and to vary as observed, such as the anti-correlation of u', v'? Second, how the coherent structures change as normal distance from the wall increases to form the above different mean-flow regions?

These questions are our main concern and will be answered below, starting back from wall shear-layer instability to fully developed wall turbulence.

11.1.3 Vorticity Equations and Statistics

The limitation of traditional statistical methods is that they shed no insight into vortical structures. As the key role of vortices in turbulent coherent structures has been revealed by abundant experimental and computational results in the past half century (pioneered by G.I. Taylor), it became natural to consider using the vorticity transport equation to predict the coherent structures or explain their evolution. In particular, experiments can withdraw information of coherent structures from turbulent field by taking the *phase-locked ensemble average*, leaving random fluctuations aside.

1. Mean-vorticity transport equations. As we have learned, the incompressible vorticity equation (2.6.6),

$$\frac{D\boldsymbol{\omega}}{Dt} = \frac{\partial \boldsymbol{\omega}}{\partial t} + \boldsymbol{u} \cdot \nabla \boldsymbol{\omega} = \boldsymbol{\omega} \cdot \nabla \boldsymbol{u} + \nu \nabla^2 \boldsymbol{\omega}, \tag{11.1.12}$$

indicates that the rate of change of the vorticity is due to stretching/tilting of the vorticity caused by the instantaneous velocity gradient, and due to viscous diffusion. To examine how (11.1.12) is applied to turbulent coherent structures, a dimensional analysis may give a simple yet important concept. Take the spanwise vortices in a free shear layer (mixing layer) as an example. Assuming that the mean velocity difference is of $O(U)$ and the thickness of the mixing layer is of $O(\delta)$, such that we have the estimates

$$|\boldsymbol{\omega}| = O(U/\delta), \quad |\nabla \boldsymbol{u}| = O(U/\delta), \quad |\boldsymbol{\omega} \cdot \nabla \boldsymbol{u}| = O(U^2/\delta^2), \quad |\nu \nabla^2 \boldsymbol{\omega}| = O(\nu U/\delta^3). \tag{11.1.13}$$

The ratio of the first term on the right side of (11.1.12) to the second term is of $O(U\delta/\nu)$, i.e., the Reynolds number based on the radial size of large spanwise vortices, which is usually a very large value. Thus, the development of large coherent structures in free shear layer is essentially an inviscid process.

Then, to study the evolution of not only the mean field but also the coherent structures, we modify Reynolds decomposition (11.1.3) to a triple decomposition, consisting of a mean field, a coherent field, and a random field of velocity and vorticity:

$$\begin{aligned} \boldsymbol{u}(x,t) &= \boldsymbol{U}(x) + \boldsymbol{u}_c(x,t) + \boldsymbol{u}_r(x,t), \\ \boldsymbol{\omega}(x,t) &= \boldsymbol{\Omega}(x) + \boldsymbol{\omega}_c(x,t) + \boldsymbol{\omega}_r(x,t), \end{aligned} \tag{11.1.14}$$

where $\boldsymbol{u}$, $\boldsymbol{\omega}$ are the instantaneous quantities, $\boldsymbol{U}$, $\boldsymbol{\Omega}$ are time mean quantities, subscripts c and r denote coherent and random constituents, respectively. Substituting (11.1.14) into (11.1.12), dropping the small viscous term, taking time average, and assuming that the coherent and random motions are uncorrelated, we obtain the mean vorticity equation

$$\frac{D\boldsymbol{\Omega}}{Dt} = (\boldsymbol{\Omega} \cdot \nabla)\boldsymbol{U} + \nu \nabla^2 \boldsymbol{\Omega} + \nabla \times \overline{(\boldsymbol{u}_c \times \boldsymbol{\omega}_c)} + \nabla \times \overline{(\boldsymbol{u}_r \times \boldsymbol{\omega}_r)}. \tag{11.1.15}$$

The first two terms on the right side of (11.1.15) are of the same form as those of (11.1.12), but the stretching and tilting here are caused by the mean velocity gradient only. Two extra nonlinear interaction terms are the curl of the turbulent force (11.1.6), solely from the solenoidal part of the mean disturbance Lamb vector:

$$\nabla \times \boldsymbol{f} = -\nabla \times \overline{(\boldsymbol{\omega}' \times \boldsymbol{u}')},$$

which is now split into coherent and random Lamb vectors and has very clear physical meaning. The third term represents the time-averaged effect of the interaction (i.e., stretching and advection) between the coherent vorticity and coherent velocity fluctuations. The fourth term is the time-mean effect of the interaction between the random vorticity and velocity fluctuations. These terms are very helpful in understanding the development of a turbulent shear flow.

2. Coherent-vorticity transport equation. The transport equation for coherent vorticity can also be derived. Assume once again that the coherent and random motions are uncorrelated. Substitute (11.1.14) into (11.1.12), take the phase-locked ensemble average (denoted by $\langle\cdot\rangle$), and neglect the higher-order quantities, the coherent vorticity equation can be obtained:

$$\begin{aligned}\frac{D\omega_c}{Dt} &= \omega_c\cdot\nabla\boldsymbol{U} + \boldsymbol{\Omega}\cdot\nabla\boldsymbol{u}_c + \nu\nabla^2\omega_c + \nabla\cdot(\omega_c\boldsymbol{u}_c - \overline{\omega_c\boldsymbol{u}_c})\\ &- \nabla\cdot(\boldsymbol{u}_c\omega_c - \overline{\boldsymbol{u}_c\omega_c}) - \nabla\cdot(\boldsymbol{u}_c\boldsymbol{\Omega})\\ &+ \nabla\cdot(\langle\omega_r\boldsymbol{u}_r\rangle - \overline{\omega_r\boldsymbol{u}_r}) - \nabla\cdot(\langle\boldsymbol{u}_r\omega_r\rangle - \overline{\boldsymbol{u}_r\omega_r}).\end{aligned} \tag{11.1.16}$$

It is of interest to notice that the above mean and coherent-vorticity equations, directly applicable to experimental studies, have a counterpart in CFD, namely the *large-eddy simulation* (LES). As a significant progress from RANS simulation, LES is required to capture large eddies (coherent structures) in turbulence of scales larger than the grid size. The small eddies, i.e., subgrid structures, are not resolvable but have to be modelled by statistic means. They are in the same position as random components $(\boldsymbol{u}_r, \omega_r)$.

11.2 Instability and Transition of Free Shear Layer

Stability analysis of free shear flows originated from the pioneering works of Helmholtz and Kelvin, and thus named *Kelvin-Helmholtz instability*.

After a free shear flow becomes unstable, there appear organized spanwise vortical structures followed by vortex pairing and tearing due to their interactions. These initially two-dimensional vortices will evolve to three-dimensional vortical structures including streamwise vortices. Even already become turbulent, spanwise vortical structures can survive up to a Reynolds number as high as 10^7; and other organized structures can also be observed in mixing layers (see Fig. 11.8).

Moreover, the interaction between spanwise and streamwise vortices may modify the structure of a shear layer, thickening it, enhancing the mixing between the fluids with different ingredients or concentrations, or enhancing chemical reactions. Thus, shear layer is often called *mixing layer* as well. The study of the stability of free shear flows can therefore provide important physical insight for flow control in aerodynamics and fluid machinery design.

11.2.1 Instability of Free Shear Layer

Consider the simplest basic flow, an inviscid incompressible flow consisting of two layers with different velocities:

$$U(y) = \begin{cases} U_2 \boldsymbol{e}_x & \text{for } y \geq 0, \\ U_1 \boldsymbol{e}_x & \text{for } y < 0, \end{cases} \tag{11.2.1}$$

which has a flat vortex sheet at $y = 0$. Assume the flows in both upper and lower regions are irrotational (reasonable for instability analysis), with potentials

$$\phi(x, y, z) = \begin{cases} \phi_2(x, y, z) & : \quad y \geq \zeta, \\ \phi_1(x, y, z) & : \quad y < \zeta, \end{cases} \tag{11.2.2}$$

where $y = \zeta(x, z, t)$ is the elevation of the disturbed vortex sheet. Substituting the normal mode decomposition

$$(\zeta, \phi_1, \phi_2) = (\hat{\zeta}, \hat{\phi}_1, \hat{\phi}_2) \exp(i(\alpha x + \beta z) + st) \tag{11.2.3}$$

into Rayleigh equation (10.1.9), and the corresponding boundary conditions, for the eigenvalue of s one obtains

$$s = ik\overline{U} \pm \frac{1}{2} k [\![U]\!], \tag{11.2.4}$$

where $\overline{U} = (U_1 + U_2)/2$ is the mean basic-flow velocity and $[\![U]\!]$ is the velocity jump across the vortex sheet. Hence,

$$(\zeta, \phi_1, \phi_2) = (\hat{\zeta}, \hat{\phi}_1, \hat{\phi}_2) \mathrm{e}^{\frac{1}{2} k [\![U]\!] t} \mathrm{e}^{i(\alpha x + \beta z + k\overline{U} t)}, \tag{11.2.5}$$

indicating that *a free flat vortex sheet is always inviscidly unstable.* Disturbance waves with any wavelength must be amplified and will propagate downstream with speed of $c = \overline{U}$.

This Kelvin-Helmholtz instability of a vortex sheet can well be explained in terms of vorticity (Batchelor 1967). Introduce transformation

$$kx' = \alpha x + \beta z, \quad kz' = -\beta x + \alpha z,$$

such that the disturbed flow becomes two-dimensional on the (x', y)-plane. Then by using (5.1.4) and $(\boldsymbol{e}'_x, \boldsymbol{e}'_z) = (\nabla x', \nabla z')$, for the vortex-sheet strength $\boldsymbol{\gamma}$ we obtain

$$\boldsymbol{\gamma} = \gamma \boldsymbol{e}'_z, \gamma = -ik [\![\hat{\phi}]\!] \, \mathrm{e}^{\frac{1}{2} k [\![U]\!] t} \mathrm{e}^{i(kx' + k\overline{U} t)}. \tag{11.2.6}$$

Notice that (11.2.5) and (11.2.6) imply a phase lag of $\pi/2$ between ζ elevation fluctuation and the sheet strength γ. In Fig. 11.6, the thick solid curve shows the periodical deformation of the vortex sheet, and its varying thickness denotes the accumulation (e.g., around point A) or rarefication (e.g., around point C) of the disturbance vorticity. Thus, according to the Biot-Savart law, the segment of vortex sheet around A with $\zeta > 0$ must move downward and will induce a negative velocity component in the x direction. This action is the most obvious at point B to make it move towards

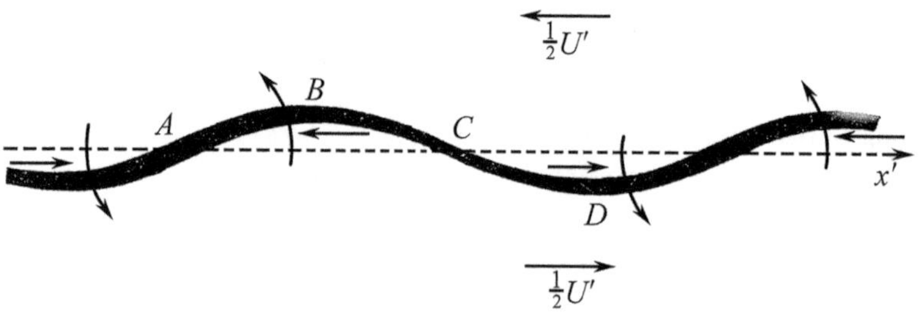

Fig. 11.6 The instability of a vortex sheet on the (x', y)-plane. From Batchelor (1967)

A and hence strengthen the sinusoidal deformation of the sheet. The more the vortex sheet deforms, the more will accumulation of vorticity disturbance around A occur. This mechanism develops a positive feedback and leads to exponential growth of disturbances.

Consider now a vortex layer with finite thickness, of which a commonly used basic-flow profile is

$$U(y) = \overline{U}\left[1 + Ra \tanh\left(\frac{y}{2\theta}\right)\right]. \tag{11.2.7}$$

Here, the velocity ratio $Ra = (U_1 - U_2)/2\overline{U}$ represents the shear level of the basic flow. The flows with $Ra = 0$ and 1 correspond to a wake and a jet, respectively. θ is the momentum thickness that is the only length scale that one can get in two-dimensional free shear layer. Calculations of spatial growth rates of disturbances, with tanh and Blasius velocity profiles, have indicated that the growth rate $(-\alpha_i\theta/Ra)$ and phase velocity $c_r/\overline{U} = \omega/(\alpha_r\overline{U})$ are both related to the Strouhal number $St = f\theta/\overline{U}$ (f is the frequency of disturbance). The most amplified wave corresponds to $St = 0.032$, with the corresponding natural frequency f_n of the mixing layer (see Fig. 11.7).

11.2.2 Free and Forced Evolutions of Spanwise Vortices

The linear stability theory of free shear flows describes the initial stage of the Kelvin-Helmholtz instability. Its prediction of the growth rate and the phase velocity of the instability waves agrees well with experiments. Downstream of the region of exponential growth, the instability waves evolve into a nonlinear region and the amplitude of the fundamental mode becomes saturated. Meanwhile, the fundamental instability wave leads to roll-up of a periodical row of spanwise vortices (Figs. 11.8 and 11.9) moving at the average velocity $\overline{U}$ with a wavelength $\lambda = \overline{U}/f_n$. The vorticity in the shear layer is gradually accumulated and concentrated through the instability mechanism. As the instability wave grows, so does the fluctuation of vorticity concentration; but the concentrated vorticity is still continuously distributed.

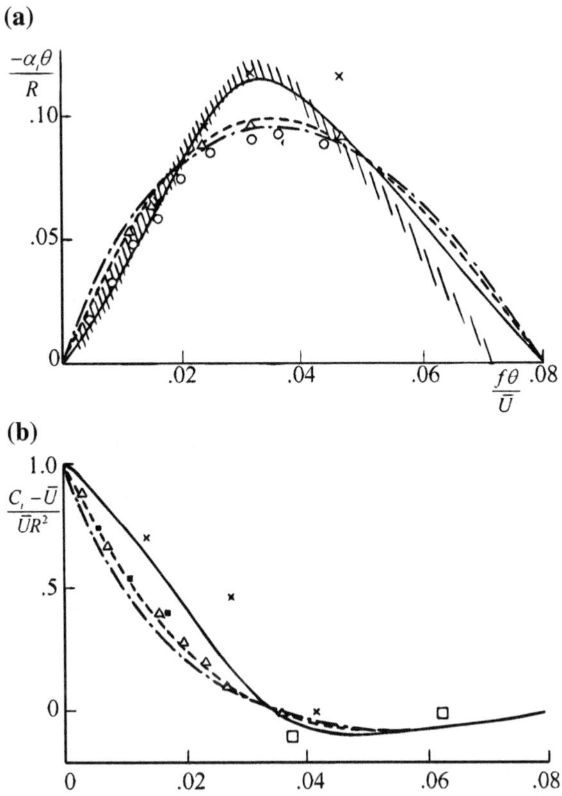

Fig. 11.7 Variation of **a** normalized amplification rate and **b** normalized phase velocity with Strouhal number $St = f\theta/\overline{U}$, from Ho and Huerre (1984)

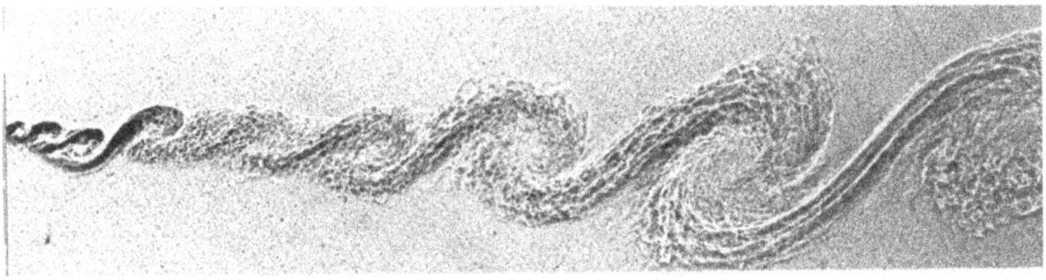

Fig. 11.8 Large-scale turbulence structures in a turbulent mixing layer. From Brown and Roshko (1974). See also Van Dyke (1982)

When the shear layer with concentrated vorticity finally rolls up, it breaks into discrete spanwise vortices as sketched in Fig. 11.9. The roll-up process is completed where the fundamental mode reaches its maximum amplitude.

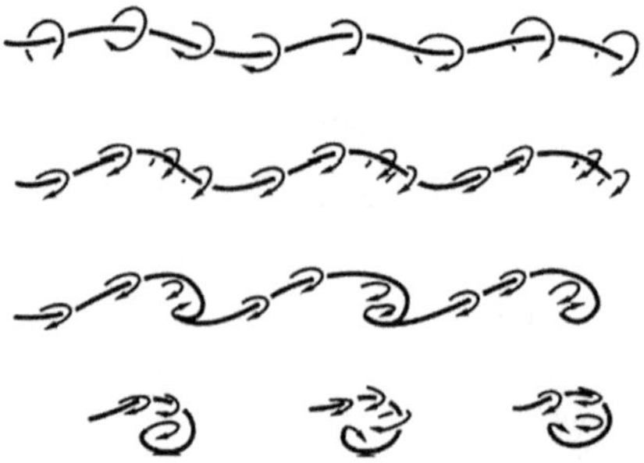

Fig. 11.9 Rolling process of a thin wavy vortex layer

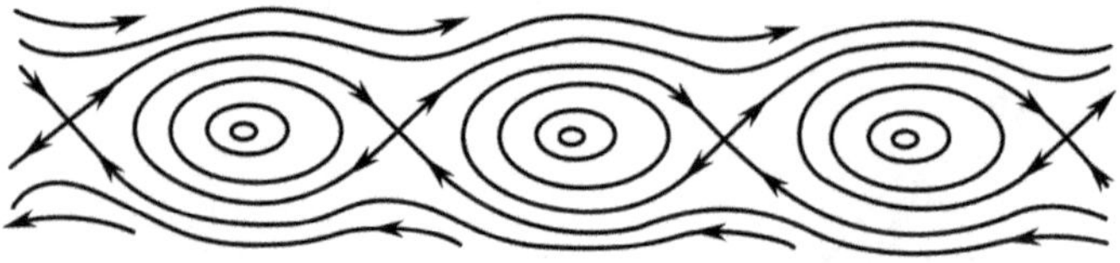

Fig. 11.10 Topology of the fundamental coherent structures in a mixing layer

Fig. 11.11 Pairing process of the spanwise vortices

A topological sketch of the mixing layer structure after roll-up, viewed in a reference frame moving at the same velocity as the average velocity of the mixing layer, is shown in Fig. 11.10, indicating a row of vortices with saddle points in between.

During the rolling up of the fundamental vortices, subharmonic components of disturbance waves also start growing. The disturbance with half of the fundamental instability frequency $f_n/2$, or with double wavelength $2\lambda_n$, produces to a pairing process (Fig. 11.11). This is one of the major physical processes in the continuing evolution of a mixing layer. Although more than two vortices involved in one merging have also been observed in unforced shear layers, the amalgamation is most often a pairing between neighboring spanwise vortices. The location where the pairing occurs can be expressed by a dimensionless streamwise distance $x^* = Rax/\lambda_0$ from the splitter plate, where $\lambda_0 = \overline{U}/f_0$ is the initial instability wavelength. Experiments in a forced mixing layer indicate that the first pairing happens at $x^* = 4$, the second at $x^* = 8$ and the third, $x^* = 16$. However, for a naturally developing mixing layer, the spatial location where the pairing occurs is fairly random. Thus, the mean velocity profiles in a naturally developing mixing layer will not show any abrupt thickening. Rather, a linear spatial growth will be observed. Note however that pairing, though important, is not the only mechanism that causes the growth of a mixing layer. The photo in Fig. 11.8 by Brown and Roshko (1974) on a naturally developing mixing layer reveals that the layer thickness may grow in streamwise direction due to vorticity diffusion without obvious pairing is involved.

In contrast to unforced shear layer, a *forced* shear layer is quite *regularized*. A rather weak periodical disturbance can dramatically influence the pairing process and result in significant modification of mixing properties. These spanwise coherent vortices evolve quite independent of small-scale random fluctuations. To see this, we consider a forced mixing layer. The mean and coherent velocity-vorticity fields can be experimentally measured, and the mean-vorticity equation (11.1.15) can be applied to analyze the flow data (Zhou and Wygnanski 2001). By assuming that the time-mean spanwise coherent motion is basically two-dimensional and that the influence of the random motion is negligible in a mixing layer under two-dimensional forcing, the first and fourth terms can also be dropped from (11.1.15). The rates of change of Ω from direct measurement and from calculation based on the third term at the righthand side of (11.1.15) are plotted in Fig. 11.12. The balance of data indicates that the above assumptions are valid. Thus, $D\Omega/Dt$, including the change in the mean-vorticity profile along the flow and the spreading of the entire mean shear field, is indeed dominated by the curl of the time-mean coherent Lamb vector. It also explains why the spreading rate depends on the variation of the forcing condition.

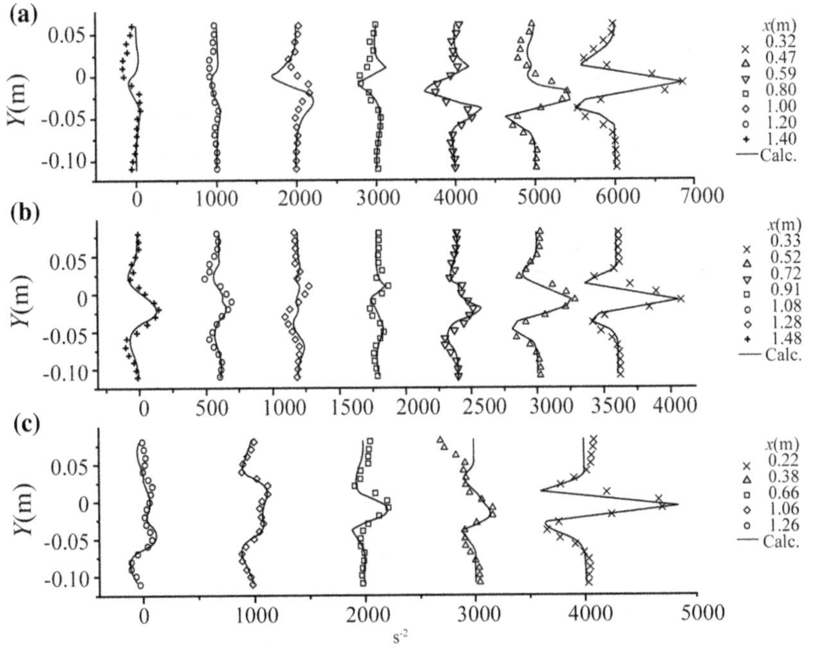

Fig. 11.12 Mean vorticity balance in a forced mixing layer between the rate of change of Ω and the curl of coherent Lamb vector. **a** Forced by single frequency; **b** forced by two frequencies (fundamental and subharmonic); **c** forced by two frequencies but with stronger amplitude. *Symbols* direct measurement of $D\Omega/Dt$; *solid lines* the third term of (11.1.15). From Zhou and Wygnanski (2001)

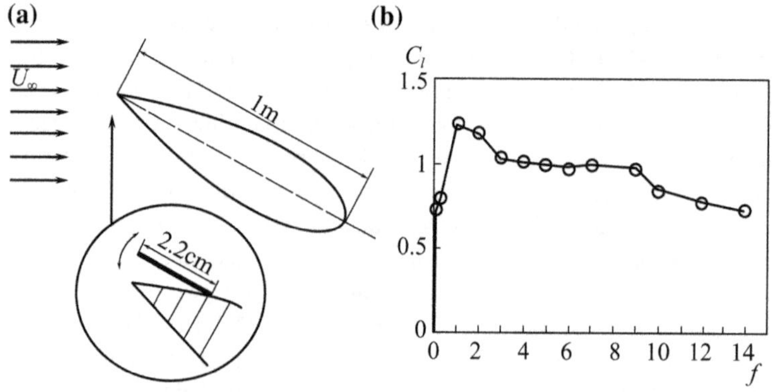

Fig. 11.13 **a** Sketch of testing model. **b** Variation of lift coefficient with forcing frequency. $\alpha = 27°$, $Re = 6.65 \times 10^5$. From Zhou et al. (1993)

The regularized behavior of forced spanwise vortices in a free shear layer and their sensitivity to the forcing have served as one of the physical backgrounds of *active shear-layer control by unsteady excitation*. This strategy has been widely explored in boundary-layer separation control. For example, since at stall angle of attack (Fig. 8.1) the boundary layer separates near the leading edge and becomes a free shear layer, as confirmed by many experiments, a weak local periodic forcing at frequency of the order of $St = 0.032$ may effectively organize the pairing process of the forced shear layer and hence delay the stall by a few degrees.

At larger post-stall angles of attack, then, even though unsteady excitation control may no longer eliminate boundary-layer separation, the same control strategy can still be applied to modulate the shear-layer evolution and thereby improve the behavior of separated flow, but at a much lower excitation frequency (of the order of wake vortex shedding, see Sect. 8.5). Physical mechanisms relevant to this kind of separated-flow control have been systematically reviewed by Wu et al. (1991), of which the key issue is the utilization of the rich spectrum of the instability and sensitivity of forced free shear layer. The first experimental realization of this *separated-flow control* strategy was made by Zhou et al. (1993), who put a thick NACA-0025 airfoil in a wind tunnel with its trailing edge toward the incoming flow at $\alpha = 27°$ (Fig. 11.13a), and used a small oscillating leading-edge flap to control the strong separated shear layer. It was found low-frequency excitation may change the massively separated flow to a closed separated bubble (in time-averaged sense), leading to a 60–70 % increment of lift (Fig. 11.13b) with little or even no penalty on drag.

Inspired by the finding of Zhou et al. (1993), Wu et al. (1998) performed a two-dimensional RANS simulation of separated-flow control by unsteady excitation on a NACA-0012 airfoil at post-stall angles of attack. They found that the massively separated and disordered unsteady flow can be effectively altered by periodic blowing-suction near the leading edge with low power input. The forcing may modulate the evolution of the separated shear layer to promote the formation of concentrated lifting vortices, which in turn interact with trailing vortices in a favorable manner

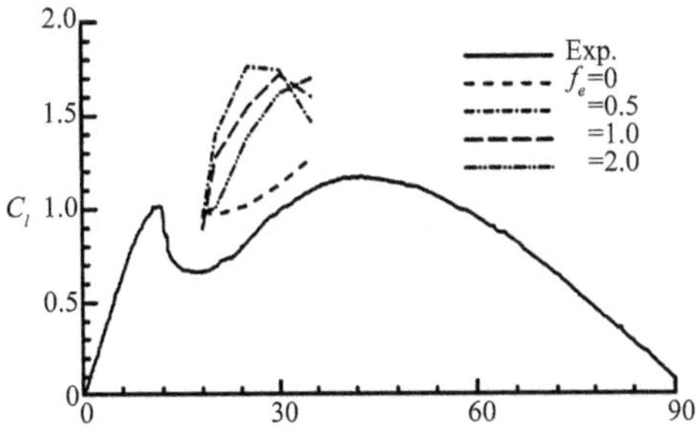

Fig. 11.14 Overall view of mean lift enhancement due to forcing. NACA-0012 airfoil. *Solid line* is experimental value (Critzos et al. 1955). From Wu et al. (1998)

and thereby alter the global deep-stall flow field. In a certain range of post-stall α and forcing frequencies, the unforced random separated flow can become periodic or quasi-periodic, associated with significant lift enhancement of upto 73 % (Fig. 11.14), some drag reduction, and alleviation of flow fluctuation.

11.2.3 Secondary Instability and Formation of Streamwise Vortices

The roll-up of spanwise vortices and their pairing always coexist with spanwise non-uniformity due to three-dimensional disturbances, which eventually leads to streamwise streaks or counter-rotating streamwise vortex pairs. Experiments indicate that these secondary vortices develop together with the spanwise structures from the very beginning of the mixing layer. The formation mechanism of the streamwise vortices in a shear layer is intuitively explained by Fig. 11.15a. If any single vortex filament in a coherent spanwise vortex is deformed by three-dimensional disturbances to form a small waviness in y direction, the humped-up portion of the vortex will be advected faster than the rest due to the shear field, thus initiating a streamwise vortex pair. The short-wave *elliptical instability* mentioned in Sect. 10.2.4 then provides a universal mechanism for the two-dimensional large-scale deformed spanwise vortices to be directly transferred to complicated three-dimensional motions. This direct transfer mechanism plays a dominant role not only in mixing layer but also in flows like wakes and vortex pairs.

Then, streamwise vortices will be wrapped onto existing large spanwise vortices, which are referred to as "braids" and sketched in Fig. 11.15b. The average spanwise spacing of these secondary structures increases in size with downstream distance and remains to be of the same order of magnitude as the local mixing layer thickness. The thickness of the braid region connecting spanwise vortex cores is of the order of Taylor microscale.

The occurrence of streamwise vortices is important in the evolution of mixing layers because it enhances the mixing efficiency. Furthermore, the interaction of spanwise and streamwise vortices has considerable influence on the transition to small scales and further development of the whole mixing layer.

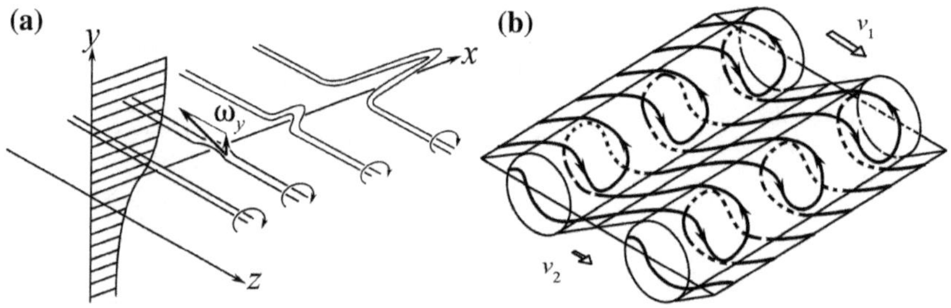

Fig. 11.15 **a** Initiation of streamwise vortices in a strain field. **b** A conceptual sketch of streamwise vortices in a mixing layer. Reproduced based on Bernal and Roshko (1986)

11.2.4 Vortex Interaction and Small-Scale Transition

The aforementioned spanwise and streamwise vortices may still be considered "laminar". With further development, including interaction of various structures, these vortices are stretched, twisted or broken down to be random, small-scale vortices or eddies, and eventually become "real" turbulence. This final stage of transition is thus referred to as *small-scale transition*. It belongs to the category of *cascade*.

Figure 11.15 suggests that stretching and compression between two neighboring spanwise vortices would be enhanced during pairing so that the streamwise vortices will be subjected to stronger strain, which may lead to small-scale transition. If one takes Kolmogorov's $-5/3$-power law (11.1.2) in energy spectrum as a criterion to identify a fully developed turbulence, this expectation has indeed been proven by experiments.

The small-scale transition involves continuous vortex interaction not only between coherent vortices, but also between coherent and random vortices. The appearance of small-scale filaments during the onset of transition due to successive interactions between a large, coherent vortex and its surrounding smaller scales has been simulated numerically, and a sequence of views is shown in Fig. 11.16. It indicates vividly how the energy cascade from large to small scales takes place.

During small scale transition, the *vortex cut-and-reconnection* mechanism due to viscosity plays a key role in the final formation of small-scale structures. This process also appears in the late evolution stage of disturbed trailing vortex pair. For the theoretical background of vortex reconnection see, e.g., Wu et al. (2006, Sect. 8.3.3). A complete process obtained by numerical simulation is shown in Fig. 11.17.

Figure 11.18 presents a global view of vortical structures in a shear layer after the small-scale transition has happened. There exist vortical structures of various scales, from the largest, with the length scale comparable to the local thickness, to the smallest, with the order of Kolmogorov dissipative length scale. The previously formed large vortical structures still exist; but they are in a turbulent rather than laminar background.

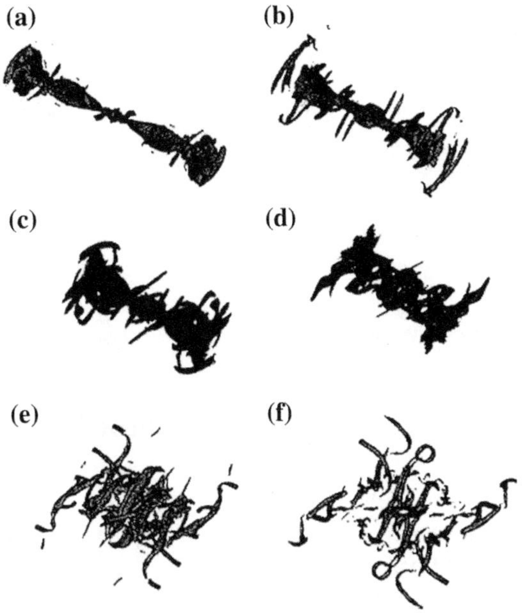

Fig. 11.16 Iso-vorticity contours illustrating appearance of small scale filaments during onset of transition. From Pradeep and Hussain (2000)

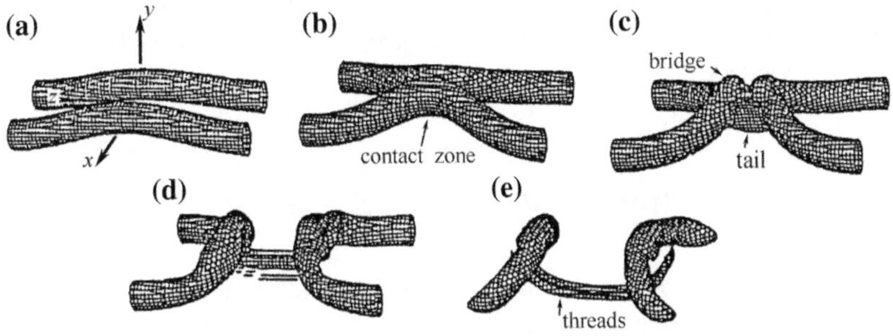

Fig. 11.17 Iso-vorticity surface to show the successive steps of small-scale formation during vortex reconnection. From Hussain and Melander (1992)

We summarize the preceding discussion as follows. In a turbulent free shear layer, two opposite physical processes are going on simultaneously, accompanied with energy transfer. One is that the large-scale vortices are gradually organized from the concentration of initially distributed vorticity through the instability mechanism. The other is that the energy contained in large-scale vortices are successively transferred to smaller vortices and eventually to small random vortices, or eddies. Both processes continue in the whole turbulent flow field so that the coexistence of vortices and waves as well as that of multi-scale structures can be observed. These coexisting waves and multi-scale vorticies continue to grow or decay, as well as interact with each other.

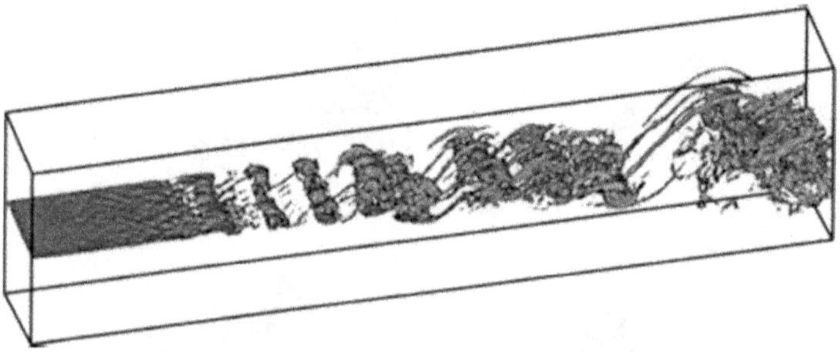

Fig. 11.18 A global view of the vortical structures in a mixing layer. From Lesieur et al. (2000)

11.3 Instability and Transition of Wall Shear Layer

In sharp contrast to the instability and transition of free shear layer (incompressible without non-conservative body force) discussed in the preceding section, where there is no solid boundary and hence no vorticity source, in a *wall-bounded shear layer* (wall shear layer for short) new vorticity is continuously produced via the boundary vorticity flux (BVF). Newly produced vorticity enters the flow field and makes its instability mechanisms and coherent structures considerably more complicated than those of a free shear layer. We now turn to these topics, going from instability to fully developed wall turbulence, focusing on the corresponding vortical structures. For recent reviews of boundary-layer turbulence, see Kachanov (1994), Smits et al. (2011) and Wallace (2013).

11.3.1 Instability Waves and Coherent Structures

1. Tollmien-Schlichting waves and Λ-waves. Reynolds was the first to realize that boundary-layer transition to turbulence is caused by its instability, preceded by a *receptivity* process to disturbance environment (Sect. 10.1.3). Consider now a Blasius boundary layer (Sect. 4.2.3) in a uniform incompressible free stream. The flow near the leading edge is laminar, and the local Reynolds number Re_x increases as x. According to the normal-mode analysis (Sect. 10.1.1), once Re_x passes a critical value, the flow will lose its stability and two-dimensional travelling waves appear, known as the *Tollmien-Schlichting waves* (T-S waves), which are the solution of the Orr-Sommerfeld equation (10.1.8) and propagate downstream with certain frequencies, wavelengths and growing amplitudes. The calculated neutral curve for the Blasius boundary layer is compared with experimental results in Fig. 11.19.

A boundary layer is an attached vortex layer, and the Tollmien-Schlichting wave is *vorticity wave*. At $Re_x \gg 1$, the vorticity in a shear layer always tends to organize itself into vortical structures. This tendency is suppressed by vorticity diffusion.

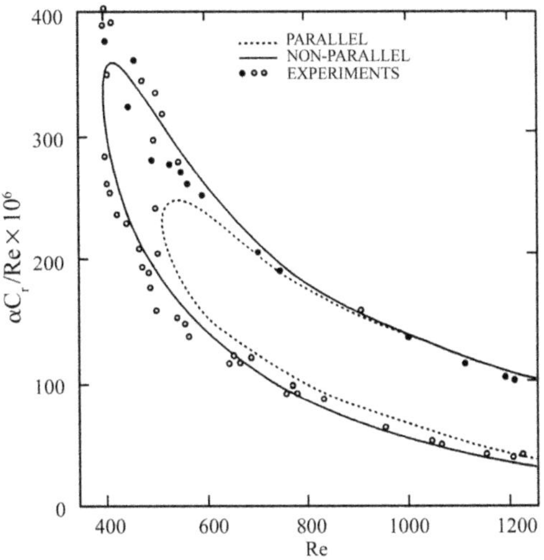

Fig. 11.19 The curve of the neutral stability of the Blasius boundary-layer profile. From Saric and Nayfeh (1975)

Toward downstream the boundary layer is thickened, making the vorticity diffusion weaker due to gradually smaller vorticity gradient. Thus, sufficiently far downstream from the leading edge, the tendency of forming organized structure will overwhelm the diffusion effect.

Note that the predicted growth rate of selective T-S waves by linear theory is effective only in the region close to the neutral curve of Fig. 11.19, and nonlinear processes must appear after the disturbances have grown to certain level. Thus, T-S wave growth does not immediately lead to wall turbulence; wall shear-layer transition involves a series of complicated processes and various coherent structures.

If there were any chance to keep a perfect two-dimensionality of the growing T-S waves, they would reach a maximum strain rate and vortex sheets would roll-up to become spanwise vortices as it happens in free shear layer. However, there has never been such a chance in a wall shear layer. Even in a laminar boundary layer on a smooth plate, the plate can never been perfectly smooth and the free stream is never perfectly disturbance-free. Since these disturbances are always three-dimensional, even if the T-S waves may shortly keep its two dimensionality during the stage of linear growth, they will never do so in nonlinear stage. Spanwise disturbance waves may occur and couple with the two-dimensional T-S wave, see (10.1.5). This results in the appearance of oblique *Λ-waves* in different arrangements as sketched in Fig. 11.20a. Of these types, the K-type (Klebanoff-type) are often observed in naturally evolved boundary layers.

Actually, in experiments and numerical simulations, an effective way of triggering a quick formation of streamwise vortices is to impose three T-S waves of proper wave vectors, two highly oblique plus a two-dimensional one or a spanwise Squire wave (see 10.1.5), to form a *three-wave resonance* (Craik 1985), of which the disturbed flow is of course highly three-dimensional and develops rapidly.

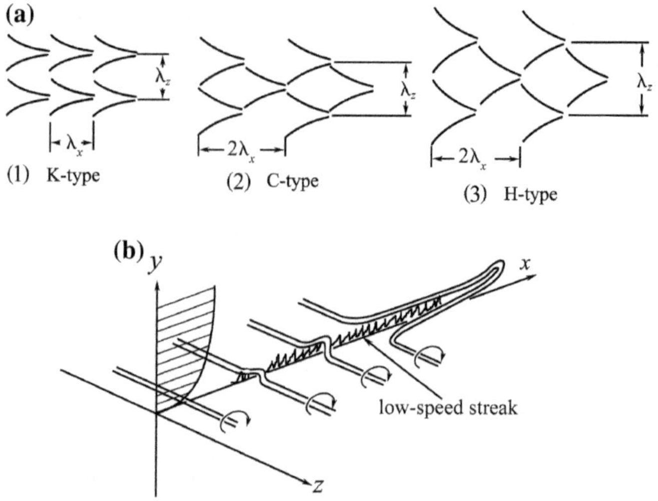

Fig. 11.20 Three-dimensional waves and structures in transitional boundary layer. **a** Λ-waves. **b** The formation of Λ-vortices, streamwise vortices, and low-speed streaks. From Wu et al. (2006)

2. Three-dimensional coherent structures. We have seen in Sect. 11.2 that the nonlinear evolution of vortical waves must result in vortical structures, both spanwise and streamwise. In a wall shear layer, by similar mechanisms, this process leads to the appearance of *Λ vortices*. Three-dimensional disturbance will make a spanwise vortex subjected to certain slight bending as sketched in Fig. 11.20b. Due to the very strong shear at the wall, the vortex segment at a little farther away from the wall will be transported much faster than the segments closer to the wall, and vortex segments connecting them will be stretched and strengthened to become Λ-shaped vortices, which are then further stretched to become *streamwise vortices* or "rolls". This mechanism is similar to the formation of streamwise vortices in a mixing layer (Fig. 11.15a). Each neighboring pair of rolls are counter-rotating, which are the dominant vortical structure in the boundary layer.

Figure 11.20b differs from Fig. 11.15a in a major aspect, namely the occurrence of *low-speed streaks* beneath the Λ-vortices/streamwise vortices. In regions where vortex pairs pump the near-wall low-speed fluid away from the wall, the flow slows down and forms low-speed streaks. On the contrary, in regions where the vortex pairs pump the outer high-speed fluid towards the wall, high-speed streaks will form. Actually, such a vortex-pair induction exists even if there is no mean shear normal to the wall. This is why, more often than not, streamwise streaks are observed coexisting with streamwise vortices.

Besides, the two legs of a Λ-vortex, once formed, will be subjected to two opposite actions. On the one hand, they will be tilted further towards downstream direction due to the mean shear and become streamwise vortices. On the other hand, the mutual induction between the two legs tends to lift both legs upward. Thus, their orientation depends on the relative intensity of the two. Obviously, the strong shear dominates in the near-wall region so that the upstream roots of the legs develop to streamwise vortices. The downstream portions of the legs farther away from the wall will be lifted up to become elongated *hairpin vortices* that asymptotically tend to the direction of maximum production of $\overline{\omega_x^2}$, i.e., $\overline{\omega_x \omega_y} \partial_y U$ or to the direction of maximum stretching (45° from the wall). After the heads of initial vortices have swept over, some portions of their legs or the streamwise vortices could be gradually dissipated by viscosity; but streaks would still be left behind.

In engineering applications, the dominant roles of streamwise vortices in mass transfer (e.g., mixing), momentum transfer (e.g., Reynolds shear stress and skin friction), and energy transfer (e.g., heat transfer) are also significant.

11.3.2 Secondary Instability and Self-sustaining Cycle of Wall Turbulence

Although streamwise vortices (or "rolls") and low-speed streaks cannot last indefinitely long but must die out, the complex interactions among coherent structures and mean shear flow in boundary layer can lead to regeneration of rolls and streaks, and hence wall turbulence has a *self-sustaining cycle* for the regeneration of coherent structures. This property is unique to wall turbulence and absent in free shear turbulence. For relevant physical mechanisms of the cycle see the detailed reviews of Panton (1997, 2001). Here we just briefly outline a few major events.

Once a low-speed streak is formed, the normal and spanwise distributions of the streamwise velocity cause an instability in both directions. This is called the *secondary instability*. Once a low-speed streak is formed, the normal and spanwise distributions of streamwise velocity $U(y, z)$ will have an inflectional instability in both directions, see the sketch of Fig. 11.21. Similar to what happens in a free shear layer, instability waves will grow around the low-speed streak and Λ-shaped vortex will roll up.

Swearingen and Blackwelder (1987) argued that, after streamwise vortices cause the formation of low-speed *streaks*, the streak instability and breakdown is responsible for the turbulence sustaining cycle. So the first issue is how streaks become unstable, and the second issue is how streamwise vortices are regenerated or re-energized.

Fig. 11.21 Sketch of normal and spanwise reflectional instabilities of streak (secondary instability)

The above mechanism of Λ-vortex formation does not work individually. Instead, it occurs successively, aligned to each other, and produce x-dependent waviness of the vortex layer around the low-speed streak. Combined with the disturbance from the fundamental T-S wave, these periodical Λ-vortices may occur with a period of (or lock-on to) the fundamental T-S wave. As a consequence, the roots of the early-produced group of Λ-vortices will not only re-energize the original rolls on both sides of the original low-speed streak, but also in turn strengthen the upward pumping of the low-momentum fluids and strengthen the low-speed streak. This fact is best described using wall unit (Sect. 11.1.2). The legs of streamwise vortices stay in the near-wall region with $x^+ \approx 200$, while the streaks occupy a region of $x^+ \approx$1,000. Experiments and numerical studies have also shown that the spanwise scale of the hairpin vortices and spanwise wavelength of the streaks are both $\lambda_z u_\tau/\nu \approx 100$ (Kline et al. 1967), where λ_z is the spansiwe wavelength. Jiménez and Moin (1991) regard this dimensionless spanwise scale as a Reynolds number, below which turbulence cannot be self-sustained. Thus, it is proposed that this may be a universal critical Reynolds number for self-sustaining of wall turbulence. Based on their DNS study of turbulent channel flow, Jiménez and Pinelli (1999) show that a streak instability cycle exists in the inner region ($20 < y^+ < 60$). It needs no interaction with the outer region and hence could be dominant at all Reynolds numbers. Waleffe and coworkers (Waleffe 1997) studied the streak instability and regeneration process in plane Couette flow. Through a normal-mode analysis of the instability of a spanwise varying shear flow $U(y, z)$, they were the first to explain theoretically how the streamwise vortices are regenerated, so that the cycle is closed. Their finding is neatly summarized as: *the steady rolls lead to steady streaks that lead to a neutral mode that generates steady rolls* (ibid).

To summarize, both normal and spanwise inflectional instabilities contribute to rolling up of vortices and breaking down of T-S waves into three-dimensional turbulence. The presence of highly inflectional profiles (high-shear regions) is the key to the secondary instabilities that lead to the structure regeneration cycle. As one of the evidences, the high-shear layer shown in numerical simulation (Rist and Fasel 1995) matches the experiments (Kachanov 1994) very well, where a hairpin vortex riding above the low-speed streak is the main conclusion of their simulation. It is also worthy mentioning that when the Reynolds number is low, the regeneration cycle could be limited in certain local regions and develop turbulent spots. In a high Reynolds-number field, this regeneration cycle, combining with further wave interaction and small-scale transition (to be shown later), will become the major physical background of the self-sustaining turbulence in the whole turbulence field.

No consensus has so far been reached, however, on what is the specific mechanism of streak instability. While the aforementioned workers explained the instability by normal-mode analysis, Schoppa and Hussain (2002) used their careful DNS data to analyze turbulent channel flow, and compared the energy growth of an unstable normal-mode and a "transient-growth disturbance". They found that the latter is much larger. The important point is that the non self-adjoint nature of the Navier-Stokes equations (Sect. 10.1.2) implies that when the disturbance is at a large level it is essentially a new flow pattern that can initiate other events (Panton 2001). In this regard, Kim and Lim (2000) made an interesting DNS test on fully developed turbulent channel flow. They calculate the flow by full Navier-Stokes equations, equations without coupling term (see 10.1.6b), and equations without nonlinear terms. They found that coupling term gives three-dimensional turbulence a large transient growth. But in absence of either term the turbulence eventually die. Thus, Kim and Lim propose that formation of near-wall vortices is essentially nonlinear, but that maintenance is linear.

11.3.3 Transient Growth and Bypass Transition

Boundary-layer transition to turbulence does not necessarily have to go through the T-S growth stage. In Sect. 10.1.2 we have seen that the non-normality of the N-S equations permits the algebraic *transient growth* of a disturbance, which indeed plays a significant role in the transition of wall shear layers, see e.g., the review of Reshotko (2001).

For example, in boundary-layer shear flow (10.1.8) possess solutions of discrete and continuous spectra. The former are of T-S type disturbances, localized inside the boundary layer and can be unstable. In contrast, the latter are decaying disturbances in space and time, but their sum may experience a transient growth. This mechanism is responsible for the *sub-critical transition* of wall layers at a Reynolds number lower than the critical value predicted by the normal-mode theory, which is widely observed to occur in a plane Poiseuille or Couette flow and boundary layer. A strong transient growth may even bypass the T-S waves and directly trigger nonlinear growth of disturbance, and hence lead to the so-called *bypass transition.*

The strongest transient growth is associated with steady *counter-rotating streamwise vortices* in boundary layer. It has been demonstrated that, on a time scale shorter than that for asymptotically growing mode predicted by normal-mode theory, some finite but small disturbances can experience algebraic linear growth of surprisingly large amplitude and thereby trigger their nonlinear growth mechanisms very rapidly. In a Poiseuille flow at Reynolds number 5000, it is found that "optimal disturbances" (for which the transient growth factor is maximized) could grow in energy by a factor 4897, although the flow is asymptotically stable at this Reynolds number (Butler and Farrel 1992). For plane Couette flow at $Re = 4000$, the optimal disturbance-energy growth can be as large as by a factor of 18000. Although optimal disturbances are hardly realizable in reality, transient-growth phenomenon has been found in several experiments of pipe flow, the earliest one was made by Kaskel in 1961 and is shown in Fig. 11.22. The phenomenon can now be interpreted by the transient growth theory (Reshotko 2001).

Travelling vortices may be detected as and expressed by waves. This is however not the case for a steady streamwise vortex. Accordingly, the mechanism of disturbance growth related to streamwise vortices cannot be expressed by the growth of normal modes either. The current understanding of the streak development is also a transient growth.

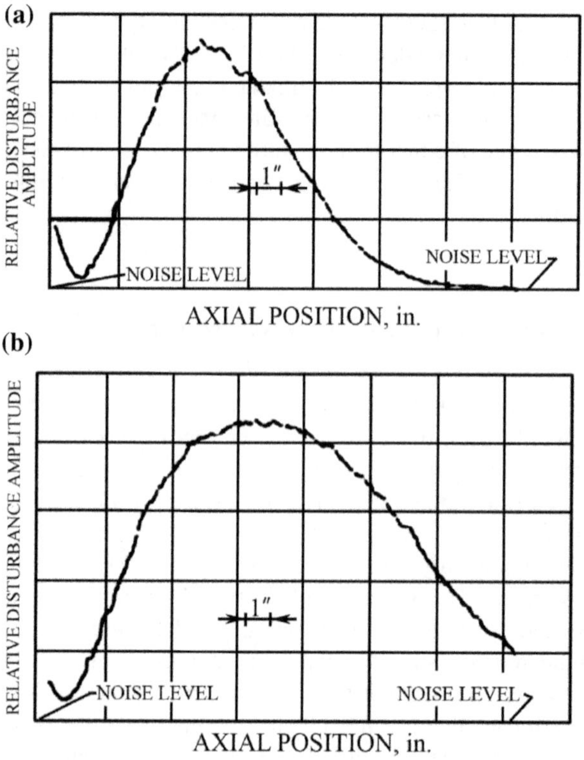

Fig. 11.22 Kaskel experimental observation of transient disturbance growth in a pipe flow at $Re_D = 7600$. Reproduced from Reshotko (2001)

Fig. 11.23 Transient growth and counter rotating vortices

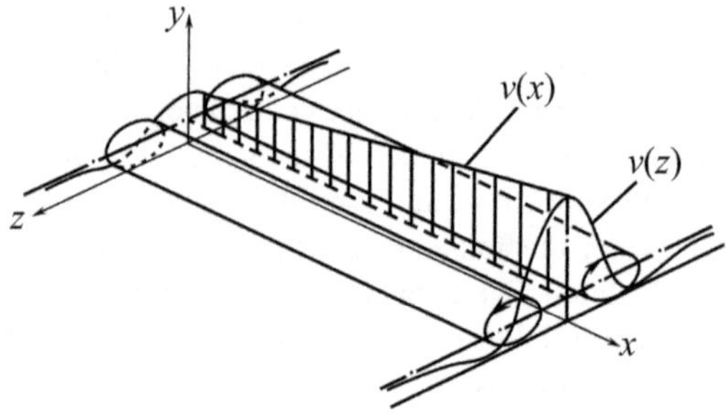

As shown in Fig. 11.23, a pair of counter rotating streamwise vortices in a boundary layer will cause wall-normal velocity disturbance that accumulates (or grows) algebraically along the streamwise direction. Even if the streamwise vortices decay along x, the normal velocity disturbance could still grow as an integrated effect. A closely related phenomenon is the occurrence of low-speed streaks and the surrounding high shear layers.

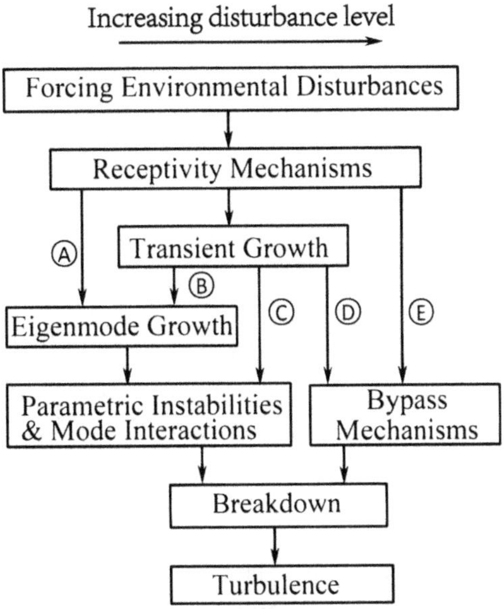

Fig. 11.24 The chart of multiple paths of boundary-layer transition to turbulence. From Morkovin et al. (1994)

The later breakdown of low-speed streaks occurs through a secondary instability, which is developed on the local shear layer between high- and low-speed streaks when a critical Reynolds number based on their size is sufficiently large. If this mechanism overwhelms the normal-mode transition, the bypass transition occurs. This kind of transition has also been observed in a boundary layer on a rough surface or in a circumstance of high turbulence intensity (e.g., a turbine blade).

In some cases transient growth may not directly leads to bypass transition, but cause some other instability mechanisms to grow. Consequently, the transition paths to wall turbulence become multiple as summarized in Fig. 11.24. There, according to Reshotko (2001), path A is the case where transient growth is weak and transition is of traditional T-S type. In path B, transient growth provides a higher amplitude to the eigenmode to cross into an exponentially unstable region. Path C occurs if eigenmode growth is absent or the transient growth is strong enough to directly excite secondary instabilities and nonlinear mode interactions. Path D is the case where transient growth results in a full disturbance spectrum (like a turbulent spectrum, but the mean profiles are still laminar). Finally, path E represents very large forcing amplitude so that no linear regime. An alternative view of path E was proposed by Lee and Wu (2008) in terms of so-called soliton-like coherent structures to be discussed in the context of Fig. 11.31.

11.3.4 Hairpin Vortices and Hairpin-Vortex Packets

So far we have encountered a few key coherent structures in transitional and turbulent wall shear layers, including Λ-vortices, streamwise vortices, hairpin vortices, and

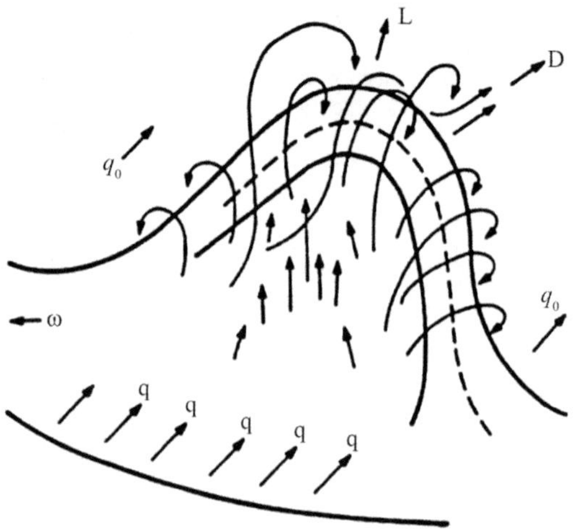

Fig. 11.25 Sketch of "horseshoe" or hairpin vortex proposed by Theodorson (1952). $\boldsymbol{q}$, $\boldsymbol{q}_0$, ω, L and D denote mean velocity, free-stream velocity, vorticity, vortex lift, and drag, respectively

long streaks. It would be a great success if a unified physical picture could be developed in which these structures can be identified as various integrated constituents, and of which the self-sustaining or auto-generating cycle is a natural consequence.

This ambitious mission was pioneered by Theodorson (1952) who, with little supporting experimental evidence, hypothesized hairpin or horseshoe vortices as the fundamental structural feature of all wall turbulence (Fig. 11.25). But since then there had been a controversy on their importance and even their presence. Head and Bandyopadhyay (1981) were the first to provide strong experimental evidence for the unambiguous nature of hairpin vortices, that the outer region with $y^+ > 100$ is full of successive hairpins vortices with their heads along a line inclined at about 20° to the wall. Perry and Chong (1982) constructed a theoretical model and demonstrate that a proper distribution of hairpins vortices can reproduce many aspects of measured statistics in turbulent boundary layer, such as mean velocity, Reynolds stress, and spectra. However, it had been impossible to gain essential breakthrough until the significant advances in experimental and computational techniques in the last two decades. The central figure of this unified picture is indeed hairpin vortices; for relevant progresses see Adrian (2007).

Then, the DNS of Wu and Moin (2009), on the whole transition process of a flat-plate boundary layer with zero pressure gradient, confirmed strikingly their unambiguous and densely populated presence, see Fig. 11.26. Similar "hairpin forests" have also been found in supersonic and hypersonic boundary layers (e.g., Wang and Lu 2012), and at larger Reynolds numbers. It has now been believed that when the "hairpin forest" travels at the same convective velocity it creates large-scale motions of the outer region (cf. Smits et al. 2011).

The unified picture in terms of hairpins vortices has not yet been completed. Some detailed issues still remain open for future studies. Nevertheless, it has more than ever accommodated so many coherent structures into a whole. Below we discuss some major relevant physical mechanisms.

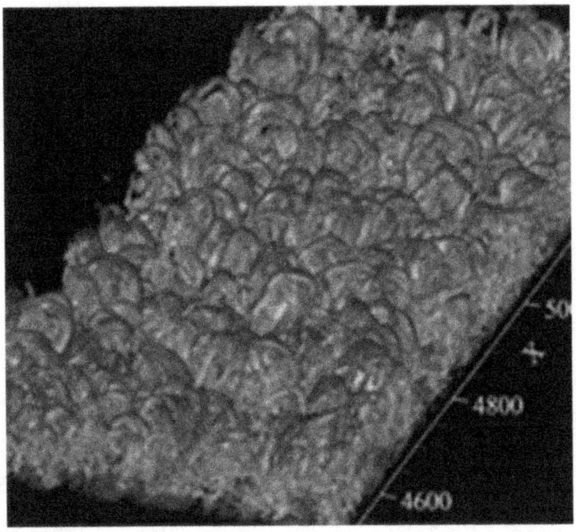

Fig. 11.26 "Hairpin vortex forest". Iso-surfaces of the second invariant of the velocity gradient tensor $\boldsymbol{Q}$. According $\boldsymbol{Q}$ criteria the vortex structures are presented. See Wu and Moin (2009)

1. Self-induced motion of a hairpin vortex. In a DNS study of channel turbulence, Zhu et al. (1999) took a conditional average of a Q2 event (ejection) in the buffer layer from earlier measured fluctuating velocity data base, and found that it is a combination of a hairpin eddy and two relatively short counter-rotating quasi-streamwise vortices. The hairpin head is first lifted up by the mean shear flow and then broadened by its self-induction as sketched in Fig. 11.27a. Then the self-induction continuously and non-uniformly deforms and stretches the vortex shown in Fig. 11.27(b–d). This process is somewhat like that of Fig. 3.10, and can be described by the local-induction approximation of vortex filament (Sect. 3.1.4),

$$\boldsymbol{u} \approx K\frac{\Gamma}{4\pi}\kappa\boldsymbol{b}, \tag{11.3.1}$$

where Γ is its circulation, κ the local curvature, and $\boldsymbol{b}$ the unit bi-normal vector. K is a constant determined by the core-vorticity distribution.

2. Formation of hairpin-vortex packets and low-speed streak. An important finding on hairpin vortex evolution is that (Smith et al. 1991; Zhu et al. 1999) a single hairpin vortex of strength above a threshold value can automatically generate new hairpins at its both upstream and downstream, forming a packet of hairpins vortex moving as a whole with the same advection velocity, see Fig. 11.28. Similar packets have been observed by PIV measurements in the outer layer of a boundary layer. Then, as a cooperative action of all hairpins aligned one behind the other along the streamwise direction, an aforementioned long *low-speed streak* forms between the legs in the near-wall region. This explains why an individual hairpin has a length of about $400\delta_v$ but the streak length can reach about $1000\delta_v$ as observed. Note that Fig. 11.28 also shows that below the hairpin packet there also appear near-wall quasi-streamwise vortices.

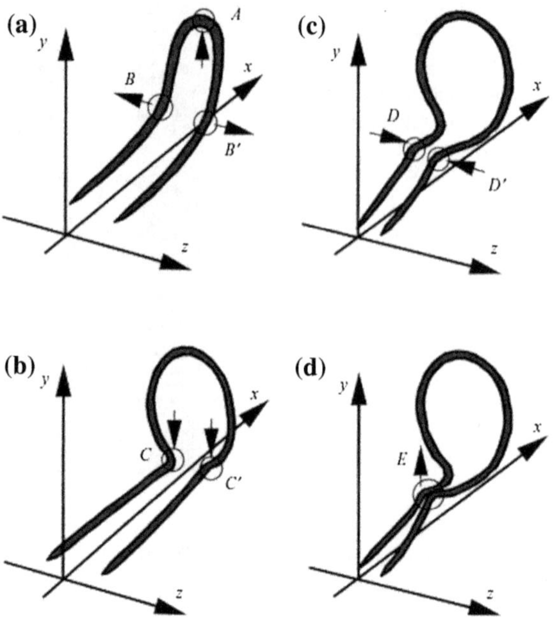

Fig. 11.27 Sketch of self-induced motion of a hairpin vortex. **a** Lifting and broadening of the head; **b** formation of the Ω-shaped head and subsequent downward motion; **c** formation of kinks in the legs and their approaching toward each other; **d** lifting of the kinks. From Zhu et al. (1999)

Further PIV-DNS studies have indicated that a hairpin packet may also generate new packets of different heights, sizes, and advection speeds, each being associated with a low-speed streak. This is conceptually sketched in Fig. 11.29. For detailed discussion on the role of this conceptual scenario in wall turbulence and its self-sustaining, see Adrian (2007) and references therein.

3. Small-scale transition. It was mentioned in Sect. 11.2.4 that the local breakdown of the "laminar" coherent structures is referred to as small-scale transition. In general, the occurrence of spikes in the velocity signals is the symbolic starting of a small-scale transition in a boundary layer. The continuous production of turbulence (or *small-scale transition*) in turbulent boundary layers was first considered as *bursts* by Kline et al. (1967). They discovered that the low-speed streaks are subjected to intermittent *turbulent bursts* such as *sweep* (Q4) and *ejection* (Q2), with an averaged period of $300\nu/u_\tau^2$.

Now, with hairpin packets as the central feature of coherent structures in transitional and turbulent boundary layers, small-scale transition can also be integrated into the unified picture. Note that hairpins vortex may also fall into category of small-scale coherent structures as the diameters of their legs are only of $O(\delta_\upsilon)$ although their lengths are of $O(100\delta_\upsilon)$; thus their head and legs must actively involve in small-scale transition. Actually, those small-scale phenomena can be explained as merely the manifestations of the passage of a hairpin packet (Adrian 2007). The head of a hairpin will produce an inrush motion on its downstream side and an adverse pressure gradient to its upstream field. The latter, combined with the shear-layer instability caused by the low-speed streak, initiates the burst (breakdown) of the

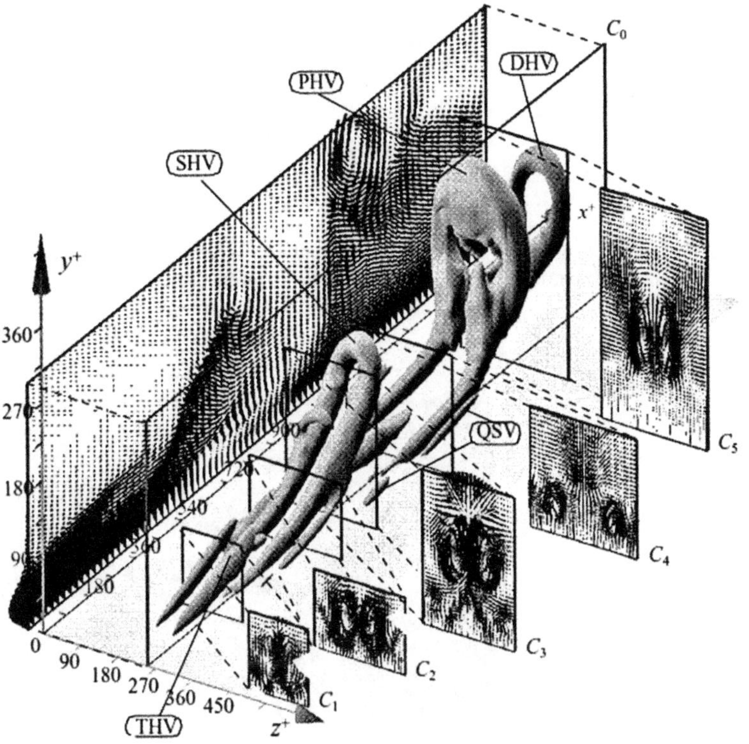

Fig. 11.28 Hairpin-vortex packets and cuts of associated velocity field by DNS of channel flow. *PHV* primary hairpin; *SHV* secondary hairpin; *THV* tertiary hairpin; *DHV* downstream hairpin; *QSV* quasi-streamwise vortices. From Zhu et al. (1999)

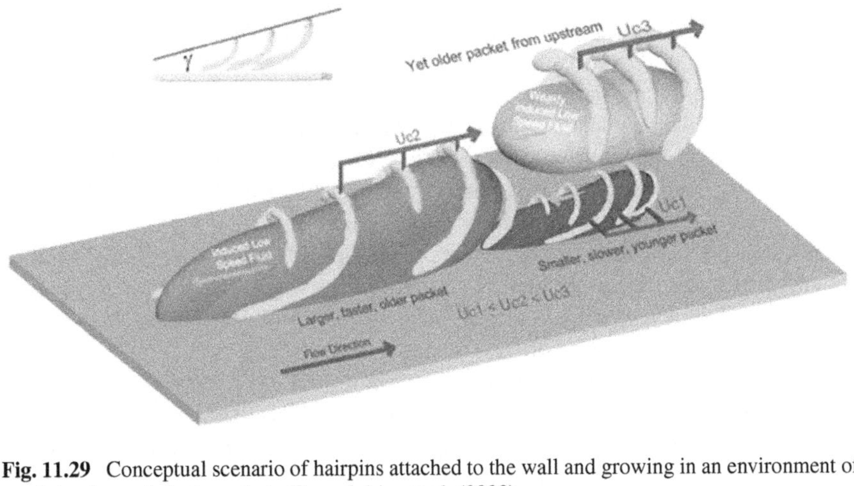

Fig. 11.29 Conceptual scenario of hairpins attached to the wall and growing in an environment of overlying larger hairpin packets. From Adrian et al. (2000)

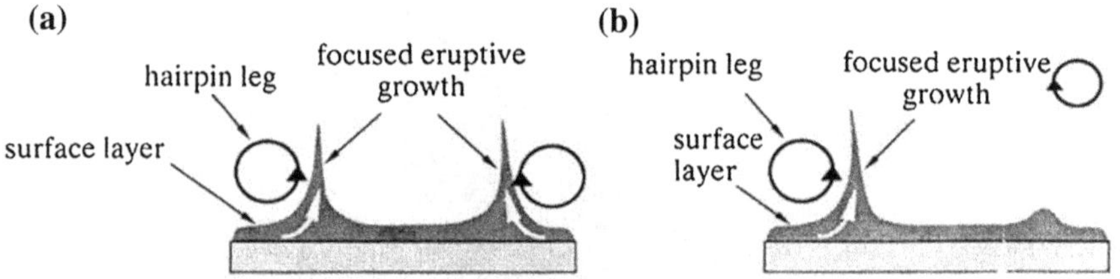

Fig. 11.30 Schematic diagram of location of vortex-induced separation of the surface layer for symmetric (**a**) and asymmetric (**b**) hairpin vortex legs. From Smith et al. (1991)

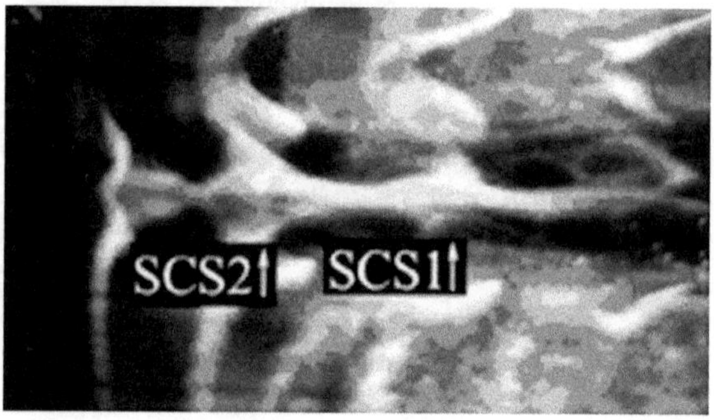

Fig. 11.31 Hydrogen bubble visualization of the Soliton-like coherent structures. From Lee and Wu (2008)

low-speed streak and breakdown of the parent vortices into small-scale structures. It has been found that almost 80 % of the turbulence production in the entire boundary layer occurs in the buffer region during intermittent, violent outward ejections of low-speed fluid and inrushes of high-speed fluid toward the wall. This near-wall turbulence production process, in particular the sweep events, is important for the generation of turbulent wall-shear stress.

While the spikes and bursts are hardly visualized by PIV and DNS with current visualization method that only captures concentrated vortices, hydrogen bubbles released from the viscous sublayer may exhibit their character. Smith et al. (1991) used this technique to identify that the rapid formation of very thin spikes (vortex sheets) in normal direction is the result of unsteady Lagrangian separation (Sect. 8.5.1) of surface layer due to the induced pressure gradient by the legs of hairpins, for which a schematic plot is shown in Fig. 11.30. The sheets then quickly roll-up into new streamwise vortices as seen in Fig. 11.28.

In the last decade, another explanation was proposed on the small-scale transition based on hydrogen-bubble visualization by Lee and coworkers, see the review of Lee and Wu (2008) and comments of Yaglom (2012). Lee found a kind of rhombus-shaped coherent structure in the near wall region, named as the *soliton-like coherent structure* (SCS) that is related to wave resonance during the development of three-dimensional disturbances. Figure 11.31 is a plane-view of several SCSs (marked by

SCS1 and SCS2) forming a long streak. These SCSs appear periodically from early transitional flow and are transported downstream to the later stages of transition. The rhombus-shaped SCS could be another view of the sub-layer phenomenon in the K-type transition (Fig. 11.20a). But Lee's observation gives more details in the later stage of vortex interaction and small-scale spike generation. The SCS causes strong periodical humping up of the low-speed streak and periodical ejections. The ejection is strong enough to cause a secondary vortex ring around it, similar to the vortex ring produced by a pulsed jet. When the secondary vortex ring is advected downstream, it will interact with existing hairpin vortices or other vortical structures and form a series of small-scale vortex loops.

11.3.5 Hypersonic Boundary-Layer Instability and Transition

We now leave incompressible flow for a while and turn to the other extremal case, the instability and transition of boundary layer in hypersonic flow. This subject is of fundamental importance and strong relevance to the safety of hypersonic vehicle flight, including significant increases in aerodynamic heating, entropy production, and drag (Fedorov 2011; Zhong and Wang et al. 2012). Compared to previously discussed incompressible boundary-layer transition, hypersonic transition is much less understood due to additional complexities.

A distinguished property of compressible boundary-layer transition is the appearance of a *second-mode instability* at very high frequency. Unlike the low-speed first-mode instability associated with transverse vortical waves, this is acoustic, longitudinal waves reflected between the relative sonic line and the solid wall, see the sketch of Fig. 11.32. The second-mode instability becomes stronger with increasing Mach number. The dynamic process of the evolutions of the two modes and their interactions play a crucial role in high Mach-number transition.

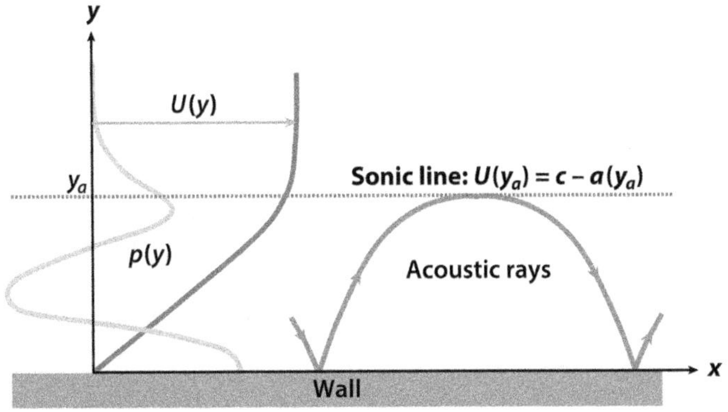

Fig. 11.32 Acoustic mode in a high-speed boundary layer, where $U(y)$ is the mean-flow profile, and $p(y)$ is the pressure disturbance profile. From Fedorov (2011)

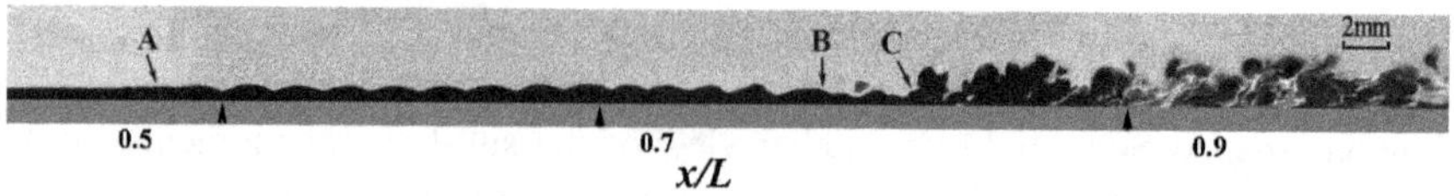

Fig. 11.33 Visualization of the whole evolution process of the second mode in transition. Formed from independent measurements over four successive sections that were spliced together. *Arrow A* appearance of second mode; *B* second mode's decay to almost zero; and *C* onset of turbulence. The *small vertical arrows* indicate where a splice occurs. The flow is from *left* to *right*. From Zhu et al. (2015)

As an example toward understanding hypersonic boundary-layer transition, in this subsection we present just one recent comprehensive experimental study and relevant physical analysis by Zhu et al. (2015). They conducted a series of flow visualizations and measurements for a hypersonic flow over a sharp-nosed flared cone of total length $L = 260\,\text{mm}$ at Mach number 6 and Reynolds number 9×10^6 per meter in a quiet wind tunnel. The gaseous medium is N_2. The flow visualization of the second mode evolution in the full boundary-layer transition process by using Rayleigh-scattering technique is shown in Fig. 11.33.[4] In the upstream region, the boundary layer is laminar. Then a linear instability results in growing second-mode patterns. The unstable waves are weak at the early stage of the destabilized laminar flow, but eventually the amplitude increases dramatically. The instability waves are formed and persist for a long distance as shown in the region between arrow A and arrow B. After further nonlinear development, the second mode decays to small value in the region between arrows B and C (to be quantified in Fig. 11.34). The picture to the right of arrow C shows that a turbulent boundary layer appears immediately after transition.

The qualitative picture of Fig. 11.33 is quantified by careful measurements. One measurement was pressure fluctuations on the model surface by fast-response piezo-electric pressure transducers (PCB). Figure 11.34 shows the detailed developments of frequencies and amplitudes of the two modes at peak frequencies of 30 and 350 kHz, respectively. The first mode grows continually to the onset of turbulence, while the second mode grows initially and then decays very quickly, eventually reaching the relatively quiet region. The second measurement was the flow-field transition process by using PIV. Figure 11.35 displays an instantaneous PIV results, of which the combination with Figs. 11.33 and 11.34 provides a clear evolution picture of the two modes.

Of the new observed phenomena, the most remarkable ones are the rapid annihilation of the second (dilatation) mode after its fast growth, and the associated very strong enhancement of the first (vortical) mode that leads to a quick transition. These peculiar phenomena demand a physical interpretation. Obviously, the splitting and

[4]Since the boundary layer is very thin compared to the cone radius, the flow was quasi two-dimensional. Here and below the experimental results are all displayed and analyzed on a meridional plane with Cartisian coordinates (x, y) with velocity components (u, v), vorticity $\omega = \partial_x v - \partial_y u$, and dilatation $\vartheta = \partial_x u + \partial_y v$.

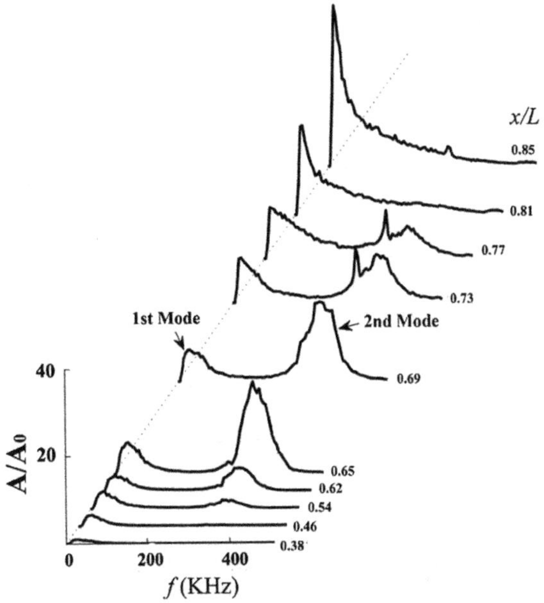

Fig. 11.34 Frequency spectrum of first and second modes at different x-positions. From Zhu et al. (2015)

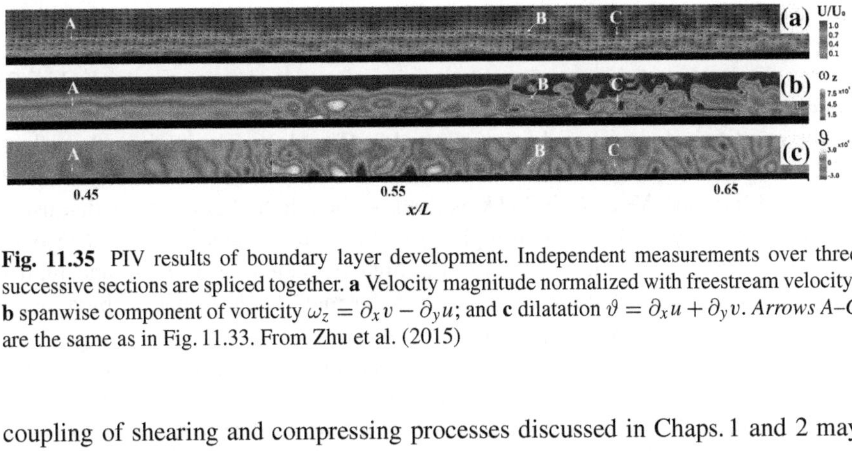

Fig. 11.35 PIV results of boundary layer development. Independent measurements over three successive sections are spliced together. **a** Velocity magnitude normalized with freestream velocity; **b** spanwise component of vorticity $\omega_z = \partial_x v - \partial_y u$; and **c** dilatation $\vartheta = \partial_x u + \partial_y v$. *Arrows A–C* are the same as in Fig. 11.33. From Zhu et al. (2015)

coupling of shearing and compressing processes discussed in Chaps. 1 and 2 may provide good insight into the relevant mechanisms.

In addition to the above evolution process of vortical mode and acoustic mode, Zhu et al. (2015) noticed that when the dilatation wave becomes sufficiently strong, it contributes considerably to the irreversible energy dissipation, see (1.2.24c). Neglecting the divergence term therein that has no definite sign, this relation reads

$$\Phi \approx \mu_\theta \vartheta^2 + \mu\omega^2, \quad \mu_\theta = \zeta + \frac{4}{3}\mu, \tag{11.3.2}$$

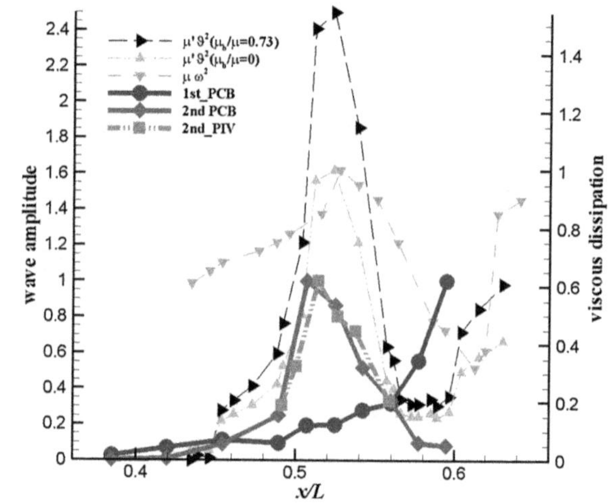

Fig. 11.36 Comparison of wave amplitudes and dissipation functions by shear and dilatation. Dissipation caused by dilatation is calculated for both $\zeta = 0.73\mu$ and $\zeta = 0$ (in the figure ζ is denoted by μ_b). From Zhu et al. (2015)

which clearly reveals the corresponding contributions of the two processes to dissipation. In particular, as remarked in Sect. 1.2.2, for diatomic and polyatomic gases the bulk viscosity ζ is nonzero. Unlike the first-mode induced dissipation $\mu\omega^2$ which is largely independent of the compressibility but affected by temperature, the second-mode induced dissipation $\mu_\theta\vartheta^2$ depends not only on ρ, T but also on the frequency of the gas motion since ζ does, and can therefore be enormously enhanced at very high frequencies. This is just the case found in the experiment, which causes considerable energy dissipation at the peak-amplitude zone of $\mu_\theta\vartheta^2$. The two terms of (11.3.2) computed from experimental data and their correlation with the spatial evolutions of the amplitudes of two modes, are plotted in Fig. 11.36. There, since the values of ζ at high frequencies are hard to determine, a conservative value $\zeta/\mu = 0.73$ measured for quite N_2 at $T = 293K$ is used. Although the figure shows that using this value already leads to much higher $\mu_\theta\vartheta^2$ than that assuming $\zeta = 0$ (Stokes hypothesis for monatomic gas), given the high frequency of 350 kHz and higher temperature, the real situation must have still far more larger (could be by a factor of 10^3) dilatation-caused dissipation.

Having obtained the evolution and dissipation characters of each mode, Zhu et al. (2015) proceeded to explain why the quick transition occurs right after the second mode rapid decay in terms of mode-coupling. As discussed in Sect. 2.3, the $(\boldsymbol{\omega}, \boldsymbol{\vartheta})$-couplings exist both in the interior of the boundary layer and at the wall. The internal coupling occurs via the compressible vorticity transport equation, which in two dimensions read

$$\frac{\partial \omega}{\partial t} + \boldsymbol{u} \cdot \nabla \omega + \vartheta \omega - \nu \nabla^2 \omega = \frac{\partial (T, s)}{\partial (x, y)}, \tag{11.3.3}$$

where the Jacobian represents baroclinic vorticity production. We consider the flow region away from the viscous sublayer, where the entropy gradient and viscous diffusion can be neglected. Let $(\boldsymbol{u}, \omega, \vartheta)$ be decomposed to a basic steady flow $(\boldsymbol{U}, \Omega, \Theta)$ and disturbance waves $(\boldsymbol{u}', \omega', \vartheta')$, and assume to the leading order $\boldsymbol{U} = U(y)\boldsymbol{e}_x$ so that $\Omega = -U'(y)$ and $\Theta = 0$. Substitute the decomposition into (11.3.3), and drop nonlinear disturbance terms in the equation for ω' but retain only $\vartheta'\omega' = \vartheta\omega'$ due to the large dilatation amplitude, one obtains

$$\frac{\partial \omega'}{\partial t} + U(y)\frac{\partial \omega'_z}{\partial x} + \vartheta\omega' = 0. \tag{11.3.4}$$

The experiment has found that in the range of $x/L \in (0.47, 0.53)$ the vortical and acoustic waves have almost the same phase speed c, so both can be approximately expressed as functions of $\eta = x - ct$ with $c \neq u$ in this region. Thus, (11.3.4) yields the following coupling relation between the phase-locked traveling ω_z-wave and ϑ-wave:

$$\omega(\eta) \propto \exp\left(-\int_0^\eta \frac{\vartheta(\zeta)d\zeta}{U - c}\right), \tag{11.3.5}$$

indicating that the ω-wave can be significantly magnified by the ϑ-wave at its peak-value zone and near the critical layer. The theory of the phase-locked nonlinear interaction has been well developed, known as *nonlinear critical-layer analysis* (cf. Wu 2004). It tells that the second mode will significantly prompt the growth of the first mode. Thus, as observed in experiment, in addition to being dissipated into heat, the kinetic energy has been transferred from the high-frequency second mode to the low-frequency first mode.

On the other hand, the boundary (ω, p)-coupling occurs via viscosity and the no-slip condition. Consider the interaction of two travelling waves of the form of $\eta = x - ct$: the normal-stress wave $\Pi \equiv p - \mu'\vartheta = \Pi(k_\theta\eta)$ and the shear-stress wave $\nu\omega(k_\omega\eta)$, with $k_\theta \gg 1$ and $k_\omega \ll k_\theta$ being the wave numbers of second and first modes, respectively. Then, by a simple extension of (2.3.7a) to compressible flow, the vorticity creation rate at the wall is measured by boundary vorticity flux caused by tangent gradient of Π:

$$\sigma(x, t) = -\nu\frac{\partial \omega}{\partial y} = \frac{k_\theta}{\rho}\Pi'(\eta) \quad \text{at } y = 0, \tag{11.3.6}$$

indicating that high-frequency ϑ-wave must create new ω-wave of the same k_θ, resulting in a high-frequency Stokes layer, which directly prepares the appearance of strong turbulent fluctuations. The extra-large bulk viscosity ζ must significantly enhance this boundary coupling.

With these (ω, p)-couplings inside the flow and at the wall, why the first-mode instability with small vorticity wave number $k_\omega \ll k_\theta$ is quickly enhanced and gains high-frequency components right after the dissipation of second mode can now be understood. It may be mentioned that, by an extension of (2.3.7b), symmetric to (11.3.6) there is

$$\frac{1}{\rho}\frac{\partial \Pi}{\partial y} = \nu k_\omega \omega'(\eta) \quad \text{at } y = 0, \tag{11.3.7}$$

which represents how the low-frequency first-mode wave alters the Neumann condition of the dilatation equation. Since $\nu k_\omega \sim Re^{-1/2} k_\omega \ll 1$, this effect is much weaker. This observation is in consistency with the aforementioned nonlinear critical-layer analysis on the one-way interaction of fast-mode and slow-mode instability waves.

In summary, the findings of Zhu et al. (2015) suggest the following route from second-mode instability to turbulence production. At first, the free-stream disturbances penetrate into the boundary layer near the tip of the cone and evolves as the linear stability theory predicts. The instability waves of both modes are amplified. When the amplitudes of these waves become large enough, nonlinear interactions may be triggered. The nonlinear coupling of dilatation waves and vortical waves inside the boundary layer and their linear coupling at the wall enable the highly localized strong peak of dilatation wave to enhance the first-mode instability and thereby cause the production of turbulence. In particular, the bulk viscosity ζ serves as a strong magnifier not only in the very strong dissipation peak of high-frequency dilatation waves but also in the viscous $(\boldsymbol{\omega}, \vartheta)$-coupling at the wall.

11.4 Two Basic Physical Processes in Turbulence

Before closing this chapter, we return to the basic properties of turbulence outlined in the beginning of the chapter. One of the properties was identified as *cascade* process associated with multiple scales, where large coherent structures are getting smaller and smaller due to vortex interaction, associated with the gradual passing of their energy to random eddies. As the cascade continues, the random energy eventually dissipates to heat. Having discussed various coherent structures in transitional and turbulent shear flows, however, we now see that in addition to cascade process there is yet another basic process in opposite direction, namely *coherent production* which starts from a laminar/locally laminar or turbulent background. Disturbances of selected modes (not necessary normal modes) are growing due to the instability mechanism and then lead to formation of vortical structures that can be grouped to larger and larger scales. These two basic physical processes in turbulence rather than only one, should be born in mind.

To gain an overall idea about the above two opposite basic processes, in what follows we consider the energy transport of coherent structures, which just complements the preceding detailed discussions on specific structures.

1. Coherent energy equation. We start from the triple decomposition (11.1.14) of the velocity field, follow the same notation as in (11.1.16), and denote the instantaneous coherent energy per unit mass and its time-average by

$$K_c \equiv \frac{1}{2}\boldsymbol{u}_c \cdot \boldsymbol{u}_c, \quad \overline{K}_c \equiv \frac{1}{2}\overline{\boldsymbol{u}_c \cdot \boldsymbol{u}_c},$$

respectively. The transport equation for mean coherent energy $\overline{K}_c$ can be derived from the coherent momentum equation. Neglecting the correlation between the coherent and random motions, and dropping the viscous diffusion as well the contribution to energy production by normal stresses (which is small in shear layers), the equation reads (Hussain 1983)

$$\begin{aligned} U_j \frac{\partial \overline{K}_c}{\partial x_j} &= -\frac{\partial}{\partial x_j}(\overline{u_{cj} p_c} + \overline{u_{cj} K_c}) - \overline{u_{ci} u_{cj}} \frac{\partial U_i}{\partial x_j} \\ &+ \overline{\langle u_{ri} u_{rj} \rangle \frac{\partial u_{ci}}{\partial x_j}} - \frac{\partial}{\partial x_j} \overline{u_{ci} \langle u_{ri} u_{rj} \rangle} - \overline{\epsilon}_c, \end{aligned} \quad (11.4.1)$$

which shows very clearly the energy transfer between mean, coherent and random motions and is helpful in understanding, predicting and possible way of controlling coherent structures.

Specifically, terms in (11.4.1) responsible for either coherent-production or structural-evolution process are easily identifiable. The left side is the advection of $\overline{K}_c$ by the mean velocity. The terms on the right side are, in turn: the mean of the advection of K_c by coherent velocity and pressure fluctuations; the coherent production by the mean shear that is usually positive, except in some limited narrow regions of certain asymmetrical turbulent shear layers where the production could be negative; the intermodal energy transfer (the rate of energy transfer from coherent motions to random ones) that is usually negative, except in some special regions where self-organization mechanism becomes dominant; the diffusion of the coherent energy by random velocity fluctuations; and lastly, the viscous dissipation of K_c that is often negligible.

2. Coherent production—the first process. This process is the physical source to generate and maintain a turbulence, without which even an existing turbulence cannot survive. For example, the turbulence generated by a grid in a uniform flow will eventually disappear due to dissipation. This process is also the source to cause anisotropy and the variety of the coherent structures in a turbulence field, without which even existing coherent structures will eventually pass their energy to isotropic random eddies.

In terms of energy transfer, this process transfers energy from the mean to coherent energy [through instability and *coherence production*—second term of (11.4.1)] and from random to coherent [negative inter-modal transfer—third term of (11.4.1)].

As a consequence of self-organization, the number of degrees of freedom (Lesieur 1990) is reduced and thus it is a procedure that leads to a negative entropy generation. The self-organization of coherent vortices from random ones can be illustrated by the case of a coherent structure embedded in the surrounding fine-scale turbulence, of which a sketch based on numerical result is shown in Fig. 11.37. The small-scale random eddies in the absence of coherent vortex are isotropic and homogeneous. The

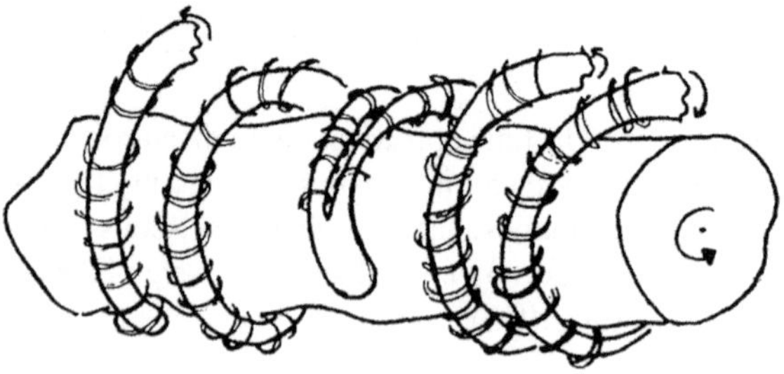

Fig. 11.37 Schematic illustration of coherent-random interaction. From Melander and Hussain (1993)

appearance of coherent vortices destroys the isotropy by aligning the random vortices to the swirl direction of the vortex, thereby giving the latter a preferred direction, and hence increases the coherent vorticity.

It should be stressed that the negative entropy generation in a turbulence field is not in conflict with the second law of thermodynamics. The latter asserts that the entropy is always increasing in an isolated system, but a given turbulence region is an open system which exchanges mass and energy with its neighboring. The given turbulence region may obtain a negative entropy flux from its neighbor so that its entropy would be locally reduced while the entropy in the neighboring region is increased. If the two regions add up to be one isolated system, the total system should still have positive entropy generation. As an interesting example from a mixing layer experiment, Huang and Ho (1990) found that the small-scale transition is first produced by the strain field of the spanwise pairing vortices imposed on the streamwise vortices. The strained streamwise vortices were unstable and initiated the random fine-scale turbulence. That is to say, the vortex merging (with negative entropy generation) is accompanied by the small-scale transition (with positive entropy generation) and the total entropy generation should still be positive.

We have seen in Sects. 11.2 and 11.3 that various instability mechanisms dominate the coherent production. In a free shear layer, one sees typically the Kelvin-Helmholtz instability followed by the formation of the spanwise vortices, the subharmonic instability and pairing etc. In a boundary layer, one sees typically the Tollmien-Schlichting instability, followed by the local inflectional instability and the formation of hairpin structures, etc. They start from laminar or locally laminar background with distributed mean vorticity (shear) and develop to organized vortices. The background can even be turbulent; e.g., a flow field with mean shear and filled with small eddies, where large vortices can also be produced by certain instability mechanism. The local inflectional instability mechanism around low speed streaks is also a good example in the self-sustaining mechanism of turbulence in fully developed turbulent boundary layers.

3. Cascading—the second process. This is an entropy generation process, including cascade, inter-modal (coherent-random) energy transfer [the third term of (11.4.1)], and dissipation. A cascade process involves complicated iterative vortex stretching, tilting and folding. But the tendency of cascading can be more easily explained by a simplified sketch. Suppose that a turbulence field is filled with many vortical

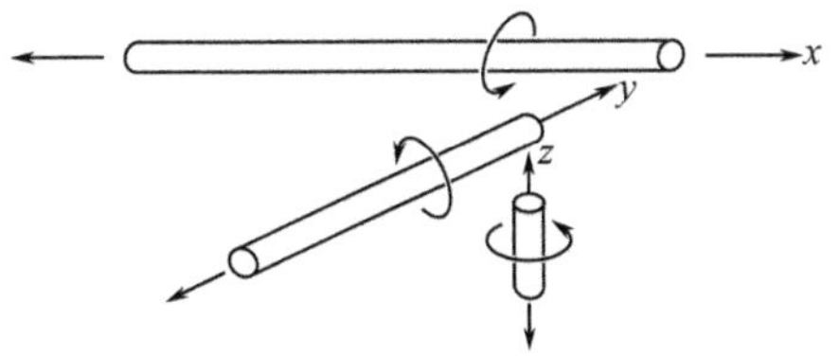

Fig. 11.38 The mutual and self induction of three vortex filament segments. Reproduced based on Chen (1986)

structures and in a local region three segments of straight vortex filaments are initially arranged as shown by Fig. 11.38. If the filament in the x-direction is somehow stretched by the induction of other segments, it will become thinner and rotate faster, which enhances the local induced velocity in the y- and z-directions. This in turn increases the local velocity gradient and causes stretching of neighboring vortices in those directions. Consequently, the latter also become thinner and rotate faster. However, as said in Sect. 3.2.3, vortex filaments cannot be only stretched without tilting, for otherwise the kinetic energy would increase unboundedly. Once the filament in x-direction is somehow tilted, say, by local-induction approximation (11.3.1) it will move in its local bi-normal direction and induce more complicated motion of the other two, making the geometry of the three-segment configuration more and more irregular. Such a procedure will continue and every step will cause further decrease of the length scale of the vortices, as well as the filament geometry. Accordingly, turbulence energy will gradually be transferred down to smaller and smaller scales.

Note that the probabilities of the cascade process as described above are uniform in all directions, and thus the turbulent structures will approach homogeneous and isotropic after several steps of cascade if there is no anisotropic influence from the first process. In fact, this process exists in all types of shear flow; and the final products of the cascade, the random eddies, are almost the same. This is why the background random eddies in turbulent shear flows are almost not dependent on the boundary conditions but coherent structures are.

If there is no influence from the first process, both *K41 scaling law* and She's universal scaling law (She and Leveque 1994) can express the cascading very well. Both revealed the physics that turbulence consists of a continuum of scales from the largest ones determined by the flow geometry to the smallest ones at the Kolmogorov dissipation scale. The largest ones are usually coherent. The smaller ones include cascaded coherent structures and random motions. If there is influence from an instability mechanism that causes production and anisotropy, a variation of the similarity parameter can be seen in She-Leveque scaling law.

The cascading process cannot continue unlimitedly. Along with the stretching, thinning and faster rotating of eddies, the dissipation rate due to the molecular viscosity is greatly enhanced. Eventually, eddies smaller than the *dissipation scale or Kolmogorov scale* $\eta = (\nu^3/\varepsilon)^{1/4}$ will be entirely dissipated with their energy being transferred to random molecular motion, the heat, and cannot be maintained in any turbulence field.

Based on Kolmogorov scale, for a given ε, a smaller ν leads to smaller dissipation scale, implying that smaller vortices can survive at higher Reynolds numbers. For example, in a high-Reynolds-number boundary layer, the order of η can be as small as tens of microns, and the corresponding timescale is of the order of microseconds. This is why DNS to date is still confined to low Reynolds numbers.

Appendix: Fields of Vectors and Tensors

A.1 Vectors and Tensors

A spatial description of the flow field is nothing but a geometrical description, of which the essence is to ensure that relevant physical quantities are invariant under artificially introduced coordinate systems. This is realized by tensor analysis (cf. Aris 1962). Here we introduce the concept of tensors in an informal way, through some important examples in fluid mechanics.

A.1.1 Scalars and Vectors

Scalars and vectors are geometric entities independent of the choice of coordinate systems. This independence is a necessary condition for an entity to represent some physical quantity. A scalar, say the fluid pressure p or density ρ, obviously has such independence. For a vector, say the fluid velocity $\boldsymbol{u}$, although its three components (u_1, u_2, u_3) depend on the chosen coordinates, say Cartesian coordinates with unit basis vectors $(\boldsymbol{e}_1, \boldsymbol{e}_2, \boldsymbol{e}_3)$, as a single geometric entity the one-form of $\boldsymbol{e}_i$,

$$\boldsymbol{u} = u_1\boldsymbol{e}_1 + u_2\boldsymbol{e}_2 + u_3\boldsymbol{e}_3 = u_i\boldsymbol{e}_i, \tag{A.1.1}$$

has to be independent of the basis vectors. Note that Einstein's convention has been used in the last expression of (A.1.1): unless stated otherwise, a repeated index always implies summation over the dimension of the space.

The inner (scalar) and cross (vector) products of two vectors are familiar. If θ is the angle between $\boldsymbol{a}$ and $\boldsymbol{b}$, the inner product generates a scalar

$$\boldsymbol{a} \cdot \boldsymbol{b} = |\boldsymbol{a}||\boldsymbol{b}| \cos\theta, \tag{A.1.2}$$

J.-Z. Wu et al., *Vortical Flows*, DOI 10.1007/978-3-662-47061-9

which is a projection operation and changing θ to $-\theta$ does not alter the result. In contrast, the cross product of $\boldsymbol{a}$ and $\boldsymbol{b}$ generates a vector

$$\boldsymbol{a} \times \boldsymbol{b} = \boldsymbol{n}(|\boldsymbol{a}||\boldsymbol{b}| \sin\theta), \tag{A.1.3}$$

where $\boldsymbol{n}$ is a unit vector perpendicular to both $\boldsymbol{a}$ and $\boldsymbol{b}$, and $|\boldsymbol{a}||\boldsymbol{b}| \sin\theta$ is the area of the parallelogram spanned by $\boldsymbol{a}$ and $\boldsymbol{b}$. Thus $\boldsymbol{a} \times \boldsymbol{b}$ determines a *vectorial area*. The positive direction of $\boldsymbol{n}$ follows the right-hand rule from $\boldsymbol{a}$ to $\boldsymbol{b}$. Since changing θ to $-\theta$ alters the sign of the result, we have $\boldsymbol{a} \times \boldsymbol{b} = -\boldsymbol{b} \times \boldsymbol{a}$.

In particular, the inner and cross products of Cartesian basis vectors satisfy

$$\boldsymbol{e}_i \cdot \boldsymbol{e}_j = \delta_{ij}, \tag{A.1.4a}$$

$$\boldsymbol{e}_i \times \boldsymbol{e}_j = \boldsymbol{e}_k, \quad i, j, k = 1, 2, 3 \text{ and cycles}, \tag{A.1.4b}$$

where

$$\delta_{ij} = \begin{cases} 1 \text{ if } i = j, \\ 0 \text{ if } i \neq j \end{cases} \tag{A.1.5}$$

is the Kroneker symbol.

The gradient operator

$$\nabla = \boldsymbol{e}_i \frac{\partial}{\partial x_i} \tag{A.1.6}$$

is also a vector, which as a single entity is invariant under any coordinate transformations. Thus, the pressure gradient ∇p is a vector; the divergence and curl of velocity,

$$\nabla \cdot \boldsymbol{u} = \vartheta, \quad \nabla \times \boldsymbol{u} = \boldsymbol{\omega}, \tag{A.1.7}$$

are the scalar *dilatation* and vector *vorticity*, respectively.

If we shift to another coordinates system, the components of $\boldsymbol{u}$ must vary according to certain rule during the coordinate transformation to ensure the invariance of vector $\boldsymbol{u}$. For example, consider two sets of Cartesian coordinates: $S(\boldsymbol{e}_i)$ and $S'(\boldsymbol{e}'_i)$. Let

$$\boldsymbol{e}_i \cdot \boldsymbol{e}'_j = c_{ij}.$$

Then the basis vectors of S' can be expressed in terms of those of S and vise versa:

$$\boldsymbol{e}'_j = c_{ij} \boldsymbol{e}_i, \tag{A.1.8a}$$

$$\boldsymbol{e}_i = c_{ij} \boldsymbol{e}'_j. \tag{A.1.8b}$$

Since

$$\boldsymbol{e}'_j \cdot \boldsymbol{e}'_k = \delta_{jk} = c_{ij} c_{lk} \boldsymbol{e}_i \cdot \boldsymbol{e}_l = c_{ij} c_{lk} \delta_{il},$$

we have

$$c_{lj}c_{lk} = \delta_{jk}, \tag{A.1.9}$$

i.e., the matrices C and C^T are orthogonal: $C^T C = I$. Then, we can expend any vector $\boldsymbol{u}$ in S and S'. Because $\boldsymbol{u}$ is *independent* of coordinates, there must be

$$\boldsymbol{u} = \boldsymbol{e}_i u_i = \boldsymbol{e}'_j u'_j = c_{ij}\boldsymbol{e}'_j u_i.$$

Therefore, using (A.1.8b) we find

$$u'_j = c_{ij}u_i. \tag{A.1.10}$$

Comparing (A.1.8a) and (A.1.10) indicates that as $\{\boldsymbol{e}_i\}$ are transformed to $\{\boldsymbol{e}'_i\}$, $\{u_i\}$ are transformed to $\{u'_i\}$ by the same transformation rule as that of the basis vectors.

Note that coordinate transformations can be either continuous or discrete. In the former c_{ij} vary continuously as a set of parameters, say the rotation angles of the basis vectors; while in the latter the direction of one of the basis vectors is reversed or one changes from right-handed to left-handed coordinate systems, like viewing the vectors through a mirror, and c_{ij} vary discretely. A *true vector* (or polar vector) is invariant under both continuous and discrete transformations, for example position vector $\boldsymbol{x}$, velocity $\boldsymbol{u}$, force $\boldsymbol{F}$, and electric field $\boldsymbol{E}$, etc. The force includes pressure force $-\nabla p$, so the vector operator ∇ is also a true vector.

On the other hand, a *pseudo vector* (or axial vector) is invariant only under continuous transformation but changes sign under discrete transformation. This happens once a cross product is involved. For example, in (A.1.4b) the unit vector $\boldsymbol{n}$ of the area, perpendicular to both $\boldsymbol{a}$ and $\boldsymbol{b}$, is an axial vector. The reason is that the positive direction of $\boldsymbol{n}$ has no objective definition but is just chosen by convention: if we turn $\boldsymbol{a}$ into $\boldsymbol{b}$ through the angle $\theta < 180°$, then we require $(\boldsymbol{a}, \boldsymbol{b}, \boldsymbol{n})$ form a right-hand triad. The direction of $\boldsymbol{n}$ will be reversed if we change the convention to the left-hand one, since $\boldsymbol{b}\times\boldsymbol{a} = -\boldsymbol{a}\times\boldsymbol{b}$. In other words, an axial vector is associated with a rotation, like the axis in which a screw with a right-hand thread will advance. Examples include angular velocity $\boldsymbol{\Omega}$, torque $\boldsymbol{x} \times \boldsymbol{F}$, and magnetic field $\boldsymbol{B}$; after all the vorticity field $\boldsymbol{\omega} = \nabla \times \boldsymbol{u}$ that concerns us most in this book.

Note however the curl of a pseudo vector is a true vector. Moreover, the inner product of a true vector and a pseudo vector, say $\boldsymbol{c} \cdot (\boldsymbol{a} \times \boldsymbol{b})$, must be a *pseudo scalar* that changes sign as we change from right-handed to left-handed coordinate systems.

A.1.2 Tensors

A physical field often involves geometric entities more complicated than vectors. For example, let the fluid velocity at a spatial point $\boldsymbol{x}$ be $\boldsymbol{u}$, and consider the velocity change at any neighboring point $\boldsymbol{x} + d\boldsymbol{x}$. To the first order of $d\boldsymbol{x}$, there is

$$du = (dx \cdot \nabla)u = dx_j u,_j ,$$

where dx_j $(j = 1, 2, 3)$ are Cartesian coefficients of dx. Here we have used a simple notation $(\cdot),_j$ to indicate the derivative with respect to x_j. So u has three directional derivatives

$$u,_j \equiv \frac{\partial u}{\partial x_j}, \quad j = 1, 2, 3,$$

each being a vector. Thus we may further expand them as

$$u,_j = e_k u_{k,j}, \quad j = 1, 2, 3.$$

Now, just as $p,_j$ is the jth component of pressure gradient vector ∇p, the vector $u,_j$ can also be viewed as the jth component of a geometric entity called *deformation tensor*, which is defined as the operator ∇ directly acting to u and equals the sum of each e_j times $u,_j$. Consequently, we obtain

$$\nabla u = e_j u,_j = e_j e_k u_{k,j}, \tag{A.1.11}$$

which can also be directly obtained from (A.1.1) and (A.1.6). In this geometric entity the basis vectors appear twice, and hence the summation is to be taken twice, once for j and once for k; so we call this entity a second-rank tensor. Note that in (A.1.11) the index j implies the components of ∇ and goes first, while the index k implies the components of u and goes after j. The order inverse in $u_{k,j}$ is only apparent, because this component of ∇u is merely an abbreviation of $\partial u_k/\partial x_j$.[1] Since ∇u is independent of the magnitude and direction of dx, in comparison with (1.1.3) it is a more general description of the velocity variation at the neighborhood of a point and will appear in many later analyses.

As another example of tensor of second rank, assume on an arbitrary surface element $dS = n dS$ there is a pressure $p > 0$, where n is the unit normal vector of the surface. Then by definition p must act as a normal pressing force on the element, i.e., a *surface stress* $t_p = -np$ or $\{t_{pi}\} = -\{n_i p\}$ per unit area. But if on dS there is a viscous force t_v, it is generically not along n but pointing to another direction, yet still depending linearly on the orientation of n. Thus, in general we may write $\{t_{vi}\} = \{n_j V_{ji}\}$ with $\{V_{ji}\}$ being the 3^2 coefficients. Since t_v and n are both geometric entities independent of the choice of coordinates, so must be $\mathbf{V} = e_j e_i V_{ji}$; it is called a *tensor of second rank*. Later we will see that the relation $\{t_{vi}\} = \{n_j V_{ji}\}$ has a vector form $t_v = n \cdot \mathbf{V}$ as the "contraction" or inner product of n and $\mathbf{V}$.

Scalars and vectors can be considered as special tensors of rank (or order) zero and one respectively. In general, as the immediate extension of vectors, a tensor $\mathbf{T}$ of rank n is a geometric entity independent of the choice of a class of coordinate systems, and has 3^n components with respect to a given coordinate in three-dimensional space.

[1] The order of j, k is a matter of convention. A single convention has to be followed throughout the whole analysis.

Like the one-form (A.1.1) for a vector, these 3^n components $T_{ij...k}$ constitute the coefficients of an n-form of the given base vectors, i.e.,

$$\mathbf{T} = T_{ij...k}\boldsymbol{e}_i\boldsymbol{e}_j \ldots \boldsymbol{e}_k.$$

The transformation rule of vector components can be extended to higher-rank tensor. For example, there is

$$\nabla\boldsymbol{u} = \boldsymbol{e}'_p\boldsymbol{e}'_q c_{jp}c_{kq}u_{k,j},$$

thus

$$u'_{q,p} = c_{jp}c_{kq}u_{k,j}.$$

This observation on the relation between tensor components and basis vectors under frame transformation suggests that a tensor can be alternatively defined in terms of the transformation behavior of its coefficients. For example:

Definition. A *Cartesian tensor* of nth rank at a point is an n-form of some Cartesian basis vectors at that point, of which the coefficients are the components of the tensor with respect to these basis vectors; under a Cartesian transformation, these components transform according to the transformation rule of basis vectors, (A.1.8a and A.1.8b) and (A.1.9), so that *the tensor remains invariant.*

Note that a Cartesian tensor may not remain invariant under a more general curvilinear transformation. For the latter we have "curvilinear tensors", which are invariant under a Cartesian transformation too. The transformation rule of curvilinear tensors is more complicated; see Aris (1962). But, using our primitive definition of tensors, their invariance under any coordinate transformation is always ensured.

In physical and numerical experiments one always chooses one or more coordinate system, and the direct output is always components of tensors. Using component operations is also convenient in deriving formulas, and sometimes only one typical component is sufficient. For example, for the velocity gradient we may just write down a representative component $u_{k,j}$. But writing the final equations in the coordinate-independent tensorial form can make the physical objectivity of these equations be seen clearly.

A.1.3 Tensor Algebra

As has been familiar, the addition (subtraction) of two vectors $\boldsymbol{a}$ and $\boldsymbol{b}$ forms a new vector whose components are the algebraic sum (difference) of those of $\boldsymbol{a}$ and $\boldsymbol{b}$. The same rule holds for tensors of any ranks: the addition (subtraction) of two tensors of the same rank forms a new tensor of that rank, whose components are the algebraic sum (difference) of those of the original tensors.

1. Tensor product. The product of a scalar and a tensor is also trivial. But the product of tensors, in general, results in a tensor of higher rank. For example, the deformation tensor $\nabla \boldsymbol{u} = \boldsymbol{e}_j \boldsymbol{e}_k \partial_j u_k$ suggests that, in addition to the inner and cross products, we may define another *dyadic multiplication* for vectors, denoted here by putting directly two (or more) vectors together, associated with increasing number of basis vectors, and thereby producing higher-rank tensors by vectors. Thus, starting from a scalar, we may write a sequence of terms like

$$\lambda, \quad \boldsymbol{a} = a_i \boldsymbol{e}_i, \quad \boldsymbol{ab} = a_i b_j \boldsymbol{e}_i \boldsymbol{e}_j, \quad \boldsymbol{abc} = a_i b_j c_k \boldsymbol{e}_i \boldsymbol{e}_j \boldsymbol{e}_k, \quad \ldots, \tag{A.1.12}$$

called *dyadics of rank 0, 1, 2, 3, ...*. Thus $\nabla \boldsymbol{u}$ is a second-rank dyadics or simply *dyad*. By using (A.1.12), one can easily verified that dyadic products satisfy the following operational laws (cf. Milne-Thomson 1968):

(1) The associative law: A dyadics can be bracketed in any manner without change of meaning. Thus,

$$\boldsymbol{abc} = (\boldsymbol{ab})\boldsymbol{c} = \boldsymbol{a}(\boldsymbol{bc}), \tag{A.1.13a}$$

$$\boldsymbol{abcd} = (\boldsymbol{ab})(\boldsymbol{cd}) = \boldsymbol{a}(\boldsymbol{bc})\boldsymbol{d}. \tag{A.1.13b}$$

(2) The scalar law: In a dyadics a vector is replaced by a scalar, which lowers the rank of the dyadics by one. For example, replacing $\boldsymbol{a}$ in (A.1.13a) by λ leads to

$$\lambda \boldsymbol{bc} = (\lambda \boldsymbol{b})\boldsymbol{c} = \lambda(\boldsymbol{bc}). \tag{A.1.13c}$$

(3) The contraction law: In a dyadics equality the dyadic product can be replaced by the inner product in one and the same position on both sides of the equality, to reduce the rank of the dyadics by two. Thus, from (A.1.13a) it follows that

$$(\boldsymbol{ab}) \cdot \boldsymbol{c} = \boldsymbol{a}(\boldsymbol{b} \cdot \boldsymbol{c}), \tag{A.1.13d}$$

by using the scalar law; while from (A.1.13b) there is

$$\boldsymbol{ab} \cdot \boldsymbol{cd} = \boldsymbol{a}(\boldsymbol{b} \cdot \boldsymbol{c})\boldsymbol{d} = (\boldsymbol{b} \cdot \boldsymbol{c})\boldsymbol{ad}. \tag{A.1.13e}$$

(4) The distributive law: If **A** is a dyadics, and **B** and **C** are dyadics of the same rank, then

$$\boldsymbol{A}(\mathbf{B} + \mathbf{C}) = \mathbf{AB} + \mathbf{AC}, \quad (\mathbf{B} + \mathbf{C})\mathbf{A} = \mathbf{BA} + \mathbf{CA}. \tag{A.1.13f}$$

The above dyadic products of vectors and associated laws can be extended to general tensors. For example, the *tensor product* of an m-rank tensor $\mathbf{P} = \{P_{i_1 i_2 \ldots i_m}\}$ and an n-rank tensor $\mathbf{Q} = \{Q_{j_1 j_2 \ldots j_n}\}$ produces an $m + n$ rank tensor $\mathbf{R} = \mathbf{PQ} = \{R_{i_1 i_2 \ldots i_m j_1 j_2 \ldots j_n}\}$. On the other hand, opposite to the tensor product, as was seen in dyadics operation the product called **contraction** reduces the rank of the resulting tensor by two. The prototype of this operation is the inner product of vectors: $\boldsymbol{a} \cdot \boldsymbol{b} = a_i b_i$, where one sums a pair of components with the same index ("dummy index"). Evidently, the contraction comes from the inner product $\boldsymbol{e}_i \cdot \boldsymbol{e}_j = \delta_{ij}$ of the two basis vectors involved in the operation. For example, the contraction or inner product of two second-rank tensors **P** and **Q** gives a second-rank tensor:

$$\mathbf{P}\cdot\mathbf{Q} = (P_{ij}\boldsymbol{e}_i\boldsymbol{e}_j)\cdot(Q_{kl}\boldsymbol{e}_k\boldsymbol{e}_l) = P_{ij}Q_{kl}\boldsymbol{e}_i\boldsymbol{e}_l\delta_{jk} = P_{ij}Q_{jl}\boldsymbol{e}_i\boldsymbol{e}_l = \mathbf{R} = R_{il}\boldsymbol{e}_i\boldsymbol{e}_l,$$

where j is the dummy index. Clearly, this contraction is the same as the multiplication of two matrices (the coefficients of any second-rank tensor can be expressed by a square matrix, but not vise versa because an arbitrary matrix does not obey the coordinate transformation rule).

The tensor product of the gradient operator ∇ and a tensor **P**, and the contraction of ∇ and tensor **Q**, say, also follow the above operation rules.

2. Tensor identification theorem. In a tensor product of any kind, say $\mathbf{A}\cdot\mathbf{B} = \mathbf{C}$, we know that if **A** and **B** are known to be geometric entities independent of the choice of coordinate systems, namely they are tensors, then evidently so must be **C**, for otherwise their independency of coordinate systems cannot be ensured. By the same reason, if **A** and **C** are known to be tensors, so must be **B**. This observation is sometimes called *tensor identification theorem*.

The theorem can be confirmed by the transformation rule of tensor components between different coordinate systems. For example, if $\boldsymbol{a} = \boldsymbol{e}_i a_i$ and $\boldsymbol{b} = \boldsymbol{e}_j b_j$ are two vectors, and we are given an equality $a_i = C_{ij}b_j$. Then from the transformation rule of a_i and b_j it can be easily inferred that C_{ij} must be the components of a second-order tensor.

3. Symmetric and antisymmetric second-order tensors. Second-order tensors are especially important in fluid dynamics. Like any square matrix, any second-rank tensor, for example the deformation tensor, can always be decomposed into a symmetric part and an anti-symmetric part, each being a second-order tensor:

$$\nabla\boldsymbol{u} = \mathbf{D} + \boldsymbol{\Omega}, \tag{A.1.14}$$

where

$$D_{ij} \equiv \frac{1}{2}(u_{j,i} + u_{i,j}) = D_{ji} \tag{A.1.15}$$

is called the *strain-rate tensor*, having six independent components; and

$$\Omega_{ij} \equiv \frac{1}{2}(u_{j,i} - u_{i,j}) = -\Omega_{ji} \tag{A.1.16}$$

is the *spin tensor*, having three independent components.

Like symmetric matrix, a real symmetric tensor **S** has three real *principal values* (eigenvalues) associated with three *principal directions*. A vector $\boldsymbol{p}$ in a principal direction is an eigenvector of **S**, defined by the homogeneous equation

$$\mathbf{S}\cdot\boldsymbol{p} = \lambda\boldsymbol{p} \quad \text{or} \quad (S_{ij} - \lambda\delta_{ij})p_j = 0, \tag{A.1.17}$$

which yields three principal values if and only if

$$\det(S_{ij} - \lambda\delta_{ij}) = -\lambda^3 + I_1\lambda^2 - I_2\lambda + I_3 = 0, \tag{A.1.18}$$

where

$$\begin{aligned} I_1 &= S_{ii} = \text{tr}\mathbf{S}, \\ I_2 &= \frac{1}{2}(S_{ii}S_{jj} - S_{ij}S_{ji}) = \frac{1}{2}[(\text{tr}\mathbf{S})^2 - \text{tr}(\mathbf{S}^2)], \\ I_3 &= \frac{1}{6}\epsilon_{ijk}\epsilon_{pqr}S_{ip}S_{jq}S_{kr} = \det\mathbf{S} \end{aligned} \tag{A.1.19}$$

are three invariants of $\mathbf{S}$. Once three roots $\lambda_1, \lambda_2, \lambda_3$ of (A.1.18) are solved, their associated principal directions $\boldsymbol{p}^{(1)}, \boldsymbol{p}^{(2)}, \boldsymbol{p}^{(3)}$ can be calculated from (A.1.17). Here, owing to the vanishing determinant, only two of the three equations are independent from which one may calculate the ratios of the magnitudes of $\boldsymbol{p}^{(1)}$ and $\boldsymbol{p}^{(2)}$ to that of $\boldsymbol{p}^{(3)}$, say, and choose any one of them be of unit length or the sum of their length squares be unity, for instance.

If the three principal values are all distinct, the principal directions are orthogonal and form the *principal axes* of $\mathbf{S}$ in which S_{ij} is diagonalized with diagonal elements being $(\lambda_1, \lambda_2, \lambda_3)$. If two of the principal values are the same, only one principal direction is unique. The other two lie in a plane perpendicular to that direction and can be chosen arbitrarily to form an orthogonal basis. If all three principal values are the same, $\lambda_1 = \lambda_2 = \lambda_3 = \lambda$, then $\mathbf{S}$ is *isotropic*, $S_{ij} = \lambda\delta_{ij}$.

Moreover, for symmetric tensor $\mathbf{S}$ and any position vector element $d\boldsymbol{x}$ from origin O, we may construct an invariant scalar

$$\phi = d\boldsymbol{x} \cdot \mathbf{S} \cdot d\boldsymbol{x} = S_{ij}dx_i dx_j \tag{A.1.20}$$

which, when taken as a constant, say $\phi = 1$, represents a center quadric surface with center at O, since changing $d\boldsymbol{x}$ to $-d\boldsymbol{x}$ does not alter ϕ. Quadric surfaces include ellipsoid, hyperboloid, paraboloid, cones, or pair of planes. This surface is known as a *tensor surface* of $\mathbf{S}$, which may help understanding the property of the tensor. For example, in principal-axis coordinates, we may write $\lambda_i = a_i^{-2}$ for $i = 1, 2, 3$ (a_i will be imaginary if $\lambda_i < 0$), so the tensor surface of $\mathbf{S}$ is

$$\frac{x_1^2}{a_1^2} + \frac{x_2^2}{a_2^2} + \frac{x_3^2}{a_3^2} = 1,$$

whose geometric property becomes obvious. An important feature of tensor surface is that, by the nature of gradient, the vector

$$2\nabla\phi = d\boldsymbol{x} \cdot \mathbf{S} \tag{A.1.21}$$

must be along the normal direction of the tensor surface.

In contrast to symmetric tensor, in three-dimensional space an antisymmetric tensor $2\boldsymbol{\Omega}$ has three independent components, which must correspond to a vector $\boldsymbol{\omega}$, in the form

$$\boldsymbol{\omega} = \begin{bmatrix} \omega_1 \\ \omega_2 \\ \omega_3 \end{bmatrix}, \quad \boldsymbol{\Omega} = \frac{1}{2}\begin{bmatrix} 0 & \omega_3 & -\omega_2 \\ -\omega_3 & 0 & \omega_1 \\ \omega_2 & -\omega_1 & 0 \end{bmatrix}. \tag{A.1.22}$$

The principal values of $2\boldsymbol{\Omega}$ are $\lambda_1 = 0$, $\lambda_{2,3} = \pm i|\boldsymbol{\omega}|$, and hence the tensor has only one principal direction as the rotation axis.

A.1.4 Unit Tensor and Permutation Tensor

The most fundamental symmetric second-rank tensor is the *unit tensor*

$$\mathbf{I} \equiv \boldsymbol{e}_i \boldsymbol{e}_j \delta_{ij}. \tag{A.1.23}$$

Under a coordinate transformation there is

$$\mathbf{I} = \boldsymbol{e}_i \boldsymbol{e}_j \delta_{ij} = c_{ip} c_{jq} \boldsymbol{e}'_p \boldsymbol{e}'_q = \delta'_{pq} \boldsymbol{e}'_p \boldsymbol{e}'_q,$$

which implies that, for the components of $\mathbf{I}$,

$$\delta'_{pq} = c_{ip} c_{jq} \delta_{ij} = c_{ip} c_{iq} = \delta_{pq}$$

by (A.1.9). Thus, the components of $\mathbf{I}$ in any Cartesian coordinates are invariably the Kronecker delta. Its contraction with itself gives

$$\delta_{ij}\delta_{jk} = \delta_{ik}, \tag{A.1.24}$$

which simply means $\mathbf{I} \cdot \mathbf{I} = \mathbf{I}$. Making contraction again, we have

$$\mathbf{I} : \mathbf{I} = \delta_{ii} = d \quad \text{(dimension of the space)}. \tag{A.1.25}$$

We mention here that the components of most of tensors are changed after an arbitrary coordinate rotation. These tensors are called *unisotropic tensors*. But there are a few tensors called *isotropic*, whose components remain unchanged during the coordinate rotation. We see the unit tensor is one of them. In fact, it can be shown that the most general *second-rank isotropic tensor* is nothing but $\lambda\delta_{ij}\boldsymbol{e}_i\boldsymbol{e}_j$, where λ is an arbitrary scalar. Moreover, one can find that the most general *fourth-rank isotropic tensor*, say $\mathbf{H} = H_{ijkl}\boldsymbol{e}_i\boldsymbol{e}_j\boldsymbol{e}_k\boldsymbol{e}_l$, is a quadratic form of the unit tensor, its components read (the proof is omitted)

$$H_{ijkl} = \lambda\delta_{ij}\delta_{kl} + \alpha\delta_{ik}\delta_{jl} + \beta\delta_{il}\delta_{jk}. \tag{A.1.26}$$

An important third-rank tensor appears whenever we make cross product (vector product) of vectors:

$$(\boldsymbol{a} \times \boldsymbol{b})_i = \boldsymbol{e}_i \cdot (\boldsymbol{e}_j a_j \times \boldsymbol{e}_k b_k) = a_j b_k \boldsymbol{e}_i \cdot (\boldsymbol{e}_j \times \boldsymbol{e}_k) = \epsilon_{ijk} a_j b_k, \tag{A.1.27}$$

where ϵ_{ijk} are components of a third-rank tensor,

$$\mathbf{E} \equiv \boldsymbol{e}_i \boldsymbol{e}_j \boldsymbol{e}_k \epsilon_{ijk}. \tag{A.1.28}$$

To find these components, we use the familiar rule of vector product:

$$\begin{aligned}\epsilon_{ijk} = \boldsymbol{e}_i \cdot (\boldsymbol{e}_j \times \boldsymbol{e}_k) &= \begin{vmatrix} \boldsymbol{e}_i \cdot \boldsymbol{e}_1 & \boldsymbol{e}_i \cdot \boldsymbol{e}_2 & \boldsymbol{e}_i \cdot \boldsymbol{e}_3 \\ \boldsymbol{e}_j \cdot \boldsymbol{e}_1 & \boldsymbol{e}_j \cdot \boldsymbol{e}_2 & \boldsymbol{e}_j \cdot \boldsymbol{e}_3 \\ \boldsymbol{e}_k \cdot \boldsymbol{e}_1 & \boldsymbol{e}_k \cdot \boldsymbol{e}_2 & \boldsymbol{e}_k \cdot \boldsymbol{e}_3 \end{vmatrix} = \begin{vmatrix} \delta_{i1} & \delta_{i2} & \delta_{i3} \\ \delta_{j1} & \delta_{j2} & \delta_{j3} \\ \delta_{k1} & \delta_{k2} & \delta_{k3} \end{vmatrix} \\ &= \begin{cases} 1 & \text{if} \quad (ijk) = (123), (231), (312), \\ -1 & \text{if} \quad (ijk) = (132), (213), (321), \\ 0 & \text{otherwise.} \end{cases}\end{aligned} \tag{A.1.29}$$

For example, by (A.1.27) we now have

$$\begin{aligned}\boldsymbol{a} \times \boldsymbol{b} &= \boldsymbol{e}_i (\boldsymbol{a} \times \boldsymbol{b})_i \\ &= \boldsymbol{e}_1 (a_2 b_3 - a_3 b_2) + \boldsymbol{e}_2 (a_3 b_1 - a_1 b_3) + \boldsymbol{e}_3 (a_1 b_2 - a_2 b_1),\end{aligned} \tag{A.1.30}$$

which is the Cartesian component form of (A.1.4b).

Therefore, the components of $\mathbf{E}$ in any Cartesian coordinates are invariably the permutation symbol ϵ_{ijk}, completely anti-symmetric with respect to the exchange of any pair of its three indices. For this reason, tensor $\mathbf{E}$ is called the *permutation tensor*, which is also isotropic. It will be used very frequently, whenever there is a cross product.

If more than one cross products appear, one need the multiplication of two permutation tensors. In a three-dimensional space, from (A.1.29) there is

$$\epsilon_{ijk} \epsilon_{lmn} = \begin{vmatrix} \delta_{il} & \delta_{im} & \delta_{in} \\ \delta_{jl} & \delta_{jm} & \delta_{jn} \\ \delta_{kl} & \delta_{km} & \delta_{kn} \end{vmatrix}. \tag{A.1.31}$$

Contraction with respect to k, n yields

$$\epsilon_{ijk} \epsilon_{lmk} = \delta_{il} \delta_{jm} - \delta_{im} \delta_{jl}. \tag{A.1.32}$$

Continuing the contraction with respect to j, m then gives

$$\epsilon_{ijk} \epsilon_{ljk} = (3 - 1) \delta_{il} = 2 \delta_{il}, \tag{A.1.33}$$

and continuing again,

$$\epsilon_{ijk}\epsilon_{ijk} = 3(3-1) = 6. \tag{A.1.34}$$

One often needs to handle two-dimensional flow. If $\boldsymbol{a}$ and $\boldsymbol{b}$ are vectors in the flow plane such that $\boldsymbol{a} \times \boldsymbol{b}$ is along the third dimension $\boldsymbol{e}_3$ normal to the plane, then in ϵ_{ijk} one of the i, j, k must be 3 due to (A.1.29), with the other two varying between 1 and 2. In this case (A.1.32) yields

$$\epsilon_{3jk}\epsilon_{3mn} = \begin{vmatrix} \delta_{jm} & \delta_{jn} \\ \delta_{km} & \delta_{kn} \end{vmatrix}, \quad j, k, m, n = 1, 2. \tag{A.1.35}$$

Contracting with respect to k, n yields

$$\epsilon_{3jk}\epsilon_{3mk} = (2-1)\delta_{jm} = \delta_{jm}, \tag{A.1.36}$$

and continuing again,

$$\epsilon_{3jk}\epsilon_{3jk} = 2(2-1) = 2. \tag{A.1.37}$$

For the permutation tensor, under a coordinate transformation there is

$$\mathbf{E} = \boldsymbol{e}_i\boldsymbol{e}_j\boldsymbol{e}_k\epsilon_{ijk} = \boldsymbol{e}'_p\boldsymbol{e}'_q\boldsymbol{e}'_r\epsilon'_{pqr},$$

which gives

$$\epsilon'_{pqr} = c_{ip}c_{jq}c_{kr}\epsilon_{ijk} = \epsilon_{pqr},$$

the last equality is from using (A.1.29) and relevant contraction operations. Thus, the components of this special tensor are also invariant.

With the aid of unit tensor and permutation tensor, all commonly encountered formulas in the algebra and differentiations of vectors and tensors can be readily derived. An important elementary use of permutation tensor is to link an antisymmetric tensor and its associated vector, as shown by (A.1.22) for the vorticity vector $\boldsymbol{\omega}$ and spin tensor $\boldsymbol{\Omega}$. This one-to-one correspondence can now be simply expressed as

$$\omega_i = \epsilon_{ijk}u_{k,j} = \epsilon_{ijk}\Omega_{jk}, \tag{A.1.38}$$

due to (A.1.16). Inversely, by using (A.1.32) it is easily seen that

$$\Omega_{jk} = \frac{1}{2}\epsilon_{ijk}\omega_i. \tag{A.1.39}$$

This pair of intimate relations between vorticity vector and spin tensor is a neat expression of (A.1.22), which shows that they have the same independent components and hence can represent, or are dual to, each other. Note that (A.1.39) also indicates that the inner product of a vector and an anti-symmetric tensor can always be conveniently expressed as the cross product of that vector and the dual vector of the anti-symmetric tensor; for instance

$$\boldsymbol{a}\cdot\boldsymbol{\Omega}=\frac{1}{2}\boldsymbol{\omega}\times\boldsymbol{a},\quad \boldsymbol{\Omega}\cdot\boldsymbol{a}=\frac{1}{2}\boldsymbol{a}\times\boldsymbol{\omega}. \tag{A.1.40}$$

Obviously, we also have

$$\nabla\cdot\boldsymbol{\Omega}=-\frac{1}{2}\nabla\times\boldsymbol{\omega}. \tag{A.1.41}$$

A.2 Integral Theorems

The key results of tensor integrations are two frequently used theorems.

The first prototype of tensor integrals is the integral of a vector $\boldsymbol{f}$ over a surface S of normal $\boldsymbol{n}$, which is primarily associated with the "net outflow" of the "$\boldsymbol{f}$-lines" in the domain V enclosed by $S=\partial V$, namely whether the $\boldsymbol{f}$-lines are generated or lost in V, or simply pass through it from one side to another. In this case one's concern is naturally the normal component of the $\boldsymbol{f}$-field on S, $\boldsymbol{n}\cdot\boldsymbol{f}$, and the key result is the familiar **Gauss theorem** that connects the integral of $\boldsymbol{f}\cdot\boldsymbol{n}$ over S to that of $\nabla\cdot\boldsymbol{f}$ over V:

$$\int_{\partial V}\boldsymbol{n}\cdot\boldsymbol{f}dS=\int_V\nabla\cdot\boldsymbol{f}dV. \tag{A.2.1}$$

The second prototype of tensor integrals is the integral of a vector $\boldsymbol{f}$ around a closed loop C of unit tangent vector $\boldsymbol{t}$, which is primarily a measure of the tendency of the $\boldsymbol{f}$-lines to "curl up". In this case one's concern is naturally the tangential component of the $\boldsymbol{f}$-field along C, $\boldsymbol{t}\cdot\boldsymbol{f}$, and the key result is the familiar **Stokes theorem** that connects the integral of $\boldsymbol{t}\cdot\boldsymbol{f}$ along C to that of $(\boldsymbol{n}\times\nabla)\cdot\boldsymbol{f}$ over any surface S spanned by C:

$$\oint_C d\boldsymbol{x}\cdot\boldsymbol{f}=\int_S\boldsymbol{n}\cdot(\nabla\times\boldsymbol{f})dS=\int_S(\boldsymbol{n}\times\nabla)\cdot\boldsymbol{f}dS, \tag{A.2.2}$$

where $d\boldsymbol{x}=\boldsymbol{t}dl$ is the vector line element of C.

In developing vorticity dynamics, there is a need to generalize the above Gauss and Stokes theorems to broader kinds of tensors and products rather than vector and inner product. We do this below based on the derivation of Milne-Thomson (1968). Note that these theorems hold only if the partial derivatives of relevant fields exist and are continuous.

A.2.1 Generalized Gauss Theorem

To generalize (A.2.1), we first examine the situation on a small volume element $V\to 0$ enclosed by boundary surface $S=\partial V$ with outward normal $\boldsymbol{n}$. Let $\boldsymbol{x}_0$ be a fixed reference point in V and $\boldsymbol{r} = \boldsymbol{x} - \boldsymbol{x}_0$ be the position vector of any point $\boldsymbol{x}$ on S with respect to $\boldsymbol{x}_0$. Let $\boldsymbol{F}$ be any vector field, of which the values on S can be approximated by $\boldsymbol{F} = \boldsymbol{F}_0 + \boldsymbol{r} \cdot (\nabla \boldsymbol{F})_0$. Thus, the surface integral of the dyad $\boldsymbol{nf}$ over S reads

$$\int_S \boldsymbol{nF} dS = \left(\int_S \boldsymbol{n} dS\right) \boldsymbol{F}_0 + \left(\int_S \boldsymbol{nr} dS\right) \cdot (\nabla \boldsymbol{F})_0,$$

where we recall a basic fact

$$\int_S \boldsymbol{n} dS = \mathbf{0} \tag{A.2.3}$$

for *any* closed S, because the projection of this integral on any direction must vanish due to the cancellation of $\boldsymbol{n}$ at opposite surface elements. Thus

$$\int_S \boldsymbol{nF} dS = \left(\int_S \boldsymbol{nr} dS\right) \cdot (\nabla \boldsymbol{F})_0. \tag{A.2.4}$$

But, it can be generally proved that

$$\int_S \boldsymbol{nr} dS = V\mathbf{I} \tag{A.2.5}$$

for any closed S and arbitrarily chosen $\boldsymbol{x}_0$, with $\mathbf{I}$ being the unit tensor. For our purpose, it suffices to demonstrate (A.2.5) for the case where $V = h^3$ is a small cubic square box in three-dimensional space and $\boldsymbol{x}_0$ is the center of the box. In this case S consists of six sides, each with area h^2 and unit normal $\pm \boldsymbol{e}_i$, $i = 1, 2, 3$. We can then use the center point of each side with $\boldsymbol{r} = \pm h\boldsymbol{e}_i/2$ to calculate the integral. This approximation gives

$$\begin{aligned}\int_S \boldsymbol{nr} dS &= \sum_{i=1}^{3} [\boldsymbol{e}_i h^2 \boldsymbol{e}_i h/2 + (-\boldsymbol{e}_i) h^2 (-\boldsymbol{e}_i) h/2] \\ &= h^3 \sum_{i=1}^{2} \boldsymbol{e}_i \boldsymbol{e}_i = \sum_{i,j=1}^{2} \boldsymbol{e}_i \boldsymbol{e}_j \delta_{ij} V = V\mathbf{I},\end{aligned}$$

which confirms (A.2.5). Therefore, (A.2.4) implies an important relation between the local gradient of a vector field $\boldsymbol{F}$ at a point $\boldsymbol{x}_0$ and the surface integral of $\boldsymbol{nF}$ over the boundary of the infinitesimal volume surrounding $\boldsymbol{x}_0$:

$$\lim_{V \to 0} \left(\frac{1}{V} \int_S \boldsymbol{nF} dS\right) = \nabla \boldsymbol{F}. \tag{A.2.6}$$

Obviously, the same argument holds true for a box of $V = h^2$ in two-dimensional space.

Then, recall that we have shown at the end of Sect. A.1.3 that a dyad has a scalar invariant and a vector invariant, we may replace the dyadic product $\boldsymbol{nr}$ in (A.2.6) by inner or cross product; and, since by the scaling law one vector in a dyad can be replaced by a scalar such that the dyadic product is replaced by conventional scalar multiplication. Moreover, noticing that the vector $\boldsymbol{F}$ in (A.2.6) can be replaced by any tensor, denoted by $\mathcal{F} = F_{ij...k}\boldsymbol{e}_i\boldsymbol{e}_j\ldots\boldsymbol{e}_k$, because only its far-left element is involved in the product, we may write (A.2.6) as a general form

$$\lim_{V\to 0}\left(\frac{1}{V}\int_S \boldsymbol{n}\circ\mathcal{F}dS\right) = \nabla\circ\mathcal{F}, \tag{A.2.7}$$

where the small circle denotes any permissible products: multiplication, scalar, vector, or dyadic.

We now return to an arbitrary finite volume V bounded by S. We can always cut V into sufficiently small boxes, for each of which (A.2.7) holds. Since at the common boundary of two neighboring boxes the outward normals are of opposite sign, the surface integrals over all boundaries shared by two boxes must cancel out, leaving only the boundary S of the whole V. This fact exhibits the *additive* property of (A.2.7) and yields the **generalized Gauss theorem**:

$$\int_{\partial V}\boldsymbol{n}\circ\mathcal{F}dS = \int_V \nabla\circ\mathcal{F}dV, \tag{A.2.8}$$

of which (A.2.1) is a special case. Taking $\mathcal{F}$ as a constant we return to (A.2.3), while taking $\mathcal{F}$ as $\boldsymbol{r}$ and $\circ$ as diadic product we return to (A.2.5). Evidently, for the total pressure force on ∂V as well as the boundary integrals of normal and tangential velocity, we have

$$\int_{\partial V}\boldsymbol{n}pdS = \int_V \nabla pdV, \tag{A.2.9}$$

$$\int_{\partial V}\boldsymbol{n}\cdot\boldsymbol{u}dS = \int_V \vartheta dV, \tag{A.2.10}$$

$$\int_{\partial V}\boldsymbol{n}\times\boldsymbol{u}dS = \int_V \boldsymbol{\omega}dV. \tag{A.2.11}$$

The generalized Gauss theorem (A.2.8) can also be read from right to left. Then it represents an extension of the integration of a total differentiation

$$\int_a^b f'(x)dx = \int_a^b df(x) = f(b) - f(a) \tag{A.2.12}$$

to multi-dimensional space: in a volume integral if all terms form a tensor function $\mathcal{F}$ acted by a gradient operator ∇, then the integrand must be a total differentiation and the integral can be simply expressed by the boundary value of $\mathcal{F}$ at ∂V.

A.2.2 Generalized Stokes Theorem

To generalize the Stokes theorem (A.2.2), we follow the same approach and first examine the line integral of dyad $d\boldsymbol{x}\boldsymbol{F}$ along the boundary of a small triangular surface element $\boldsymbol{n}\Delta$. Let the three edges of Δ be $\boldsymbol{a}_i$, $i = 1, 2, 3$, that form a directional loop $C = \partial\Delta$ with line element $d\boldsymbol{x} = \boldsymbol{t}dl$, and let $\boldsymbol{x}_0$ be the center point O of the triangle, see Fig. A.1 such that to the leading order $\boldsymbol{F} = \boldsymbol{F}_0 + \boldsymbol{r} \cdot (\nabla\boldsymbol{F})_0$ on C. Then by using the evident fact similar to (A.2.3),

$$\oint d\boldsymbol{x} = \boldsymbol{0}, \tag{A.2.13}$$

the integral of $d\boldsymbol{x}\boldsymbol{F}$ along the loop or the "dyad circulation" reads

$$\oint_C d\boldsymbol{x}\boldsymbol{F} = \left(\oint_C d\boldsymbol{x}\boldsymbol{r}\right) \cdot (\nabla\boldsymbol{F})_0.$$

We approximate this integral by

$$\oint_C d\boldsymbol{x}\boldsymbol{F} = \sum_{i=1}^{3} \boldsymbol{a}_i\boldsymbol{F}_i = \boldsymbol{a}_1(\boldsymbol{F}_1 - \boldsymbol{F}_3) + \boldsymbol{a}_2(\boldsymbol{F}_2 - \boldsymbol{F}_3) \tag{A.2.14}$$

due to $\boldsymbol{a}_3 = -(\boldsymbol{a}_1 + \boldsymbol{a}_2)$, where $\boldsymbol{F}_i$ $(i = 1, 2, 3)$ is the value of $\boldsymbol{F}$ at the midpoint A_i of each edge $\boldsymbol{a}_i$. Now since $\boldsymbol{r}_i = \overrightarrow{OA_i}$ (Fig. A.1), in the above equation there is

$$\boldsymbol{F}_1 - \boldsymbol{F}_3 = (\boldsymbol{r}_1 - \boldsymbol{r}_3) \cdot (\nabla\boldsymbol{F})_0, \quad \boldsymbol{F}_2 - \boldsymbol{F}_3 = (\boldsymbol{r}_2 - \boldsymbol{r}_3) \cdot (\nabla\boldsymbol{F})_0.$$

But, from Fig. A.1 and the elementary plane geometry, we know $\boldsymbol{r}_1 - \boldsymbol{r}_3 = -\boldsymbol{a}_2/2$ and $\boldsymbol{r}_2 - \boldsymbol{r}_3 = \boldsymbol{a}_1/2$. Hence (A.2.14) becomes

$$\oint_C d\boldsymbol{x}\boldsymbol{F} = -\frac{\boldsymbol{a}_1}{2}[\boldsymbol{a}_2 \cdot (\nabla\boldsymbol{F})_0] + \frac{\boldsymbol{a}_2}{2}[\boldsymbol{a}_1 \cdot (\nabla\boldsymbol{F})_0],$$

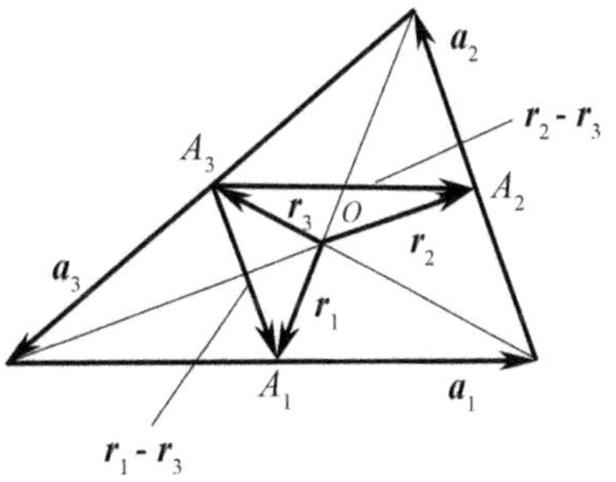

Fig. A.1 A triangle formed by vectors $\boldsymbol{a}_1$, $\boldsymbol{a}_2$, and $\boldsymbol{a}_3 = -(\boldsymbol{a}_1 + \boldsymbol{a}_2)$. Shown in the figure are the midpoints and lines that intersect at the center point O, vectors $\boldsymbol{r}_i = \overrightarrow{OA_i}$, $i = 1, 2, 3$, as well as $\boldsymbol{r}_1 - \boldsymbol{r}_3 = -\boldsymbol{a}_2/2$ and $\boldsymbol{r}_2 - \boldsymbol{r}_3 = \boldsymbol{a}_1/2$

which by using the identity

$$\frac{1}{2}(\boldsymbol{a}_1 \times \boldsymbol{a}_2) \times \nabla = \Delta \boldsymbol{n} \times \nabla = \frac{1}{2}[-\boldsymbol{a}_1(\boldsymbol{a}_2 \cdot \nabla) + \boldsymbol{a}_2(\boldsymbol{a}_1 \cdot \nabla)]$$

is cast to (similar to the proof of (A.2.7), we insert a general product $\circ$ and replace $\boldsymbol{F}$ by a general tensor $\mathcal{F}$)

$$\lim_{\Delta \to 0} \left(\frac{1}{\Delta} \oint_{\partial \Delta} d\boldsymbol{x} \circ \mathcal{F} \right) = (\boldsymbol{n} \times \nabla) \circ \mathcal{F}. \tag{A.2.16}$$

Now, for a finite loop enclosing a finite surface S, we can cut S into many triangles, of which the line integrals along a boundary shared by two neighboring triangles are again cancelled. Thus, (A.2.16) is also additive and we finally obtain the **generalized Stokes theorem**

$$\oint_{\partial S} d\boldsymbol{x} \circ \mathcal{F} = \int_S (\boldsymbol{n} \times \nabla) \circ \mathcal{F} dS, \tag{A.2.17}$$

of which (A.2.2) is a special case. The classic application of the latter in fluid dynamics is Kelvin's identification of the velocity circulation along ∂S as the vorticity flux through S:

$$\oint_{\partial S} \boldsymbol{u} \cdot d\boldsymbol{x} = \int_S \boldsymbol{\omega} \cdot \boldsymbol{n} dS. \tag{A.2.18}$$

Other examples include

$$\oint_{\partial S} \phi d\boldsymbol{x} = \int_S \boldsymbol{n} \times \nabla \phi dS, \tag{A.2.19}$$

$$\oint_{\partial S} d\boldsymbol{x} \times \boldsymbol{\psi} = \int_S (\boldsymbol{n} \times \nabla) \times \boldsymbol{\psi} dS. \tag{A.2.20}$$

The generalized Stokes theorem is another extension of the elementary calculus formula (A.2.12) to multiple-dimensional space, stating that if in an open-surface integral all terms form a tensor $\mathcal{F}$ acted by the tangential gradient operator $(\boldsymbol{n} \times \nabla)$, then the integrand must be a total differentiation and the integral can be expressed by values of $\mathcal{F}$ along ∂S.[2]

An immediate corollary of the generalized Stokes theorem is

$$\int_S (\boldsymbol{n} \times \nabla) \circ \mathcal{F} dS = 0 \;\text{ on closed } S, \tag{A.2.21}$$

[2]Thus, it is the second expression on the right-hand side of (A.2.2), rather than the first one, that is suitable to be generalized.

since a closed surface has no boundary. An example of (A.2.21) is

$$\int_S \boldsymbol{n} \times \nabla\phi dS = 0 \text{ on closed } S. \tag{A.2.22}$$

A.2.3 Derivative Moment Transformation

Standard force formulas can be transformed to different but equivalent forms where local dynamics appears explicitly. The formulas to appear below are all from extending the elementary one-dimensional integral formula

$$\int_a^b d[xf(x)] = [xf(x)]_a^b = \int_a^b xf'(x)dx + \int_a^b f(x)dx$$

to higher dimensions by using the Gauss and Stokes theorems. This trick enables expressing the integral of a vector by that of its proper derivatives. For example, in two or three dimensions, from the vector identities

$$\nabla \cdot (\boldsymbol{x}\boldsymbol{f}) - \nabla(\boldsymbol{x} \cdot \boldsymbol{f}) = k\boldsymbol{f} - \boldsymbol{x} \times (\nabla \times \boldsymbol{f})$$
$$\nabla \cdot (\boldsymbol{f}\boldsymbol{x}) = (\nabla \cdot \boldsymbol{f})\boldsymbol{x} + \boldsymbol{f},$$

where and below $k = n - 1$ and $n = 2, 3$ denotes the spatial dimensionality,[3] and applying the generalized Gauss theorem (A.2.8), there is

$$\int_V \boldsymbol{f} dV = \frac{1}{k}\int_V \boldsymbol{x} \times (\nabla \times \boldsymbol{f})dV - \frac{1}{k}\int_{\partial V} \boldsymbol{x} \times (\boldsymbol{n} \times \boldsymbol{f})dS, \tag{A.2.23}$$

$$\int_V \boldsymbol{f} dV = -\int_V \boldsymbol{x}(\nabla \cdot \boldsymbol{f})dV + \int_{\partial V} \boldsymbol{x}(\boldsymbol{n} \cdot \boldsymbol{f})dS. \tag{A.2.24}$$

It is easily seen that the total-vorticity formula (3.3.2) is just a special application of identity (A.2.24). These integrations by parts cast the integral of a vector field to that of the moment of its curl or divergence (spatial derivatives). We thus name these kind of transformations *derivative moment transformation* (DMT).

Some more DMT identities can be deduced from the Gauss and Stokes theorems. For example, from the generalized Stokes theorem (A.2.17), for the integral of any normal vector $\phi\boldsymbol{n}$ over an open surface S bounded by closed loop C, there is[4]

$$\int_S \phi\boldsymbol{n} dS = -\frac{1}{k}\int_S \boldsymbol{x} \times (\boldsymbol{n} \times \nabla\phi)dS + \frac{1}{k}\oint_C \phi\boldsymbol{x} \times d\boldsymbol{x}. \tag{A.2.25}$$

---

[3]When $n = 2$, V is a deck of unit thickness.

[4]When S is a closed surface, the second term of (A.2.25) vanishes. This special case can also be derived by setting $\boldsymbol{f} = \nabla\phi$ in (A.2.23).

And, for $n = 3$ only, the surface integral of a tangent vector can be cast to

$$\int_S \boldsymbol{n} \times \boldsymbol{A} dS = -\int_S \boldsymbol{x} \times [(\boldsymbol{n} \times \nabla) \times \boldsymbol{A}] dS + \oint_{\partial S} \boldsymbol{x} \times (d\boldsymbol{x} \times \boldsymbol{A}). \tag{A.2.26}$$

Moreover, if the original integral is already a moment, the DMT may cast it to an integral of a second-order moment. Examples include:

$$\int_V \boldsymbol{x} \times \boldsymbol{f} dV = -\frac{1}{2}\int_V x^2 \nabla \times \boldsymbol{f} dV + \frac{1}{2}\int_{\partial V} x^2 \boldsymbol{n} \times \boldsymbol{f} dS, \tag{A.2.27}$$

$$\int_S \boldsymbol{x} \times \boldsymbol{n}\phi dS = \frac{1}{2}\int_S x^2 \boldsymbol{n} \times \nabla\phi dS - \frac{1}{2}\oint_C x^2 \phi d\boldsymbol{x} \tag{A.2.28a}$$

$$= -\frac{1}{3}\int_S \boldsymbol{x} \times [\boldsymbol{x} \times (\boldsymbol{n} \times \nabla\phi)] dS + \frac{1}{3}\oint_C \phi \boldsymbol{x} \times (\boldsymbol{x} \times d\boldsymbol{x}). \tag{A.2.28b}$$

For more DMT identities see Wu et al. (2006). All these identities are invariant with respect to the change of the origin of the position vector $\boldsymbol{x}$. Therefore, the origin of $\boldsymbol{x}$ can be arbitrarily chosen, usually depending on the convenience.

A.3 Intrinsic Decompositions of Vector Fields

In a three-dimensional space, a vector field say $\boldsymbol{f}(\boldsymbol{x}, t)$ can be decomposed into three independent scalar functions of $\boldsymbol{x}$ and t. If they are components of $\boldsymbol{f}$ in a coordinate system, they do not have intrinsic meaning and are cumbersome in practical analysis. What is desired here is to seek an intrinsic and coordinate-independent decomposition of $\boldsymbol{f}$, such that each scalar component has clear physical implication as nature itself exhibits. This is of critical importance in understanding the physics of the fundamental processes in fluid motion. A systematic and thorough approach in studying this subject is the classic Helmholtz decomposition and its modern sharpening to be introduced below. While the theory applies to any vector field, we use vector fields in fluid mechanics as illustration.

A.3.1 Longitudinal and Transverse Vector Fields

In Sect. A.1 we have learned that the vector operator ∇ can produce two kinds of vectors. Its product with a scalar φ produces a curl-free true vector, and its cross product with a vector produces a divergence-free vector:

$$\boldsymbol{f}_{\parallel} = \nabla\varphi, \quad \boldsymbol{f}_{\perp} = \nabla \times \boldsymbol{A}. \tag{A.3.1}$$

These two vector fields have very different behavior. For example, if $\boldsymbol{f}_{\parallel}$ and $\boldsymbol{f}_{\perp}$ represent different forces, by the generalized Gauss theorem (A.2.8) their integrals over a volume V yield a pair of boundary integrals:

$$\int_V \boldsymbol{f}_{\parallel} dV = \int_{\partial V} \boldsymbol{n}\varphi dS, \quad \int_V \boldsymbol{f}_{\perp} dV = \int_{\partial V} \boldsymbol{n} \times \boldsymbol{A} dS. \tag{A.3.2}$$

Moreover, if φ and $\boldsymbol{A}$ have wave behavior, say a plane travelling wave of the form of $f(\boldsymbol{x}, t) = f_0 e^{i(\boldsymbol{k}\cdot\boldsymbol{x}-nt)}$ with constant amplitude f_0, then we have

$$\boldsymbol{f}_{\parallel} = i\boldsymbol{k}\varphi_0 e^{i(\boldsymbol{k}\cdot\boldsymbol{x}-nt)}, \quad \boldsymbol{f}_{\perp} = i\boldsymbol{k} \times \boldsymbol{A}_0 e^{i(\boldsymbol{k}\cdot\boldsymbol{x}-nt)}, \tag{A.3.3}$$

so that the waves of $\boldsymbol{f}_{\parallel}$ and $\boldsymbol{f}_{\perp}$ propagate along and perpendicular to the direction of $\boldsymbol{k}$, respectively: the former is associated with *longitudinal* wave, while the latter is associated with *transverse* wave.

These simple examples suggest that the curl-free and divergence-free fields represent two physically distinct vectorial processes, which are mutually independent and complementary in many aspects. Owing to their different wave propagation direction, we call the curl-free process the *longitudinal process* and the divergence-free process the *transverse process*.[5]

Thinking of the mutually independent and complementary behavior of longitudinal and transverse vectors, one may further conjecture that these specific vector fields may coexist in a general vector field with both nonzero curl and nonzero divergence, or a general vector field could be intrinsically decomposed into such two parts, say $\boldsymbol{f} = \boldsymbol{f}_{\parallel} + \boldsymbol{f}_{\perp}$. This conjecture has been well confirmed by the following famous *Helmholtz decomposition* and its later sharpening.

A.3.2 Helmholtz Decomposition

We start from considering an intrinsic geometrical decomposition of a vector, say velocity $\boldsymbol{u}$, to the component along the direction defined by a given vector, say $\boldsymbol{k}$,

[5]We shall use these terms in a broader sense than that implied by (A.3.3). If in a bounded domain the Fourier expansion is not feasible, the simple geometric relations between wave oscillating directions and $\boldsymbol{k}$ will no longer exist; but (A.3.1) will still be used as the definition of longitudinal and transverse vectors.

and the components perpendicular to it. Using these component vectors to recover the original vector is easily achieved by the well-known identity:

$$|\boldsymbol{k}|^2\boldsymbol{u} = \boldsymbol{k}(\boldsymbol{k}\cdot\boldsymbol{u}) - \boldsymbol{k}\times(\boldsymbol{k}\times\boldsymbol{u}) \equiv |\boldsymbol{k}|^2(\boldsymbol{u}_\parallel + \boldsymbol{u}_\perp), \tag{A.3.4}$$

where $\parallel$ and $\perp$ denote the component velocity vectors parallel and perpendicular to $\boldsymbol{k}$, respectively.

Splitting a vector into parallel and perpendicular parts as in (A.3.4) is the prototype of a more advanced intrinsic decomposition of vectors, which can be conveniently observed by replacing $\boldsymbol{k}$ by ∇ in (A.3.4), yielding a differential identity

$$\begin{aligned}\nabla^2\boldsymbol{u} &= \nabla(\nabla\cdot\boldsymbol{u}) - \nabla\times(\nabla\times\boldsymbol{u}) \\ &= \nabla\vartheta - \nabla\times\boldsymbol{\omega}.\end{aligned} \tag{A.3.5}$$

This is an example of the *Helmholtz decomposition* (HD for short), characterized by the fact that the first part is curl-free (irrotational) while the second part is divergence-free (solenoidal). The functions $\nabla\cdot\boldsymbol{u}$ and $\nabla\times\boldsymbol{u}$ are said to be the scalar and vector *Helmholtz potentials* of $\nabla^2\boldsymbol{u}$, respectively.

In (A.3.5) the Helmholtz potentials themselves are both physical entities with natural implications. More generally, a theorem by Helmholtz asserts that *for a finite and continuous vector field* $\boldsymbol{u}(\boldsymbol{x}, t)$ *defined in a domain* V*, the following decomposition exists globally*:

$$\boldsymbol{u} = \nabla\phi + \nabla\times\boldsymbol{\psi}. \tag{A.3.6a}$$

Of the three independent components of $\boldsymbol{u}$, ϕ represents one of them and $\boldsymbol{\psi}$ should represent the remaining two although it has three components. This is possible because in (A.3.6a) the curl-free part of $\boldsymbol{\psi}$ (a scalar gradient that takes one component) is useless. Thus it suffices to impose a constraint or *gauge condition* to specify the curl-free part, and the most convenient choice for fluid dynamics is to set

$$\nabla\cdot\boldsymbol{\psi} = 0. \tag{A.3.6b}$$

Note that in (A.3.5) this gauge condition is automatically satisfied.

To prove that for a prescribed field $\boldsymbol{u}(\boldsymbol{x}, t)$ with associated known ϑ and $\boldsymbol{\omega}$ fields, the Helmholtz potentials ϕ and $\boldsymbol{\psi}$ do exist globally, we take divergence and curl of (A.3.6a), and using (A.3.6b). This yields

$$\nabla\cdot\boldsymbol{u} = \vartheta = \nabla^2\phi, \tag{A.3.7a}$$

$$\nabla\times\boldsymbol{u} = \boldsymbol{\omega} = -\nabla^2\boldsymbol{\psi}. \tag{A.3.7b}$$

Then the existence of ϕ and $\boldsymbol{\psi}$ amounts to that of solutions for Poisson equations (A.3.7a) and (A.3.7b), respectively, for which the answer has long been known. Later in Sect. A.3.3 we shall give a pair of analytical solutions of (A.3.7) in integral form.

Unlike unbounded domain, in an arbitrary bounded domain V the Helmholtz decomposition (A.3.6) has not yet exhausted the intrinsic decomposition of a vector field to its full extent. Some issues need to be further addressed. First, the Helmholtz potentials in (A.3.6a) are not unique, since $\boldsymbol{\psi} + \nabla\eta$ for an arbitrary scalar η is also a vector potential due to $\nabla \times \nabla\eta = 0$. This arbitrariness cannot be completely removed by (A.3.6b), which only implies $\nabla^2\eta = 0$.

Second, in mathematic term, as a generalization of geometric orthogonality, two vector fields $\boldsymbol{u}_1$ and $\boldsymbol{u}_2$ defined on a domain V are said *functionally orthogonal* in the inner-product integral sense, if and only if they are square-integrable and the integral of $\boldsymbol{u}_1 \cdot \boldsymbol{u}_2$ over V vanishes. Now, we have seen that the properties of the longitudinal and transverse parts of $\boldsymbol{u}$ differ radically from each other and mutually complementary in certain sense; but it is not yet clear whether or under what condition the orthogonality condition holds:

$$\int_V \nabla\phi \cdot (\nabla \times \boldsymbol{\psi}) dV = 0. \tag{A.3.8}$$

When (A.3.8) holds, we use the same notation $\|$ and $\perp$ to denote

$$\boldsymbol{u}_{\|} = \nabla\phi, \quad \boldsymbol{u}_{\perp} = \nabla \times \boldsymbol{\psi}, \tag{A.3.9}$$

and call $\boldsymbol{u}_{\|}$ and $\boldsymbol{u}_{\perp}$ the *longitudinal* and *transverse* components of $\boldsymbol{u}$, respectively. A full clarification of this fundamental problem is given by the following theorem (Chorin and Marsden 1992):

Helmholtz-Hodge Theorem. *A vector field $\boldsymbol{f}$ on V can be uniquely and orthogonally decomposed in the form $\boldsymbol{f} = \nabla\phi + \boldsymbol{f}_{\perp}$, where $\boldsymbol{f}_{\perp}$ has zero divergence and is parallel to ∂V.*

This result sharpens the Helmholtz decomposition (A.3.6) and is called *Helmholtz-Hodge decomposition*. We now prove the theorem under the assumption that V is *simply-connected* for simplicity.

First, since

$$\boldsymbol{n} \cdot \boldsymbol{f}_{\perp} = 0 \text{ or } \boldsymbol{n} \cdot \boldsymbol{f} = \frac{\partial\phi}{\partial n} \quad \text{on } \partial V, \tag{A.3.10}$$

by the Gauss theorem there is

$$\int_V \nabla\phi \cdot \boldsymbol{f}_{\perp} dV = \int_{\partial V} \nabla \cdot (\phi \boldsymbol{f}_{\perp}) dS = \int_{\partial V} \phi \boldsymbol{n} \cdot \boldsymbol{f}_{\perp} dS = 0,$$

which proves the orthogonality.

Second, consider two splittings $\boldsymbol{f} = \nabla\phi_1 + \boldsymbol{f}_{\perp 1} = \nabla\phi_2 + \boldsymbol{f}_{\perp 2}$, both satisfying (A.3.10). Then

$$\boldsymbol{0} = \boldsymbol{f}_{\perp 1} - \boldsymbol{f}_{\perp 2} + \nabla(\phi_1 - \phi_2).$$

Taking the inner product with $\boldsymbol{f}_{\perp 1} - \boldsymbol{f}_{\perp 2}$ and integrating, we have

$$\begin{aligned} 0 &= \int_V \{|\boldsymbol{f}_{\perp 1} - \boldsymbol{f}_{\perp 2}|^2 + (\boldsymbol{f}_{\perp 1} - \boldsymbol{f}_{\perp 2}) \cdot \nabla(\phi_1 - \phi_2)\} dV \\ &= \int_V |\boldsymbol{f}_{\perp 1} - \boldsymbol{f}_{\perp 2}|^2 dV, \end{aligned}$$

due to the Gauss theorem and (A.3.10). Thus, there must be $\boldsymbol{f}_{\perp 1} = \boldsymbol{f}_{\perp 2}$ and hence $\nabla\phi_1 = \nabla\phi_2$. This guarantees the uniqueness of the decomposition and excludes any scalar potentials from $\boldsymbol{f}_\perp$. Moreover, (A.3.7a) and the second equation of (A.3.10) form a well-posed Neumann problem for ϕ, of which the solution exists and is unique up to an additive constant. Hence so does $\boldsymbol{f}_\perp = \boldsymbol{f} - \nabla\phi$. Thus the proof of the uniqueness and existence of the decomposition is completed.

Note that for unbounded domain the existence, uniqueness, and orthogonality of the longitudinal and transverse parts of $\boldsymbol{f}$ are all ensured provided that one of $\nabla\phi$ and $\boldsymbol{f}_\perp$ decays fast enough as $|\boldsymbol{x}| \to \infty$.

A.3.3 Integral Expressions of Helmholtz Potentials

In the Helmholtz decomposition (A.3.5) of $\nabla^2\boldsymbol{u}$, the Helmholtz potentials ϑ and $\boldsymbol{\omega}$ themselves are important physical fields and will appear in dynamic equations. We call such an HD *natural*. In contrast, in (A.3.6a) the potentials ϕ and $\boldsymbol{\psi}$ do not have independent physical meaning nor appear in physical laws (unless under very idealized and simplified conditions), but serve as auxiliary functions. In this case they have to be calculated from the given $\boldsymbol{u}$-field by solving the Poission equations (A.3.7). We do this now.

The idea comes from a comparison of (A.3.6a) and (A.3.5), which suggests that if we set

$$\boldsymbol{u} = -\nabla^2\boldsymbol{F} = -\nabla(\nabla \cdot \boldsymbol{F}) + \nabla \times (\nabla \times \boldsymbol{F}), \tag{A.3.11a}$$

then there will be

$$\phi = -\nabla \cdot \boldsymbol{F}, \quad \boldsymbol{\psi} = \nabla \times \boldsymbol{F}. \tag{A.3.11b}$$

Thus, computing these potentials for given $\boldsymbol{u}$ amounts to solving the vector Poisson equation (A.3.11a), which can be done by using the fundamental solution or Green's function $G(\boldsymbol{x})$ of Poisson equation in free space, defined by

$$\nabla^2 G(\boldsymbol{x}) = \delta(\boldsymbol{x}). \tag{A.3.12}$$

In n-dimensional space, $n = 2, 3$, this solution is known as

$$G(\boldsymbol{x}) = \begin{cases} \dfrac{1}{2\pi} \log |\boldsymbol{x}| \text{ if } n = 2, \\ -\dfrac{1}{4\pi |\boldsymbol{x}|} \quad \text{if } n = 3. \end{cases} \tag{A.3.13}$$

We will often use the vector

$$\nabla G = \frac{\boldsymbol{x}}{2(n-1)\pi |\boldsymbol{x}|^n}. \tag{A.3.14}$$

Now assume $\boldsymbol{u}$ is given in a domain V and $\boldsymbol{u} = \boldsymbol{0}$ outside V. Then the vector field $\boldsymbol{F}$ satisfying (A.3.11b) can be written by

$$\boldsymbol{F}(\boldsymbol{x}) = -\int_V G(\boldsymbol{x} - \boldsymbol{x}')\boldsymbol{u}(\boldsymbol{x}')dV',$$

where $dV' = dV(\boldsymbol{x}')$. Indeed, acting both sides by ∇^2, and noticing that $\boldsymbol{x}$ is fixed and $\boldsymbol{x}'$ is the integration variable, so that ∇^2 can be shifted into the integral and acts on $G(\boldsymbol{x} - \boldsymbol{x}')$ only, by (A.3.12) we have

$$-\nabla^2 \boldsymbol{F} = \int_V \delta(\boldsymbol{x} - \boldsymbol{x}')\boldsymbol{u}(\boldsymbol{x}')dV' = \begin{cases} \boldsymbol{u} \text{ if } \boldsymbol{x} \in V, \\ \boldsymbol{0} \text{ if } \boldsymbol{x} \notin V. \end{cases} \tag{A.3.15}$$

Therefore, owing to (A.3.11b), denoting the gradient operator with respect to the integration variable $\boldsymbol{x}'$ by ∇' so that $\nabla G = -\nabla' G$ and $\nabla^2 G = \nabla'^2 G$, when $\boldsymbol{x}$ is in V we obtain the desired integral expressions of $(\phi, \boldsymbol{\psi})$ by $\boldsymbol{u}$

$$\phi = -\int \nabla' G \cdot \boldsymbol{u} dV', \tag{A.3.16a}$$

$$\boldsymbol{\psi} = \int \nabla' G \times \boldsymbol{u} dV'. \tag{A.3.16b}$$

It should be stressed that for bounded domain (A.3.16) only provides *one of all possible pairs* of Helmholtz potentials of $\boldsymbol{u}$ because it does not care any boundary condition for $\nabla\phi$ and $\nabla \times \boldsymbol{\psi}$. In particular, the velocity field $\boldsymbol{u}$ may often be nonzero outside V but is both divergence-free and curl-free. This is seen clearly by applying integration by parts to (A.3.16) and using the Gauss theorem, yielding

$$\phi = \int_V G\vartheta dV' - \int_{\partial V} G\boldsymbol{n} \cdot \boldsymbol{u} dS', \tag{A.3.17a}$$

$$\boldsymbol{\psi} = -\int_V G\boldsymbol{\omega} dV' + \int_{\partial V} G\boldsymbol{n} \times \boldsymbol{u} dS', \tag{A.3.17b}$$

where the sum of two surface integrals gives a harmonic field $\boldsymbol{u} = \nabla\varphi$, say, with $\nabla^2\varphi = 0$ and prescribed boundary condition $\partial\varphi/\partial n$ at ∂V. In order to obtain the unique Helmholtz-Hodge decomposition, the simplest way is to solve the scalar boundary-value problem (A.3.7a) and (A.3.10).

When $\boldsymbol{x}$ is outside V, these integrals vanish. Therefore, we have

$$\boldsymbol{u} = \nabla\phi + \nabla\times\boldsymbol{\psi} \quad \text{for } \boldsymbol{x}\in V, \tag{A.3.18a}$$

$$\boldsymbol{0} = \nabla\phi + \nabla\times\boldsymbol{\psi} \quad \text{for } \boldsymbol{x}\notin V. \tag{A.3.18b}$$

For unbounded domain, the Helmholtz decomposition is still valid provided that the above integrals converge. This is the case if $(\boldsymbol{\omega}, \vartheta)$ vanish outside some finite region or decay sufficiently fast (Phillips 1933). Therefore, (A.3.17a) and (A.3.17b) provide a constructive proof of the global existence of the Helmholtz decomposition for any differentiable vector field. Moreover, using the same notation as above, the split vectors $\nabla\phi$ and $\nabla\times\boldsymbol{\psi}$ can be expressed in terms of dilatation and vorticity, respectively:

$$\nabla\phi = \int_V \vartheta\nabla G dV' - \int_{\partial V}(\boldsymbol{n}\cdot\boldsymbol{u})\nabla G dS', \tag{A.3.19}$$

$$\nabla\times\boldsymbol{\psi} = \int_V \boldsymbol{\omega}\times\nabla G dV' - \int_{\partial V}(\boldsymbol{n}\times\boldsymbol{u})\times\nabla G dS'. \tag{A.3.20}$$

These formulas not only show the non-local nature of the decomposition but also, via (A.3.14), tell how fast the influence of $\boldsymbol{\omega}$ and ϑ from a distant point $\boldsymbol{x}'$ on the field point $\boldsymbol{x}$ decays.

A.4 Problems

A.1. Let f be a sufficiently smooth scalar field and $\boldsymbol{F}, \boldsymbol{G}, \boldsymbol{H}$ be sufficiently smooth vector fields. Use component notation of vectors along with the properties of δ_{ij} and ϵ_{ijk}, prove the following identities. (1) Algebraic identities

$$(\boldsymbol{F}\times\boldsymbol{G})\cdot\boldsymbol{H} = (\boldsymbol{G}\times\boldsymbol{H})\cdot\boldsymbol{F}, \tag{A.4.1}$$

$$\boldsymbol{F}\times(\boldsymbol{G}\times\boldsymbol{H}) = (\boldsymbol{F}\cdot\boldsymbol{H})\boldsymbol{G} - (\boldsymbol{F}\cdot\boldsymbol{G})\boldsymbol{H}. \tag{A.4.2}$$

(2) Differential identities

$$\nabla\cdot(f\boldsymbol{F}) = (\nabla f)\cdot\boldsymbol{F} + f\nabla\cdot\boldsymbol{F}, \tag{A.4.3}$$

$$\nabla\times(f\boldsymbol{F}) = (\nabla f)\times\boldsymbol{F} + f\nabla\times\boldsymbol{F}, \tag{A.4.4}$$

$$\nabla\cdot(\boldsymbol{F}\times\boldsymbol{G}) = \boldsymbol{G}\cdot(\nabla\times\boldsymbol{F}) - \boldsymbol{F}\cdot(\nabla\times\boldsymbol{G}), \tag{A.4.5}$$

$$\nabla\times(\boldsymbol{F}\times\boldsymbol{G}) = \boldsymbol{F}(\nabla\cdot\boldsymbol{G}) - \boldsymbol{G}(\nabla\cdot\boldsymbol{F}) + (\boldsymbol{G}\cdot\nabla)\boldsymbol{F} - (\boldsymbol{F}\cdot\nabla)\boldsymbol{G}, \tag{A.4.6}$$

$$\nabla(\boldsymbol{F}\cdot\boldsymbol{G}) = \boldsymbol{F}\times(\nabla\times\boldsymbol{G}) + (\boldsymbol{F}\cdot\nabla)\boldsymbol{G} + \boldsymbol{G}\times(\nabla\times\boldsymbol{F}) + (\boldsymbol{G}\cdot\nabla)\boldsymbol{F}, \quad (A.4.7)$$

$$\nabla\times(\nabla f) = \boldsymbol{0}, \quad (A.4.8)$$

$$\nabla\cdot(\nabla\times\boldsymbol{F}) = 0, \quad (A.4.9)$$

$$\nabla\times(\nabla\times\boldsymbol{F}) = \nabla(\nabla\cdot\boldsymbol{F}) - \nabla^2\boldsymbol{F}, \quad (A.4.10)$$

$$(\boldsymbol{F}\times\nabla)\times\boldsymbol{G} = (\nabla\boldsymbol{G})\cdot\boldsymbol{F} - \boldsymbol{F}(\nabla\cdot\boldsymbol{G}) \quad (A.4.11)$$

$$\boldsymbol{F}\times(\nabla\times\boldsymbol{G}) = (\nabla\boldsymbol{G})\cdot\boldsymbol{F} - \boldsymbol{F}\cdot\nabla\boldsymbol{G} \quad (A.4.12)$$

$$(\boldsymbol{F}\times\nabla)\times\boldsymbol{G} - \boldsymbol{F}\times(\nabla\times\boldsymbol{G}) = \boldsymbol{F}\cdot\nabla\boldsymbol{G} - \boldsymbol{F}(\nabla\cdot\boldsymbol{G}). \quad (A.4.13)$$

A.2. Prove all the DMT identities given in Sect. A.2.3.

A.3. Consider a position vector $\boldsymbol{x} = (x_1, x_2, x_3)$ and a fixed position vector $\boldsymbol{x}_0 = (x_{01}, x_{02}, x_{03})$, so that their distance vector is $\boldsymbol{r} = \boldsymbol{x} - \boldsymbol{x}_0$. Let $r = |\boldsymbol{r}|$ be the length of $\boldsymbol{r}$. Prove:

$$\boldsymbol{F}\cdot\nabla\boldsymbol{r} = \boldsymbol{F}, \quad (A.4.14)$$

and the following identities for $\boldsymbol{x}' \neq \boldsymbol{0}$:

$$\nabla(1/r) = -\boldsymbol{r}/r^3, \quad (A.4.15a)$$

$$\nabla\cdot[\nabla(1/r)] = \nabla^2(1/r) = 0, \quad (A.4.15b)$$

$$\nabla\times[\nabla(1/r)] = \boldsymbol{0}. \quad (A.4.15c)$$

Hint: Start your algebra from $r^2 = \boldsymbol{r}\cdot\boldsymbol{r}$.

A.4. Consider the solution of a scalar Poisson equation

$$\nabla^2 F(\boldsymbol{x}) = f(\boldsymbol{x}), \quad (A.4.16)$$

where $f(\boldsymbol{x})$ is the source distributed in domain V enclosed by boundary S with outward unit vector $\boldsymbol{n}$.

(1) Prove

$$\int_V f(\boldsymbol{x})dV = \int_S \boldsymbol{n}\cdot\nabla F dS = \int_S \frac{\partial F}{\partial n} dS, \quad (A.4.17)$$

where the right-hand side represents the total flux of F.

(2) In spherical coordinates (r, θ, ϕ), let $F(r) = -1/(4\pi r)$ be a spherically symmetric source decaying as $1/r$ from the origin. Prove that this $F(r)$ represents a unit-strength point source, i.e., it produces a unit total flux for any S which encloses the origin, but zero flux for any S which does not.

(3) Prove that for the above case we have

$$\nabla^2\phi(r) = \delta(r), \quad (A.4.18)$$

where $\delta(r)$ is a delta function with properties

$$\delta = \begin{cases} \infty \text{ if } r = 0, \\ 0 \text{ if } r \neq 0, \end{cases} \tag{A.4.19a}$$

and

$$\int_V \delta(r) dV = 1. \tag{A.4.19b}$$

(4) Assume that the outer boundary S extends to infinity, and extend $f(r)$ to a two-point function (Green's function) $G(\boldsymbol{x}, \boldsymbol{x}') = -1/(4\pi r)$, where $\boldsymbol{x} \in V$ is the point to estimate $\phi(\boldsymbol{x})$, $\boldsymbol{x}' \in V$ the point to estimate the source, and $\boldsymbol{r} = \boldsymbol{x} - \boldsymbol{x}'$. Prove that the integral expression of the solution of (A.4.16) with a general source $f(\boldsymbol{x}')$ is

$$F(\boldsymbol{x}) = \int_V G(\boldsymbol{x}, \boldsymbol{x}') f(\boldsymbol{x}') dV(\boldsymbol{x}'), \tag{A.4.20}$$

where the integration variable is $\boldsymbol{x}'$.

(5) Check that the solution (A.4.20) indeed satisfies (A.4.16).

4. Let V be a volume enclosed by boundary S with outward unit normal $\boldsymbol{n}$, and ϕ and ψ be scalar functions. Prove the following Green's identities:

$$\int_V (\phi \nabla^2 \psi + \nabla\phi \cdot \nabla\psi) dV = \int_S \phi \frac{\partial \psi}{\partial n} dS, \tag{A.4.21a}$$

$$\int_V (\phi \nabla^2 \psi - \psi \nabla^2 \phi) dV = \int_S \left(\phi \frac{\partial \psi}{\partial n} - \psi \frac{\partial \phi}{\partial n} \right) dS. \tag{A.4.21b}$$

Moreover, if $\nabla^2 \phi = 0$, prove

$$\int_V \nabla\phi \cdot \nabla\phi dV = \int_S \phi \frac{\partial \phi}{\partial n} dS. \tag{A.4.21c}$$

References

Adrian, R. J., Meinhart, C. D., & Tomkins, C. D. (2000). Vortex organization in the outer region of the turbulent boundary layer. *Journal of Fluid Mechanics*, *422*, 1–54.

Adrian, R. J. (2007). Hairpin vortex organization in wall turbulence. *Physics of Fluids*, *19*, 041301.

Akhmetov, D. G. (2009). *Vortex rings*. New York, NY: Springer.

Alekseenko, S. V., Kuibin, P. A., & Okulov, V. L. (2007). *Theory of concentrated vortices*. New York, NY: Springer.

Archer, P. J., Thomas, T. G., & Coleman, G. N. (2008). Direct numerical simulation of vortex ring evolution from laminar to the early turbulent regime. *Journal of Fluid Mechanics*, *598*, 201–226.

Aris, R. (1962). *Vectors, tensors and the basic equations of fluid mechanics*. New Jerse: Dover.

Barton, G. (1989). *Elements of Green's function and propagation*. Oxford: Oxford University Press.

Batchelor, G. K. (1956). On steady laminar flow with closed streamlines at large Reynolds number. *Journal of Fluid Mechanics*, *1*, 177–190.

Batchelor, G. K. (1964). Axial flow in trailing line vortices. *Journal of Fluid Mechanics*, *20*, 645–658.

Batchelor, G. K. (1967). *An introduction to fluid mechanics*. Cambridge, UK: Cambridge University Press.

Benjamin, T. B. (1976). *The alliance of practical and analytical insights into the nonlinear problems of fluid mechanics*. Lecture Notes in Mathematics Number 503, Springer, pp. 8–29.

Bernal, L. P., & Roshko, A. (1986). Streamwise vortex structure in plane mixing layers. *Journal of Fluid Mechanics*, *170*, 499–525.

Betz, A. (1950). Wie entsteht ein Wirbel in einer wenig žahen Flüssigkeit. *Naturwissenschaften*, *37*, 193–196.

Bloor, D. (2011). *The Enigma of the Airfoil*. Chicago: The University of Chicago Press.

Brown, G. L., & Lopez, J. M. (1990). Axisymmetric vortex breakdown. Part II. Physical mechanism. *Journal of Fluid Mechanics*, *221*, 553–576.

Brown, G. L., & Roshko, A. (1974). On density effects and large structure in turbulent mixing layers. *Journal of Fluid Mechanics*, *64*, 775–816.

Burgers, J. M. (1921). On the resistance of fluid and vortex motion. *Proceedings of the Koninklijke Nederlandse Akademie van Wetenschappen*, *23*, 774–782.

Butler, K. M., & Farrell, B. F. (1992). Three-dimensional optimal perturbations in viscous shear flow. *Physics of Fluids A*, *4*, 1637–1650.

Caswell, B. (1967). Kinematics and stress on a surface of rest. *Archives of Review of Fluid Mechanics*, *26*, 385–399.

Chen, M. Z. (1986). *Turbulence and the related enginnering calculations*. Beijing: Beijing Aeronautical Institute Press (in Chinese).

Chen, Q. N., Chen, S. Y., & Eyink, G. L. (2003). The joint cascade of energy and helicity in three-dimensional turbulence. *Physics of Fluids*, *15*, 361–374.

J.-Z. Wu et al., *Vortical Flows*, DOI 10.1007/978-3-662-47061-9

Chernyshenko, S. I. (1998). Asymptotic theory of global separation. *ASME Applied Mechanics Reviews, 51*, 523–536.

Chorin, A. J., & Marsdon, J. E. (1992). *A mathematical introduction to fluid mechanics*. New York, NY: Springer.

Craik, A. D. D. (1985). *Wave interactions and fluid flows*. Cambridge: Cambridge University Press.

Critzos, C. C., Heyson, H. H., & Boswinkle, R. W. (1955). Aerodynamic characteristics of NACA0012 airfoil section at angles of attack from 0° to 180°. NACA TN-3361.

Crow, S. C. (1970). Stability theory for a pair of trailing vortices. *AIAA Journal, 8*, 2172–2179.

Dabiri, J. O. (2009). Optimal vortex formation as a unifying principle in biological propulsion. *Annual Review of Fluid Mechanics, 41*, 17–33.

Darrigol, O. (2005). *Worlds of flow*. Oxford: Oxford University Press.

Dazin, A., Dupont, P., & Stanislas, M. (2006a). Experimental characterization of the instability of the vortex ring. *Part I: Linear Phase Experiments in Fluids, 40*, 383–399.

Dazin, A., Dupont, P., & Stanislas, M. (2006b). Experimental characterization of the instability of the vortex ring. *Part II: Nonlinear Phase Experiments in Fluids, 41*, 401–413.

Delbende, I., Chomaz, J. M., & Huerre, P. (1998). Absolute/convective instabilities in the Batchlor vortex: A numerical study of the linear impulse response. *Journal of Fluid Mechanics, 355*, 229–254.

Délery, J. (2001). Robert Legendre and Henri Werlé: Toward the elucidation of three-dimensional separation. *Annual Review of Fluid Mechanics, 33*, 129–154.

Didden, N. (1979). On the formation of vortex rings: Rolling-up and production of circulation. *Journal of Applied Mathematics and Physics (ZAMP), 30*, 101–116.

Drazin, P. G. (2002). *Introduction to Hydrodynamic stability*. Cambridge: Cambridge University Press.

Du, Y., & Karniadakis, G. E. (2000). Suppressing wall-turbulence by means of a transverse traveling wave. *Science, 288*, 1230–1234.

Earnshaw, P. B. (1961). An experimental investigation of the struncture of a leading edge vortex. *Aeronautical Research Council, 22876*, R& M. 3281.

Emanuel, K. A. (1984). A note on the instability of columnar vortices. *Journal of Fluid Mechanics, 145*, 235–238.

Ertel, H. (1942). Ein neuer hydrodynamischer Erhaltungssatz. *Naturwissenschaften, 30*(36), 543–544.

Fabre, D., Cossu, C., & Jacqiun, L. (2000a). Spatio-temporal development of the long and short-wave vortex-pair instabilities. *Physics of Fluids, 12*(5), 1247–1250.

Fabre, D., Cossu, C., & Jacqiun, L. (2000b). Absolute/convective instabilities and spatial growth in a vortex pair. In A. Maurel & P. Petitjeans (Eds.), *Dynamics and structures of vortices* (pp. 162–172). New York, NY: Springer.

Faddy, J. M., & Pullin, D. I. (2005). Flow structure in a model of aircraft trailing vortices. *Physics of Fluids, 17*, 085106.

Fedorov, A. (2011). Transition and stability of high-speed boundary layers. *Annual Review of Fluid Mechanics, 43*, 79–95.

Feys, J., & Maslowe, A. (2014). Linear stability of the Moore-Saffman model for a trailing wingtip vortex. *Physics of Fluids, 26*, 024108.

Fiabane, L., Gohlke, M., & Cadot, O. (2011). Characterization of flow contributions to drag and lift of a circular cylinder using a volume expression of the fluid force. *European Journal of Mechanics–B/Fluids, 30*, 311–315.

Flandro, G. A., Perry, E. H., & French, J. C. (2006, January). *Nonlinear rocket motor stability computation: Understanding the Brownlee-Marble observation*. AIAA paper 2006–539. AIAA Aerospace Sciences Meeting, Reno, NV.

Fornberg, B. (1985). Steady viscous flow past a circular cylinder up to Reynolds number 600. *Journal of Computational Physics, 61*, 297–320.

Fraenkel, L. E. (1970). On steady vortex rings with small cross-section in an ideal fluid. *Proceedings of the Royal Society of London A, 316*, 29–62.

Fraenkel, L. E., & Burgers, M. S. (1974). A global theory of steady vortex rings in an ideal fluid. *Acta Mathematica, 132*, 13–51.
Fukumoto, Y. (2010). Global time evolution of viscous vortex rings. *Theoretical and Computational Fluid Dynamics, 24*, 335–347.
Fukumoto, Y., & Kaplanski, F. (2008). Global time evolution of an axisymmetric vortex ring at low Reynolds numbers. *Physics of Fluids*, 053103-1–053103-13.
Fukumoto, Y., & Moffatt, H. K. (2008). Kinematic variational principle for motion of vortex rings. *Physica D: Nonlinear Phenomena, 237*, 2210–2237.
Gad-el-Hak, M. (2000). *Flow control*. Cambridge: Cambridge University Press.
Galdi, G. P. (2011). *An introduction to the mathematical theory of the Navier-Stokes equations* (2nd ed.). New York, NY: Springer.
Gan, L., Nickels, T. B., & Dawson, J. R. (2011). An experimental study of a turbulent vortex ring: A three-dimensional representation. *Experiments in Fluids, 51*, 1493–1507.
Gerrard, J. H. (1966). The mechanics of the formation region of vortices behind bluff bodies. *Journal of Fluid Mechanics, 25*, 401–413.
Gharib, M., Rambod, E., & Shariff, K. (1998). A universal time scale for vortex ring formation. *Journal of Fluid Mechanics, 360*, 121–140.
Giacomelli, R., & Pistolesi, E. (1934). Historical Sketch. In W. F. Durand (Ed.), *Aerodynamic theory* (Vol. I, pp. 305–394). New York, NY: Dover.
Glauert, H. (1926). *The elements of Airfoil and Airscrew theory*. Cambridge: Cambridge University Press.
Goldstein, S. (1948). On laminar boundary-layer flow near a position of separation. *Quarterly Journal of Mechanics and Applied Mathematics, 1*, 43–69.
Greenspan, H. P. (1968). *The theory of rotating fluids*. Cambridge: Cambridge University Press.
Hall, M. G. (1966). The structure of concentrated vortex cores. *Progress in Aeronautical Science, I*, 53–110.
Haller, G. (2004). Exact theory of unsteady separation for two-dimensional flows. *Journal of Fluid Mechanics, 512*, 257–311.
Hartman, P. (1964). *Ordinary differential equations, Chapter VII*. New York, NY: Wiley.
Head, M. R., & Bandyopadhyay, P. (1981). New aspects of turbulent boundary layer structure. *Journal of Fluid Mechanics, 107*, 297.
Helmholtz H. (1858). Über Integrale der hydrodynamischen Gleichungen, welche den Wirbelbewegungen ensprechen. *J. Reine Angew. Math., 55*, 25–55. English translation: On intergal of the hydrodynamical equations which express vortex motion (1867). *Phil. Mag., 33*, 485–512.
Ho, C. M., & Huerre, P. (1984). Perturbed free shear layers. *Annual Review of Fluid Mechanics, 16*, 365–424.
Hou, T. Y., Stredie, V. G., & Wu, T. Y. (2007). Mathematical modeling and simulation of aquatic and aerial animal locomotion. *Journal of Computional Physics, 225*, 1603–1631.
Howard, L. N., & Gupta, A. S. (1962). On the hydrodynamic and hydromagnetic stability of swirling flows. *Journal of Fluid Mechanics, 14*, 463–476.
Huang, L. S., & Ho, C. M. (1990). Small-scale transition in a plane mixing layer. *Journal of Fluid Mechanics, 210*, 475–500.
Huerre, P., & Monkewitz, P. A. (1990). Local and global instabilities in spatially developing flows. *Annual Review of Fluid Mechanics, 22*, 473–537.
Hunt, J. C. R., & Vassilicos, J. C. (Eds.). (2000). *Turbulence structure and vortex dynamics*. Cambridge: Cambridge University Press.
Hussain, A. K. M. F. (1983). Coherent structures–reality and myth. *Physics of Fluids, 26*, 2816–2850.
Hussain, M. F., & Melander, M. V. (1992). Understanding turbulence via vortex dynamics. In T. B. Gatski, S. Sarkar, & C. G. Speziale (Eds.), *Studies in turbulence* (pp. 157–178). New York, NY: Springer.
Izydorek, M., Rybicki, S., & Szafaniec, Z. (1996). A note on the Poincare-Bendixson index theorem. *Kodai Mathematical Journal, 19*, 145–156.

Jeong, J., & Hussain, F. (1995). On the identification of a vortex. *Journal of Fluid Mechanics, 285*, 69–94.
Jiménez, J. (2004). *Turbulence and vortex dynamics*. Paris: Ecolepolytechnique.
Jiménez, J., & Moin, P. (1991). The minimal flow unit in near-wall turbulence. *Journal of Fluid Mechanics, 225*, 213–240.
Jiménez, J., & Pinelli, A. (1999). The autonomous cycle of near-wall turbulence. *Journal of Fluid Mechanics, 389*, 335–359.
Kachanov, Y. S. (1994). Physical mechanisms of laminar-boundary-layer transition. *Annual Review of Fluid Mechanics, 26*, 411–482.
Kaden, H. (1931). Aufwicklung einer unstabilen unstetigkeitsfläche. *Ingenieur Archive, 2*, 140–168.
von Kármán T. (1954). On the foundation of high speed aerodynamics. In: W.R. Sears (Ed.), *General theory of high speed aerodynamics*. Princeton, NJ: Princeton University Press.
von Kármán, T., & Burgers, J. M. (1935). General aerodynamic theory–perfect fluids. In W. F. Durand (Ed.), *Aerodynamic theory* (Vol. II). New York, NY: Dover.
von Kármán, T., & Sears, W. R. (1938). Airfoil theory for non-uniform motion. *Journal of Aerosol Science, 5*(10), 379–390.
Katz, J., & Plotkin, A. (2001). *Low-speed aerodynamics* (2nd ed.). Cambridge: Cambridge University Press.
Kelvin L. (1869). On vortex motion. *Trans. R. Soc. Edinb., 25*, 217–260.
Kerswell, R. R. (2002). Elliptical instability. *Annual Review of Fluid Mechanics, 34*, 83–113.
Khorrami, M. R. (1991). On the viscous modes of instability of a trailing line vortex. *Journal of Fluid Mechanics, 225*, 197–212.
Kim, J., & Lim, J. (2000). A linear process in wall-bounded turbulent shear flows. *Physics of Fluids, 12*(8), 1885–1888.
Kleckner, D., & Irvine, T. M. (2013). Creation and dynamics of knotted vortices. *Nature Physics*, doi:10.1038/nphys2560.
Kline, S. J., Reynolds, W. C., Schraub, F. A., & Runstadler, P. W. (1967). The structure of turbulent boundary layers. *Journal of Fluid Mechanics, 30*, 741–773.
Krasny, R. (1987). Computation of vortex sheet roll-up in the Trefftz plane. *Journal of Fluid Mechanics, 184*, 123–155.
Krasny, R. (1988). Numerical simulation of vortex sheet evolution. *Fluid Dynamics Research, 3*, 93–97.
Küchemann, D. (1978). *The aerodynamic design of aircraft*. New York, NY: Pergamon.
Küchemann, D., & Weber, J. (1965). Vortex motions. *ZAMM, 45*, 457–474.
Kurosaka, M., & Sundaram, P. (1986). Illustrative examples of streaklines in unsteady vortices: Interpretational difficulties revisited. *Physics of Fluids, 29*, 3474–3477.
Lagerstrom, P. A. (1964). *Laminar flow theory*. Princeton: Princeton University Press.
Lagerstrom, P. M. (1975). Solutions of the Navier-Stokes equation at large Reynolds number. *SIAM Journal on Applied Mathematics, 28*, 202–214.
Lamb, H. (1932). *Hydrodynamics*. Cambridge: Cambridge University Press.
Lanchester, F. W. (1907). *Aerodynamics*. London: Constable.
Landau, L. D., & Lifshitz, E. M. (1959). *Fluid mechanics*. Oxford: Pergamon.
Lautenschlager, M., Eppel, D. P., & Thacker, W. C. (1988). Subgrid-parameterization in helical flows. *Beitraege zur Physik der Atmosphaer, 61*, 87.
Lee, C. B., & Wu, J. Z. (2008). Transition in wall-bounded flows. *Applied Mechanics Reviews, 61*, 030802.
Lee, C. B., Su, Z., Zhong, H. J., Chen, S. Y., Zhou, M. D., & Wu, J. Z. (2013). Experimental investigation of freely falling thin disks. Part 2. Transition of three-dimensional motion from zigzag to spiral. *Journal of Fluid Mechanics, 732*, 77–104.
Lee, T., & Pereira, J. (2010). Nature of wakelike and jetlike axial tip vortex flows. *Journal of Aircraft, 47*, 1946–1954.
Leibovich, S. (1984). Vortex stability and breakdown: Survey and extension. *AIAA Journal, 22*(1), 1192–1206.

Leibovich, S., & Stewartson, K. (1983). A sufficient condition for the instability of columnar vortices. *Journal of Fluid Mechanics*, *126*, 335–356.

Lesieur, M. (1990). *Turbulence in fluids*. Dordrecht Netherlands: Kluwer.

Lesieur, M., Comte, P., & Metais, O. (2000). LES and vortex topology in shear and rotating flows. In J. C. R. Hunt & J. C. Vassilicos (Eds.), *Turbulence structure and vortex dynamics* (pp. 269–288). Cambridge: Cambridge University Press.

Lessen, M., Singh, P. J., & Paillet, F. (1974). The stability of a trailing line vortex. Part I. Inviscid theory. *Journal of Fluid Mechanics*, *63*, 753–763.

Leweke, T., & Williamson, C. H. K. (1998). Coopr erative elliptic instability of a vortex pair. *Journal of Fluid Mechanics*, *360*, 85–119.

Li, G. J., & Lu, X. Y. (2012). Force and power of flapping plates in a fluid. *Journal of Fluid Mechanics*, *712*, 598–613.

Liepmann, H. W., & Roshko, A. (1957). *Elements of Gas dynamics*. New York, NY: Wiley.

Lighthill, M. J. (1956). Viscosity effects in sound waves of finite amplitude. In G. K. Batchelor & R. M. Davies (Eds.), *Surveys in mechanics* (pp. 250–351). Cambridge: Cambridge University Press.

Lighthill, M. J. (1963). Introduction of boundary layer theory. In L. Rosenhead (Ed.), *Laminar Boundary Layers* (pp. 46–113). Oxford: Oxford University Press.

Lighthill, M. J. (1978). Acoustic streaming. *Journal of Sound and Vibration*, *61*, 391–418.

Lighthill, M. J. (1986). *An informal introduction to theoretical fluid mechanics*. Oxford: Clarendon Press.

Lighthill, M. J. (1995). Fluid mechanics. In L. M. Brown, A. Pais, & B. Pippard (Eds.), *Twentieth centry physics, Chapter 10*. New York, NY: Institute of Physics Publishing and American Institute of Physics Press.

Lim, T. T., & Nickels, T. B. (1995). Vortex rings. In S. I. Green (Ed.), *Fluid vortices*. Dordrecht Netherlands: Kluwer.

Lin, C. C. (1957). Motion in the boundary layer with a rapidly oscillating external flow. *Proceedings of the 9th International Congress of Applied Mechanics, Brussels*, *4*, 155–167.

Liu, L. Q., Zhu, J. Y., & Wu, J. Z. (2015). *Lift and drag in two-dimensional steady viscous and compressible flow* (in revision).

Liu, L. Q., Wu, J. Z., Shi, Y. P., & Zhu, J. Y. (2014). A dynamic counterpart of the Lamb vector in viscous compressible aerodynamics. *Fluid Dynamics Research*, *46*(6), 061417.

Liu, T. (2013). Extraction of skin-friction fields from surface flow visualizations as an inverse problem. *Measurement Science and Technology*, *24*, 124004.

Liu, T., Woodiga, S., & Ma, T. (2011). Skin friction topology in a region enclosed by penetrable boundary. *Experiments in Fluids*, *51*, 1549–1562.

Liu, T., Wang, B., & Choi, D. (2012). Flow structures of Jupiter's Great Red Spot extracted by using optical flow method. *Physics of Fluids*, *24*, 096601–096613.

Liu, T., Wu, J. Z., Zhu, J. Y., & Liu, L. Q. (2015). The origin of lift revisited: I. A complete physical theory. In 45th AIAA Fluid Dynamics Conference, Dallas, Texas, 22–26 June 2015.

Loiseleux, T., Chomaz, J. M., & Huerre, P. (1998). The effects of swirl on jets and wakes: Linear instability of Rankine vortex with axial flow. *Physics of Fluids*, *10*(5), 1120–1134.

Mao, F., Shi, Y. P., & Wu, J. Z. (2010). On a general theory for compressing process and aeroacoustics: Linear analysis. *Acta Mechanica*, *26*, 355–364.

Mao, F., Shi, Y. P., Xuan, L. J., Su, W. D., & Wu, J. Z. (2011). On the governing equations for the compressing process and its coupling with other processes. *Science in China*, *54*, 1154–1167.

Marongiu, C., & Tognaccini, R. (2010). Far-field analysis of the aerodynamic force by Lamb vector integrals. *AIAA Journal*, *48*, 2543–2555.

Marongiu, C., Tognaccini, R., & Ueno, M. (2013). Lift and lift-induced drag computation by Lamb vector integration. *AIAA Journal*, *51*, 1420–1430.

McCune, J. E., & Tavares, T. S. (1993). Perspective: unsteady wing theory–the Kármán/Sears legacy. *Journal of Fluids Engineering*, *115*, 548–560.

McLean, D. (2012). *Understanding aerodynamics*. West Sussex: Wiley.

Melander, M. V., & Hussain, F. (1993). Coupling between a coherent structure and fine-scale turbulence. *Physical Review E, 48*(4), 2669–2689.

Meleshko, V. V., Gourjii, A. A., & Krasnopolskaya, T. S. (2012). Vortex rings: History and state of the art. *Journal of Mathematical Sciences, 187*, 772–808.

Milne-Thomson, L. M. (1968). *Theoretical hydrodynamics*. New York, NY: Dover.

Moffatt, H. K. (1969). Degree of knottedness of tangled vortex lines. *Journal of Fluids Mechanics, 35*, 117–129.

Moore, D. W., & Saffman, P. G. (1973). Axial flow in laminar trailing vortices. *Proceedings of the Royal Society of London A, 333*, 491–508.

Morse, P. M., & Feshbach, H. (1953). *Methods of theoretical physics* (Vol. I). New York, NY: McGrall-Hill.

Morkovin, M. V., Reshotko, E., & Herbert, T. (1994). Transition in open flow systems: A reassessment. *Bulletin of the American Physical Society, 39*(9), 1–31.

Nitsche, M., & Krasny, R. (1994). A numerical study of vortex ring formation at the edge of a circular tube. *Journal of Fluid Mechanics, 276*, 139–161.

Noack, B. R. (1999). On the flow around a circular cylinder. Part I: laminar and transitional regime, Part II: turbulent regime. *Journal of Applied Mathematics and Mechanics, 79*, S223–S226, S227–S230.

Norbury, J. (1973). A family of steady vortex rings. *Journal of Mechanics, 57*, 417–431.

Oertel, H. (2004). *Prandtl's essentials of fluid mechanics*. New York, NY: Springer.

Olendraru, C., & Sellier, A. (2002). Viscous effects in the absolute-convective instability of the Batchelor vortex. *Journal of Fluid Mechanics, 459*, 371–396.

Olendraru, C., Sellier, A., Rossi, M., & Huerre, P. (1999). Inviscid instability of the Batchelor vortex: Absolute-convective transition and spatial branches. *Physics of Fluids, 11*(7), 1805–1820.

Panton, R. L. (1997). *Advances in fluid mechanics: Self-sustaining mechanisms of wall turbulence*. Boston: Computational Mechanics Publications.

Panton, R. L. (2001). Overview of the self-sustaining mechanisms of wall turbulence. *Prog. Aerosp. Sci., 37*, 341–83.

Panton, R. L. (2013). *Incompressible flow*. Hoboken: Wiley.

Perry, A. E., & Chong, M. S. (1982). On the mechanism of wall turbulence. *Journal of Fluid Mechanics, 119*, 173–217.

Perry, A. E., Chong, M. S., & Lim, T. T. (1982). The vortex-shedding process behind two-dimensional bluff bodies. *Journal of Fluid Mechanics, 116*, 77–90.

Phillips, H. B. (1933). *Vector analysis*. New York, NY: Wiley.

del Pino, C., Parras, L., Felli, M., & Fernandez-Feria, R. (2011). Structure of trailing vortices: Comparison between particle image velocimetry measurements and theoretical models. *Physics of Fluids, 23*, 013602.

Pope, S. B. (2000). *Turbulence flows*. Cambridge: Cambridge University Press.

Pradeep, D., & Hussain, F. (2000). Core dynamics of a coherent structure: A prototypical physical-space cascade mechanism. In J. C. R. Hunt & J. C. Vassilicos (Eds.), *Turbulence structure and vortex dynamics* (pp. 51–82). Cambridge: Cambridge University Press.

Prandtl, L. (1904). *Über Flüssigkeitsbewegung bei sehr kleiner Reibung*. Heidelberg: Proceedings of the III International Mathematics Congress.

Prandtl, L. (1918). Tragflügeltheorie. I. Mitteilung, Nachrichten der Gesellschaft der Wissenschaften zu Göttingen. *Mathematisch-Physischen Klasse der*, 151–177.

Prandtl, L., & Tietjens, O. G. (1934). *Applied Hydro- and aeromechanics*. New York, NY: Springer.

Pritchard, P. (2011). *Fox and McDonalds introduction to fluid mechanics* (8th ed.). New York, NY: Wiley.

Ren, H., & Lu, X. Y. (2013). Large eddy simulation of a vortex ring impinging on a three-dimensional circular cylinder. *Theoretical and Applied Mechanics Letters, 3*, 032007.

Reshotko, E. (1984). *Environment and reseptivity*. AGARD R-709, Paper 4.

Reshotko, E. (2001). Transient growth: A factor in bypass transition. *Physics of Fluids, 13*, 1067–1075.

Richardson, L. F. (1922). *Weather prediction by numerical process*. Cambridge: Cambridge University Press.
Rist, U., & Fasel, H. (1995). Direct numerical simulation of controlled transition in a flat-plate boundary layer. *Journal of Fluid Mechanics, 298*, 211–248.
Rosenhead, L. (1963). *Laminar Boundary Layer*. New York, NY: Dover.
Roshko, A. (1993). Perspectives on bluff body aerodynamics. *Journal of Wind Engineering and Industrial Aerodynamics, 49*, 79–100.
Saffman, P. G. (1970). The velocity of viscous vortex rings. *Studies in Applied Mathematics, 49*, 371–380.
Saffman, P. G. (1978). The number of waves on unstable vortex rings. *Journal of Fluid Mechanics, 84*, 625–639.
Saffman, P. G. (1992). *Vortex dynamics*. Cambridge: Cambridge University Press.
Salmon, R. (1988). Hamiltonian fluid mechanics. *Annual Review of Fluid Mechanics, 20*, 225–256.
Salmon, R. (1998). *Lecture on Geophysical fluid dynamics*. Oxford: Oxford University Press.
Saric, W. S., & Nayfeh, A. H. (1975). Nonparallel stability of boundary layer flows. *Physics of Fluids, 18*, 945–950.
Schlichting, H., & Gersten, K. (2000). *Boundary layer theory* (8th Revised and enlarged Edition). London: Springer.
Schmid, P. J., & Henningson, D. S. (2001). *Stability and transition in shear flows*. New York, NY: Springer.
Schoppa, W., & Hussain, F. (2002). Coherent structure generation in near-wall turbulence. *Journal of Fluid Mechanics, 453*, 57–108.
Sears, W. R. (1941). Some aspects of non-stationary airfoil theory and its practical application. *Journal of the Aeronautical Sciences, 8*, 104–108.
Serrin, J. (1959). Mathematical principles of classic fluid dynamics. In S. Flügge (Ed.), *Handbuch der Physik VIII/1* (pp. 125–263). New York, NY: Springer.
Shariff, K., & Leonard, A. (1992). Vortex rings. *Annual Review of Fluid Mechanics, 24*, 235–279.
She, Z. S., & Leveque, E. (1994). Universal scaling laws in fully developed turbulence. *Physical Review Letters, 72*, 336–339.
Sheldahl, R. E., & Klimas, P. C. (1981). *Aerodynamic characteristics of seven airfoil sections through 180 degrees angle of attack for use in aerodynamic analysis of vertical axis wind turbines. SAND80-2114*. Albuquerque, NM: Sandia National Laboratories.
Siggia, E. D. (1985). Collapse and amplification of a vortex filament. *Physics of Fluids, 28*, 794–805.
Smith, F. T. (1977). Laminar separation of an incompressible fluid streaming past a smooth surface. *Proceedings of the Royal Society of London A, 356*, 443–463.
Smith, C. R., Walker, J. D. A., Haidari, A. H., & Sobrun, U. (1991). On the dynamics of near-wall turbulence. *Philosophical Transactions of the Royal Society London A, 336*, 131–175.
Smits, A. J., McKeon, B. J., & Marusic, I. (2011). High-Reynolds number wall turbulence. *Annual Review of Fluid Mechanics, 43*, 353–375.
Sun, M., & Wu, J. H. (2004). Large aerodynamic forces on a sweeping wing at low Reynolds numbers. *Acta Mechanica Sinica, 20*, 24–31.
Surana, A., Grunberg, O., & Haller, G. (2006). Exact theory of three-dimensional flow separation. Part I: Steady separation. *Journal of Fluid Mechanics, 564*, 57–103.
Swearingen, J. D., & Blackwelder, R. F. (1987). The growth and breakdown of streamwise vortices in the presence of a wall. *Journal of Fluid Mechanics, 182*, 255–290.
Sychev, V. V., Rubin, A. I., Sychev, V. V., & Korolev, G. L. (1998). *Asymptotic theory of separated flows*. Cambridge: Cambridge University Press.
Synge, J. L. (1933). The stability of heterogeneous liquids. *Transactions of the Royal Society of Canada, 3*, 1–18.
Taneda, S. (1985). Flow field visualization. In F. I. Niordson & N. Olhoff. (Eds.), *Theoretical and applied mechanics* (pp. 399–410). Amsterdam: Elsevier Science.
Tennekes, H., & Lumley, J. L. (1972). *A first course in turbulence*. Cambridge: MIT Press.

Theodorson, T. (1935). *General theory of aerodynamic instability and the mechanism of flutter* (p. 496). Rept: NACA Tech.

Theodorson, T. (1952). Mechanism of turbulence. *Proceedings of the 2nd Midwestern Conference in Fluid Mechanics*.

Ting, L. (1959). On the mixing of two parallel streams. *Journal of Mathematical Physics, 38*, 153–165.

Tobak, M., & Peake, D. J. (1982). Topology of 3-dimensional separated flows. *Annual Review of Fluid Mechanics, 14*, 61–85.

Tong, B. G., Yin, X. Y., & Zhu, K. Q. (2009) *Theory of vortex motion* (2nd ed.). China: University of Science and Technology of China Press (in Chinese).

Truesdell, C. (1954). *The kinematics of vorticity*. Bloomington: Indiana University Press.

Tung, C., & Ting, L. (1967). Motion and decay of a vortex ring. *Physics of Fluids, 10*, 901–910.

Van Dyke, M. (1975). *Perturbation methods in fluid mechanics*. New York, NY: Academic Press.

Van Dyke, M. (1982). *An album of fluid motion*. Stanford: The Parabolic.

Van Dommenlen, L. L., & Shen, S. F. (1982). The genesis of separation. In T. Cebici (Ed.), *Numerical and physical aspects of aerodynamic flows* (pp. 293–311). New York, NY: Springer.

Wagner, H. (1925). Über die Entstehung des dynamischen Auftriebes von Tragflügeln. *ZAMM Journal of Applied Mathematics and Mechanics, 5*, 17–35.

Waleffe, F. (1997). On a self-sustaining process in shear flows. *Physics of Fluids, 9*(4), 883–900.

Walker, J. D. A., Smith, C. R., Cerra, A. W., & Doligalski, T. L. (1987). The impact of a vortex ring on a wall. *Journal of Fluid Mechanics, 181*, 99–140.

Wallace, J. M. (2013). Highlights from 50 years of turbulent boundary layer research. *Journal of Turbulence, 13*(53), 1–70.

Wang, J. C., Shi, Y. P., Wang, L. P., Xiao, Z. L., He, X. T., & Chen, S. Y. (2012). Scaling and statistics in three-dimensional compressible turbulence. *Physical Review Letters, 108*, 214505.

Wang, J. C., Yang, Y., Shi, Y. P., Xiao, Z. L., He, X. T., & Chen, S. Y. (2013). Cascade of kinetic energy in three-dimensional compressible turbulence. *Physical Review Letters, 110*, 214505.

Wang, K. C. (1982). New development about open separation. In H. H. Fernholz & E. Krause (Eds.), Three-dimensional Turbulent Boundary Layers, *Proceedings IUTAM Symposium* (pp. 94–105). New York, NY: Springer.

Wang, L., & Lu, X. Y. (2012). Flow topology in compressible turbulent boundary layer. *Journal of Fluid Mechanics, 703*, 255–278.

Wang, K. C., Zhou, H. C., Hu, C. H., & Harrington, S. (1990). Three-dimensional separated flow structure over prolate spheroids. *Proceedings of Royal Society of London A, 421*, 73–90.

Widnall, S. E., Bliss, D. B., & Tsai, C. Y. (1974). Instability of short waves on a vortex ring. *Journal of Fluid Mechanics, 66*, 35–47.

Widnall, S. E., & Tsai, C. Y. (1977). Instability of thin vortex ring of constant vorticity. *Philosophical Transactions of the Royal Society London A, 287*, 273–305.

Woodiga, S. A., & Liu, T. S. (2009). Skin frictions fields on delta wings. *Experiments in Fluids, 47*, 897–911.

Wu, C. J., Wang, L., & Wu, J. Z. (2007). Suppression of the von Karman vortex street behind a circular cylinder by a traveling wave generated by a flexible surface. *Journal of Fluid Mechanics, 574*, 365–391.

Wu, J. C. (1981). Theory for aerodynamic force and moment in viscous flows. *AIAA Journal, 19*, 432–441.

Wu, J. C. (2005). *Elements of vorticity aerodynamics*. Beijing: Tsinghua University Press.

Wu, J. Z. (1987). Incompressible theory of the interaction between moving bodies and vorticity field. *Acta Aerodynamica Sinica, 5*, 22–30 (in Chinese).

Wu, J. Z. (1995). A theory of three-dimensional interfacial vorticity dynamics. *Physics of Fluids, 7*, 2375–2395.

Wu, J. Z., & Wu, J. M. (1993). Interactions between a solid surface and a viscous compressible flow field. *Journal of Fluid Mechanics, 254*, 183–211.

Wu, J. Z., & Wu, J. M. (1996). Vorticity dynamics on boundaries. *Advances in Applied Mechanics*, *32*, 119–275.

Wu, J. Z., Vakili, A. D., & Wu, J. M. (1991). Review of the physics of enhancing vortex lift by unsteady excitation. *Progress in Aerospace Sciences*, *28*, 73–131.

Wu, J. Z., Ma, H. Y., & Zhou, M. D. (1993). *Introduction to vorticity and vortex dynamics*. China: High Education Publications (in Chinese).

Wu, J. Z., Lu, X. Y., Denny, A. G., Fan, M., & Wu, J. M. (1998). Post-stall flow control on an airfoil by local unsteady forcing. *Journal of Fluid Mechanics*, *371*, 21–58.

Wu, J. Z., Zhou, Y., & Fan, M. (1999). A note on kinetic energy, dissipation, and enstrophy. *Physics of Fluids*, *11*, 503–505.

Wu, J. Z., Tramel, R. W., Zhu, F. L., & Yin, X. Y. (2000). A vorticity dynamic theory of three-dimensional flow separation. *Physics of Fluids*, *12*, 1932–1954.

Wu, J. Z., Ma, H. Y., & Zhou, M. D. (2006). *Vorticity and vortex dynamics*. New York, NY: Springer.

Wu, J. Z., Wu, H., & Li, Q. S. (2009). Boundary vorticity flux and enginnering flow management. *Advances in Applied Mathematics and Mechanics*, *1*(3), 353–366.

Wu, J. Z., Lu, X. Y., Yang, Y. T., & Zhang, R. K. (2010, December). Vorticity dynamics in complex flow diagnosis and management. *Proceedings of the 13th Asia Congress Fluid Mechanics* (pp. 17–21). Dhaka, Bangladesh.

Wu, T. Y. (1956). Small perturbations in the unsteady flow of a compressible, viscous and heat-conducting fluid. *Journal of Mathematical Physics*, *35*, 13–27.

Wu, T. Y. (2007). A nonlinear theory for a flexible unsteady wing. *Journal of Engineering Mathematics*, *58*, 279–287.

Wu, X. H., & Moin, P. (2009). Direct numerical simulation of turbulence in a nominally zero-pressure-gradient flat-plate boundary layer. *Journal of Fluids Mechanics*, *630*, 5–41.

Wu, X. S. (2004). Non-equilibrium, nonlinear critical layers in laminar-turbulent transition. *Acta Mechanica Sinica*, *20*, 327–339.

Xia, X. J., & Deng, X. Y. (1991). *Engineering separated flow dynamics*. Beijing: Beijing University of Aeronautical and Astronautical Press (in Chinese).

Xiao, Z., Wan, M., Chen, S., & Eyink, G. L. (2009). Physical mechanism of the inverse energy cascade of two-dimensional turbulence: a numerical investigation. *Journal of Fluids Mechanics*, *619*, 1–44.

Xuan, L. J., Mao, F., & Wu, J. Z. (2012). Water hammer prediction and control: The Green's function method. *Acta Mechanica*, *28*, 266–273.

Yaglom, A. M. (2012). *Hydrodynamic instability and transition to turbulence* (pp. 552–553). New York, NY: Springer.

Yang, Y. T. (2004). *The effect of wave pattern of flexible wall on the drag reduction in turbulent channel flow*. BS Thesis, Peking University, Beijing (in Chinese).

Yang, Y. T., Su, W. D., & Wu, J. Z. (2010). Helical-wave decomposition and applications to channel turbulence with streamwise rotation. *Journal of Fluids Mechanics*, *662*, 91–122.

Yang, Y. T., Zhang, R. K., An, Y. R., & Wu, J. Z. (2007). Steady vortex force theory and slender-wing flow diagnosis. *Acta Mechanics*, *23*, 609–619.

Yin, X. Y., & Sun, D. J. (2003). *Vortex stability*. Beijing: National Defence Industry Press (in Chinese).

Yin, X. Y., Sun, D. J., Wei, M. J., & Wu, J. Z. (2000). Absolute and convective instability character of slender viscous vortices. *Physics of Fluids*, *12*(5), 1062–1072.

Zhang, H. X. (1985). The separation criteria and flow behavior for three-dimensional steady separated flow. *Acta Aerodynamica Sinica*, *1*(1), 1 (in Chinese).

Zhang, R. K., Mao, F., Wu, J. Z., Chen, S. Y., Wu, Y. L., & Liu, S. H. (2009). Characteristics and control of the draft-tube flow in part-load Francis turbine. *Journal of Fluids Engineering*, *131*, 021101-1–021101-13.

Zhao, H., Wu, J. Z., & Luo, J. S. (2004). Turbulent drag reduction by traveling wave of flexible wall. *Fluid Dynamics Research*, *34*, 175–198.

Zhong, H. J., Chen, S. Y., & Lee, C. B. (2011). Experimental investigation of freely falling disks: Transition from zigzag to spiral. *Physics of Fluids*, *23*, 9–12.

Zhong, H. J., Lee, C. B., Su, Z., Chen, S. Y., Zhou, M. D., & Wu, J. Z. (2013). Experimental investigation of freely falling thin disks. Part 1. The flow structures and Reynolds number effects on the zigzag motion. *Journal of Fluid Mechanics*, *716*, 228–250.

Zhong, X. L., & Wang, X. W. (2012). Direct numerical simulation on the receptivity, instability, and transition of hypersonic boundary layers. *Annual Review of Fluid Mechanics*, *44*, 527–561.

Zhou, M. D., Fernholz, H. H., Ma, H. Y., Wu, J. Z., & Wu, J. M. (1993). *Vortex capture by a two dimensional airfoil with a small oscillating leading edge flap*. AIAA 93–3266.

Zhou, M. D., & Wygnanski, I. (2001). The response of a mixing layer formed between parallel streams to a concomitant excitation at two frequencies. *Journal of Fluids Mechanics*, *441*, 139–168.

Zhu, F. L. (2000). *Applications of boundary vorticity dynamics to flow simulation, airfoil design, and flow control*. Ph.D. Dissertation, University of Tennessee, Knoxville.

Zhu, J., Adrian, R. J., Balachandar, S., & Kendall, T. M. (1999). Mechanisms for generating coherent packets of hairpin vortices in channel flow. *Journal of Fluids Mechanics*, *387*, 353–396.

Zhu, J. Y., Zou, S. F., Liu, L. Q., Wu, J. Z., & Liu, T. (2015). The Origin of Lift Revisited: II. Physical Processes of Airfoil-Circulation Formation in Starting Flow. In 45th AIAA Fluid Dynamics Conference, Dallas, Texas, 22–26 June 2015.

Zhu, Y. D., Zhang, C. H., Tang, Q., Yuan, H. J., Wu, J. Z., & Chen, S. Y., et al. (2015). *Transition in hypersonic boundary layers: Role of dilatational waves* (in revision).

Zhuang, L. X., Yin, X. Y., & Ma, H. Y. (2009). *Fluid mechanics* (2nd ed.). Beijing: University of Science and Technology of China Press (in Chinese).

Index

J.-Z. Wu et al., *Vortical Flows*, DOI 10.1007/978-3-662-47061-9

CPSIA information can be obtained
at www.ICGtesting.com
Printed in the USA
LVHW081249151219
640572LV00013B/632/P

9 783662 470602